20th Anniversary of *Pharmaceuticals*—Advances in Pathophysiology, Pharmacology and Neuroprotection in Glaucoma

20th Anniversary of *Pharmaceuticals*—Advances in Pathophysiology, Pharmacology and Neuroprotection in Glaucoma

Guest Editor

Najam A. Sharif

MDPI

Basel • Beijing • Wuhan • Barcelona • Belgrade • Novi Sad • Cluj • Manchester

Guest Editor
Najam A. Sharif
Nanoscope Therapeutics, Inc.
Dallas
USA

Editorial Office
MDPI AG
Grosspeteranlage 5
4052 Basel, Switzerland

This is a reprint of the Special Issue, published open access by the journal *Pharmaceuticals* (ISSN 1424-8247), freely accessible at: https://www.mdpi.com/journal/pharmaceuticals/special_issues/915H07E9K8.

For citation purposes, cite each article independently as indicated on the article page online and as indicated below:

Lastname, A.A.; Lastname, B.B. Article Title. *Journal Name* **Year**, *Volume Number*, Page Range.

ISBN 978-3-7258-5335-9 (Hbk)
ISBN 978-3-7258-5336-6 (PDF)
https://doi.org/10.3390/books978-3-7258-5336-6

Contents

About the Editor

Najam A. Sharif

Najam A. Sharif is currently Vice-President of Global Research and Development at Nanoscope Therapeutics Inc (Dallas, TX, USA). He holds BSc, PhD and DSc degrees from Southampton University (UK) and has 40 years' pharmaceutical drug/device/biologics/gene therapy experience gained through long-term tenures at Syntex/Roche, Alcon-Novartis, Santen Pharmaceuticals and now Nanoscope Therapeutics. Dr. Sharif has been praised as a "drug hunter" for the treatment of glaucoma, ocular allergies, inherited retinal diseases, and neurological disorders with some global regulatory approvals and marketed medicines (e.g., Travatan, Patanol, Simbrinza, and Betaxon). For such efforts he has been recognized through a number of awards and prizes (Dr. Roger Vogel Award for pharmaceutical research; Sir James Black Award for drug discovery contributions; Professor Ernst H. Barany Prize for exemplary contributions to ocular pharmacology). He is an Adjunct Professor at multiple global universities and a faculty committee member for MS and PhD research candidates, advising and mentoring the same. Dr. Sharif has published >260 scientific articles, edited four books and has been granted 26 patents. His research interests encompass pharmacology, pathophysiology, drug discovery and optogenetic gene therapy, amongst many others.

pharmaceuticals

MDPI

Article

Tauroursodeoxycholic Acid Protects Retinal Ganglion Cells and Reduces Inflammation in Mice Following Optic Nerve Crush

Nan Zhang [1,2,3], Ying Li [1,2], Xian Zhang [1,2,4], Micah A. Chrenek [2], Jiaxing Wang [2,5], Preston E. Girardot [1,2], Jana T. Sellers [1,2], Eldon E. Geisert [2], John M. Nickerson [2] and Jeffrey H. Boatright [1,2,*]

1 Atlanta Veterans Administration Center for Visual and Cognitive Rehabilitation, Decatur, GA 30033, USA; nan.zhang@emory.edu (N.Z.); ying.li@emory.edu (Y.L.); xianz0707@csu.edu.cn (X.Z.); preston.girardot@emory.edu (P.E.G.); jana.t.sellers@emory.edu (J.T.S.)

2 Department of Ophthalmology, School of Medicine, Emory University, Atlanta, GA 30322, USA; micah.chrenek@emory.edu (M.A.C.); eldon.e.geisert@emory.edu (E.E.G.); jx.wang@med.miami.edu (J.W.); litjn@emory.edu (J.M.N.)

3 Department of Ophthalmology, Zhongnan Hospital, Wuhan University, Wuhan 430071, China

4 Department of Ophthalmology, Second Xiangya Hospital of Central South University, Changsha 410011, China

5 Bascom Palmer Eye Institute, University of Miami Miller School of Medicine, Miami, FL 33136, USA

* Correspondence: jboatri@emory.edu; Tel.: +1-404-778-4113; Fax: +1-404-778-2231

Abstract: Purpose: The aim of this study was to investigate the protective effects of systemically administered tauroursodeoxycholic acid (TUDCA) in an optic nerve crush (ONC) mouse model of retinal ganglion cell (RGC) death. **Methods:** C57BL/6J mice were injected intraperitoneally (i.p.) three times per week with TUDCA (500 mg/kg) for two weeks, after which unilateral ONC was performed. A control cohort was identically treated with a drug vehicle (phosphate buffered saline; PBS). A separate cohort did not undergo any injections or surgeries (this was termed the "Naïve" group). Pattern electroretinography (PERG) was recorded 3 days after ONC. Retinas were harvested for whole-mount immunofluorescence staining with an antibody against RGC marker Brn3a and imaged by fluorescent confocal microscopy. Apoptotic cells in the ganglion cell layer (GCL) were detected by Terminal Deoxynucleotidyl Transferase-Mediated dUTP Nick End Labeling (TUNEL) performed on fixed retina sections. Glial fibrillary acidic protein (GFAP) immunostaining on fixed retina sections was conducted to detect the activation of Müller cells. Total RNA was extracted from retinas and expression of interleukin (IL)-1β, IL-6, tumor necrosis factor (TNF)-α, and IL-10 was determined by digital droplet PCR (ddPCR). **Results:** TUDCA treatment preserved visual function as assessed by PERG. P1 and N2 amplitudes from the PBS-treated ONC group were significantly diminished compared to those of the Naïve group ($p < 0.001$). TUDCA treatment prevented this diminution. The amplitudes of P1 and N2 in the TUDCA-treated ONC group were statistically indistinguishable from those of the Naïve group and were higher than the PBS-treated ONC group (TUDCA+ONC vs. PBS+ONC, P1: 6.99 ± 0.89 μV vs. 3.60 ± 0.69 μV, $p < 0.01$; N2: −9.30 (IQR: −13.43–−6.44) μV vs. −4.47 (IQR: −10.26–−2.17) μV). TUDCA treatment preserved RGCs. The ONC-vehicle-only group had 25% fewer RGCs (Brn3a-positive cells) than Naïve eyes ($p < 0.0001$). TUDCA treatment nearly completely prevented this loss, preserving all but 7.7% of the RGCs, and the number of RGCs in the TUDCA-treated ONC group was significantly higher than in the PBS-treated ONC group (TUDCA+ONC vs. PBS+ONC, 1738.00 ± 14.43 cells per field vs. 1454.00 ± 6.55 cells per field, $p < 0.0001$). The number of TUNEL-positive cells in the GCL (Naïve vs. PBS+ONC group: 1.00 (IQR: 0.00–2.00) % vs. 37.00 (IQR: 8.50–48.50) %, $p < 0.05$) and GFAP-positive fibers transversing retina sections (Naïve vs.

PBS+ONC group: 33.00 ± 1.15 vs. 185.70 ± 42.37 fibers/retina, $p < 0.05$), and the expression of IL-6, TNF-α were significantly greater in the PBS-treated ONC group compared to that of the Naïve group (Naïve vs. PBS+ONC group, IL-6: 0.07 (IQR: 0.06–0.31) vs. 0.99 (IQR: 0.56–1.47), $p < 0.05$, TNF-α: 0.19 ± 0.069 vs. 1.39 ± 0.23; $p < 0.01$), an increase not observed with TUDCA treatment. **Conclusions:** Systemic TUDCA treatment significantly preserved RGC function and survival in the mouse ONC model of RGC damage. TUDCA treatment prevented RGC apoptosis, Müller glial cell activation, and retinal expression of several inflammatory cytokines. These data suggest that TUDCA is a promising therapeutic candidate for preserving RGC numbers and function.

Keywords: tauroursodeoxycholic acid; bile acid; retinal ganglion cell; inflammation; optic nerve crush; mus musculus; mouse; mice

1. Introduction

Glaucoma blinds more than 75 million people worldwide [1]. It causes the death of retinal ganglion cells (RGCs), leading to visual field loss and blindness [2,3]. Elevated intraocular pressure is a significant risk factor for the disease [4]. However, even with well-managed intraocular pressure (IOP), RGC loss can progress, eventually resulting in vision loss [5]. As a chronic and progressive neurodegenerative disease, there is increasing interest in neuroprotective strategies and treatments that are independent of IOP management [6], including anti-glaucomatous agents, antibiotics, dietary supplements, novel small-molecule neuroprotectants, and neurotrophic factors [7]. However, as of now, no clinically recognized medication exists that effectively protects RGCs and the optic nerve in glaucoma treatment.

Optic nerve crush (ONC) is a surgical compression injury to the optic nerve, usually performed in rodents [8]. ONC causes axonal injury followed by retrograde degeneration and apoptosis of RGCs and has been widely used to model aspects of RGC damage [9–11]. Inflammatory retinal responses, including reactive gliosis and increased expression of pro-inflammatory cytokines after ONC, also contribute to RGC death and may mimic secondary RGC degeneration that occurs in glaucoma [12–15].

Tauroursodeoxycholic acid (TUDCA) is an endogenous bile acid and the taurine conjugate of ursodeoxycholic acid (UDCA) [16]. Recently, hydrophilic bile acids, notably TUDCA, have been reported to have profound anti-apoptotic and neuroprotective effects [17]. Systemic treatment with TUDCA is protective in models of diseases such as diabetes, kidney disease, Parkinson's Disease, Huntington's Disease, Alzheimer's Disease, and retinal degeneration [18–29]. TUDCA has good bioavailability and is well tolerated when taken orally [30,31]. Further, clinical studies report that it is protective in neurodegenerative diseases [17], including increased survival in amyotrophic lateral sclerosis (ALS) patients [30]. Our previous studies reported that TUDCA offers protection in multiple retinal degeneration models, including diabetic retinopathy [32], retinitis pigmentosa (rd10 [33], and rd1 mice [34]). Additionally, TUDCA treatment is protective in animal models of RGC damage [35–38], though effects specifically on RGC function and in murine models are heretofore untested. The present study addressed this gap in knowledge.

Among several potential mechanisms of action [39,40], TUDCA has anti-inflammatory effects. It exhibits neuroprotective and anti-inflammatory effects in animal models of stroke by binding to the GPBAR1/TGR5 receptor in microglial cells [41]. It also reduces glial cell activation in multiple disease models including fatty liver disease, lipopolysaccharide-induced neuroinflammation, and retinitis pigmentosa [25,42–44], though this is relatively unexplored in models of RGC damage. Beyond its anti-inflammatory effects, extensive in vitro research on cultured neuronal cells and in vivo experiments on animal disease models have deeply explored and defined the anti-apoptotic properties of TUDCA [45]. TUDCA has been reported to prevent apoptosis by disrupting the E2F-1/p53/Bax apoptotic pathway and lowering nuclear fragmentation and caspase activity in Alzheimer's disease [46]. TUDCA also prevents apoptosis by suppressing ER stress activation and oxidative stress on tunicamycin-induced apoptosis in primary cultured rat DRG neurons [47]. TUDCA reduced neuronal apoptosis and improved neurological functions after subarachnoid hemorrhage by targeting the TGR5/SIRT3 signaling pathway [48]. Based on the neuroprotective effects of TUDCA in neurodegenerative and retinal degenerative diseases, we hypothesize that TUDCA may exert potential neuroprotective effects on RGCs. Although a few studies have reported the protective effects of TUDCA in other animal models of RGC damage, they have not yet explored the impact on RGC function or the mechanisms underlying TUDCA's protective action.

This study aimed to explore the protective effects of systemically administered TUDCA in a mouse model of RGC death induced by ONC, and whether TUDCA prevents or reduces RGC retinal inflammation. The rationale is a combination of the aforementioned need to develop IOP-independent treatments to protect RGC cells in glaucomatous disease, the oral bioavailability and tolerance in human subjects, the neuroprotective, anti-apoptotic, and anti-inflammatory properties we and others find with TUDCA treatment, including in ALS trials. We expanded our understanding of the protective effects that TUDCA treatment has on RGCs, including treatment effects on RGC function, apoptosis, and survival as assessed by retinal electrophysiology, TUNEL of retina section, and immunofluorescence labeling of RGCs in retinal flatmounts. Further, we assessed treatment effects on retinal inflammation via immunofluorescence staining of mouse retinal sections to observe Müller glia cell activation and digital drop PCR on retinal RNA to test for expression of pro-inflammatory genes. The results indicate that systemic TUDCA treatment significantly preserved RGC survival and function in an ONC mouse model of RGC damage. TUDCA treatment also significantly suppressed ONC-induced RGC apoptosis, activation of Müller glial cells, and overexpression of pro-inflammatory cytokines.

2. Results

2.1. *TUDCA Treatment Preserves RGC Function After Optic Nerve Injury*

To test whether TUDCA can protect RGC function from injury, we performed PERG on mice 3 days after ONC. As shown in Figure 1, functional diminution was observed in eyes that underwent ONC in PBS-treated mice ("PBS+ONC"). These eyes exhibited markedly diminished P1 and N2 amplitudes compared with the contralateral eye that did not undergo ONC (62% decrease and 59% decrease separately, PBS+ONC vs. Contralateral, P1: 3.60 ± 0.70 μV vs. 9.59 ± 0.35 μV, $p < 0.0001$, one-way ANOVA with Tukey's test; N2: −4.47 (IQR: −10.26–−2.17) μV vs. −14.89 (IQR: −15.82–−12.63) μV, $p < 0.001$, Kruskal–Wallis test with Dunn's multiple comparison test) or compared to eyes from Naïve mice (61% decrease and 55% decrease separately, PBS+ONC vs. Naïve, P1: 3.60 ± 0.70 μV vs. 9.27 ± 0.46 μV, $p < 0.0001$, one-way ANOVA with Tukey's test; N2: −4.47 (IQR: −10.26–−2.17) μV vs.−14.05 (IQR: −14.68–−11.77) μV, $p < 0.001$, Kruskal–Wallis test with Dunn's multiple comparison test). Conversely, functional diminution was largely prevented in the eyes of TUDCA-treated mice that underwent ONC. TUDCA treatment preserved both P1 and N2 amplitudes compared with crushed eye from PBS-treated mice (P1: 6.99 ± 0.89 μV vs. 3.60 ± 0.69 μV, $p < 0.01$, one-way ANOVA with Tukey's test; N2: −9.30 (IQR: −13.43–−6.44) μV vs. −4.47 (IQR: −10.26–−2.17) μV, Kruskal–Wallis test with

Dunn's multiple comparison test). There was no statistically significant difference in PERG responses between the eyes of Naïve mice and the Contralateral eyes of PBS-treated mice or between the eyes of Naïve mice and ONC eyes of TUDCA-treated mice (Figure 1B,C), though functional preservation was not absolute.

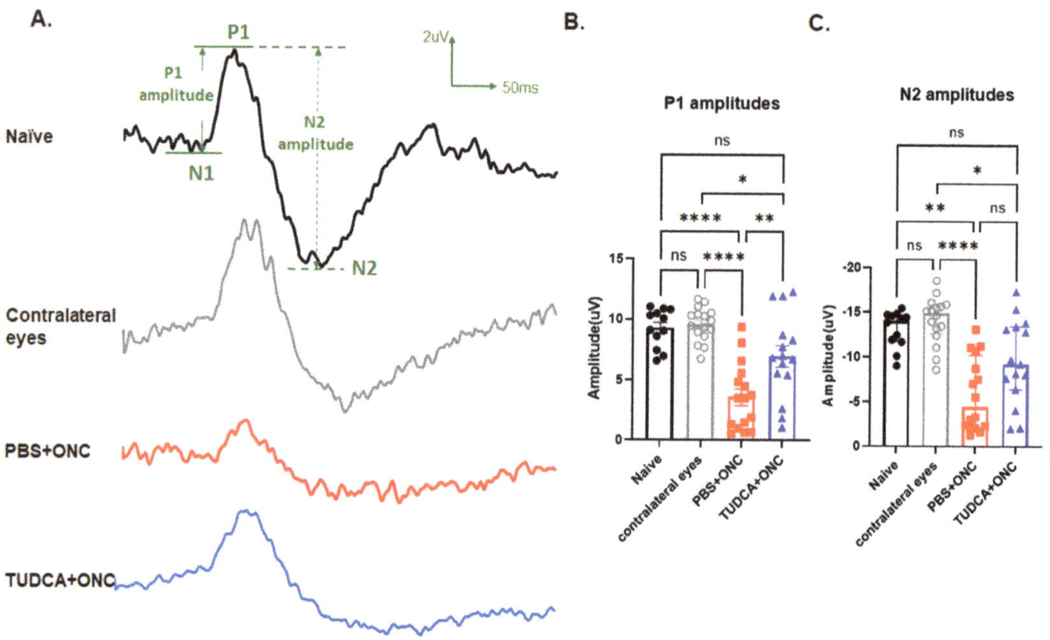

Figure 1. TUDCA treatment preserved RGC function 3 days after optic nerve crush. (**A**) Representative PERG waveforms from each group. (**B**) Quantification of P1 amplitudes 3 days after ONC. P1 amplitudes from PBS+ONC group were significantly diminished compared to those of the Naïve or Contralateral eye groups. TUDCA treatment preserved the P1 amplitudes significantly ($p < 0.01$). (**C**) Quantification of N2 amplitudes 3 days after ONC. N2 amplitudes from PBS-treated ONC group were significantly diminished compared to those of the Naïve group or Contralateral eye groups. TUDCA treatment preserved the N2 amplitudes significantly ($p < 0.05$). The results are represented as mean ± SEM (**B**), median ± IQR (**C**). A one-way ANOVA with Tukey's multiple comparisons test (**B**) and Kruskal–Wallis test with Dunn's multiple comparison test (**C**) was conducted between the mean cells/field in all combinations. ns: no statistical difference, * $p < 0.05$, ** $p < 0.01$, **** $p < 0.0001$. n = 12 mice in Naïve group; n = 16 mice in PBS+ONC group and for the Contralateral eyes of that group; n = 15 mice in TUDCA + ONC group. PERG: pattern electroretinogram; RGC: retinal ganglion cell; PBS: phosphate buffered solution; ONC: optic nerve crush; TUDCA: tauroursodeoxycholic acid.

2.2. TUDCA Treatment Maintains RGC Survival After Optic Nerve Injury

RGC loss was observed within 3 days of ONC (Figure 2). As shown in Figure 2, the number of immunofluorescent Brn3a-positive cells, assumed to be RGCs, declined by about 25% in retinas from eyes that had undergone ONC in PBS-treated mice compared to retinas from Naïve mice (1454.00 ± 6.55 vs. 1883.00 ± 41.02 cells per field; n = 4/group, one-way ANOVA with Tukey's test, $p < 0.0001$). Retinas from mice treated with TUDCA had significantly more RGCs compared to the PBS-treated mice after ONC (1738.00 ± 14.43 cells per field vs. 1454.00 ± 6.54 cells per field, n = 4, one-way ANOVA with Tukey's test, $p < 0.0001$) (Figure 2B). TUDCA treatment did not perfectly preserve RGC numbers. The number of RGCs declined slightly but statistically significantly in retinas from eyes that had undergone ONC in TUDCA-treated mice compared to retinas from Naïve mice (1883.00 ± 41.02 vs. 1738.00 ± 14.43 cells per field; n = 4/group, one-way ANOVA with Tukey's test, $p < 0.01$).

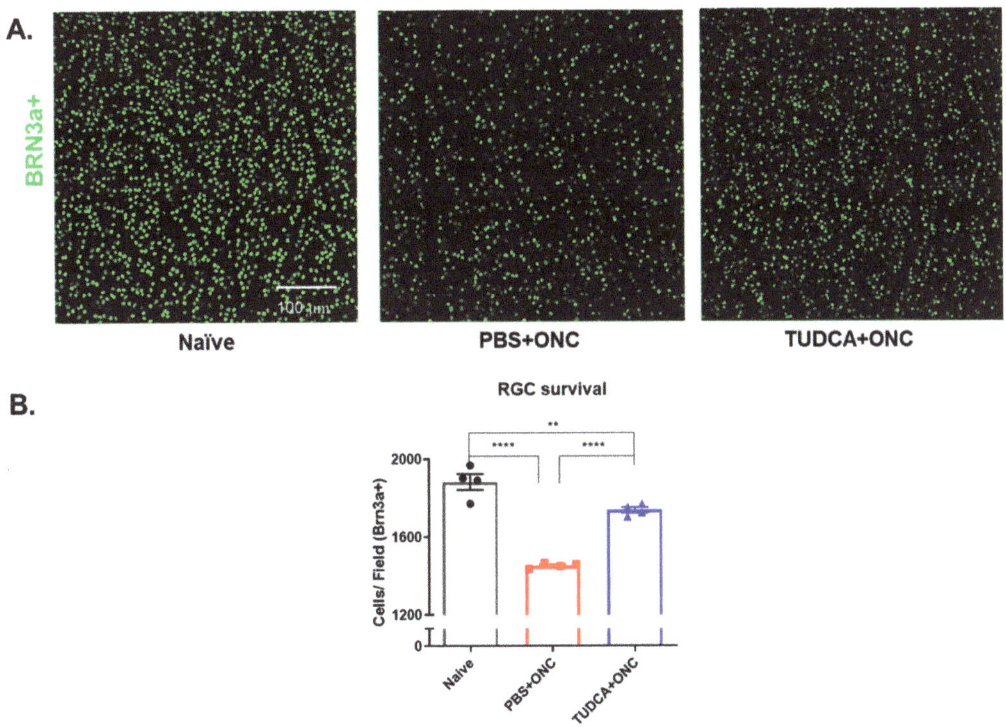

Figure 2. TUDCA treatment protects against ONC-induced RGC loss. (**A**) Representative images of retina flatmounts stained against Brn3a from different groups. (**B**) Quantification of Brn3a-positive cells, assumed to be RGCs, 3 days after ONC. The number of Brn3a-positive cells was significantly less in the PBS-treated ONC group 3 days after ONC compared to Naïve group (25% decrease). TUDCA treatment partially but significantly prevented this loss. The number of Brn3a-positive cells in TUDCA-treated ONC group was slightly lower than Naïve group (1883 ± 41.0 vs. 1738 ± 14.4 cells/field). The results are represented as mean ± SEM. A one-way ANOVA with Tukey's multiple comparisons test was conducted between the mean cells/field in all combinations. ** $p < 0.01$, **** $p < 0.0001$. n = 4 mice per group. Scale bar = 100 microns. RGC: retinal ganglion cell; PBS: phosphate buffered solution; ONC: optic nerve crush; TUDCA: tauroursodeoxycholic acid.

2.3. TUDCA Treatment Prevents ONC-Induced RGC Apoptosis

TUNEL-positive cells in the GCL, presumed to be RGCs affected by axon crush [49,50], were observed and counted across entire retina sections. Compared with the Naïve group, which had few to no TUNEL-positive cells in the GCL (Figure 3A,D), the PBS-treated ONC group had a significant percentage of TUNEL-positive cells (Figure 3B,D) (Naïve vs. PBS+ONC group: 1.00 (IQR: 0.00–2.00) % vs. 37.00 (IQR: 8.50–48.50), $p < 0.05$, Kruskal–Wallis test with Dunn's multiple comparison test). In contrast, sections from the TUDCA-treated ONC group had few TUNEL-positive cells in the GCL (Figure 3C), an amount statistically indistinguishable from the Naïve group (Figure 3D).

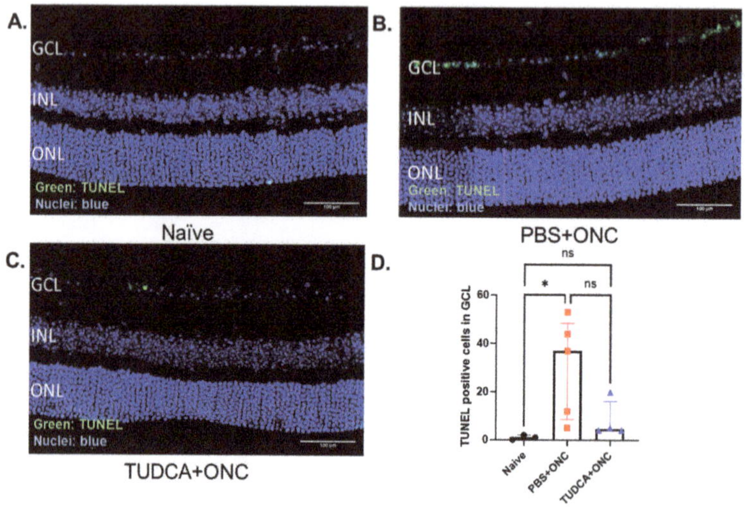

Figure 3. TUDCA treatment prevented ONC-induced RGC apoptosis. (**A–C**) Representative morphologic images of TUNEL staining from the region of 250–750 µm from the optic nerve. (**D**) Quantification of percentage of TUNEL positive RGC in GCL. RGC apoptosis significantly increased 3 days after ONC ($p < 0.05$). TUDCA treatment prevented the increase in TUNEL-positive cells of the GCL observed in retinas from eyes that had undergone ONC. TUNEL-positive cells in GCL were counted by masked observers across entire sections, which include optic nerve head. The results are represented as median ± IQR. Kruskal–Wallis test with Dunn's multiple comparison test was conducted between the mean percent in all combinations. ns: no statistical difference, * $p < 0.05$. n = 3 mice in Naïve group; n = 5 mice in PBS group; n = 4 mice in TUDCA group. Scale bar = 100 microns. GCL: ganglion cell layer; INL: inner nuclear layer; ONL: outer nuclear layer; TUNEL: terminal deoxynucleotidyl transferase dUTP nick end labeling; PBS: phosphate buffered solution; ONC: optic nerve crush; TUDCA: tauroursodeoxycholic acid.

2.4. TUDCA Treatment Prevents Müller Cell Activation After Optic Nerve Injury

GFAP is widely used to assess the gliosis of Müller cells [51–53]. Here, we used retinal sections stained with GFAP to test whether TUDCA treatment might reduce Müller cell activation in our ONC model. The GFAP signal in Naïve retina sections was mainly in the inner retina (Figure 4A). In the ONC+PBS condition, the number of GFAP-positive cells increased, and the signal appeared to span the inner retina out to the outer plexiform layer (Figure 4B). Treatment with TUDCA prevented these ONC-induced changes in the GFAP signal (Figure 4C). The number of GFAP-positive fibers spanning from the ganglion cell layer (GCL) to the inner nuclear layer (INL) were counted by masked observers across entire sections for each cohort [54]; these quantified outcomes are presented in Figure 4D. ONC injury markedly increased the number of GFAP-positive fibers in PBS-treated mice compared with retinas from Naïve mice (Naïve vs. PBS+ONC group: 33.00 ± 1.5 vs. 185.70 ± 42.37 fibers/retina, n = 3 in Naïve group, n = 6 in PBS+ONC group, $p < 0.05$, one-way ANOVA with Tukey's test), indicating that activation of Müller cells occurred after crush injury. TUDCA treatment prevented the Müller cell gliosis observed in retinas from eyes that had undergone ONC in PBS-treated mice, with the number of GFAP-positive strands being statistically indistinguishable between Naïve and TUDCA+ONC groups. These data suggest that ONC stressed the retina, leading to apparent Müller cell activation. TUDCA treatment prevented this stress and/or activation.

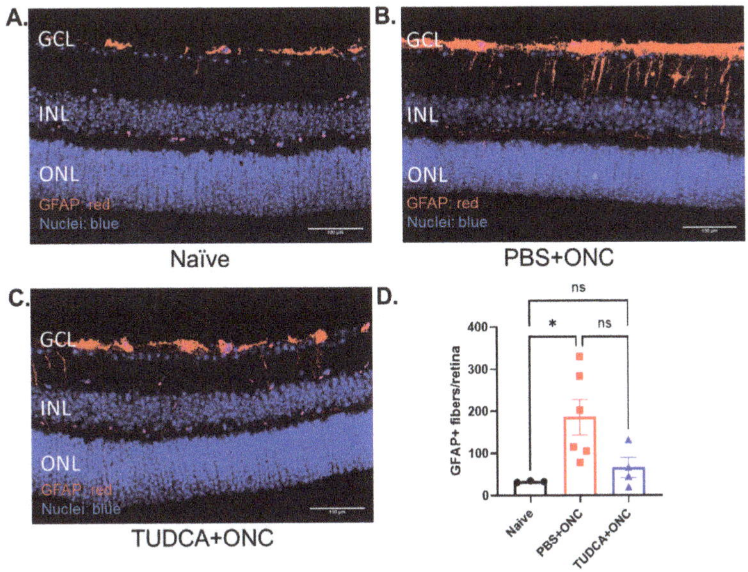

Figure 4. TUDCA treatment suppresses ONC-induced Müller cell activation. (**A–C**) Representative fluorescent images of GFAP staining from the region of 250–750 μm from the optic nerve. (**D**) Quantification of GFAP-positive fibers per retina. GFAP-positive fibers significantly increased 3 days after ONC. TUDCA treatment prevented the increase in number of GFAP-positive fibers in mouse retina caused by ONC damage. No significant difference was been found between Naïve and TUDCA treated groups. GFAP-positive fibers fully penetrating the INL were counted across entire sections by masked observers. The results are represented as mean ± SEM. A one-way ANOVA with Tukey's multiple comparisons test was conducted between the mean cells/field in all combinations. ns: no statistical difference, * $p < 0.05$, ns: no significant difference. n = 3 mice in Naïve group; n = 6 mice in PBS group; n = 4 mice in TUDCA group. Scale bar = 100 microns. GCL: ganglion cell layer; INL: inner nuclear layer; ONL: outer nuclear layer; GFAP: glial fibrillary acidic protein; PBS: phosphate buffered solution; ONC: optic nerve crush; TUDCA: tauroursodeoxycholic acid. Control cohort "PBS+ONC" was shared across the experiment reported in Figure 4 in [55].

2.5. TUDCA Inhibits Inflammatory Response After Injury

TUDCA treatment appears to suppress Müller cell activation induced by ONC (Figure 4). To further investigate whether TUDCA might reduce inflammation, we measured retinal mRNA expression levels of proinflammatory factors interleukin (IL)-1β, IL-6, and tumor necrosis factor-α (TNF-α). As shown in Figure 5, the mRNA expression levels of IL-6, and TNF-α were higher in the PBS-treated ONC group compared with the Naïve group (Naïve vs. PBS+ONC group, IL-6: 0.07 (IQR: 0.06–0.31) vs. 0.99 (IQR: 0.56–1.47), Kruskal–Wallis test with Dunn's multiple comparison test, $p < 0.05$, TNF-α: 0.19 ± 0.069 vs. 1.39 ± 0.23, one-way ANOVA with Tukey's test, $p < 0.01$; n = 4–5 for each group; data reported as mRNA ratio to HPRT multiplied by 1000). There was no significant difference between Naïve and TUDCA-treated ONC groups in IL-6 and TNF-α mRNA levels, indicating that treatment with TUDCA prevented an ONC-induced increase in expression of these genes (Naïve vs. TUDCA+ONC group, IL-6: 0.07 (IQR: 0.06–0.31) vs.0.48 (IQR: 0.17–1.14), Kruskal–Wallis test with Dunn's multiple comparison test; TNF-α: 0.19 ± 0.069 vs. 0.88 ± 0.24, one-way ANOVA with Tukey's test; n = 4–5 for each group; data reported as mRNA ratio to HPRT multiplied by 1000). Different from others, IL-1β and IL-10, did not show significant differences among groups (Figure 5D).

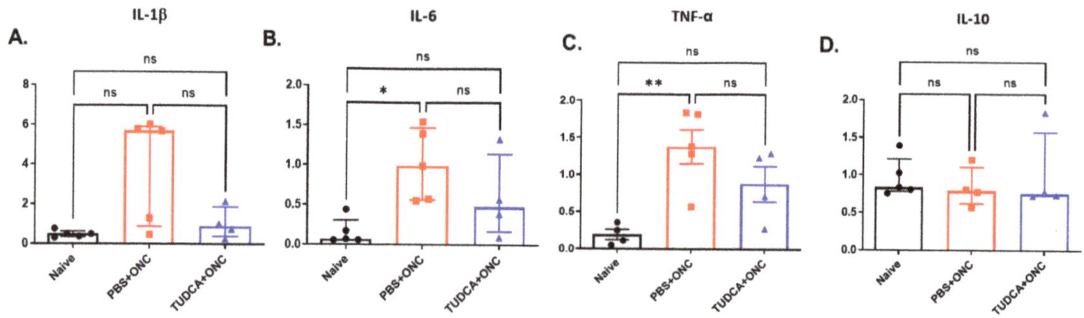

Figure 5. TUDCA treatment prevented ONC-induced overexpression of pro-inflammatory factors. The mRNA levels of IL-1β (**A**), IL-6 (**B**), TNF-α (**C**), IL-10 (**D**) were examined 3 days after ONC. *Y*-axis in all channels represent the mRNA ratio to HPRT multiplied by one thousand. The results are represented as mean ± SEM (**C**), median ± IQR (**A**,**B**,**D**). Kruskal–Wallis test with Dunn's multiple comparison test was conducted between the mean ratio to HPRT in all combinations (**A**,**B**,**D**). One-way ANOVA with Tukey's tests was conducted between the mean ratio to HPRT in all combinations (**C**). ns: no statistical difference, * $p < 0.05$, ** $p < 0.01$, ns: no significant difference, n = 5 mice in Naïve group/PBS+ONC group; n = 4 mice in TUDCA+ONC group. IL-1β: interleukin 1β; IL-6: interleukin 6; IL-10: interleukin 10; TNF-α: tumor necrosis factor-α; PBS: phosphate buffered solution; ONC: optic nerve crush; TUDCA: tauroursodeoxycholic acid.

3. Discussion

In this study, we observed that TUDCA treatment significantly preserved RGC function enhanced RGC survival, and suppressed inflammation and apoptosis in an ONC mouse model of RGC damage. Prior to this, the protective effects of systemic TUDCA treatment on RGCs had not been tested in mice, nor had effects on RGC function or retinal inflammatory responses been assessed following ONC, regardless of species.

We found that systemic treatment with TUDCA partially preserved the number of RGCs in mouse eyes that had undergone ONC (Figure 2) and prevented apoptosis in the cells of the GCL (Figure 3). This confirms results in an ONC rat model [35] and extends findings into a new species. Further expanding on this, we found that systemic TUDCA treatment nearly completely protected against RGC functional loss caused by ONC, as shown by the preservation of amplitudes of P1 and N2 elements in PERGs (Figure 1). These functional findings may have clinical relevance as it is known that changes in PERG amplitude occur before visual field loss in glaucoma patients, and this has been used as a sensitive biomarker for monitoring RGC function in both glaucoma patients and animal models of glaucoma [56].

We observed that Müller glia cell activation and elevated production of proinflammatory factors occurred after ONC, but that both were attenuated with TUDCA treatment (Figures 4 and 5). Inflammation plays an important role in the pathogenesis of glaucoma [57,58] in both glaucoma patients and animal models [59,60]. Upregulation of proinflammatory cytokines is observed in multiple glaucoma models [12,61–63]. Müller cell gliosis, assessed by GFAP upregulation [52,54,64], that occurs after ONC and other retinal damage can result in the release of glutamate, cytokines, and other mediators of inflammation and RGC injury [54,65,66]. We speculate that TUDCA treatment might promote RGC survival by inhibiting Müller cell activation and proinflammatory cytokine production to reduce neuroinflammation. An additional explanation is that TUDCA treatment prevented ONC-induced RGC apoptosis (Figure 3) via cell-autologous mechanisms (e.g., suppressing ER stress [67]), which would preclude signals from RGC-invoking inflammatory responses [13].

ONC mouse model serves as an acute RGC damage model, mimicking both glaucoma and acute optic nerve trauma.

The effects of systemic TUDCA treatment we report here in protecting RGC density and function and in diminishing retinal inflammation compare well with recently reported effects of other neuroprotective agents in the ONC mouse model. For instance, i.p. injection of the RXR agonist, 9-cis-13,14-dihydroretinoic acid (9CDHRA) diminished RGC apoptosis and retinal inflammation but had little or no protective effect on function [68]. A combination of ocular topical and i.p. injection of brimonidine modestly preserved RGC numbers (13% loss across retina without treatment, 10% loss with treatment), but had minimal effect on retinal inflammation [69]. Intravitreal injection of Brain-Derived Neurotrophic Factor (BDNF) modestly preserved RGC density but did not affect visual function loss [70]. Conversely, i.p. injections of the NAD+ precursor nicotinamide riboside (NR) significantly protected against losses in RGC density and function and suppressed retinal inflammation [55]. Thus, compared to other RGC neuroprotectants actively being studied, systemic treatment with TUDCA appears to be a candidate treatment worth pursuing.

A limitation of this study lies in its focus solely on the short-term protective effects of TUDCA, specifically examining its protective efficacy only three days after ONC injury, without addressing the long-term protective effects of TUDCA. Off-target effects of TUDCA were not observed (e.g., no gross changes were observed in behavior or in mouse weights with treatment). Another potential limitation of this study is that the effects of anesthesia and other animal handling were directly compared to outcomes of naïve, PBS+ONC, or TUDCA-ONC groups in experiments assessing RGC function via PERG (Figure 1), but not in other experiments. However, in addition to data from that experiment, we previously reported no observable differences between such sham treatment (contralateral, uninjured eyes from mice that underwent ONC) and naïve groups [55] and so did not repeat that comparison in every instance of the present study. Lastly, this study did not investigate the molecular mechanisms underlying TUDCA's anti-inflammatory properties and its inhibition of RGCs apoptosis. To further elucidate the neuroprotective mechanisms of TUDCA, we aimed to explore earlier time points to identify immediate early gene responses following injury. The mechanisms of RGC protection for TUDCA are subjects of our future research. In addition, the additive or synergistic therapeutic effects of TUDCA combined with other glaucoma treatment medications, such as IOP-lowering drugs and other neuroprotective agents, will also be explored in our future studies to prevent RGC damage in glaucoma effectively.

4. Materials and Methods

4.1. Experimental Design

Mice were divided into three groups: the Naïve group, the PBS+ONC group, and the TUDCA+ONC group. The Naïve group (12 mice) received no treatments. The PBS+ONC group (16 mice) received intraperitoneal (i.p.) injections of phosphate-buffered saline (PBS) and underwent unilateral optic nerve crush (ONC) surgery. The TUDCA + ONC group (15 mice) received i.p. injections of tauroursodeoxycholic acid (TUDCA) followed by unilateral ONC surgery. The experimental timeline is illustrated in Figure 6. Injections of PBS or TUDCA were administered on Monday, Wednesday, and Friday between 9:00 and 10:00 AM for two weeks. ONC surgery was performed 30 to 45 min after the sixth injection of either PBS or TUDCA. A seventh injection was given two days after the surgery. One day later, which was three days after the ONC, pattern electroretinograms (ERGs) were conducted, followed by euthanasia and tissue collection.

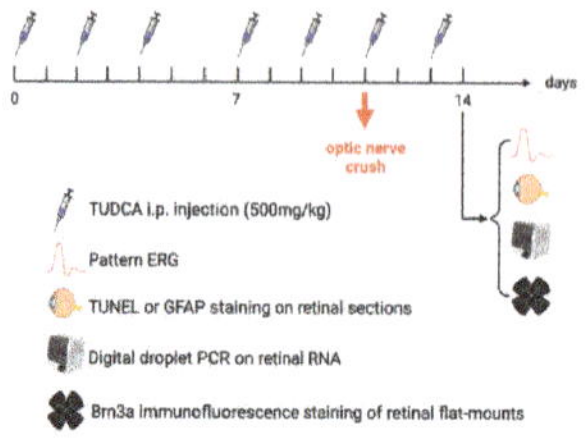

Figure 6. Experimental timeline. Details in text of Section 4.1.

4.2. Animals

C57BL/6J mice (2–3 months old during experiments, weighing 20–25 g) were obtained from Jackson Laboratory (Stock# 000664, Bar Harbor, ME, USA). We maintained them on a 12 h light–dark cycle (7 a.m. on and 7 p.m. off) with food and water ad libitum. Mice were euthanized with CO_2 gas or with cardiac perfusion. Husbandry and experimental procedures followed an approved protocol (Emory Institutional Animal Care and Use Committee Protocol ID PROTO201800248) and the "ARVO Statement for the Use of Animals in Ophthalmic and Vision Research".

4.3. Drugs

We prepared and injected TUDCA as previously detailed [32,33,71]. Paraphrasing briefly from our prior studies, we dissolved TUDCA (Lot# 3310670, MilliporeSigma, Billerica, MA, USA) in phosphate-buffered saline (PBS; VWRVK813, Cat#97063-660, VWR International, LLC, Radnor, PA, USA). The 1× solution has a composition of 137 mM NaCl, 2.7 mM KCl, and 9.5 mM phosphate buffer. We prepared sterile TUDCA (5% (w/v)) solution just before injection, and adjusted the pH to 7.4, as previously detailed [71]. We injected TUDCA solution (i.p., 500 mg/kg mouse body weight) into mice in the morning 3 times per week as per our previous studies [32,33]. We injected "Control mice" with an equivalent volume of sterile PBS based on body weight.

4.4. Optic Nerve Crush

The ONC procedure was performed unilaterally following protocols we previously developed [72]. During the initial three days after the surgery, they were checked for potential infection, bleeding, and muscle control issues. A group of age-matched C57BL/6J mice that had no surgeries served as the Naïve group.

4.5. Pattern ERG

Three days after ONC, mice were dark-adapted overnight before pattern electroretinograms (PERG) were performed, as previously described [70]. Summarizing and paraphrasing this procedure briefly, once mice were anesthetized (as described in the last section), 0.5% proparacaine and 1% tropicamide eye drops (Akorn Inc., Lake Forest, IL, USA) were administered to reduce eye sensitivity and dilate the pupils. Mice were placed on a heating pad (37 °C) under a dim red light provided by the overhead lamp of the Celeris-Diagnosys system (Diagnosys, LLC, Lowell, MA, USA). Transient PERG responses were recorded using black and white vertical stimuli. The pattern stimulator was placed on the tested eye, with the flash stimulator for the contralateral eye acting as a reference electrode. Pattern stimuli of 50 cd·s/m^2 were presented. Six hundred averaged signals, with cut-off filter frequencies of 1 to 300 Hz were recorded under scotopic conditions. The P1 value was defined as the N1 trough to the peak of P1. N2 was defined as the preceding P1 peak to the trough of N2 (similar to what is termed "pERG amplitude" [73] or "P1N2" [74]). Mice were euthanized after PERG measurement by IACUC-approved CO_2 asphyxiation.

4.6. RGC Quantification

Retinal ganglion cells (RGCs) were quantified after immunostaining whole flatmounts of the retina for Brn3a immunofluorescence, as we previously detailed [72]. Briefly summarizing our protocol, mice were deeply anesthetized and perfused with 4% paraformaldehyde. Eyes were removed and fixed in 4% paraformaldehyde for one hour at room temperature. Dissected retinas were washed in PBS containing 1% Triton X-100 (Sigma-Aldrich, Cat. # 9002-93-1, St. Louis, MO, USA) and blocked with 10% normal donkey serum and 5% BSA. The retinas were then placed in the primary antibody, Brn3a (1:1000; Santa Cruz; SC-31984; Dallas, TX, USA), and incubated overnight at 4 °C. Following this, each retina was washed three times for 10 min in PBST (PBS with 0.1% Tween-20) and incubated overnight at 4 °C with the secondary antibody (1:1000, Alexa Fluor 488 AffiniPure Donkey Anti-Goat, Invitrogen, Cat# ABS82, Eugene, OR, USA). After another three washes with PBST, the retinas were flat-mounted with coverslips using Fluoromount-G (Southern Biotech, Cat. # 0100-01, Birmingham, AL, USA). A Nikon C1 confocal microscope (Nikon, Inc., Melville, NY, USA) was used to image the RGCs. The analysis of RGC survival was performed as previously reported [72]. For each retina, four fields, each measuring 636.5 µm × 636.5 µm and located 1.0 mm from the optic nerve, were randomly sampled from the middle regions of each retina under 20× magnification. The images were processed using a custom pipeline in CellProfiler v.6 [75] (https://cellprofiler.org/citations [accessed on 16 April 2021]) to automatically count the number of Brn3a-positive cells in a masked manner [76]. To evaluate RGC survival for each retina, the number of RGCs counted from the four regions was averaged.

4.7. Immunochemistry

Mice were euthanized with CO_2 gas after PERG. Eyes were fixed in fixation solution (97% methanol, Cat#: BDH20291GLP, VWR International, LLC, USA; 3% acetic acid, Cat#: Fisher BP2401-500, Thermo Fisher, Waltham, MA, USA) at −80 °C for 4 days, dehydrated in a series of methanol and xylenes baths, embedded in paraffin, and sectioned through the sagittal plane on a microtome at thickness of 5 µm, with minor variation of the freeze-substitution method of Ying et al. [77,78].

Immunofluorescence was used to detect glial fibrillary acidic protein (anti-GFAP, #Z0334, Dako, CA, USA) to assess Müller cell reactivity as we have previously described [79]. To ensure consistent examination across animals, sections encompassing the optic nerve head and the central corneal were chosen. One section from the retina of each animal was selected for the subsequent immunofluorescence staining.

Nikon Ti2 inverted microscope with A1R-HD25 confocal scanner was used for the imaging of the slides immunostained for GFAP (Nikon Instruments Inc., Melville, NY, USA) using a 20× objective lens, resonance scanning with 2× zoom at 1024 × 1024 per field, processed using Nikon's Denoise.ai algorithm, and tiled using NIS Elements. GFAP quantification was achieved by manually counting GFAP-positive fibers fully penetrating the inner nuclear layer (INL) across whole sections as others have reported [54]. Control cohort "PBS+ONC" was shared across the experiment reported in Figure 4 of [55].

4.8. TUNEL Assay

One section from the retina of each animal that traverses the optic nerve was used for TUNEL (Terminal deoxynucleotidyl transferase dUTP nick end labeling) staining. TUNEL assay was performed according to the protocol for the DeadEnd Fluorometric TUNEL Kit (Cat: #G3250, Promega, Fitchburg, WI, USA) as we previously described [78]. To ensure consistent examination across animals, sections encompassing the optic nerve head and the central corneal were chosen. One section from the retina of each animal was selected for the subsequent immunofluorescence staining. The slides stained for TUNEL were imaged using a Nikon Ti2 inverted microscope with an A1R-HD25 confocal scanner (Nikon Instruments Inc., Melville, NY, USA), as described immediately above. TUNEL-positive cells in the GCL were manually counted across whole sections for each retina using Adobe Photoshop CS6.

4.9. RNA Extraction and Digital Droplet PCR

The retinal gene expression of IL-1β, IL-6, TNF-α, and IL-10 were assayed relative to hypoxanthine-guanine phosphoribosyltransferase (HPRT) expression by digital droplet PCR (ddPCR) as previously described [80]. Mice were euthanized, and retinal tissues were collected and flash-frozen on dry ice. Total RNA was isolated from each individual retina. RNA extraction and cDNA synthesis were achieved by following the standard protocols of Qiagen RNeasy mini kit (Cat: # 74106, Qiagen, Mississauga, ON, Canada) and Qiagen QuantiNova RT kit (Cat: #205410, Qiagen, Mississauga, ON, Canada) respectively [81]. 6-fluorescein amidite (FAM) hydrolysis probe sets for ddPCR were ordered from Integrated DNA Technologies (Coralville, IA, USA) or Biorad (Hercules, CA, USA) or for each target and normalized to a hexachlorofluorescein (HEX) HPRT hydrolysis probe set (Integrated DNA Technologies). ddPCR was performed using 5 ng of cDNA and FAM hydrolysis probe sets for IL-1β (Mm.PT.58.41616450; Integrated DNA Technologies, IA, USA), IL-6 (dMmuCPE5095532; Biorad, CA, USA), TNF-α (Mm.PT.58.12575861; Integrated DNA Technologies) and IL-10 (dMmuCPE5094026; Bio-Rad, Hercules, CA, USA), and a HEX hydrolysis probe set for HPRT (Mm.PT.39a22214828; Integrated DNA Technologies, IA, USA) in Biorad ddPCR Supermix for Probes (no deoxyuridine triphosphate [dUTP]). The droplets were prepared using an Automated Droplet Generator model AutoDG (Biorad, CA, USA), plates were sealed with a PX1 plate sealer, PCR was performed using a C1000 Touch thermal cycler (Bio-Rad, Hercules, CA, USA) with deep well block, and read using a QX200 droplet reader (Bio-Rad, Hercules, CA, USA). Data were analyzed using Bio-Rad QuantaSoft software v1.7.4.0917 and Microsoft (Redmond, WA, USA) Excel 2016.

4.10. Masking and Statistical Analyses

Trained observers conducting assessments that required judgments were masked to treatment group identities. This included marking PERG peaks and nadirs and counting of Brn3a-positive cells in retina flatmounts and of TUNEL-positive cells and GFAP-positive fibers in fixed retina sections. Statistical analyses were conducted using Prism 8.4.2 Software (GraphPad Software Inc., La Jolla, CA, USA). A Shapiro–Wilk test was used to check the normality of the distribution of results. Data that passed this check were tested for significant differences in treatment outcomes by one-way ANOVA with Tukey's post hoc testing; data that did not undergo a Kruskal–Wallis test with Dunn's multiple comparison testing. The statistical test used for each result is identified in the figure captions. Every possible comparison between the three groups was considered. For all analyses, results were considered statistically significant when $p < 0.05$. Data are presented as mean ± SEM or median ± IQR.

5. Conclusions

We demonstrated that systemic TUDCA treatment significantly preserved RGC survival and RGC function in an ONC mouse model of RGC damage. TUDCA treatment also significantly suppressed activation of Müller glial cells, overexpression of some proinflammatory cytokines, and RGC apoptosis. These data suggest that TUDCA is a promising therapeutic candidate for preserving RGC function and delaying vision loss in glaucoma patients. However, more preclinical and clinical studies are needed to confirm the neuroprotective potential of TUDCA in glaucoma.

Author Contributions: Conceptualization, J.H.B. and J.M.N.; methodology, M.A.C.; software, J.W.; validation, M.A.C. and E.E.G.; investigation, N.Z. and J.T.S.; data curation, P.E.G. and J.T.S.; writing—original draft preparation, N.Z.; writing—review and editing, J.H.B. and Y.L.; visualization, N.Z.; supervision, J.H.B.; project administration, N.Z., X.Z. and P.E.G.; funding acquisition, J.H.B. All authors have read and agreed to the published version of the manuscript.

Funding: Funded by the Abraham J. and Phyllis Katz Foundation (J.H.B.); The Foundation for Atlanta Veterans Education and Research (J.H.B.); BrightFocus Foundation (G2020286, J.H.B. & Y.L); BrightFocus grant (G2022012S, J.W.). The joint training program between Emory University School of Medicine and Xiangya School of Medicine, Central South University; NIH R01EY028859 (J.H.B.); NIH R01EY021592 (J.M.N.); NIH R01EY028450 (J.M.N.); NIH R01 EY031042 (E.E.G.); VA I01RX002806 (J.H.B.); VA I21RX001924 (J.H.B.); VARR&D C9246C (Atlanta VAMC); and NIH P30EY06360 (Emory).

Institutional Review Board Statement: The animal study protocol was approved by the Emory Institutional Animal Care and Use Committee (Protocol ID PROTO201800248; approved 19 November 2018 through 18 November 2021) and followed the ARVO Statement for the Use of Animals in Ophthalmic and Vision Research.

Informed Consent Statement: Not applicable.

Data Availability Statement: All data are contained within the submitted manuscript. The data presented in this study are available in this submission.

Conflicts of Interest: The authors declare no conflicts of interest.

References

1. Tham, Y.C.; Li, X.; Wong, T.Y.; Quigley, H.A.; Aung, T.; Cheng, C.Y. Global prevalence of glaucoma and projections of glaucoma burden through 2040: A systematic review and meta-analysis. *Ophthalmology* **2014**, *121*, 2081–2090. [CrossRef]
2. Gupta, D.; Chen, P.P. Glaucoma. *Am. Fam. Physician* **2016**, *93*, 668–674. [PubMed]
3. Almasieh, M.; Levin, L.A. Neuroprotection in Glaucoma: Animal Models and Clinical Trials. *Annu. Rev. Vis. Sci.* **2017**, *3*, 91–120. [CrossRef]
4. McMonnies, C.W. Glaucoma history and risk factors. *J. Optom.* **2017**, *10*, 71–78. [CrossRef] [PubMed]
5. Quigley, H.A. Glaucoma. *Lancet* **2011**, *377*, 1367–1377. [CrossRef] [PubMed]
6. He, S.; Stankowska, D.L.; Ellis, D.Z.; Krishnamoorthy, R.R.; Yorio, T. Targets of Neuroprotection in Glaucoma. *J. Ocul. Pharmacol. Ther.* **2018**, *34*, 85–106. [CrossRef]
7. Shen, J.; Wang, Y.; Yao, K. Protection of retinal ganglion cells in glaucoma: Current status and future. *Exp. Eye Res.* **2021**, *205*, 108506. [CrossRef]
8. Sarikcioglu, L.; Demir, N.; Demirtop, A. A standardized method to create optic nerve crush: Yasargil aneurysm clip. *Exp. Eye Res.* **2007**, *84*, 373–377. [CrossRef]
9. Nuschke, A.C.; Farrell, S.R.; Levesque, J.M.; Chauhan, B.C. Assessment of retinal ganglion cell damage in glaucomatous optic neuropathy: Axon transport, injury and soma loss. *Exp. Eye Res.* **2015**, *141*, 111–124. [CrossRef]
10. Cameron, E.G.; Xia, X.; Galvao, J.; Ashouri, M.; Kapiloff, M.S.; Goldberg, J.L. Optic Nerve Crush in Mice to Study Retinal Ganglion Cell Survival and Regeneration. *Bio Protoc.* **2020**, *10*, e3559. [CrossRef]
11. Chen, Z.; Wang, R.; Wu, J.; Xia, F.; Sun, Q.; Xu, J.; Liu, L. Low-dose carbon monoxide inhalation protects neuronal cells from apoptosis after optic nerve crush. *Biochem. Biophys. Res. Commun.* **2016**, *469*, 809–815. [CrossRef] [PubMed]

12. Qijun, Z.; Huan, Z.; Ling, G.; Kaijian, C.; Wei, L.; Shuxing, J.; Xiang, C.; Rongdi, Y.; Jian, Y. The levels and significance of inflammasomes in the mouse retina following optic nerve crush. *Int. Immunopharmacol.* **2019**, *71*, 313–320. [CrossRef] [PubMed]
13. Mac Nair, C.E.; Schlamp, C.L.; Montgomery, A.D.; Shestopalov, V.I.; Nickells, R.W. Retinal glial responses to optic nerve crush are attenuated in Bax-deficient mice and modulated by purinergic signaling pathways. *J. Neuroinflamm.* **2016**, *13*, 93. [CrossRef]
14. Li, H.Y.; Ruan, Y.W.; Ren, C.R.; Cui, Q.; So, K.F. Mechanisms of secondary degeneration after partial optic nerve transection. *Neural Regen. Res.* **2014**, *9*, 565–574. [CrossRef]
15. Tang, F.; Xu, F.; Cui, L.; Huang, W.; Jiang, L.; Chen, L.; Yan, W.; He, W.; Shen, C.; Huang, H.; et al. The expression and role of PIDD in retina after optic nerve crush. *J. Mol. Histol.* **2020**, *51*, 89–97. [CrossRef] [PubMed]
16. Kusaczuk, M. Tauroursodeoxycholate-Bile Acid with Chaperoning Activity: Molecular and Cellular Effects and Therapeutic Perspectives. *Cells* **2019**, *8*, 1471. [CrossRef]
17. Khalaf, K.; Tornese, P.; Cocco, A.; Albanese, A. Tauroursodeoxycholic acid: A potential therapeutic tool in neurodegenerative diseases. *Transl. Neurodegener.* **2022**, *11*, 33. [CrossRef]
18. Keene, C.D.; Rodrigues, C.M.; Eich, T.; Chhabra, M.S.; Steer, C.J.; Low, W.C. Tauroursodeoxycholic acid, a bile acid, is neuroprotective in a transgenic animal model of Huntington's disease. *Proc. Natl. Acad. Sci. USA* **2002**, *99*, 10671–10676. [CrossRef]
19. Castro-Caldas, M.; Carvalho, A.N.; Rodrigues, E.; Henderson, C.J.; Wolf, C.R.; Rodrigues, C.M.; Gama, M.J. Tauroursodeoxycholic acid prevents MPTP-induced dopaminergic cell death in a mouse model of Parkinson's disease. *Mol. Neurobiol.* **2012**, *46*, 475–486. [CrossRef]
20. Pardue, M.T.; Allen, R.S. Neuroprotective strategies for retinal disease. *Prog. Retin. Eye Res.* **2018**, *65*, 50–76. [CrossRef]
21. Daruich, A.; Picard, E.; Boatright, J.H.; Behar-Cohen, F. Review: The bile acids urso- and tauroursodeoxycholic acid as neuroprotective therapies in retinal disease. *Mol. Vis.* **2019**, *25*, 610–624. [PubMed]
22. Zangerolamo, L.; Vettorazzi, J.F.; Solon, C.; Bronczek, G.A.; Engel, D.F.; Kurauti, M.A.; Soares, G.M.; Rodrigues, K.S.; Velloso, L.A.; Boschero, A.C.; et al. The bile acid TUDCA improves glucose metabolism in streptozotocin-induced Alzheimer's disease mice model. *Mol. Cell. Endocrinol.* **2021**, *521*, 111116. [CrossRef] [PubMed]
23. Lee, J.H.; Yoon, Y.M.; Lee, S.H. TUDCA-Treated Mesenchymal Stem Cells Protect against ER Stress in the Hippocampus of a Murine Chronic Kidney Disease Model. *Int. J. Mol. Sci.* **2019**, *20*, 613. [CrossRef]
24. Gao, X.; Fu, L.; Xiao, M.; Xu, C.; Sun, L.; Zhang, T.; Zheng, F.; Mei, C. The nephroprotective effect of tauroursodeoxycholic acid on ischaemia/reperfusion-induced acute kidney injury by inhibiting endoplasmic reticulum stress. *Basic Clin. Pharmacol. Toxicol.* **2012**, *111*, 14–23. [CrossRef]
25. Wang, W.; Zhao, J.; Gui, W.; Sun, D.; Dai, H.; Xiao, L.; Chu, H.; Du, F.; Zhu, Q.; Schnabl, B.; et al. Tauroursodeoxycholic acid inhibits intestinal inflammation and barrier disruption in mice with non-alcoholic fatty liver disease. *Br. J. Pharmacol.* **2018**, *175*, 469–484. [CrossRef] [PubMed]
26. Drack, A.V.; Dumitrescu, A.V.; Bhattarai, S.; Gratie, D.; Stone, E.M.; Mullins, R.; Sheffield, V.C. TUDCA slows retinal degeneration in two different mouse models of retinitis pigmentosa and prevents obesity in Bardet-Biedl syndrome type 1 mice. *Investig. Ophthalmol. Vis. Sci.* **2012**, *53*, 100–106. [CrossRef]
27. Oveson, B.C.; Iwase, T.; Hackett, S.F.; Lee, S.Y.; Usui, S.; Sedlak, T.W.; Snyder, S.H.; Campochiaro, P.A.; Sung, J.U. Constituents of bile, bilirubin and TUDCA, protect against oxidative stress-induced retinal degeneration. *J. Neurochem.* **2011**, *116*, 144–153. [CrossRef]
28. Fernandez-Sanchez, L.; Bravo-Osuna, I.; Lax, P.; Arranz-Romera, A.; Maneu, V.; Esteban-Perez, S.; Pinilla, I.; Puebla-Gonzalez, M.D.M.; Herrero-Vanrell, R.; Cuenca, N. Controlled delivery of tauroursodeoxycholic acid from biodegradable microspheres slows retinal degeneration and vision loss in P23H rats. *PLoS ONE* **2017**, *12*, e0177998. [CrossRef]
29. Tao, Y.; Dong, X.; Lu, X.; Qu, Y.; Wang, C.; Peng, G.; Zhang, J. Subcutaneous delivery of tauroursodeoxycholic acid rescues the cone photoreceptors in degenerative retina: A promising therapeutic molecule for retinopathy. *Biomed. Pharmacother.* **2019**, *117*, 109021. [CrossRef]
30. Ketabforoush, A.; Faghihi, F.; Azedi, F.; Ariaei, A.; Habibi, M.A.; Khalili, M.; Ashtiani, B.H.; Joghataei, M.T.; Arnold, W.D. Sodium Phenylbutyrate and Tauroursodeoxycholic Acid: A Story of Hope Turned to Disappointment in Amyotrophic Lateral Sclerosis Treatment. *Clin. Drug Investig.* **2024**, *44*, 495–512. [CrossRef]
31. Setchell, K.D.; Rodrigues, C.M.; Podda, M.; Crosignani, A. Metabolism of orally administered tauroursodeoxycholic acid in patients with primary biliary cirrhosis. *Gut* **1996**, *38*, 439–446. [CrossRef] [PubMed]
32. Fu, J.; Aung, M.H.; Prunty, M.C.; Hanif, A.M.; Hutson, L.M.; Boatright, J.H.; Pardue, M.T. Tauroursodeoxycholic Acid Protects Retinal and Visual Function in a Mouse Model of Type 1 Diabetes. *Pharmaceutics* **2021**, *13*, 1154. [CrossRef] [PubMed]
33. Phillips, M.J.; Walker, T.A.; Choi, H.Y.; Faulkner, A.E.; Kim, M.K.; Sidney, S.S.; Boyd, A.P.; Nickerson, J.M.; Boatright, J.H.; Pardue, M.T. Tauroursodeoxycholic acid preservation of photoreceptor structure and function in the rd10 mouse through postnatal day 30. *Investig. Ophthalmol. Vis. Sci.* **2008**, *49*, 2148–2155. [CrossRef] [PubMed]
34. Lawson, E.C.; Bhatia, S.K.; Han, M.K.; Aung, M.H.; Ciavatta, V.; Boatright, J.H.; Pardue, M.T. Tauroursodeoxycholic Acid Protects Retinal Function and Structure in rd1 Mice. *Adv. Exp. Med. Biol.* **2016**, *854*, 431–436. [CrossRef]

35. Kitamura, Y.; Bikbova, G.; Baba, T.; Yamamoto, S.; Oshitari, T. In vivo effects of single or combined topical neuroprotective and regenerative agents on degeneration of retinal ganglion cells in rat optic nerve crush model. *Sci. Rep.* **2019**, *9*, 101. [CrossRef]
36. Gomez-Vicente, V.; Lax, P.; Fernandez-Sanchez, L.; Rondon, N.; Esquiva, G.; Germain, F.; de la Villa, P.; Cuenca, N. Neuroprotective Effect of Tauroursodeoxycholic Acid on N-Methyl-D-Aspartate-Induced Retinal Ganglion Cell Degeneration. *PLoS ONE* **2015**, *10*, e0137826. [CrossRef]
37. Xia, H.; Nan, Y.; Huang, X.; Gao, J.; Pu, M. Effects of Tauroursodeoxycholic Acid and Alpha-Lipoic-Acid on the Visual Response Properties of Cat Retinal Ganglion Cells: An In Vitro Study. *Investig. Ophthalmol. Vis. Sci.* **2015**, *56*, 6638–6645. [CrossRef]
38. Boatright, J.H.; Boyd, A.P.; Garcia–Valenzuela, E. Effect of Tauroursodeoxycholic Acid on Mouse Retinal Ganglion Cell Degeneration Following Optic Nerve Transection. *Investig. Ophthalmol. Vis. Sci.* **2005**, *46*, 3180.
39. Li, P.; Fu, D.; Sheng, Q.; Yu, S.; Bao, X.; Lv, Z. TUDCA attenuates intestinal injury and inhibits endoplasmic reticulum stress-mediated intestinal cell apoptosis in necrotizing enterocolitis. *Int. Immunopharmacol.* **2019**, *74*, 105665. [CrossRef]
40. Deng, T.; Xie, J.; Ge, H.; Liu, Q.; Song, X.; Hu, L.; Meng, L.; Zhang, C. Tauroursodeoxycholic acid (TUDCA) enhanced intracytoplasmic sperm injection (ICSI) embryo developmental competence by ameliorating endoplasmic reticulum (ER) stress and inhibiting apoptosis. *J. Assist. Reprod. Genet.* **2020**, *37*, 119–126. [CrossRef]
41. Yanguas-Casas, N.; Barreda-Manso, M.A.; Nieto-Sampedro, M.; Romero-Ramirez, L. TUDCA: An Agonist of the Bile Acid Receptor GPBAR1/TGR5 With Anti-Inflammatory Effects in Microglial Cells. *J. Cell. Physiol.* **2017**, *232*, 2231–2245. [CrossRef] [PubMed]
42. Jiao, P.; Ma, J.; Feng, B.; Zhang, H.; Diehl, J.A.; Chin, Y.E.; Yan, W.; Xu, H. FFA-induced adipocyte inflammation and insulin resistance: Involvement of ER stress and IKKbeta pathways. *Obesity* **2011**, *19*, 483–491. [CrossRef]
43. Yanguas-Casas, N.; Barreda-Manso, M.A.; Nieto-Sampedro, M.; Romero-Ramirez, L. Tauroursodeoxycholic acid reduces glial cell activation in an animal model of acute neuroinflammation. *J. Neuroinflamm.* **2014**, *11*, 50. [CrossRef] [PubMed]
44. Noailles, A.; Fernandez-Sanchez, L.; Lax, P.; Cuenca, N. Microglia activation in a model of retinal degeneration and TUDCA neuroprotective effects. *J. Neuroinflamm.* **2014**, *11*, 186. [CrossRef]
45. Xing, C.; Huang, X.; Wang, D.; Yu, D.; Hou, S.; Cui, H.; Song, L. Roles of bile acids signaling in neuromodulation under physiological and pathological conditions. *Cell Biosci.* **2023**, *13*, 106. [CrossRef]
46. Ramalho, R.M.; Ribeiro, P.S.; Sola, S.; Castro, R.E.; Steer, C.J.; Rodrigues, C.M. Inhibition of the E2F-1/p53/Bax pathway by tauroursodeoxycholic acid in amyloid beta-peptide-induced apoptosis of PC12 cells. *J. Neurochem.* **2004**, *90*, 567–575. [CrossRef] [PubMed]
47. Chen, F.; Ge, Z.; Li, N.; Yu, Z.; Wu, R.; Zhao, Y.; He, X.; Cai, G. TUDCA protects against tunicamycin-induced apoptosis of dorsal root ganglion neurons by suppressing activation of ER stress. *Exp. Ther. Med.* **2022**, *24*, 509. [CrossRef]
48. Wu, H.; Yu, N.; Wang, X.; Yang, Y.; Liang, H. Tauroursodeoxycholic acid attenuates neuronal apoptosis via the TGR5/ SIRT3 pathway after subarachnoid hemorrhage in rats. *Biol. Res.* **2020**, *53*, 56. [CrossRef]
49. Garcia-Valenzuela, E.; Gorczyca, W.; Darzynkiewicz, Z.; Sharma, S.C. Apoptosis in adult retinal ganglion cells after axotomy. *J. Neurobiol.* **1994**, *25*, 431–438. [CrossRef]
50. Donahue, R.J.; Maes, M.E.; Grosser, J.A.; Nickells, R.W. BAX-Depleted Retinal Ganglion Cells Survive and Become Quiescent Following Optic Nerve Damage. *Mol. Neurobiol.* **2020**, *57*, 1070–1084. [CrossRef]
51. Lewis, G.P.; Fisher, S.K. Up-regulation of glial fibrillary acidic protein in response to retinal injury: Its potential role in glial remodeling and a comparison to vimentin expression. *Int. Rev. Cytol.* **2003**, *230*, 263–290. [CrossRef] [PubMed]
52. Bringmann, A.; Pannicke, T.; Grosche, J.; Francke, M.; Wiedemann, P.; Skatchkov, S.N.; Osborne, N.N.; Reichenbach, A. Muller cells in the healthy and diseased retina. *Prog. Retin. Eye Res.* **2006**, *25*, 397–424. [CrossRef]
53. Zhao, T.T.; Tian, C.Y.; Yin, Z.Q. Activation of Muller cells occurs during retinal degeneration in RCS rats. *Adv. Exp. Med. Biol.* **2010**, *664*, 575–583. [CrossRef]
54. Augustine, J.; Pavlou, S.; Ali, I.; Harkin, K.; Ozaki, E.; Campbell, M.; Stitt, A.W.; Xu, H.; Chen, M. IL-33 deficiency causes persistent inflammation and severe neurodegeneration in retinal detachment. *J. Neuroinflamm.* **2019**, *16*, 251. [CrossRef]
55. Zhang, X.; Zhang, N.; Chrenek, M.A.; Girardot, P.E.; Wang, J.; Sellers, J.T.; Geisert, E.E.; Brenner, C.; Nickerson, J.M.; Boatright, J.H.; et al. Systemic Treatment with Nicotinamide Riboside Is Protective in Two Mouse Models of Retinal Ganglion Cell Damage. *Pharmaceutics* **2021**, *13*, 893. [CrossRef] [PubMed]
56. Bach, M.; Hoffmann, M.B. Update on the pattern electroretinogram in glaucoma. *Optom. Vis. Sci.* **2008**, *85*, 386–395. [CrossRef] [PubMed]
57. Madeira, M.H.; Boia, R.; Santos, P.F.; Ambrosio, A.F.; Santiago, A.R. Contribution of microglia-mediated neuroinflammation to retinal degenerative diseases. *Mediat. Inflamm.* **2015**, *2015*, 673090. [CrossRef]
58. Adornetto, A.; Russo, R.; Parisi, V. Neuroinflammation as a target for glaucoma therapy. *Neural Regen. Res.* **2019**, *14*, 391–394. [CrossRef]
59. Wei, X.; Cho, K.S.; Thee, E.F.; Jager, M.J.; Chen, D.F. Neuroinflammation and microglia in glaucoma: Time for a paradigm shift. *J. Neurosci. Res.* **2019**, *97*, 70–76. [CrossRef]

60. Vohra, R.; Tsai, J.C.; Kolko, M. The role of inflammation in the pathogenesis of glaucoma. *Surv. Ophthalmol.* **2013**, *58*, 311–320. [CrossRef]
61. Wilson, G.N.; Inman, D.M.; Dengler Crish, C.M.; Smith, M.A.; Crish, S.D. Early pro-inflammatory cytokine elevations in the DBA/2J mouse model of glaucoma. *J. Neuroinflamm.* **2015**, *12*, 176. [CrossRef] [PubMed]
62. Cai, X.F.; Lin, S.; Geng, Z.; Luo, L.L.; Liu, Y.J.; Zhang, Z.; Liu, W.Y.; Chen, X.; Li, X.; Yan, J.; et al. Integrin CD11b Deficiency Aggravates Retinal Microglial Activation and RGCs Degeneration After Acute Optic Nerve Injury. *Neurochem. Res.* **2020**, *45*, 1072–1085. [CrossRef] [PubMed]
63. Yang, X.; Hondur, G.; Tezel, G. Antioxidant Treatment Limits Neuroinflammation in Experimental Glaucoma. *Investig. Ophthalmol. Vis. Sci.* **2016**, *57*, 2344–2354. [CrossRef] [PubMed]
64. Zhu, X.; Wu, S.; Zeng, W.; Chen, X.; Zheng, T.; Ren, J.; Ke, M. Protective Effects of Rapamycin on Trabecular Meshwork Cells in Glucocorticoid-Induced Glaucoma Mice. *Front. Pharmacol.* **2020**, *11*, 1006. [CrossRef]
65. Vecino, E.; Rodriguez, F.D.; Ruzafa, N.; Pereiro, X.; Sharma, S.C. Glia-neuron interactions in the mammalian retina. *Prog. Retin. Eye Res.* **2016**, *51*, 1–40. [CrossRef]
66. Kumar, A.; Pandey, R.K.; Miller, L.J.; Singh, P.K.; Kanwar, M. Muller glia in retinal innate immunity: A perspective on their roles in endophthalmitis. *Crit. Rev. Immunol.* **2013**, *33*, 119–135. [CrossRef]
67. Zhang, T.; Baehr, W.; Fu, Y. Chemical chaperone TUDCA preserves cone photoreceptors in a mouse model of Leber congenital amaurosis. *Investig. Ophthalmol. Vis. Sci.* **2012**, *53*, 3349–3356. [CrossRef]
68. Parrilla, G.E.; Vander Wall, R.; Chitranshi, N.; Basavarajappa, D.; Gupta, V.; Graham, S.L.; You, Y. RXR agonist, 9-cis-13,14-dihydroretinoic acid (9CDHRA), reduces damage and protects from demyelination in transsynaptic degeneration model. *Neuroscience* **2024**, *559*, 91–104. [CrossRef]
69. Maciulaitiene, R.; Kalesnykas, G.; Pauza, D.H.; Januleviciene, I. A combination of topical and systemic administration of brimonidine is neuroprotective in the murine optic nerve crush model. *PLoS ONE* **2024**, *19*, e0308671. [CrossRef]
70. Taniguchi, T.; Sharif, N.A.; Ota, T.; Farjo, R.A.; Rausch, R. Assessment of Brain-Derived Neurotrophic Factor on Retinal Structure and Visual Function in Rodent Models of Optic Nerve Crush. *Pharmaceuticals* **2024**, *17*, 798. [CrossRef]
71. Boatright, J.H.; Moring, A.G.; McElroy, C.; Phillips, M.J.; Do, V.T.; Chang, B.; Hawes, N.L.; Boyd, A.P.; Sidney, S.S.; Stewart, R.E.; et al. Tool from ancient pharmacopoeia prevents vision loss. *Mol. Vis.* **2006**, *12*, 1706–1714. [PubMed]
72. Li, Y.; Struebing, F.L.; Wang, J.; King, R.; Geisert, E.E. Different Effect of Sox11 in Retinal Ganglion Cells Survival and Axon Regeneration. *Front. Genet.* **2018**, *9*, 633. [CrossRef] [PubMed]
73. Talla, V.; Koilkonda, R.; Porciatti, V.; Chiodo, V.; Boye, S.L.; Hauswirth, W.W.; Guy, J. Complex I subunit gene therapy with NDUFA6 ameliorates neurodegeneration in EAE. *Invest. Ophthalmol. Vis. Sci.* **2015**, *56*, 1129–1140. [CrossRef] [PubMed]
74. Miura, G.; Wang, M.H.; Ivers, K.M.; Frishman, L.J. Retinal pathway origins of the pattern ERG of the mouse. *Exp. Eye Res.* **2009**, *89*, 49–62. [CrossRef]
75. Carpenter, A.E.; Jones, T.R.; Lamprecht, M.R.; Clarke, C.; Kang, I.H.; Friman, O.; Guertin, D.A.; Chang, J.H.; Lindquist, R.A.; Moffat, J.; et al. CellProfiler: Image analysis software for identifying and quantifying cell phenotypes. *Genome Biol.* **2006**, *7*, R100. [CrossRef]
76. Dordea, A.C.; Bray, M.A.; Allen, K.; Logan, D.J.; Fei, F.; Malhotra, R.; Gregory, M.S.; Carpenter, A.E.; Buys, E.S. An open-source computational tool to automatically quantify immunolabeled retinal ganglion cells. *Exp. Eye Res.* **2016**, *147*, 50–56. [CrossRef]
77. Mary-Sinclair, M.N.; Wang, X.; Swanson, D.J.; Sung, C.Y.; Mendonca, E.A.; Wroblewski, K.; Baumer, S.H.; Goldowitz, D.; Jablonski, M.M.; Skapek, S.X. Varied manifestations of persistent hyperplastic primary vitreous with graded somatic mosaic deletion of a single gene. *Mol. Vis.* **2014**, *20*, 215–230.
78. Zhang, X.; Henneman, N.F.; Girardot, P.E.; Sellers, J.T.; Chrenek, M.A.; Li, Y.; Wang, J.; Brenner, C.; Nickerson, J.M.; Boatright, J.H. Systemic Treatment With Nicotinamide Riboside Is Protective in a Mouse Model of Light-Induced Retinal Degeneration. *Investig. Ophthalmol. Vis. Sci.* **2020**, *61*, 47. [CrossRef]
79. Sellers, J.T.; Chrenek, M.A.; Girardot, P.E.; Nickerson, J.M.; Pardue, M.T.; Boatright, J.H. Initial Assessment of Lactate as Mediator of Exercise-Induced Retinal Protection. *Adv. Exp. Med. Biol.* **2019**, *1185*, 451–455.
80. Mees, L.M.; Coulter, M.M.; Chrenek, M.A.; Motz, C.T.; Landis, E.G.; Boatright, J.H.; Pardue, M.T. Low-Intensity Exercise in Mice Is Sufficient to Protect Retinal Function During Light-Induced Retinal Degeneration. *Investig. Ophthalmol. Vis. Sci.* **2019**, *60*, 1328–1335. [CrossRef]
81. Getz, T.E.; Chrenek, M.A.; Papania, J.T.; Shelton, D.A.; Markand, S.; Iuvone, P.M.; Kozmik, Z.; Boatright, J.H.; Nickerson, J.M. Conditional Knockouts of Interphotoreceptor Retinoid Binding Protein Suggest Two Independent Mechanisms for Retinal Degeneration and Myopia. *Investig. Ophthalmol. Vis. Sci.* **2024**, *65*, 32. [CrossRef] [PubMed]

pharmaceuticals

MDPI

Article

Kinin B1- and B2-Receptor Subtypes Contract Isolated Bovine Ciliary Muscle: Their Role in Ocular Lens Function and Intraocular Pressure Reduction

Najam A. Sharif [1,2,3,4,5,6,7,*], Madura Kulkarni-Chitnis [1], Anthonia Okolie [1], Ya Fatou Njie-Mbye [1] and Sunny E. Ohia [1]

1 Department of Pharmaceutical Sciences, College of Pharmacy and Health Sciences, Texas Southern University, Houston, TX 75207, USA; yafatou.njie-mbye@tsu.edu (Y.F.N.-M.); sunny.ohia@tsu.edu (S.E.O.)
2 Singapore Eye Research Institute (SERI), Singapore 169856, Singapore
3 Institute of Ophthalmology, University College London (UCL), London WC1E6BT, UK
4 Department of Ophthalmology, Imperial College of Science and Technology, St. Mary's Campus, London SW72AZ, UK
5 Department of Pharmacy Sciences, Creighton University, Omaha, NE 68178, USA
6 Department of Pharmacology and Neuroscience, University of North Texas Health Sciences Center, Fort Worth, TX 76107, USA
7 Department of Ophthalmology, Eye-ACP Duke-National University of Singapore Medical School, Singapore 169857, Singapore
* Correspondence: nsharif@nanostherapeutics.com; Tel.: +1-(817)-565-7686

Abstract: Background: Bradykinin is an endogenously produced nonapeptide with many physiological and pathological functions that are mediated by two pharmacologically defined receptor subtypes, B1- and B2-receptors. Current studies sought to characterize the functional bradykinin (BK) receptors present in freshly isolated bovine ciliary muscle (BCM) using an organ-bath tissue contraction system. **Methods:** Cumulative longitudinal isometric tension responses of BCM strips (4–5 mm) were recorded before and after the addition of test compounds to BCM strips hooked up to an isometric strain gauge transducer system. **Results:** BK and its analogs (7–11 concentrations) contracted BCM in a biphasic concentration-dependent manner. The first high affinity/potency phase accounted for 40–60% of the maximal contraction by each of BK (potency, EC_{50} = 0.9 ± 0.3 nM), Lys-BK (EC_{50} = 0.7 ± 0.1 nM), Met-Lys-BK (EC_{50} = 1 ± 0.1 nM), Hyp3-BK (EC_{50} = 1 ± 0.2 nM), RMP-7 (EC_{50} = 3.5 ± 0.5 nM), and Des-Arg^9-BK (EC_{50} = 10 ± 0.4nM) (mean ± SEM, n = 3–8). The second lower activity phase of contraction potency values for these peptides ranged between 100 nM and 3 μM. In the presence of a selective B1-receptor antagonist (R715; 0.1–10 μM), the concentration–response curves to Des-Arg9-BK (B1-receptor agonist) were still observed, indicating activation of B2-receptors by this kinin. Likewise, when B2-receptors were completely blocked by using a B2-selective antagonist (WIN-64338; 1–10 μM), BK still induced BCM contraction, now by stimulating B1-receptors. **Conclusions:** This agonist/antagonist profile of BCM receptors indicated the presence of both B1- and B2-receptor subtypes, both being responsible for contracting this smooth muscle. The BCM kinin receptors may be involved in changing the shape of the ocular lens to influence accommodation, and since the ciliary muscle is attached to the trabecular meshwork through which aqueous humor drains, endogenously released kinins may regulate intraocular pressure.

Keywords: ciliary muscle; bradykinin; receptor subtypes; B1 receptors; B2 receptors; kinin; pharmacology; IOP regulation

1. Introduction

The nine amino acid-containing peptide bradykinin (BK) and the octapeptide Des-Arg^9-BK are naturally produced and released in many tissues of the mammalian body where they perform numerous functions [1–3]. Amongst the deleterious roles, BK and

Des-Arg^9-BK release inflammatory cytokines and induce edematous conditions and pain. However, this BK is also involved in beneficial activities such as tissue contraction, hormone secretion, cell proliferation, and cellular protection in a direct or indirect manner [1–3]. BK interacts with two major receptor subtypes, B1 and B2, which possess a higher affinity for and potency at the B2 receptor whose activity can be selectively blocked by the B2-selective antagonist WIN-64338 [1–3]. Conversely, Des-Arg^9-BK is a selective agonist for the B1 receptor whereas R715 behaves as a B1 receptor selective antagonist [1–3]. While the mammalian eye is an immune-privileged organ, a robust kallikrein/kinase and kininergic system exists, which locally generates BK and other fragments of the peptide as determined by immunohistochemical and molecular biological techniques [2,4–7]. Furthermore, the latter endogenous peptides interact with B1- and B2-receptors, with the B2-type being constitutively expressed while B1-type receptors are present at a low level in most cells but are inducible in the presence of pro-inflammatory cytokines, endotoxins, or after tissue injury or trauma [1–3]. Many ocular tissues express B2-receptors in the membranes of kinin-responsive cells such as corneal and conjunctival epithelial cells, non-pigmented ciliary epithelial cells, ciliary muscle cells, and trabecular meshwork cells [2,4–8], although the mRNAs for the B1-receptors have also been reported in some of these cells [4,6]. However, both the B1- and B2-subtype receptors mediate the biological functions of BK and its analog peptides primarily via Gq-coupled signaling pathways, which utilize inositol phosphates, intracellular Ca^{2+} ($[Ca^{2+}]_i$), and diacylglycerol as second messengers within the cytoplasm, which can be quantitated as indices of receptor activation and utilized for biochemical/pharmacological characterization [2,5–8]. Additional receptor-mediated functional readouts include early response kinase (ERK1/2) and matrix metalloproteinase (MMP) gene and protein up-regulation and/or prostaglandin release in response to BK and its analogs [2,5–8].

Bradykinin has been reported to contract isolated smooth muscles from several mammalian species [9,10]. In the present study, our aim was to use a different functional readout to confirm the presence of and to pharmacologically characterize the kinin receptors in freshly isolated strips of bovine ciliary muscle (BCM) using a tissue contraction bioassay system. Such an organ-bath/tissue response system was previously utilized to characterize muscarinic cholinergic receptors [11–13], serotonin-2 receptors [14], endothelin receptors [15], and various other receptor types in isolated strips or rings of ocular tissues [16].

2. Results

All of the compounds tested yielded concentration-dependent contraction of BCM strips in vitro. The positive control agent, carbachol, always contracted the tissue, and its actions were concentration-dependently blocked by increasing concentrations of the muscarinic receptor antagonist, atropine, shifting the concentration-response curves of carbachol to the right (Figure 1). This profile confirmed the competitive antagonism properties of atropine, as has been documented in prior studies in various tissues, including BCM [11–13]. Furthermore, the concentration-response curve for carbachol, in the absence and presence of atropine, was monophasic, yielding a potency (EC_{50}) value of 1 ± 0.07 μM ($n = 6$) for atropine (Figure 1) [11], with this being like values obtained in CM-derived cells and tissues [11–13]. The use of carbachol-induced BCM contraction was deemed necessary to ensure tissue viability in each experiment before any of the kinins were tested.

Regarding the actions of BK and its analogs (Figure 2), all produced concentration-dependent contractions of the BCM over the range of peptide concentrations used (0.1 nM to 10 μM) (Figure 3). All agonist kinin peptides displayed rather shallow concentration-response curves (e.g., Figure 3) as compared to the muscarinic receptor agonist, carbachol (Figure 1). The biphasic activity of the peptides was akin to that reported by Field et al. [17,18] in the guinea pig colon and trachea. The delineation of the high- (indicated by the dotted line representing the top of the first phase of the contraction, e.g., Figure 3) and low-potency components of the kinin-induced concentration-response curves yielded the potency values

(EC_{50}, half-maximal contractile concentration) for each of the two receptor sites. The data obtained from 3–8 experiments for each of the test compounds are shown in Table 1. The apparent rank order of potency of the peptides contracting the BCM strips at the high-affinity/potency site was Lys-BK ≥ BK ≥ Met-Lys-BK = Hyp^3-BK > RMP-7 >> Des-Arg^9-BK. Conversely, the compound rank order of potency at the low-affinity/potency site from the biphasic concentration-response curves (representing interaction with the B1-receptors) was BK = Met-Lys-BK = > Hyp^3-BK > Lys-BK. However, due to the relative receptor subtype selectivity of Des-Arg^9-BK, its interaction with the B1-receptor was represented by the high-potency site (EC_{50} = 30 nM), whereas the low-potency site representing the B2-receptor exhibited a low-potency component of the concentration-response curve (EC_{50} = 3000 nM) (Table 1).

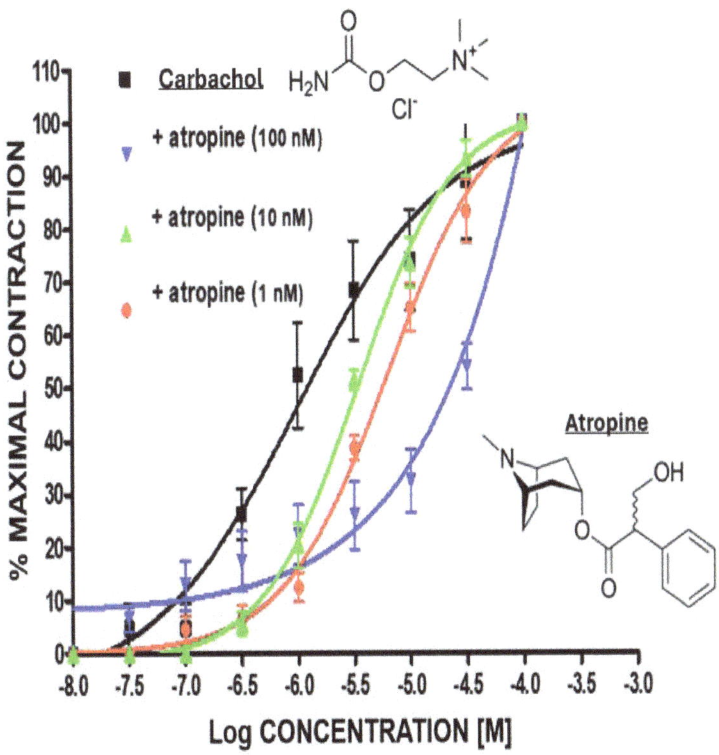

Figure 1. Carbachol contracts bovine ciliary muscle (BCM) and atropine inhibits the carbachol-induced contraction in vitro. The monophasic concentration-dependent profile of BCM contraction by carbachol was progressively shifted to the right in the presence of different concentrations of the muscarinic receptor antagonist, atropine. Mean ± SEM of n = 6–8 separate experiments using freshly isolated BCM strips.

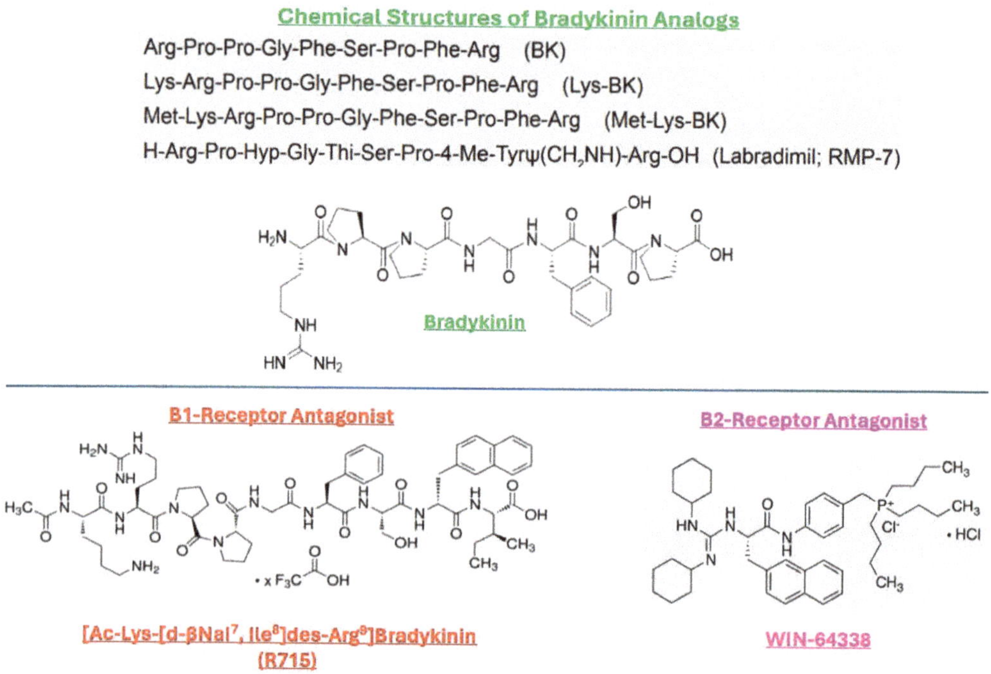

Figure 2. Chemical structures of bradykinin and its key analogs and those of the selective B1- and B2-receptor antagonists. The amino acid composition and linkage to form bradykinin and its analogs (linear peptides) are depicted in the upper panel. The lower panel shows the peptidic nature of R715 (B1-receptor antagonist; [17]) and the non-peptidic B2-receptor antagonist, WIN-64338 [1–3,8,18–20]. It must be noted that while WIN-64338 generally acts as a specific B2-receptor-selective antagonist, it has been shown to also block the cGMP-producing effects of Des-Arg^9-BK, the B1-receptor agonist, in bovine aortic endothelial cells [21].

Table 1. Functional potencies of BK and BK-based analogs at contracting bovine ciliary muscle strips in vitro.

Kinin Agonist	Reported Receptor Selectivity [1–3,18,20]	Max. Tension Generated During Contraction (gm Tension)	Functional Contractile Activity Potency (EC_{50})	
			High Affinity and Potency Phase	Low Affinity and Potency Phase
Bradykinin (BK)	B2 > B1	0.07 ± 0.01	0.9 ± 0.3 nM	100 ± 2.4 nM
Met-Lys-BK	B2 > B1	0.06 ± 0.01	1.0 ± 0.1 nM	100 ± 8.2 nM
Lys-BK	B2 ≥ B1	0.06 ± 0.01	0.7 ± 0.1 nM	500 ± 12.8 nM
Hyp^3-BK	B2 > B1	0.04 ± 0.01	1.0 ± 0.2 nM	150 ± 9.0 nM
RMP-7	B2 > B1	0.07 ± 0.01	3.5 ± 0.5 nM	N/A
Des-Arg^9-BK	B1 > B2	0.06 ± 0.01	30 ± 1.9 nM	3000 ± 72.0 nM

Data are mean ± SEM from 3–8 independent experiments using freshly obtained BCM strips and from tissue contraction concentration–response curves using 7–11 different concentrations (0.1 nM–10 μM) of each compound. N/A = not applicable.

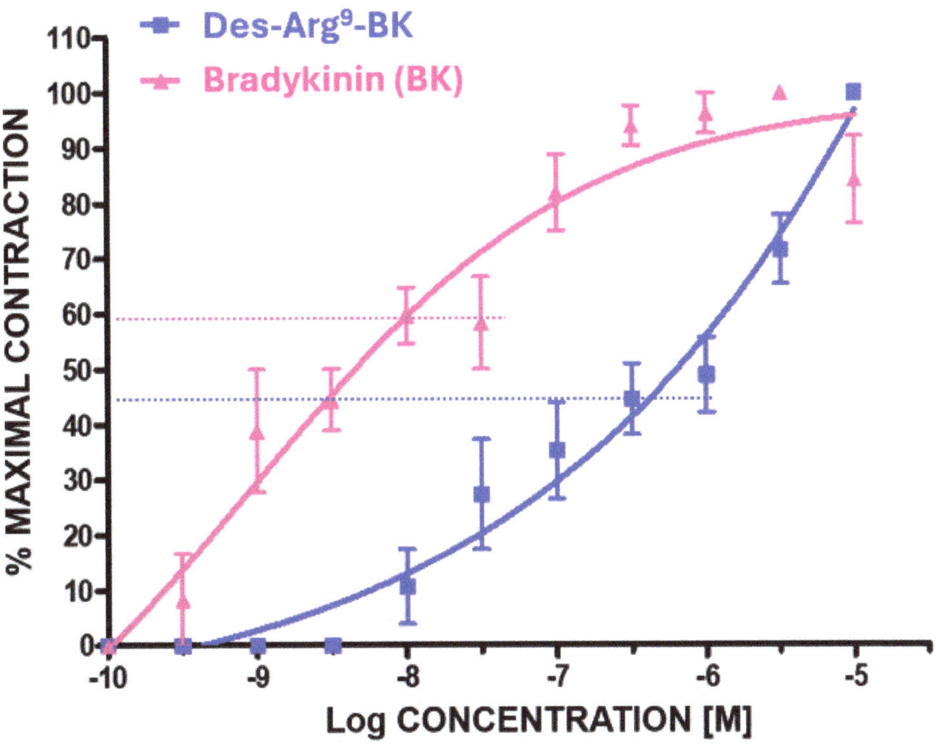

Figure 3. BCM contractions were induced by both the selective B1-receptor agonist (Des-Arg^9-BK) and the B2-receptor agonist, bradykinin (BK), with different profiles of activity. Both peptides induced bi-phasic concentration-dependent contractile responses, yielding rather shallow graphs. These become apparent when the top of each high-potency phase of the graphs is delineated with a dotted line, as shown. Data derived from each phase of the concentration–response (C-R) curve yielded the relative functional potencies of the peptides at the two receptor sub-types. Thus, for example, BK exhibited a half-maximal response (EC_{50}) at 0.9 nM concentration at the B2 receptor, while its potency at the B1 receptor was EC_{50} = 100 nM. Conversely, Des-Arg^9-BK possessed an EC_{50} of 30 nM at the B1 receptor and an EC_{50} of 3 μM at the B2 receptor. Mean ± SEM of n = 3–8 independent experiments using strips of BCM isolated from fresh bovine eyes.

By completely blocking the B1-receptor with a high affinity/potency B1-antagonist, R715 (0.1–10 μM [22]; Figure 2), Des-Arg^9-BK was still allowed to stimulate the contraction of BCM strips (Figure 4). Likewise, completely blocking the B2 receptor with an antagonist for this subtype (WIN-64338 (1–10 μM [1–3,18]; Figure 2)) did not prevent BK from concentration-dependently contracting the BCM strips (Figure 5).

The functional contractile efficacies/potencies of BK and Des-Arg^9-BK, two very receptor-subtype-differentiating kinins [1–3,19,20,23–25], in the BCM contraction bioassay, were compared with their activities in numerous tissues of different species (Table 2), either where the assays revealed biphasic actions of the peptides (e.g., guinea pig taenia caeci and trachea [17,18]) as in the BCM; or where only a single contractile receptor site was evident (e.g., B1-receptor in rat ileum [25]; or in rat aortic endothelium [22]; where Des-Arg^9-BK exhibited the highest potency); or where only a B2-receptor-mediated contraction was observed (e.g., human umbilical vein, rat uterus, rat whole bladder) [1–3,18,20,23]; or where an undefined kinin receptor was described (e.g., in mouse/human bladder strips); or where BK had a low potency at the high-affinity binding site [10]. Additional functional assay data such as kinin-induced $[Ca^{2+}]_i$ mobilization, obtained from human ciliary muscle (hCM) cells and human trabecular meshwork (hTM) cells, which contain a single B2-receptor subtype, was also tabulated for comparison purposes (Table 2) [2,8].

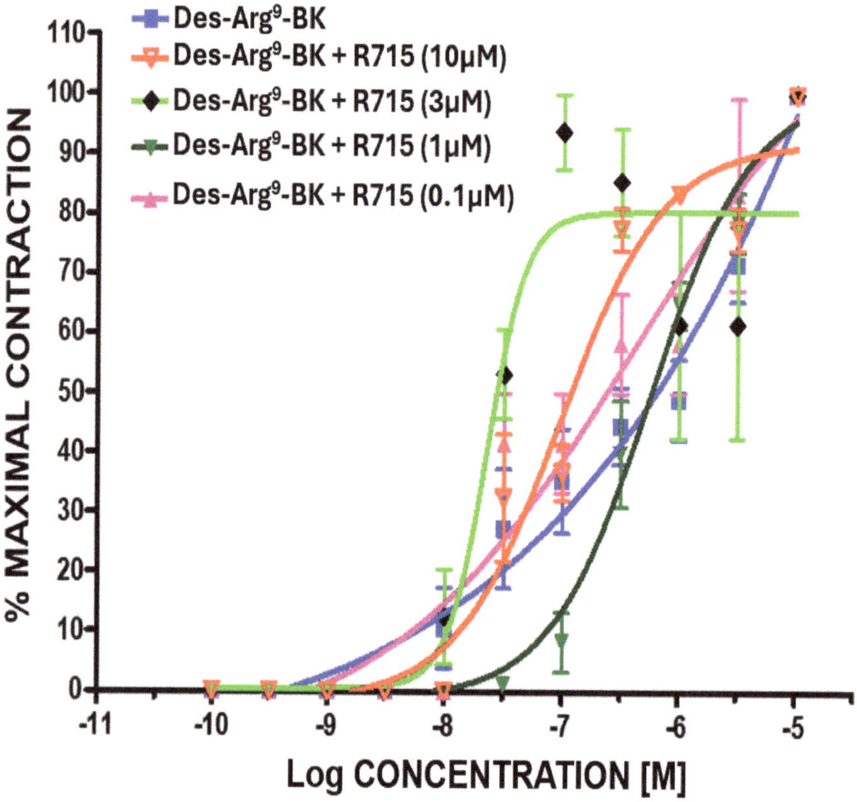

Figure 4. Complex effects of the selective B1-receptor antagonist (R715) on the concentration-response (C-R) curves of the B1-receptor agonist, Des-Arg^9-BK, in the BCM contraction assay. Despite complete blockage of the B1-receptor by R715 (0.1–10 μM), Des-Arg^9-BK still contracted the BCM strips but now by activating the B2-receptor. Mean ± SEM from up to 8 independent experiments using fresh strips of BCM isolated from bovine eyes.

Table 2. Ability of BK and Des-Arg^9-BK to Contract Various Tissues of Different Species or to Mobilize Intracellular Ca^{2+} in human Ocular Cells.

Functional Potency Values in Various Tissues and Cells																		
Compound	**BCM**		**GPTC**		**GPT**		**HUV**		**RUT**		**RWB**		**M/H BSTP**		**RI**		**hCM/hTM Cells**	
	HP Site	LP Site	HP Site	LP Site	HP Site	LP Site	HP Site	LP Site	HP Site	LP Site	HP Site	LP Site	HP Site	LP Site	HP Site	LP Site	HP Site	LP Site
BK	0.9 nM	100 nM	3 nM	1000 nM	20 nM	4000 nM	9 nM	N/A	3 nM	N/A	90 nM	N/A	1200/ 5100 nM	N/A	4890 nM	N/A	2.4/1 nM	N/A
Des-Arg^9-BK	30 nM	3000 nM	N/A	N/A	N/A	N/A	>1000 nM	N/A	>1000 nM	N/A	ND	ND	ND	ND	18.6 nM	N/A	4200/ 3600 nM	N/A

Data shown represent the mean functional potencies of the major kinins in contracting the smooth muscle contained in various tissues of multiple species, or their abilities to induce mobilization of intracellular calcium in human ciliary muscle (hCM) and human trabecular meshwork (hTM) cells in vitro, obtained from multiple experiments. BK is B2-receptor-selective, while Des-Arg^9-BK is B1-receptor-selective. BCM = bovine ciliary muscle; GPTC = guinea pig tania caeca; GPT = guinea pig trachea; RUT = rat uterus strips; RWB = rat whole bladder; M/H BSTP = mouse/human bladder strips; RI = rat ileum; hCM = human ciliary muscle; hTM = human trabecular meshwork. HP site = high potency binding site; LP site = low potency binding site.

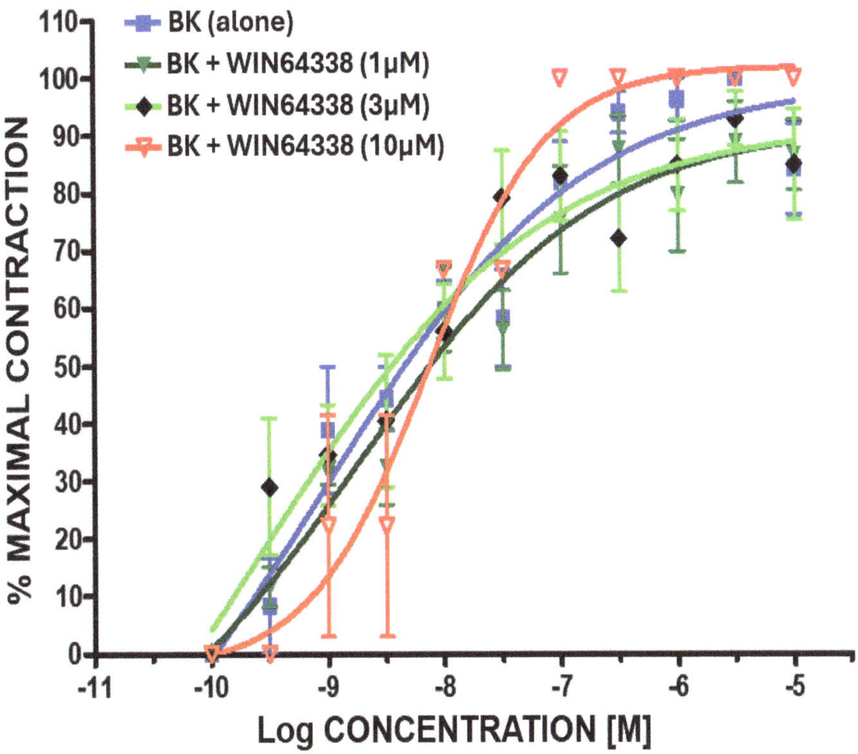

Figure 5. Complex effects of the selective B2-receptor antagonist (WIN-64338) on the concentration–response (C-R) curves of the B2-receptor agonist, BK, in the in vitro BCM contraction assay system. Even in the presence of the B2-receptor antagonist (1–10 µM), BK continued to contract the muscle, now by activating the B1-receptors. Mean ± SEM from up to 8 independent experiments using new strips of BCM isolated from freshly obtained bovine eyes.

3. Discussion

BK and Des-Arg^9-BK are highly B2- and B1-receptor selective agonists, respectively, which have been used to pharmacologically characterize the subtypes of receptors activated by these peptides to mediate many functional activities in numerous diverse cells and tissues [1–3,7,8,19,20,23–25]. Additional subtype classification has been aided by specific B1- (e.g., R715) and B2-antagonists (e.g., WIN-64338) [1–3,7,8,20,23]. The agonist peptides have been reported to induce contractions of many tissues of different species and certain bioassays have been validated using agonist/antagonist profiles. For example, B1-receptors exist in the vascular smooth muscle of humans, rats, and rabbit aorta, renal artery, coronary artery [22,24], rabbit bladder [19], and rat ileum [25]. B2 receptors have been localized and characterized in a multitude of tissues and cells (e.g., in the brain: neurons within the thalamus, cerebral cortex, and hypothalamus; in the cardiovascular system: endothelial cells of the aorta, carotid arteries, and mesenteric arteries; other tissues: eye (e.g., ciliary muscle, trabecular meshwork, corneal and conjunctival epithelial cells), kidney, liver, lung, prostate, small intestine, uterus, etc.) [1–3,7,8,19,20,23–25]. Likewise, some tissues express both B1 and B2 receptors (e.g., guinea pig colon and trachea), as is evident from biphasic tissue contraction concentration–response curves for BK [17,18] (Table 2). The latter publications by Field et al. [17,18] clearly showed the low- and high-potency components of the contractile responses to bradykinin and thus directly support our observations of biphasic responses in the BCM preparation.

Mechanistically, smooth muscle contraction is induced by kinin receptor-activated generation of inositol trisphosphate, amongst other inositol phosphates, that mobilizes and elevates $[Ca^{2+}]_i$, which binds to calmodulin and in turn activates myosin light chain (MLC) kinase. The latter process then leads to phosphorylation of the head of MLC, which eventually leads to cross-bridge cycling between actin and myosin ([16,26]; Figure 6). Myosin light chain phosphorylation is also regulated by MLC phosphatase, which removes the phosphate from MLC and promotes relaxation. Gα12/13-coupled receptor signaling involves activation of the small G-protein RhoA and consequently Rho-kinase, which inactivates MLC phosphatase leading to a sustained contraction ([10,20,26,27]; Figure 6). Although BK contracts smooth muscles in the body as described and referenced above, vasorelaxation of certain vascular beds by BK and Des-Arg^9-BK has also been reported but this requires a relatively high basal tone of the tissues often due to the type(s) of innervation they receive and/or the prevailing local cellular environmental conditions [1,3].

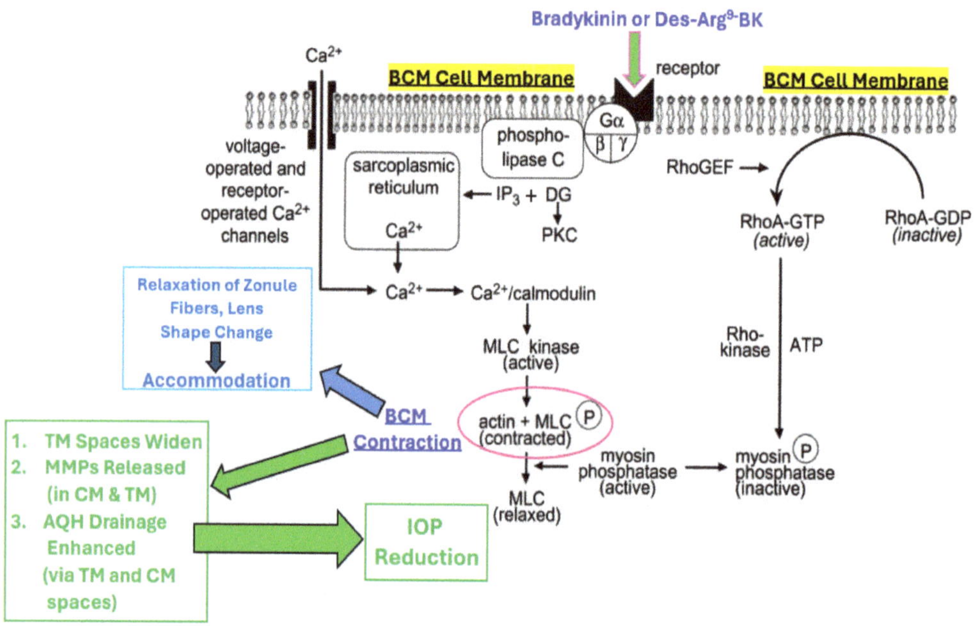

Figure 6. Schematic depiction of BCM smooth muscle cell contraction involves multiple intracellular mechanisms upon B1/B2 kinin receptor activation. BK receptor activation leads to the production of intracellular second messengers and consequential enhancement of various enzyme activities within the CM cells and connected tissues such as the TM and lens. These collective activities impact accommodative functions of the ocular lens and increase AQH dynamics to lower and control IOP.

In the current studies, both BK and Des-Arg^9-BK contracted the BCM in a biphasic fashion, indicating the involvement of both B1 and B2 receptors in the smooth muscle contraction process. Other agonist peptides related to BK also elicited similar profiles of contractility, thereby substantiating the effects of BK and Des-Arg^9-BK in this specific tissue. Furthermore, the role of both kinin receptor subtypes being responsible for contracting BCM in vitro was confirmed by sequentially recording the functional responses of the tissue in the presence of B1- and B2-receptor antagonists in numerous independent experiments. These collective observations are unique and novel and have been characterized for the first time in BCM. The agonist potency values for BK, Lys-BK, Hyp^3-BK, Met-Lys-BK, and RMP-7 observed for the high-affinity BK-binding/potency site in the BCM contraction bioassay (0.7 to 3.5 nM EC_{50} values), corresponding to stimulation of B2-receptors (Table 1), compared well to those observed for other tissue contractile responses [1,3,17–19,22–25] (Table 2) and for the stimulation of second messenger production in various ocular cell-types involved in human CM and TM contraction (Table 2; [1–3,5–8]). Similarly, the ability of Des-Arg^9-BK to induce BCM contraction via the high-affinity/potency binding site (1st phase of the contractile profile; Figure 3), and thus the B1-receptor (EC_{50} = 30 nM) (Tables 1 and 2), compared closely with the rat ileum contraction potency of 18.6 nM (Table 2) [25]. Conversely, the second phase BCM contraction potency of Des-Arg^9-BK (EC_{50} = 3000 nM), corresponding to its activation of the B2 receptor via the low-affinity/potency site, was closely similar to that obtained for human and rat uterus and for mobilizing $[Ca^{2+}]_i$ in human TM and human CM cells (Table 2) [2,5,6,8,26]. For contextual purposes, in recombinant B1 receptors expressed in host cells, BK exhibited a binding affinity (inverse binding inhibition constant; K_i values) of 200 nM (mouse), 5000 nM (rabbit), and 10,000 nM (human). Conversely, BK displayed K_i values at the cloned B2 receptor of 0.48 nM (mouse), 4.5 nM (rabbit), and 0.54 nM. Similarly, the K_i values of Des-Arg^9-BK at cloned B1 receptors were 0.7 nM (mouse), 32 nM (rabbit), and 1930 (human) [1]. The K_i values of Des-Arg^9-BK at the cloned B2-receptors were >1000 nM (rabbit), 6400 nM (mouse), and 8100 nM (human) [1]. Taken together, the agonist (and antagonist) profiles of kinin-mediated contraction of isolated BCM closely resemble the profiles of pharmacologically defined B1 and B2 receptors as reported for several tissues and cells expressing kinin receptors [1–3,7,8,17–20,22–27].

As to the functions of the B1- and B2-kinin receptor subtypes present in the BCM, several possibilities exist. The mammalian CM is composed of longitudinal and circular muscle fibers and is an important part of the eye that contributes to a person's ability to view objects clearly at varying distances [16,28]. The contraction of the CM circular fibers primarily causes relaxation of the zonule fibers connected to the lens, rendering the latter more spherical [16,28]. Under these conditions, light refraction permits near vision. Obviously, the relaxation of the circular ciliary fibers allows far vision, and thus a change from contraction to relaxation of these fibers is responsible for accommodation performed by the lens located in the middle of the eye [16,28]. On the other hand, the muscle bundles of the radial part of the CM and the inner bundles of the longitudinal portion of the CM form tendons in their anterior insertion that are continuous with the extracellular matrix (ECM) of the TM beams and connect with the ECM of the scleral spur [16,28,29]. The latter contains elastic fibers and collagen, which are continuous with those of the core of the corneoscleral TM beams. Hence, due to this structural arrangement of these anterior uveal tissues, CM contraction pulls the scleral spur posteriorly and widens the spaces within the TM, thereby allowing the aqueous humor (AQH) to flow out of the anterior chamber and thereby lowering the IOP [16,29]. Therefore, endogenously released BK and its metabolism to Des-Arg^9-BK within the CM can lead to contraction of the CM via B2 and B1 receptors, respectively, and thus regulation of intraocular pressure can be achieved by influencing the egress of AQH via the conventional outflow pathway [8,16,29]. The latter outflow pathway engaged by the CM accounts for 75–80% of the total drainage of AQH from the anterior chamber of the eye under normal conditions, which is progressively reduced in ocular hypertension and under glaucomatous circumstances [16,29]. Hence, locally released kinins can help regulate IOP and prevent glaucomatous optic neuropathy, a prevalent potentially blinding disease [2,4–6,8].

In addition to regulating conventional AQH outflow through CM contraction, the eye has developed another unconventional AQH drainage system that takes advantage of increasing the gaps between CM muscle bundles to affect IOP control [16,29]. In animals and human subjects that experience OHT/glaucoma due to blockage of the conventional outflow pathway by aberrantly depositing ECM in the TM and Schlemm's canal area [16,29], AQH drainage via the CM spaces becomes more important. However, this requires the local generation and release of matrix metalloproteinases, which can digest the ECM in between the CM bundles to permit AQH efflux [16,29]. The BK receptors, in particular the B2 subtype present in the CM cells, can initiate the up-regulation of genes to enhance MMP production and thus help eliminate some of the ECM to increase uveoscleral outflow of AQH to lower IOP [2,8]. Indeed, such mechanisms have been shown to operate in response to the delivery of B2-receptor agonists to the anterior chamber (e.g., in rabbits after intravitreal injection of BK) [2,8] and in cynomolgus monkeys following topical ocular instillation of a non-peptide B2-receptor agonist whose IOP-lowering activity could be blocked by a B2-receptor antagonist [2,8]. Whether such IOP-lowering effects of BK and other kinins occur in cattle remains to be determined. Interestingly, however, Webb et al. [5] have demonstrated that BK was able to stimulate conventional outflow of AQH from anterior segments of bovine eyes in organ culture in a time-dependent manner that required the generation of MMPs. Such data strongly support the possibility that BK and its analogs would modulate IOP in vivo in cattle, and such studies are eagerly awaited.

In conclusion, the current studies have pharmacologically defined the presence of both B1- and B2-receptor subtypes in BCM using a variety of BK analogs and selective antagonists of each of the receptors [1–3,23]. These kinin receptor subtypes potently contract the smooth muscle within the BCM, which may be involved in accommodative and AQH dynamic/IOP regulatory functions. Since BK also contracts the pupillary sphincter muscle in rabbits [30] and lowers IOP in the same species [2,8] and given that non-peptide BK agonist lowers and controls IOP in monkeys via the B2-receptor [2,8], it is anticipated that the kinin receptors in BCM serve similar functions. However, further investigations are needed to demonstrate these effects of kinin receptor activation via the CM in cattle.

4. Materials and Methods

The methods utilized in the current studies were adapted from Lograno and Romano [13] and Ohia et al. [14], with some minor modifications. Fresh bovine eyeballs were obtained from a local abattoir (Fisher Ham and Meat Company, Houston, TX, USA) and transported to the laboratory in ice. Whole eyeballs were placed in warm oxygenated Krebs buffer solution for 30 min to ensure equilibration before dissection of the ciliary muscle. Briefly, the anterior segment was dissected out of whole eyeballs after an incision was made at the limbus. The anterior segment containing the cornea, anterior uvea, and lens was then inverted and the lens dislodged with a probe along with the attached vitreous humor. After removal of the vitreous humor and lens, the whole ciliary muscle (comprising both longitudinal and circular portions) was quickly isolated and dissected from the scleral spur and choroid. Muscle strips of 4–5 mm in length were cut and mounted in 25 mL organ baths containing oxygenated Krebs solution at 37 °C using a silk thread linked to an isometric strain gauge under a constant load of 1g tension. The tissues were then allowed to equilibrate for 30 min. Two ciliary muscle strips were prepared from each eyeball and four to eight strips were used for each series of experiments. The composition of the Krebs solution used (mM) was NaCl, 118; KCl, 4.8; $CaCl_2$, 2.5; KH_2PO_4, 1.2; $NaHCO_3$, 25; $MgSO_4$, 2.0; dextrose, 10 and flurbiprofen, 0.003 (pH 7.4), all purchased from Sigma-Aldrich, St. Louis, MO (USA). Before any drug additions, the tissues were challenged with carbachol (10 μM) at least once to assess the functional state of the ciliary muscle strips and only those that yielded robust responses were subsequently utilized in the experiments. A robust response was characterized as a single contraction that consistently yielded a tension in the muscle induced by carbachol of 0.15–0.3 gm. Tissues that failed to elicit this level of response were excluded in the contractile studies of the bradykinin compounds. Tissue strips were then thoroughly rinsed with fresh buffer solution via a series of buffer additions and drainage activities to remove any residual muscarinic agonist, and the preload was readjusted prior to starting the actual study with kininergic ligands. Concentration–response studies employed 7–11 compound concentrations (0.1 nM to 10 μM) and data from multiple experiments were collected. Longitudinal isometric tension responses were recorded through an FTO3-transducer and displayed on a Polyview computer software 4.3 analyzer [14].

4.1. Chemicals

All of the compounds used were of the highest purity and were purchased from reputable vendors as described ahead. Bradykinin, Met-Lys-Bradykinin, Lys-Bradykinin, Hyp^3-Bradykinin, RMP-7, Des-Arg^9-Bradykinin, R715, and WIN-64338 were purchased from Tocris, Bio-Techne Corporation, Minneapolis, MN, USA. Atropine and carbachol were purchased from Sigma-Aldrich, Inc., St Louis, MO, USA. All other commonly utilized reagents/chemicals were also purchased from the latter company.

4.2. Data Analysis

The contractile responses obtained were expressed as gram tension developed and as percentages of the maximum response of each of the compounds. GraphPad Prism 4.0 Software (San Diego, CA, USA) was then used to construct the concentration–response curves for the kinins and to derive the half-maximal stimulation potencies (EC_{50}) of each phase of the bi-phasic curves and the values tabulated as previously described [14].

5. Concluding Remarks

The data reported in this article represent novel findings pertaining to the possible functions of the bovine ciliary muscle when activated by bradykinin and its close structural peptidic analogs, including its metabolite Des-Arg^9-BK. The availability and utility of the latter agonists and two well-known B1- and B2-receptor antagonists permitted the characterization of BK receptor subtypes involved in the contraction of the ciliary muscle. Unexpectedly, we discovered that BCM is the other unique tissue, along with the guinea pig colon and esophagus, that expresses functionally active B1- and B2-receptors, which promote the contractile process. These observations have extended our knowledge about BK receptor heterogeneity in an animal species that is not often utilized for ocular research purposes and thus opens up new avenues for future research using BCM preparations. The possible roles of endogenous or exogenously administered BK agonists for modulating intraocular lens function, perhaps combating myopia [31], and for treating elevated IOP, not just in monkeys [32] but in human subjects, warrant further studies in vitro and in vivo.

Author Contributions: Conceptualization, N.A.S. and S.E.O.; Methodology, N.A.S., S.E.O. and Y.F.N.-M.; Software, Formal analysis and investigation, M.K.-C. and Y.F.N.-M.; Data curation, M.K.-C. and Y.F.N.-M.; Writing, N.A.S., S.E.O. and Y.F.N.-M.; Review and editing, N.A.S., S.E.O., Y.F.N.-M., M.K.-C. and A.O.; Supervision, Y.F.N.-M. and S.E.O.; Project administration and funding acquisition, S.E.O. and Y.F.N.-M. All authors have read and agreed to the published version of the manuscript.

Funding: This research was made possible by internal departmental funding at TSU and was also partially supported by grant number 1R15EY022215-01 from the National Institute of Health and the National Eye Institute.

Institutional Review Board Statement: The work describes work performed on isolated tissues of bovine ciliary muscle, which is derived from eyes of cattle post-mortem obtained from a local slaughterhouse. Hence, no institutional board approvals are needed.

Informed Consent Statement: Not applicable.

Data Availability Statement: The original contributions presented in the study are included in the article, further inquiries can be directed to the corresponding author.

Conflicts of Interest: Authors declare no conflicts of interest.

References

1. Leeb-Lundberg, L.M.; Marceau, F.; Müller-Esterl, W.; Pettibone, D.J.; Zuraw, B.L. International union of pharmacology. XLV. Classification of the kinin receptor family: From molecular mechanisms to pathophysiological consequences. *Pharmacol. Rev.* **2005**, *57*, 27–77. [CrossRef] [PubMed]
2. Sharif, N.A. Novel potential treatment modalities for ocular hypertension: Focus on angiotensin and bradykinin system axes. *J. Ocul. Pharmacol. Ther.* **2015**, *31*, 131–145. [CrossRef] [PubMed]

3. Regoli, D.; Gobeil, F. Kinins and peptide receptors. *Biol. Chem.* **2016**, *397*, 297–304. [CrossRef] [PubMed]
4. Ma, J.X.; Song, Q.; Hatcher, H.C.; Crouch, R.K.; Chao, L.; Chao, J. Expression and cellular localization of the kallikrein-kinin system in human ocular tissues. *Exp. Eye Res.* **1996**, *63*, 19–26. [CrossRef]
5. Webb, J.G.; Husain, S.; Yates, P.W.; Crosson, C.E. Kinin modulation of conventional outflow facility in the bovine eye. *J. Ocul. Pharmacol. Ther.* **2006**, *22*, 310–316. [CrossRef]
6. Webb, J.G. The kallikrein/kinin system in ocular function. *J. Ocul. Pharmacol. Ther.* **2011**, *27*, 539–543. [CrossRef]
7. Wiernas, T.K.; Griffin, B.W.; Sharif, N.A. The expression of functionally coupled bradykinin receptors in human corneal epithelial cells and their pharmacological characterization with agonists and antagonists. *Br. J. Pharmacol.* **1997**, *121*, 649–656. [CrossRef]
8. Sharif, N.A. Prospects of treating ocular hypertension and glaucoma with peptidic and non-peptide kinin mimetic drugs. *JOJ Ophthalmol.* **2017**, *3*, 555601. [CrossRef]
9. Ribeiro, A.S.; Fernandes, V.S.; Martínez, M.P.; López-Oliva, M.E.; Barahona, M.V.; Recio, P.; Martínez, A.C.; Blaha, I.; Orensanz, L.M.; Bustamante, S.; et al. Pre-and post-junctional bradykinin B2 receptors regulate smooth muscle tension to the pig intravesical ureter. *Neurourol. Urodynam.* **2016**, *35*, 115–121. [CrossRef]
10. Borsodi, K.; Balla, H.; Molnár, P.J.; Lénárt, Á.; Kenessey, I.; Horváth, A.; Keszthelyi, A.; Romics, M.; Majoros, A.; Nyirády, P.; et al. Signaling pathways mediating bradykinin-induced contraction in murine and human detrusor muscle. *Front. Med.* **2022**, *8*, 745638. [CrossRef]
11. Masuda, H.; Goto, M.; Tamaoki, S.; Kamikawatoko, S.; Tokoro, T.; Azuma, H. M3-type muscarinic receptors predominantly mediate neurogenic quick contraction of bovine ciliary muscle. *Gen. Pharmacol.* **1998**, *30*, 579–584. [CrossRef] [PubMed]
12. Sugawara, R.; Takai, Y.; Miyazu, M.; Ohinata, H.; Yoshida, A.; Takai, A. Agonist and antagonist sensitivity of non-selective cation channel currents evoked by muscarinic receptor stimulation in bovine ciliary muscle cells. *Auton. Autacoid Pharmacol.* **2006**, *26*, 285–292. [CrossRef]
13. Lograno, M.D.; Daniele, E.; Trabucchi, M.; Govoni, S. Evidence for protein kinase C modulation of the ciliary muscle response to carbachol and desensitization. *Eur. J. Pharmacol.* **1991**, *204*, 49–53. [CrossRef]
14. Ohia, S.E.; Njie-Mbye, Y.F.; Robinson, J.; Mitchell, L.; Mckoy, M.; Opere, C.A.; Sharif, N.A. Serotonin-2B/2C receptors-mediate bovine ciliary muscle contraction: Role in IOP regulation. *J. Ocular Pharmacol. Ther.* **2018**, *34*, 70–75. [CrossRef]
15. Millar, C.; Poyer, J.F.; Gabelt, B.T.; Kaufman, P.L. Endothelin subtypes: Effect on isolated rhesus monkey ciliary muscle. *J. Pharmacol. Exp. Ther.* **1995**, *275*, 1143–1147.
16. Wiederholt, M.; Schäfer, R.; Wagner, U.; Lepple-Wienhues, A. Contractile response of the isolated trabecular meshwork and ciliary muscle to cholinergic and adrenergic agents. *Ger. J. Ophthalmol.* **1996**, *5*, 146–153.
17. Field, J.L.; Hall, J.M.; Morton, I.K. Bradykinin receptors in the guinea pig taenia caeci are similar to proposed BK3 receptors in the guinea pig trachea and are blocked by HOE 140. *Br. J. Pharmacol.* **1992**, *105*, 293–296. [CrossRef]
18. Field, J.L.; Butt, S.K.; Morton, I.K.; Hall, J.M. Bradykinin B2 receptors and coupling mechanisms in the smooth muscle of the guinea pig taenia caeci. *Br. J. Pharmacol.* **1994**, *113*, 607–613. [CrossRef]
19. Butt, S.K.; Dawson, L.G.; Hall, J.M. Bradykinin B1 receptors in the rabbit urinary bladder: Induction of responses, smooth muscle contraction, and phosphatidylinositol hydrolysis. *Br. J. Pharmacol.* **1995**, *114*, 612–617. [CrossRef]
20. Regoli, D.; Gobeil, F.; Nguyen, Q.T.; Jukic, D.; Seoane, P.R.; Salvino, J.M.; Sawutz, D.G. Bradykinin receptor types and B2 subtypes. *Life Sci.* **1994**, *55*, 735–749. [CrossRef]
21. Wirth, K.J.; Schölkens, B.A.; Wiemer, G. The bradykinin B2 receptor antagonist WIN 64338 inhibits the effect of des-Arg^9-bradykinin in endothelial cells. *Eur. J. Pharmacol.* **1994**, *288*, R1–R2. [CrossRef] [PubMed]
22. Batista, C.; Sales, V.M.; Merino, V.F.; Bader, M.; Feres, T.; Pesquero, J.B. Role of endothelial kinin B1 receptor on the membrane potential of transgenic rat aorta. *Physiol. Res.* **2022**, *71*, 477–487. [CrossRef] [PubMed]
23. Marceau, F.; Bachelard, H. A robust bioassay of the human bradykinin B2 receptor that extends molecular and cellular studies: The isolated umbilical vein. *Pharmaceuticals* **2021**, *14*, 177. [CrossRef]
24. McLean, P.G.; Perretti, M.; Ahluwalia, A. Kinin B(1) receptors and the cardiovascular system: Regulation of expression and function. *Cardiovasc. Res.* **2000**, *48*, 194–210. [CrossRef]
25. Meini, S.; Lecci, A.; Maggi, C.A. The longitudinal muscle of rat ileum as a sensitive monoreceptor assay for bradykinin B1 receptors. *Br. J. Pharmacol.* **1996**, *117*, 1619–1624. [CrossRef]
26. Husain, S.; Abdel-Latif, A.A. Protein kinase C isoforms in iris sphincter smooth muscle: Differential effects of phorbol ester on contraction and cAMP accumulation are species specific. *Curr. Eye Res.* **1996**, *15*, 329–334. [CrossRef]
27. Webb, R.C. Smooth muscle contraction and relaxation. *Am. J. Physiol. Adv. Physiol. Educ.* **2003**, *27*, 201–206. [CrossRef]
28. Tamm, E.R.; Lütjen-Drecoll, E. Ciliary body. *Microsc. Res. Tech.* **1996**, *33*, 390–439. [CrossRef]
29. Acott, T.S.; Vranka, J.A.; Keller, K.E.; Raghunathan, V.; Kelley, M.J. Normal and glaucomatous outflow regulation. *Prog. Retin. Eye Res.* **2021**, *82*, 100897. [CrossRef]
30. Bynke, G.; Håkanson, R.; Hörig, J.; Leander, S. Bradykinin contracts the pupillary sphincter and evokes ocular inflammation through release of neuronal substance P. *Eur. J. Pharmacol.* **1983**, *91*, 469–475. [CrossRef]

31. Sharif, N.A.; Choudry, N.; Chaudhry, S.D.; Abraham, A. Eye disease genetics and therapeutics. In *Advances in Vision Research, Volume IV: From Basic to Translations Research—Developing Therapeutics and Diagnostics for Genetic Eye Diseases, Essentials in Ophthalmology*; Prakash, G., Iwata, T., Eds.; Springer: Singapore, 2024; Chapter 15, pp. 225–255. [CrossRef]
32. Sharif, N.A.; Katoli, P.; Scott, D.; Li, L.; Kelly, C.R.; Xu, S.; Husain, S.; Toris, C.; Crosson, C. FR-190997, a non-peptide bradykinin B_2-receptor partial agonist, is a potent and efficacious intraocular pressure lowering agent in ocular hypertensive cynomolgus monkeys. *Drug Dev. Res.* **2014**, *5*, 211–223. [CrossRef] [PubMed]

pharmaceuticals

MDPI

Article

In Silico Modeling of Myelin Oligodendrocyte Glycoprotein Disulfide Bond Reduction by Phosphine-Borane Complexes

Raheem Remtulla [1], Sanjoy Kumar Das [2,3] and Leonard A. Levin [1,4,*]

1 Department of Ophthalmology and Visual Sciences, McGill University, Montreal, QC H3A 0G4, Canada; raheem.remtulla@mail.mcgill.ca
2 Division of Experimental Medicine, McGill University, Montreal, QC H3A 0G4, Canada; sanjoy_das@videotron.ca
3 Drug Discovery Core, Research Institute of the McGill University Health Centre, Montreal, QC H4A 3J1, Canada
4 Department of Neurology and Neurosurgery, McGill University, Montreal, QC H3A 0G4, Canada
* Correspondence: leonard.levin@mcgill.ca

Abstract: Background: Neurodegenerative diseases can cause vision loss by damaging retinal ganglion cells in the optic nerve. Novel phosphine-borane compounds (PBs) can protect these cells from oxidative stress via the reduction of disulfide bonds. However, the specific targets of these compounds are unknown. Proteomic evidence suggests that myelin oligodendrocyte glycoprotein (MOG) is a potential target. MOG is of significant interest due to its role in anti-MOG optic neuritis syndrome. Methods: We used in silico modeling to explore the structural consequences of cleaving the extracellular domain MOG disulfide bond, both in isolation and in complex with anti-MOG antibodies. The potential binding of PBs to this bond was examined using molecular docking. Results: Cleaving the disulfide bond of MOG altered the structure of MOG dimers and reduced their energetic favorability by 46.13 kcal/mol. The energy profiles of anti-MOG antibody complexes were less favorable when the disulfide bond of MOG was reduced in the monomeric state by 55.21 kcal/mol, but the reverse was true in the dimeric state. PBs exhibited reducing capabilities with the MOG extracellular disulfide bond, with this best-scoring compound binding with an energy of −28.54 kcal/mol to the MOG monomer and −24.97 kcal/mol to the MOG dimer. Conclusions: These findings suggest that PBs can affect the structure of MOG dimers and the formation of antibody complexes by reducing the MOG disulfide bond. Structural changes in MOG could have implications for neurodegenerative diseases and anti-MOG syndrome.

Keywords: phosphine-borane complexes; myelin oligodendrocyte glycoprotein; disulfide bonds; bond cleaving; optic neuritis; neurodegeneration

1. Introduction

Vision loss is a frequent result of neurodegenerative diseases affecting the optic nerve. Typical presentations include a loss of visual acuity, visual field defects, and dyschromatopsia [1,2]. The mechanism of action of these diseases is usually due to damage to retinal ganglion cell (RGC) axons within the optic nerve [1,2]. Damage to RGC axons induces RGC soma death through a variety of mechanisms. One such mechanism is via signaling from reactive oxygen species (ROS), particularly superoxide species [2–5]. Intracellular superoxide bursts have been identified as a critical event in RGC death and may act as a pre-apoptotic biomarker [5]. Evidence for this mechanism includes studies demonstrating increased levels of superoxide species in RGCs after injury, and the specified imaging of superoxide demonstrated that the level of superoxide increases before other markers for apoptosis in damaged RGCs [2–4]. Furthermore, superoxide scavengers that decrease superoxide signals have been able to prevent cell death in traumatic optic neuropathy models [2–5]. These studies suggest that superoxide species are upstream signals for RGC death [2–5]. Fortunately, disulfide reducing agents have been able to protect RGCs from ROS signal-related injuries [2–5]. Based on the neuroprotective properties of disulfide reducing agents, both in vitro and in vivo, we hypothesized that the mechanism by which ROS signal axonal injury is through disulfide formation [6,7]. Highly oxidative environments increase the formation of disulfide bonds between cysteine side chains and cause cellular damage [6,7].

Phosphine-borane compounds are a novel class of compounds recognized for their reactive organophosphine group and protective borane group, which have demonstrated promise in the field of neuroprotection [7–9]. In vitro and in vivo studies demonstrated that these compounds protect RGCs from oxidative stress [7–9]. Studies using triphenyl phosphine and aryl disulfides have suggested that phosphorus atoms in phosphine-borane molecules act as a nucleophile in an S_N2 nucleophilic substitution reaction in disulfide bond cleavage [7–10]. However, the exact target of novel phosphine-borane compounds has not been determined. Oxidative isotope-coded affinity tag data from our laboratory was used to identify one of the protein targets for the bis(3-propionic acid methyl ester) phenylphosphine borane complex (PB1) as myelin oligodendrocyte glycoprotein (MOG), with consistent disulfide bond cleavage.

MOG is a clinically relevant protein for neurodegeneration because of anti-MOG-related neurologic disease, which partially overlaps with multiple sclerosis. Anti-MOG syndromes result from damage to MOG [11,12], which is expressed on oligodendrocyte cell surfaces and on the outermost surface of myelin sheaths [11]. Although early studies found that 41% of multiple sclerosis (MS) patients had anti-MOG antibodies in serum, as did patients with non-MS neurologic disease and healthy patients [12–17], it is now known that anti-MOG antibodies are rare in patients with typical MS phenotypes [11] and that anti-MOG syndromes are clinically distinct entities from MS. Unlike MS, anti-MOG patients present more frequently with bilateral optic neuropathy, steroid dependence, and a lower female predominance [11,18,19]. Interestingly, the clinical presentations of anti-MOG syndrome differ between adults and children [11]. In children, the most common presenting symptom is acute disseminated encephalomyelitis (36%), whereas the most common presentation for anti-MOG syndromes in adults is optic neuritis (88%) [11,19,20]. Unlike in MS, bilateral optic neuritis is common, with one study finding 42% of anti-MOG syndromes had bilateral optic neuritis and 31% had unilateral optic neuritis [19,20]. Also unlike MS, optic disc swelling is seen in the majority of patients. Anti-MOG optic neuritis is comparatively severe, with visual acuity worse than 20/200 [19,20]. Anti-MOG optic neuritis is highly steroid-responsive, with rapid improvement in symptoms after intravenous methyprednisone. Unfortunately, the cessation of steroids often results in relapses of optic neuritis, and patients often must continue long-term steroid-sparing immunosuppressants to prevent recurrences. Although immunosuppression can prevent recurrent optic neuritis, it can be associated with systemic side effects, including recurrent infections, lymphocytopenia, hepatotoxicity, nephrotoxicity, and cardiac disease. Patients on immunosuppression have to be monitored for these adverse effects [11,18].

MOG is a minor myelin component (~0.05% of all myelin proteins), which is exclusively expressed in the central nervous system (CNS) of mammals, on the outermost lamella of myelin sheaths and oligodendrocyte processes [21,22]. MOG is a type I integral transmembrane protein with a single extracellular Ig variable domain [22–24]. A schematic of the MOG structure is presented in Figure 1. There are 245 amino acids, of which the first 126 are extracellular [23,25–27]. The protein is relatively small, with dimensions of $40 \times 35 \times 30$ Å [23,25–27] and a ß sandwich domain with one antiparallel ß sheet, with the N and C terminals on opposite ends of the monomer. Like many Ig-V domains, MOG has core packing residues, with a disulfide bond between Cys24 and Cys98 and a buried salt bridge between Arg68 and Asp92. There are three loops at the end, distal to the membrane, including an F-G strand that has been implicated in autoimmunity. The structure also has a significant strip of electronegative charge that runs about half the length of the molecule [25].

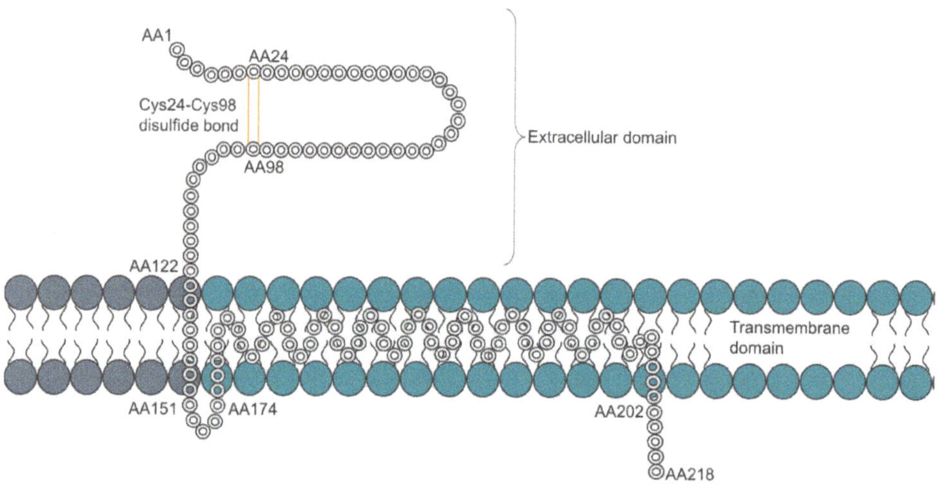

Figure 1. Simplified schematic of MOG based on Kroepfl et al. proposed model [28,29].

Native MOG exists as a mixture of monomeric and dimeric species [25]. The extracellular domain has the ability to form a homodimer in situ within the CNS [30]. A crystal structure of the dimer was described by Clements et al. The dimer forms an antiparallel head-to-tail dimer. The dimer interface involves interactions between residues 8–11 interacting with opposing residues 95–112, and residues 40–46 and opposing residues 40–46. The interaction is stabilized by 10 hydrogen bonding and van der Waals interactions [25]. Native MOG has been identified in both monomeric and dimeric forms both in and outside the central nervous system [22,30,31]. The amino acids 35–55 are buried in dimer formation, and have also been implicated in autoimmunity to MOG. Given that gel filtration identified more MOG existing as monomers than as dimers, one could infer that the dimer interface is weak and favors monomer formation over dimer formation [25].

In order to determine the nature of antibody binding to MOG, the crystal structure extracellular domain of MOG from rat was docked with the chimeric antigen-binding fragment FAb of the demyelinating MOG specific monoclonal antibody 8-18C5 [21,31]. The MOG-FAb complex showed that the antibody bound to the upper membrane distal surface of MOG, binding to three loops located at the membrane distal side of MOG, with the major contribution made by the FG loop. The FG loop formed the major interactions of the complex, and the FG loop includes amino acids 101–108 which became completely buried within the 8-18C5 antibody [21]. The sequence I of AA 101–108 is unique to MOG, which is only in the CNS. Therefore, the body does not develop immune tolerance to that sequence.

Other encephalitogenic epitopes of MOG include AA 1–22, AA 35–55 and AA 92–105 [25,32–34]. MOG AA 35–55 is of particular interest, being able to cause inflammatory and demyelinating immune responses [32–34]. Interestingly, the epitopes of mAb 8-18C5 and other demyelinating mAbs to MOG are conformation-dependent, and linear epitopes to MOG do not induce widespread demyelination [35]. This suggests that disruption in MOG's baseline structure may prevent autoimmunity. Both the monomer and the dimer species have been recognized by 8-18C5 antibodies [25].

Using in silico methodologies and molecular docking, we tested whether phosphine-borane compounds cleave the disulfide bond in the extracellular domain of MOG. Such data would be valuable in determining the mechanism of these novel compounds, provide insight into the mechanism of anti-MOG-associated optic neuritis, and identify a potential new target of neuroprotection for demyelinating disease.

2. Results

2.1. Preprocessing of MOG Structure

An X-ray diffraction of the monomer structure of MOG from Breithaupt et al. [21] was accessed as a PBD file on the RCSB protein data bank (PBD 1PKO; resolution 1.45 Å) and imported directly into Schrödinger. The PBD validation from the RCSB demonstrated that the protein was an appropriate structure compared to other X-ray diffraction structures on the database, with the only metric below average being real space R-value Z-score RSRZ outliers. Both the Ramachandran and sidechain outliers were appropriate. A protein reliability report was generated after importing it to Schrödinger. Only 2 of the 22 metrics evaluated were flagged. The first was missing atoms, where only two amino acids were involved. The second included three water molecules that did not form hydrogen bonds with the protein structure. These issues were managed with the protein preparation tool and the resultant structure utilized as the ground truth for the monomer structure with the Cys24-Cys98 disulfide bond intact (MOG-DI).

Breithaupt et al. also developed an X-ray diffraction of the MOG dimer in complex with anti-MOG antibodies (PBD 1PKQ; resolution 3 Å) [21]. However, the PBD validation from the RCSB suggested that the structure was relatively poor, with no metrics of structure quality above the 50th percentile for all PBD X-ray diffraction structures other than RSRZ outliers. Seven of the other 22 metrics evaluated in the Schrödinger protein reliability report for 1PKQ were out of the appropriate range. The most significant out-of-range metric was steric clashing, of which there were 89 clashes between amino acids on the structure. Finally, the resolution of the structure was considered unacceptable. These structures were used in antibody docking; however, the lower quality of these structures should be noted.

2.2. Impact of Disulfide Bond on MOG Monomer Structure

The structure of MOG-DI was evaluated in the workspace and the disulfide bond between Cys24 and Cys98 cleaved manually. The post-cleaved MOG structure was processed with the protein preparation wizard without automatic disulfide formation. Minimization was performed and the resultant structure named MOG-DR (MOG with disulfide bond reduced).

MOG-DI was then compared to MOG-DR. There was little difference between the two, with only a 0.19 Å root mean squared deviation (RMSD) between the FG loops on MOG-DI and MOG-DR. There was a 55.21 kcal/mol difference in minimized total complex energy favoring MOG-DR monomers over MOG-DI monomers. Table 1 demonstrates the complex energies of all structures in the proteomic analysis.

Table 1. MOG structures with total complex energy values in both disulfide bond intact and disulfide bond reduced structures.

MOG Structures	Total Complex Energy Values (kcal/mol)	
	Disulfide Bond Intact	**Disulfide Bond Reduced**
Monomer	−527.04	−582.26
Dimer	−1218.89	−1172.76
Monomer antibody complex	−1332.73	−1333.79
Dimer antibody complex	−2029.64	−2032.65

2.3. Impact of Disulfide Bond on MOG Dimer Structure

Using MOG-DI, protein–protein docking was performed to generate a computational model of the MOG-DI dimer (Figure 2). This resulted in a total of 22 poses, of which one fit the description of the dimer described by Clements et al. [25], with a total complex energy of −1218.89 kcal/mol and an RMSD of 0.05 (Table 1). The dimer interface had a total interface interaction of 1023 Å^2 and 1037 Å^2. There were also 8 salt bridges and 15 hydrogen bonds identified at the interface (Table 2).

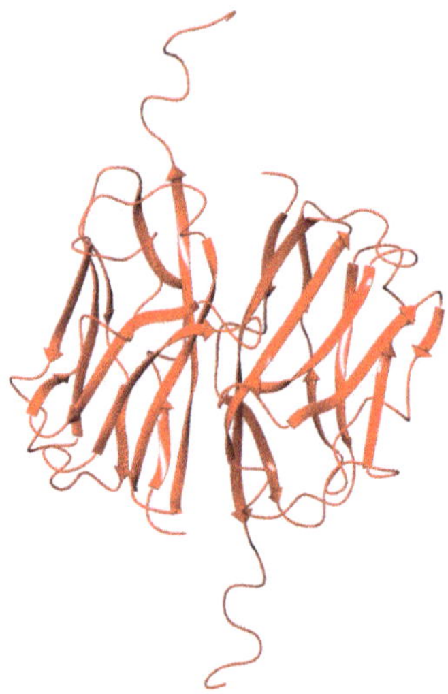

Figure 2. MOG dimer protein ribbon structure of Schrödinger protein–protein docking of extracellular MOG monomers with disulfide bond intact (PBD 1PKO).

Table 2. Summary of protein interaction interfaces between the disulfide intact and reduced structures generated by ProFunc.

		Interface Residuals	Interface Area (Å^2)	Salt Bridges	Hydrogen Bonds	Nonbond Contacts
Dimer	Intact	57	2060	8	15	174
	Reduced	42	2020	-	6	177
Monomer antibody complex	Intact	24	1169	3	9	106
	Reduced	21	1116	3	5	74
Dimer antibody complex	Intact	23	1356	3	7	76
	Reduced	34	1609	1	7	134

The process of dimer formation with protein–protein docking was repeated using MOG-DR. A total of 21 poses resulted, of which none were in keeping with the structure suggested by Clements et al. [25], with one structure closely similar to the known dimer between MOG monomers (Figure 3). The latter structure had a complex energy of −1172.76 kcal/mol and an RMSD of 0.10 (Table 1). There were no salt bridges at the interface and only six hydrogen bonds (Table 2). There was significant rotation of the DR monomers in the MOG-DR dimer structure compared to the MOG-DI dimer, with Arg126 displaced by 36.23 Å in the MOG-DR dimer compared to the MOG-DI dimer. Furthermore, the antigenic chain was shifted, with an RMSD of 11.95 Å between the MOG-DI dimer and the MOG-DR dimer [25].

There was a 46.13 kcal/mol difference in complex energies favoring the MOG-DI dimer over the MOG-DR dimer (Table 1). This was further exemplified by the presence of a greater number of residues, a larger interface area, and a greater number of salt bridges and hydrogen bonds. This suggests that the Cys24–Cys98 disulfide bond of the MOG monomer is important for MOG dimer formation.

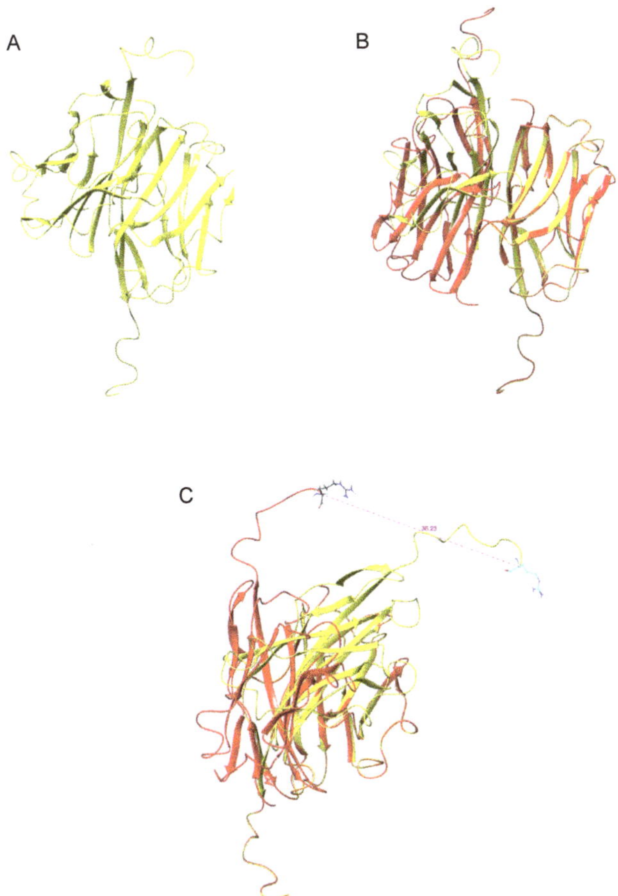

Figure 3. (**A**) MOG dimer protein ribbon structure of Schrödinger protein–protein docking of extracellular MOG monomers with disulfide bond cleaved pre dimer formation (PBD 1PKO). (**B**) Overlapping structures of MOG dimer protein ribbon structure with disulfide bond intact (RED) and MOG dimer protein ribbon structure with disulfide bond cleaved (YELLOW-GREEN) from Schrödinger protein–protein docking (**C**) Lateral view of overlapping structures of MOG dimer protein ribbon structure with disulfide bond intact (RED) and MOG dimer protein ribbon structure with disulfide bond cleaved (YELLOW-GREEN) from Schrödinger protein–protein docking. Measurement between arginine 126 on MOG dimer protein ribbon structure with disulfide bond intact and MOG dimer protein ribbon structure with disulfide bond cleaved in Ångstrom units.

2.4. Impact of Disulfide on Interactions Between Anti-MOG Antibody and MOG Monomer

Given the role of the disulfide on MOG dimer formation, the nature of the binding of anti-MOG antibodies to MOG was explored. The MOG monomer in complex with the antibody [21] (PBD 1PKQ) was processed by isolating the antibody from this structure and minimizing it to its lowest energy state using the protein preparation tool. After minimization, the heavy chain from the antibody isolated from 1PKQ was docked with the protein–protein docking tool to three structures: the monomer remnant from 1PKQ, the MOG-DI monomer, and the MOG-DR monomer.

First, the top 30 poses resulting from docking the MOG monomer remnant (derived from 1PKQ) to the isolated heavy chain of the anti-MOG antibody (derived from 1PKQ) were analyzed, one of which matched the monomer–heavy chain antibody complex represented by 1PKQ. This pose, being closest in structure to 1PKQ, had a total complex energy of −1276.54 kcal/mol (Table 1). The RMSD between the MOG FG loop in the docked results and the FG loop in 1PKQ was 1.97 Å. This suggests that despite the lower resolution of 1PKQ, the protein–protein docking tool was able to redemonstrate the interactions between MOG and anti-MOG antibodies.

Second, the top 30 poses resulting from the docking of the MOG-DI monomer (derived from 1PKO) to the isolated heavy chain of the anti-MOG antibody (derived from 1PKQ) were analyzed, one of which matched the monomer–heavy chain antibody complex represented by 1PKQ. The pose had a total complex energy of −1332.73 kcal/mol and an RMSD of 0.05 (Table 1). The presented structure had 24 residues at the antibody–monomer interface, a total interface area of 1169 Å^2, 3 salt bridges, and 9 hydrogen bonds (Table 2). The modeled antibody complex was compared with 1PKQ. There was an insignificant difference in the location of antigenic chain binding to the antibody between the two structures, with an RMSD of 1.19 Å between the backbones of the FG loops of the two structures.

Third, the MOG-DR monomer from 1PKO was docked to the isolated heavy chain of the anti-MOG antibody from 1PKQ, and the top 30 poses analyzed. There were significant differences between the complex of the MOG-DR monomer and the antibody vs. the monomer–heavy chain antibody complex represented by 1PKQ. Similarly, there were significant differences between the complex of the MOG-DR monomer and the antibody vs. the complex of the MOG-DI monomer and the antibody. The MOG DR monomer–antibody complex that was closest in structure to the 1PKQ ground truth structure had a total complex energy of −1333.79 kcal/mol and an RMSD of 0.05 (Table 1). There were 21 residues at the MOG-DR monomer–antibody interface, a total interface area of 1116 Å^2, 3 salt bridges, and 5 hydrogen bonds (Table 2). The RMSD between the FG loops of the MOG-DR monomer–antibody complex and the complex of 1PKQ was 3.84 Å. In comparison, the RMSD between the FG loops of the MOG-DR monomer–antibody complex and the MOG-DI monomer–antibody complex was 3.02 Å.

There was only a 1.06 kcal/mol difference favoring the MOG-DR monomer–antibody complex over the MOG-DI monomer–antibody complex. The MOG-DR monomer had a 55.21 kcal/mol lower complex energy than the MOG-DI counterpart (Table 1). When considering the intrinsic difference in monomer–antibody complex energies, there is an approximately 54 kcal/mol complex energy difference favoring the MOG-DI monomer–antibody complex formation vs. the reduced form. This is further exemplified by increased interface residues, interface area, salt bridges, and hydrogen bonds with the MOG-DI monomer–antibody complex structure.

2.5. Impact of Disulfide on Interactions Between Anti-MOG Antibody and MOG Dimer

The analyses above established that the Cys24-Cys98 disulfide bond played a pertinent role in forming MOG antibody complexes. To explore the same process with the dimeric form of MOG, dimers of MOG-DI monomers and MOG-DR monomers were formed using the protein–protein docking tool, and then docked to the heavy chain antibody isolated from 1PKQ, as described previously.

Protein–protein docking between the MOG-DI dimer and the heavy chain of 1PKQ resulted in 30 poses. The X-ray crystallography of the dimer–antibody complex was not available, and therefore the structure analyzed was the one with the most structural similarity to the monomer–antibody complex (Figure 4). This structure had a complex energy of −2029.64 kcal/mol and an RMSD of 0.05 Å (Table 1). There were 23 residues at the interface, 3 salt bridges, and 7 hydrogen bonds, and the area of the interface was 1356 Å^2 (Table 2). The RMSD between the FG loops on 1PKQ and the MOG-DI dimer–antibody complex was 5.55 Å. The RMSD between the FG loops on the MOG-DI dimer–antibody complex and the MOG-DI monomer–antibody complex was 4.92 Å.

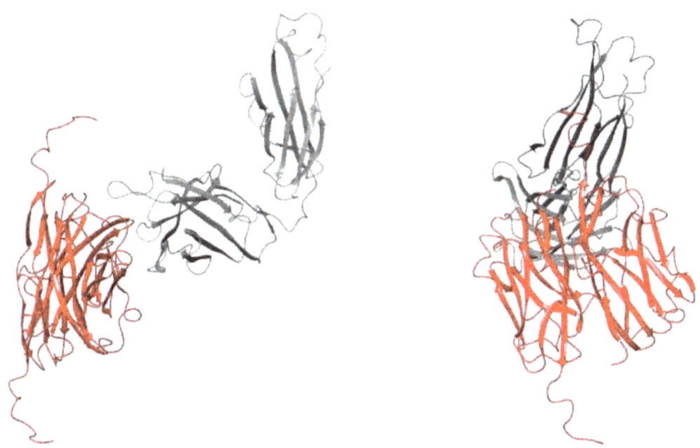

Figure 4. Protein ribbon structure of Schrödinger protein–protein docking-generated MOG dimer presented in the red structure (PBD 1PKO) with disulfide bond intact in complex with anti-MOG antibody presented in the grey structure(PBD 1PKQ). Two lateral orientations of the protein interaction are depicted.

Similarly, the MOG-DR dimer was docked to the heavy chain of the antibody from 1PKQ, resulting in 30 displayed poses. None of the presented structures were similar to the MOG-DI dimer–antibody complex. The structure that was most similar to the MOG-DI dimer–antibody complex had a total energy of −2032.65 kcal/mol and an RMSD of 0.05 Å (Table 1, Figure 5). There were 34 residues at the interface, one 1 salt bridge, and 7 hydrogen bonds, and the area of the interface was 1609 Å^2 (Table 2). The RMSD between the FG loops on the MOG-DR dimer–antibody complex and the MOG-DI dimer–antibody complex was 4.71 Å, and the distance between the Arg126 on the MOG-DI–antibody complex and that of the MOG-DR dimer–antibody complex was 36.94 Å. The RMSD between the FG loops on the intact 1PKQ complex and the MOG-DR dimer–antibody complex was 6.87 Å, while the RMSD between the FG loops on the MOG-DR dimer–antibody complex and the MOG-DI monomer–antibody complex was 6.17 Å (Figure 6).

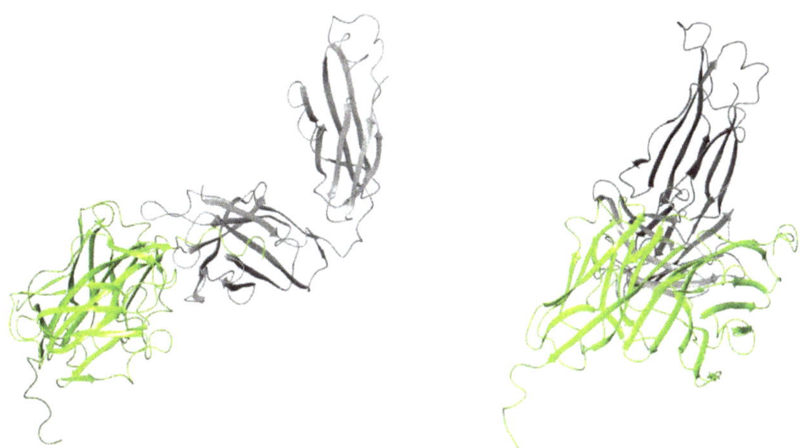

Figure 5. Protein ribbon structure of Schrödinger protein–protein docking-generated MOG dimer presented in the yellow-green structure (PBD 1PKO) with disulfide bond cleaved pre dimer formation in complex with anti-MOG antibody presented in the grey structure (PBD 1PKQ). Two lateral orientations of the protein interaction are depicted.

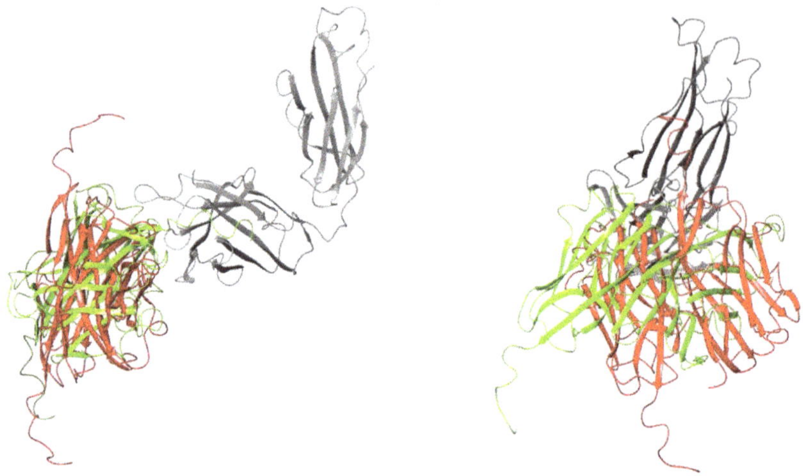

Figure 6. Overlapping structures of MOG dimer with disulfide bond intact presented in the red structure in complex with anti-MOG antibody presented in the grey structure (PBD 1PKQ), and MOG dimer with disulfide bond cleaved pre dimer formation presented in the yellow-green structure in complex with anti-MOG antibody presented in the grey structure (PBD 1PKQ). Two lateral orientations of the protein interaction are depicted.

The dimer–antibody complex models demonstrated that there was a 3.01 kcal/mol complex energy difference favoring dimer–antibody complex formation when the Cys24-Cys98 disulfide was reduced, compared to a 46.13 kcal/mol difference in energy favoring dimer formation when the disulfide was intact (Table 1). These findings suggest a more favorable energy for a dimer–antibody complex with the disulfide bond reduced vs. the disulfide bond intact. This was further exemplified by a greater number of interface residues (34 vs. 23) and a larger interface area (1609 vs. 1356) when the disulfide bond was reduced, although more salt bridges were present (1 vs. 3) when the disulfide bond was intact (Table 2).

2.6. Covalent Docking of Phosphine-Borane Compounds to the MOG Monomer Disulfide

Using the covalent docking tool, the deboronated phosphine-borane compounds and phosphine controls were docked to Cys24 and Cys98 on the MOG monomer 1PKO with the disulfide bond intact. The deboronated phosphine-borane compounds included bis(3-propionic acid methyl ester) phenylphosphine (P1) and (3-propionic acid methyl ester)diphenylphosphine. These compounds contain an ester group that can be hydrolyzed. As such, their metabolized ester bonds form acids. Since P1 has two ester bonds, both an R and S enantiomer exist for P1 (P1-monoacid-R and P1-monoacid-S). As well, a diacid form also exists known as P1-diacid. There is only one ester bond on P2, and when hydrolyzed it is known as P2-acid. Phosphine controls include tris(2-carboxyethyl)phosphine (TCEP) phosphine (P-H), trimethylphosphine (P-Me), and triphenylphosphine (P-Ph). Only isolated phosphine was able to dock to Cys98, and none of the deboronated phosphine-borane compounds studied were able to dock to that cysteine sulfur. In addition, trimethylphosphine, triphenylphosphine, and TCEP were unable to dock to Cys98. All tested compounds were able to dock to Cys24 (see ligand interaction diagrams in Figure 7). All compounds docked with an RMSD less than 1 Å, with negative docking scores and negative molecular mechanics with a generalized Born and surface area solvation (MM-GBSA). The mean docking score of the control compounds was −1.48, compared to −2.06 among the test compounds ($p = 0.12$). The mean MM-GBSA of the controls was −9.10 kcal/mol, compared to −20.83 kcal/mol ($p = 0.03$) (Table 3).

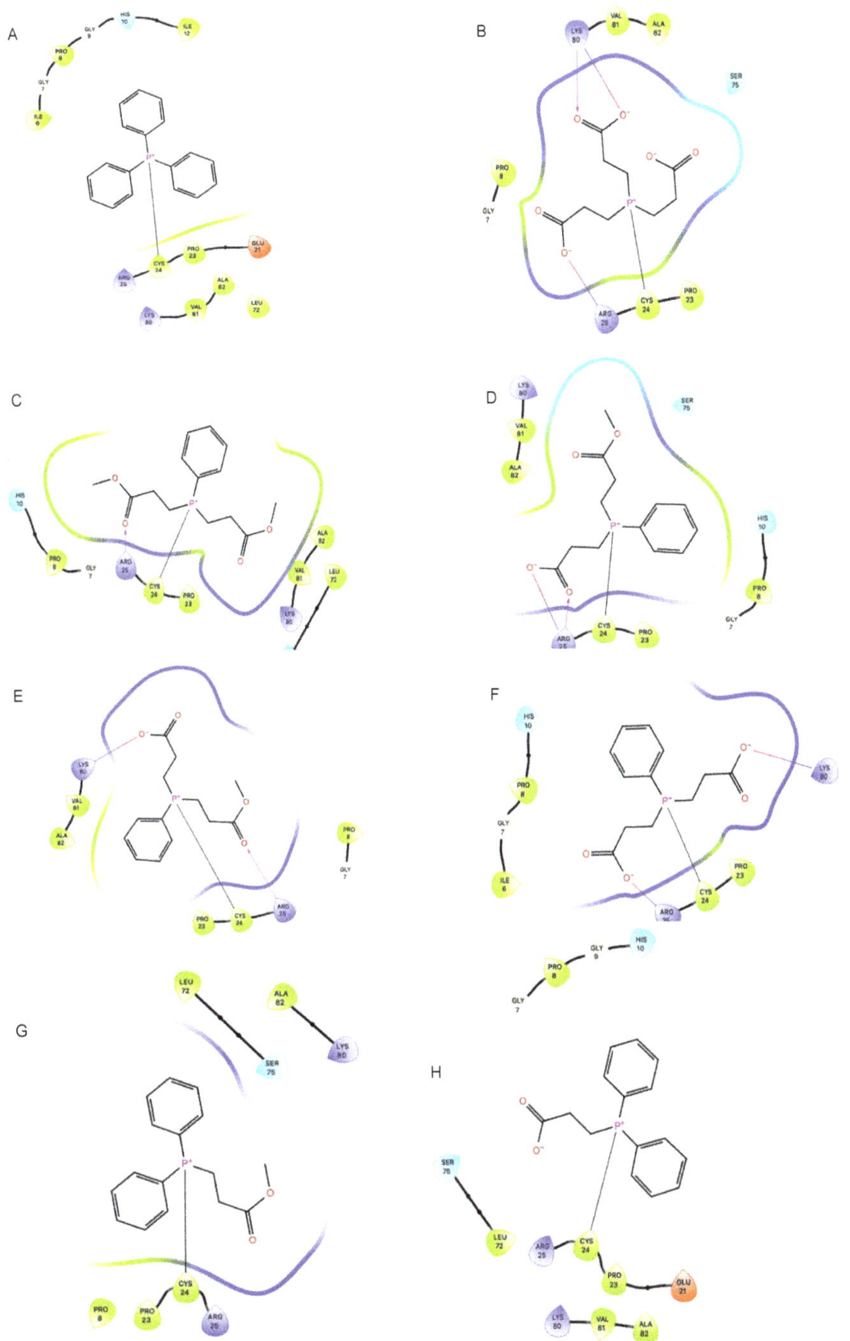

Figure 7. (**A**) Ligand interaction diagram of covalent docking of triphenylphosphine to cystine 24 on myelin oligodendrocyte glycoprotein. (**B**) Ligand interaction diagram of covalent docking of tris(2-carboxyethyl)phosphine to cystine 24 on myelin oligodendrocyte glycoprotein. (**C**) Ligand interaction

diagram of covalent docking of P1 to cystine 24 on myelin oligodendrocyte glycoprotein. (**D**) Ligand interaction diagram of covalent docking of P1 monoacid R to cystine 24 on myelin oligodendrocyte glycoprotein. (**E**) Ligand interaction diagram of covalent docking of P1-monoacid S to cystine 24 on myelin oligodendrocyte glycoprotein. (**F**) Ligand interaction diagram of covalent docking of P1-diacid to cystine 24 on myelin oligodendrocyte glycoprotein. (**G**) Ligand interaction diagram of covalent docking of P2 to cystine 24 on myelin oligodendrocyte glycoprotein. (**H**) Ligand interaction diagram of covalent docking of P2 acid to cystine 24 on myelin oligodendrocyte glycoprotein.

Table 3. Covalent docking results of phosphine compounds to cystine 24 on myelin oligodendrocyte protein monomer.

	CovDock Score	MM-GSBA
P1	−1.30	−28.54
P1-Monoacid-R	−2.55	−20.98
P1-Monoacid-S	−2.47	−24.99
P1-Diacid	−1.59	−15.72
P2	−1.98	−18.61
P2-Acid	−2.49	−16.16
P-Ph	−1.35	−21.88
P-Me	−0.90	−7.96
P-H	−1.53	−4.96
TCEP	−2.15	−1.60

TCEP was the control compound with the most negative docking score (−2.15), and three of the test compounds docked with a more negative docking score than TCEP. Furthermore, all test compounds had an MM-GBSA change in free energy of binding less than −15 kcal/mol, with three test compounds having an MM-GBSA less than −20 kcal/mol. Only isolated phosphine had an MM-GBSA less than −15 kcal/mol, with no other controls meeting this threshold. Of the test compounds, the monoacid metabolite of PB1 in both the S and R enantiomer had the most suitable docking profiles. The S enantiomer had a docking score of −2.46 and an MM-GBSA of −24.99 kcal/mol, and the R enantiomer a docking score of −2.55 and a MM-GBSA of −20.98 kcal/mol. The docking scores and MM-GBSA of the docked compounds to Cys24 in the 1PKO monomer are listed in Table 3.

The amino acids Arg25 and Lys80 were the only amino acids other than Cys24 to have direct interactions with the docked phosphine compounds (Figure 7). Arg25 interacted with all P1 compounds and TCEP. Both the ester groups and carboxylic acid groups of the P1 compounds and the carboxylic acid groups of TCEP interacted with the guanidino group of arginine. Salt bridges were formed between nitrogen and carboxylic acids. Despite the presence of ester groups and carboxylic acid on the P2 compounds, there was no interaction with Arg25. The P2 compounds have an additional phenyl group, a larger effective molecular diameter, and are more rigid than the carboxylic acid chains on P1 compounds. The compounds that interacted with Arg25 had a mean docking score of −2.01 and MM-GBSA of −18.37 kcal/mol, whereas those that did not interact with Arg25 had a docking score of −1.65 and an MM-GBSA of −13.91 kcal/mol.

Lys80 interacted with P1-diacid, P1-monoacid(S), and TCEP through hydrogen bonding and salt bridges. Lys80 was only able to interact with carboxylic acid groups, and not with ester groups. The compounds that interacted with Lys80 has a mean docking score of −2.06 and an MM-GBSA of −14.10 kcal/mol, whereas those that did not had a mean docking score of −1.73 and an MM-GBSA of −17.01 kcal/mol. All tested compounds that interacted with Lys80 also interacted with Arg25.

To ensure that the Cys24 binding site was accessible, the Schrödinger binding site mapping tool was used to determine which aspects of the protein surfaces were accessible binding sites. These binding site maps were overlapped with the output complexes from the covalent docking procedure to determine if the bound phosphine compounds were docking at a predicted binding site. The binding site mapping tool was first employed without evaluating shallow binding sites, and then with shallow binding sites. Two binding sites were identified when only deep binding sites were evaluated, neither of which included the exposed sulfur of Cys24. However, when shallow binding sites were evaluated, four additional sites were identified. One of these sites included the exposed sulfur on Cys24. This site met all the parameters expected of a valid binding site, with a SiteScore of 0.89, a Dscore of 1.02, and a volume of 421.55 Å. The binding site map over Cys24 in complex with P1-diacid is demonstrated in Figure 8.

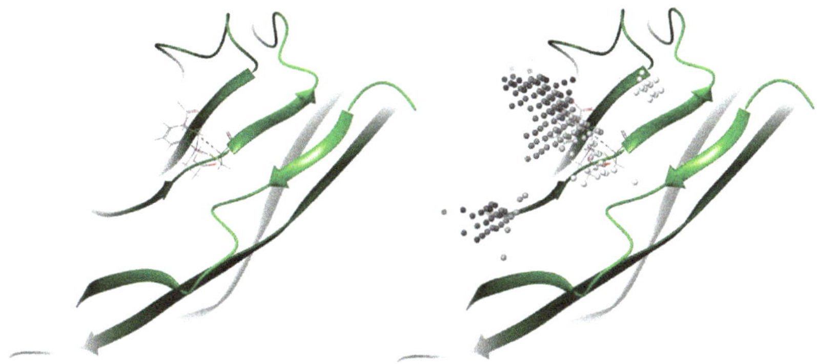

(a) (b)

Figure 8. (**a**) Covalent docking results of P1-diacid (ball and stick) in complex with cystine 24 on ribbon diagram of MOG monomer (PBD 1PKO). (**b**) Binding site map on MOG monomer (PBD 1PKO) overlayed with covalent docking results of P1-diacid (ball and stick) in complex with cystine 24 on ribbon diagram of MOG monomer (PBD 1PKO).

2.7. *Covalent Docking of Phosphine-Borane Compounds to the MOG Dimer Disulfide*

The deboronated phosphine-borane compounds and phosphine controls were docked to the MOG-DI dimer. Cys24 and Cys98 in the monomers of the MOG dimer were set as the reactive amino acid and covalently docked, as described previously.

As with the MOG monomer, only phosphine was able to dock to Cys98. None of the deboronated PB1 or PB2 metabolites, trimethylphosphine, triphenylphosphine, or TCEP docked to Cys98. All tested compounds were able to dock to Cys24. All compounds docked with an RMSD less than 1, with negative docking scores and negative prime energies. The mean docking score of the control compounds was −1.25, compared to −2.39 among test compounds ($p < 0.05$), and the mean MM-GBSA of the controls was −11.77 kcal/mol, compared to −21.80 kcal/mol ($p < 0.01$). Of the control compounds, the one with the most negative docking score was isolated phosphine, with a score of −1.55; five of the test compounds were able to dock with a more negative docking score than isolated phosphine. All the test compounds had a prime energy less than −15 kcal/mol, with five having a prime energy less than −20 kcal/mol. Only triphenylphosphine had an MM-GBSA less than −15 kcal/mol, with no other controls meeting this threshold. The docking scores and MM-GBSA for docking of all the studied phosphine compounds to Cys24 in the 1PKO dimer are listed in Table 4.

Similar to the monomer docking results, the amino acids Arg25 and Lys80 both interacted with the docked compounds. However, unlike when docking to the monomer, the amino acid His10 also interacted with docked compounds. Arg25 only interacted through hydrogen bonding and salt bridges with four of the six test compounds and none of the control compounds. Only carboxylic acid groups appeared to interact with the guanidino group of Arg25. All phosphine-borane compounds with a hydrolyzed ester group formed interactions with Arg25. The compounds that interacted with Arg25 had an average docking score of −2.69 and an MM-GBSA of −21.65 kcal/mol, whereas those that did not interact with Arg25 had a docking score of −1.43 and an MM-GBSA of −15.21 kcal/mol.

Table 4. Covalent docking results of phosphine compounds to cystine 24 on myelin oligodendrocyte protein disulfide intact dimer.

	CovDock Score	MM-GSBA
P1	−1.22	−24.97
P1-Monoacid-R	−2.37	−22.45
P1-Monoacid-S	−3.11	−20.80
P1-Diacid	−2.48	−21.60
P2	−2.34	−19.21
P2-Acid	−2.82	−21.76
P-Ph	−1.37	−21.93
P-Me	−0.61	−9.71
P-H	−1.55	−5.34
TCEP	−1.48	−10.11

Interactions of Lys80 were with P1-diacid, P-Ph, and TCEP. The carboxylic acid groups on P1-diacid and TCEP formed hydrogen bonds and salt bridges with the primary amine group on Lys80. The phenyl group of P-Ph formed a Pi-cat interaction with the primary amine group of Lys80. This interaction was not found in any other compound docked. Compounds that interacted with Lys80 had a mean docking score of −1.78 and an MM-GBSA of −17.88 kcal/mol, while those that did not interact with Lys80 had a docking score of −2.00 and an MM-GBSA of −17.75 kcal/mol.

Only P1-monoacid(R) and TCEP interacted with His10. In these interactions, the imidazole ring of His10 formed hydrogen bonds with the carbonyl oxygen of carboxylic acid groups on these compounds. The two compounds that interacted with His10 had a mean docking score of −2.29 and an MM-GBSA of −15.46 kcal/mol, while and those that did not interact with His10 had a docking score of −1.84 and an MM-GBSA of −18.37 kcal/mol.

As with the monomer, the binding site tool was applied to the MOG-DI dimer and then was overlapped with the MOG-DI dimer covalent docking results. A total of six docking sites were identified, of which three were shallow docking sites. One of the shallow binding sites overlapped with the binding site of the phosphine compounds in complex with Cys24. This shallow binding site had a SiteScore of 0.89, a Dscore of 1.02, and a volume of 421.55, and was compatible with a valid binding site.

3. Discussion

The molecular modeling demonstrated that the disulfide bond between Cys24 and Cys98 has an impact on the structure of the MOG dimer. Although reducing the disulfide bond made limited structural differences in the MOG monomer, it resulted in a more energetically stable structure. Most of the existing research on MOG evaluates the MOG monomer with the disulfide bond intact. It is unclear if the increased stability in the disulfide-reduced monomer has a clinical or biochemical significance. However, it is possible that favoring a more energetically stable monomer has positive effects, and possibly even relates to the neuroprotection seen with phosphine-borane complexes [8,36]. Cleaving the disulfide bond appeared to impair dimer formation. The total complex energy value of the MOG-DR dimer was nearly 50 kcal/mol lower than the MOG-DI dimer, suggesting that the monomer disulfide bond plays a vital role in dimer formation. Previous studies in computational modeling have suggested that differences between 10 and 30 kcal/mol may suggest greater affinities in docking [37,38]. Furthermore, there was more structural similarity between the MOG-DI dimer and the dimer described by Clements et al. [25] than between the MOG-DR dimer and the Clements et al. dimer. Clements et al. used gel electrophoresis to demonstrate that there is a higher proportion of MOG monomer than MOG dimer [25]. It is possible that reducing environments lead to the cleavage of disulfide bonds and reduced dimer formation, and that in non-pathological states MOG monomers are formed preferentially over MOG homodimers. It has been well known that cleaving disulfide bonds has protective effects on neurons, and one possibility is that this is mediated through preventing the dimer formation of MOG monomers.

The findings in the present study may provide insights into the interaction of phosphine-borane complexes and MOG, with respect to three areas opposing neurodegeneration: remyelination, neurotrophin-related neuroprotection, and the prevention of complement-dependent neuronal loss or synaptic pruning. First, the MOG dimer, as demonstrated previously and replicated in these findings, is an antiparallel head to tail homodimer [25]. This is consistent with the theory that MOG plays a role in adhesion between opposing myelin sheaths [21,25]. Considering that the disulfide bond of the extracellular domain of MOG produces a more stable dimer, it is conceivable that the function of MOG is to promote opposing myelin adhesion in oxidative environments through dimer formation. This may relate to neurodegenerative diseases where oxidative stress plays a role, and by promoting disulfide bond formation (favoring dimer over monomer formation), there may be an increase in opposing myelin sheath adhesion.

There is evidence that myelin adhesion molecules play a role in regulating myelin growth [39]. Elazar et al. found that knockout models of cell adhesion molecules, including Cadm4, Mag, and Caspr, increase the number of axons surrounded by multiple myelin sheaths and identified that in that way neighbouring myelin sheaths overgrew into adjacent territories [39,40]. They suspected that in knockout models, myelin sheath growth may be disinhibited and suggested that cell adhesion molecules play a role in myelin growth regulation [39,40]. This is further supported by studies demonstrating that domains of Cahm4 restricted myelin sheath growth in the CNS and Schwann cell myelination [39,40].

It is possible that MOG dimers behave as a cell adhesion molecule and inhibit myelin sheath growth. In states of oxidative stress, disulfide bond formation (and by extension, dimer formation) may be favored. Such a dimer formation could contribute to the restriction of myelin sheath growth through cell adhesion. Although oxidative stress is associated with neurodegeneration in both the CNS and PNS, there is a stronger implication for oxidative stress playing a role in CNS diseases [41–44]. This is pertinent, as MOG is absent in the PNS [21,22]. As evidenced by the favorable covalent docking scores that were identified, it is possible that phosphine-borane compounds cleave the disulfide bond of MOG and decrease dimer formation. This may decrease opposing myelin adhesion in oxidate environments and promote myelination and myelin sheath growth.

Second, the reduction of the MOG Cys24-Cys98 disulfide bond may be neuroprotective via the promotion of neurotrophin activity. Büdingen et al. suggested that MOG's extracellular domain has a similar structure to TrkA. They suggested that MOG binds to nerve growth factor (NGF), the ligand for TrkA, and sequesters NGF from TrkA. They explicitly evaluated the monomer structure of MOG but did not explicitly study the dimer or the impact of the disulfide bond on NGF binding to MOG [45]. It is possible that reducing the disulfide bond of MOG decreases the affinity for NGF, thus sequestering less NGF from TrkA [45,46]. It is therefore possible that phosphine-borane complexes could promote NFG binding to TrkA by reducing MOG's disulfide bonds.

Third, there is evidence that MOG may act as an activator of the complement cascade by binding to complement component C1q near the IgG binding site [22,47–49]. As described by Stevens et al. [50], C1q is elevated in retinal ganglion cells and the inner plexiform layer in glaucoma models compared to controls. It has been suggested that C1q plays a role in synaptic pruning, with C1q knockout models demonstrating a greater number of overlapping projections compared to controls [45]. In a prospective study involving 12 pediatric patients with anti-MOG acquired demyelinating syndromes, 8 were found to have C1q associated to myelin and phagocytic cells in brains with EAE [51]. The studies evaluating MOG–C1q docking did not discuss whether the C1q was binding to MOG as a monomer or as a dimer [22,47,48,52]. However, these studies did prepare their proteins at physiologic conditions where disulfide bonds are formed between Cys24 and Cys98 [25]. Given these findings, it is possible that C1q binds preferentially to the MOG-DI monomer over the MOG-DR monomer. If this is the case, disulfide reduction may be protective from complement binding, and this could explain how phosphine-borane compounds protect neurons from an autoimmune complement cascade.

In all three of the above presented mechanisms, the inhibition of MOG through phosphine-borane compounds would protect retinal ganglion cells from damage. It is therefore possible that phosphine-borane compounds could be used in the treatment of neuro-ophthalmic diseases beyond anti-MOG syndrome.

The major MOG-associated pathophysiology is the anti-MOG antibody syndrome. We found that docking MOG-DI and MOG-DR monomers to the 8-18C5 antibody favored MOG-DI monomer antibody complex formation. This is unsurprising given that the previously determined structure of the MOG antibody complex was between the 8-18C5 antibody and a MOG monomer with the Cys24-Cys98 disulfide bond intact. Our modeling suggests a higher complex energy in the disulfide-reduced structure and increased structural dissimilarity between the MOG-DR monomer complex and the previously determined structure by Breithaupt et al. (2003) [21]. These findings suggest that reducing the MOG disulfide bond could prevent anti-MOG antibody–MOG complex formation. However, there was a lower complex energy between the 8-18C5 antibody and the MOG-DR dimer than the MOG-DI dimer. There has been no previous determination of the MOG dimer–antibody complex, and it is unclear if such a complex is formed. It is possible that if such a complex did exist, then the reduction in the disulfide bond would increase antibody binding. However, given that reducing the disulfide bond decreases dimer formation compared to the MOG-DR monomer, which has decreased antibody binding compared to the disulfide intact structure, it is unclear if the disulfide bond would influence autoimmunity toward the dimer.

In addition, there is more structural similarity between the MOG-DI dimer–antibody complex and the previously determined MOG monomer–antibody complex than the MOG-DR dimer–antibody complex. The structure of the MOG-Fab complex was based on X-ray crystallography of the MOG monomer. It is possible that antibodies preferentially bind MOG monomers vs. MOG dimers [21]. In the previous PBD structure of antibody–MOG complexes (PBD 1PKQ), the disulfide bond on MOG's extracellular domain was intact. It would be surprising to find that cleaving the disulfide bond would have no impact on antibody–MOG complex formation, given that monoclonal antibodies to MOG are conformation-dependent and antibodies to linear epitopes of MOG do not induce widespread demyelination [35]. It should be noted that the protein quality of the antibody–MOG complex was suitable to analyze the presented structure, but it had insufficient resolution for modeling protein–protein docking. Given the clinical relevance of anti-MOG syndromes, it would be helpful to develop a crystallography structure of an anti-MOG antibody with a resolution and quality more compatible to molecular docking.

It should be noted that dynamic molecular modeling was not employed, with the authors instead using restrained minimization with OLPS3e [53]. The latter approach allowed for a more conservative assessment of the disulfide bond and was more appropriate with the computational power available. However, it should be noted that a dynamic approach could have allowed a possibly wider range of possible structures [54].

The covalent docking results suggested that phosphine-active compounds were able to bind at Cys24 and follow a S_N2 nucleophilic substitution reaction. The docking scores for all compounds were unimpressive, but these scores typically rely on previous docking patterns. Since this site had never before been targeted for binding and no crystallography of any compound in complex with Cys24 has been determined, these docking scores may be unreliable. However, the MM-GBSAs were highly suggestive that phosphine-borane compound derivatives were compatible with covalent docking to Cys24 in both the monomer and dimer. The site map also provides evidence that the exposed surface of Cys24 was compatible with an acceptable shallow binding site. It should be noted that no negative controls were used in this study. Since there has been no previous crystallography or modeling of this site, there have been no confirmed negative controls for the binding site on the MOG protein. TCEP is known to cleave disulfide bonds indiscriminately and for that reason was used as a positive control [9,10]. Furthermore, the other non PB phosphine molecules all had exposed phosphorus with the ability to complete a S_N2 nucleophilic substitution reaction with cysteine bonds [10]. However, it should be noted that CovDock in Schrödinger did not dock all phosphine compounds to all disulfide bonds. This is evidenced by the fact that only phosphine was able to cleave Cys98, while all other tested molecules were not able to. This suggests that a negative control would not necessarily be warranted, as CovDock was able to eliminate unsuitable targets.

These findings suggest that Cys24's disulfide bond could be an appropriate target for drug discovery with respect to anti-MOG syndrome. Salt bridge and hydrogen bond formation with Arg25 had a significantly positive effect on docking. The development of phosphine-borane compounds that could more specifically interact with Arg25 could help create a MOG-targeted reducing agent. In particular, P1 and its metabolites had favorable MM-GBSAs. This may indicate that the targeted hydrolysis of phosphine-borane ester bonds may improve the binding of phosphine-borane to the MOG Cys24-Cys98 disulfide bond because carboxylic acid groups are more likely to make hydrogen bonds with the amino acid than the intact ester group.

It is important to note that this study did not involve in vivo or in vitro methodologies. The isolation of MOG and anti-MOG antibodies is both a challenging and costly process, and there have been relatively few studies assessing ligand docking to these proteins. The current in silico evaluation was necessary to identify potential ligands and protein binding sites, providing a preliminary proof of concept. Before these molecules can be considered for clinical trials, comprehensive in vivo and in vitro studies will be required.

Patients with anti-MOG syndrome are often subjected to long-term immunosuppression and the potential adverse effects of these medications [11,18]. Unlike systemic immunosuppression, a phosphine-borane compound could theoretically be synthesized that would be specific to the MOG protein, thereby potentially mitigating any adverse effects.

4. Materials and Methods

4.1. Compounds

Phosphine-borane compounds were synthesized at the Keck–University of Wisconsin Comprehensive Cancer Center Small Molecule Screening Facility (Madison, WI). PB1 and the (3-propionic acid methyl ester) diphenylphosphine–borane complex PB2, and their deprotected metabolites bis(3-propionic acid methyl ester) phenylphosphine (P1) and (3-propionic acid methyl ester)diphenylphosphine (P2), cleave intracellular disulfide reporters and protect RGCs from oxidative stress. The structures of select structures are presented in Figure 9. These compounds have ester bonds (two in P1 and one in P2) that can be hydrolyzed, as previously identified by in vivo NMR [7–9], resulting in six P1 and P2 compounds with exposed phosphorus. These compounds were used as test compounds for molecular docking. Given that the disulfide bond on MOG has never been targeted previously, positive controls for its specific reduction could not be used. Instead, tris(2-carboxyethyl)phosphine (TCEP), a reducing agent known to cleave generic disulfide bonds, was used. TCEP, like phosphine-borane compounds, cleaves disulfide bonds through an S_N2 nucleophilic substitution reaction, with the exposed phosphorus acting as a nucleophile [10]. Phosphine (PH_3), trimethylphosphine ($P(CH_3)_3$), and triphenylphosphine ($P(C_6H_5)_3$) all have exposed phosphorus and could potentially cleave MOG's solitary disulfide bond. TCEP, phosphine, trimethylphosphine, and triphenylphosphine therefore served as theoretical positive controls for molecular docking.

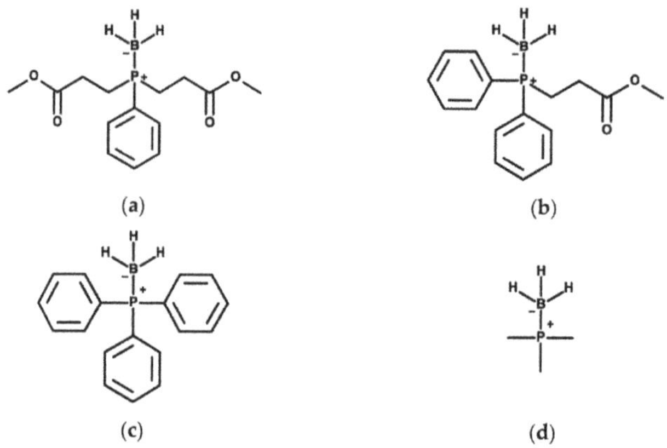

Figure 9. Phosphine-borane structures. Chemical structures of (**a**) PB1, (**b**) PB2 (**c**) PB-Ph, and (**d**) PB-Me.

4.2. Protein Structures

The extracellular structure domain of the monomer and dimer of MOG was described by Clements et al. [25]. The monomeric structure of MOG (PBD 1PKO) was also generated through X-ray diffraction at 1.4 Å resolution, R-value free of 0.22, R-value work of 0.19, and total atom count of 1160) [21]. The latter structure was imported into Schrödinger 2021-1 (Schrödinger LLC, New York, NY, USA; version 12.6.144) and prepared using the Protein Preparation wizard [55].Bond orders were assigned using the CCD database. All missing hydrogens were added, missing side chains were added using prime and het, and stats were generated using the Epik function of Schrödinger at pH 7 ± 2. Hydrogen bond assignment was optimized from sample water orientations at pH 7. Water molecules more than 5 Å from het groups were removed, as well as those with less than 3 hydrogen bonds to non-water molecules. Finally, restrained minimization was performed on all atoms within the structure using the OLPS3e force field, to determine the lowest energy structure to an RMSD of 0.3 Å [53]. The result of these steps was the MOG monomer with the disulfide bond intact ("MOG-DI monomer").

The disulfide-reduced monomer of MOG was created by identifying the disulfide bond on MOG through the Schrödinger Maestro platform, and then the disulfide bond between Cys24 and Cys98 was manually cleaved. The post-cleaved structure was processed using the Protein Preparation wizard as described previously, but without allowing the creation of disulfide bonds. The result was the MOG monomer with the disulfide reduced ("MOG-DR monomer").

Breithaupt et al. (2003) also generated a PBD structure (1PKQ; resolution 3 Å, R-value free 0.32, R-value work 0.25, total atom count 8517) of the extracellular domain of MOG in the Fab-complex, for the purposes of understanding MOG's role in autoimmunity [21]. This was achieved by determining the complex between MOG's extracellular domain and the anti-MOG antibody 8-18C5. This structure was imported into Schrödinger and prepared using the Protein Preparation wizard, as described above. The structure included two MOG monomers and two antibodies, with both heavy and light chains. We separated and isolated antibodies and MOG monomers into separate entities in Schrödinger. The isolated antibodies and MOG monomers were then prepared using the Protein Preparation wizard, as described above.

4.3. Protein Docking

Using the Schrödinger protein–protein docking tool, the MOG-DI and DR monomers were docked in order to determine the structure of the MOG-DI and DR homodimers. This was accomplished using the dimer mode in the Prime Protein–Protein docking function.

The docking tool available in Schrödinger has been previously validated in vivo, allowing for a suitable evaluation of the protein modifications [56,57]. Amino acid constraints were not labeled as attractive or repulsive. However, the monomer tail (AA 118–126) was assigned the buried state because these amino acids do not form the primary extracellular structure, are not involved in the dimer interface, and are located nearest to the myelin interface. Setting amino acids as buried results in increased van der Waals radii and repulsive potentials for buried residues and sets the attractive potential for residues to zero. The number of protein orientations to dock was set to the maximum-allowed 70,000 orientations. This represents a sampling of every 5° in the space of Euler angles. The number of poses to return was set to 30, so that only the highest-scoring 30 poses would be evaluated. The most energetically favorable pose of each complex was used for comparative analysis. Protein energetics was measured using total complex energy in kcal/mol. Total complex energies were determined by the determination of bond and molecular mechanism force fields through electrostatic and van der Waals energies and solvation free energy contributions through polar electrostatic energies and the nonelectrostatic solvation component, as implemented in the Schrödinger protein–protein docking tool [57,58].

Using the Schrödinger protein-protein docking tool, the MOG antibody complex was generated against the MOG-DI monomer, MOG-DR monomer, MOG-DI dimer, and MOG-DR dimer. The 8-18C5 anti-MOG antibody was isolated from 1PKQ, as described above. No amino acids were constrained as attraction, repulsion, or buried. The heavy chain of the antibody was detected using the default system in the Schrödinger protein–protein docking tool available in Schrödinger. The number of orientations was set to 70,000 and the number of poses to 30. The protein metrics evaluated prior to protein docking included PBD resolution, binding sites, bond angle length and angles, protein backbone and sidechain structure steric clashes, and protein site maps [59]. Given that MOG is a membrane-bound protein and approaches of the antibody or MOG monomer from the membrane side could not be completely eliminated with constraints, the protein chosen for analysis was the one with the most structural similarity to the structures presented in Clements et al. and Breithaupt et al. [21,25] Such a methodology makes fewer assumptions regarding the final protein structure than more liberal approaches [60]. These methodologies have been previously validated and compared to other in silico models in syntaxin superfamily proteins and peptide ligand complexes [61,62].

To analyze the interface of the docked proteins, ProFunc was employed to evaluate the number of residues involved at the interface, the area of the interface, the number of salt bridges at the interface, and the number of hydrogen bonds at the interface [55].

4.4. Covalent Docking

Using the Covalent Docking (CovDock) feature of Schrödinger, all compounds described above were covalently docked to Cys24 and Cys98 in the MOG-DI monomer and MOG-DI dimer. CovDock has been effectively demonstrated to exhibit consistency with in vivo modes of interaction, indicating its reliability and relevance in predicting molecular interactions, with the CovDock RMSD ranging between 1.5 and 1.9 Å [63–67]. To simulate the S_N2 nucleophilic substitution reaction [10], the following code was used to describe this reaction:

```
REACTION_NAME                 CUSTOM CODE
CUSTOM_CHEMISTRY              ('<1>', ('charge', 0, 1))
CUSTOM_CHEMISTRY              ('<1>|<2>', ('bond', 1, (1, 2)))
CUSTOM_CHEMISTRY              ('<2>', ('charge', 1, 1))
LIGAND_SMARTS_PATTERN         1,[PX4+1,PX3+0]
RECEPTOR_SMARTS_PATTERN       2,[C,c]-[S,O;H1,-1]
```

Docked ligands were confined to enclosing boxes centered around the reactive residue of interest (either Cys24 or Cys98). No constraints were applied to any amino side chain. Docking was completed with the Schrödinger thorough pose prediction methodology. Molecular mechanics energies combined with a generalized Born and surface area continuum solvation (MM-GBSA) scoring and docking scores were calculated as the primary metrics of the docking results. MM-GBSA scoring is a force field-based assessment of docking with the calculated binding energy dependant on force field predictions, charge models, continuum solvation methods, sampling methods, interior dielectric constants, and conformation entropy. Such scoring methodologies have become accepted practice as they demonstrate high accuracy and are less computationally demanding than other force field-based calculations [67]. The docking scoring in Schrödinger is calculated empirically based on lipophilic–lipophilic terms, hydrogen bond terms, a rotatable bond penalty, contributions from protein–ligand coulomb-vdW energies, and hydrophobic encounters [68,69]. The most energetically favorable pose of each compound was used for comparative analysis. The RMSD was determined for all molecular docking results as a metric for validity and metric for the reproduction of findings. The RMSD is calculated by comparing the multiple binding poses to each other [70]. The metrics evaluated in covalent docking included evaluating the structure of the ligand and protein at the site map as well as electrostatic, lipophilic, and aromatic interactions, H-bonds, rotational entropy, ligand flexibility, and the stability of the produced covalent bond [66,71].

4.5. Site Map Determination

To determine if the site of the compound binding was compatible with an appropriate receptor site, the SiteMap tool in Schrödinger was employed to determine the top ranked potential receptor binding sites. A minimum of 15 site points were required in order to be considered a site. The most restrictive definition of hydrophobicity was used, and sites less than 4 Å from the nearest site point were combined into a single site. A site map was determined for both the MOG-DI monomer and MOG-DI dimer. SiteMap was run twice for each structure, first without allowing for the detection of shallow binding sites and then with the detection of shallow binding sites.

5. Conclusions

Based on the in silico modeling presented above, we suggest that the Cys24-Cys98 disulfide bond in the extracellular domain of MOG could play an important role in MOG dimer formation. It is plausible that the neuroprotective effects of disulfide-cleaving phosphine-borane compounds are mediated by preventing dimer formation, which subsequently regulates myelin formation, the sequestration of nerve growth factor, and the activation of autoimmune complement cascades. Disruption of the Cys24-Cys98 bond also reduces the stability of anti-MOG antibody complexes, providing a potential avenue to treat anti-MOG-related disease. Although the presented computational modeling is promising, it should be noted that experimental validation is required. If verified, phosphine-borane compounds may provide clinicians with a new approach to treating one of the most serious optic neuropathies.

Author Contributions: Conceptualization, R.R. and L.A.L.; methodology, R.R. and S.K.D.; resources, S.K.D. and L.A.L.; writing—original draft preparation, R.R.; writing—review and editing, R.R., S.K.D. and L.A.L.; project administration, L.A.L.; funding acquisition, L.A.L. All authors have read and agreed to the published version of the manuscript.

Funding: The McGill Drug Discovery Platform at the McGill University Health Center Research Institute kindly provided access to the Schrödinger software platform. L.A.L has received support from Canada Research Chairs and Canadian Institutes for Health Research CIHR PJT-162396.

Institutional Review Board Statement: Not applicable.

Informed Consent Statement: Not applicable.

Data Availability Statement: The data presented in this study are available within the article.

Conflicts of Interest: L.A.L. is a consultant to Annexon, Prilenia, Janssen, Roche, Perfuse, Genentech, UNITY, Eyevensys, and Santen; is a member of the Scientific Advisory Board for the Gilbert Family Foundation and the Steering Committee for the National Eye Institute Audacious Goal Initiative;

and holds patents assigned to the Wisconsin Alumni Research Foundation. The funders had no role in the design of the study; in the collection, analyses, or interpretation of data; in the writing of the manuscript; or in the decision to publish the results.

References

1. Behbehani, R. Clinical approach to optic neuropathies. *Clin. Ophthalmol.* **2007**, *1*, 233–246. [PubMed]
2. Almasieh, M.; Wilson, A.M.; Morquette, B.; Cueva Vargas, J.L.; Di Polo, A. The molecular basis of retinal ganglion cell death in glaucoma. *Prog. Retin. Eye Res.* **2012**, *31*, 152–181. [CrossRef] [PubMed]
3. Syc-Mazurek, S.B.; Fernandes, K.A.; Wilson, M.P.; Shrager, P.; Libby, R.T. Together JUN and DDIT3 (CHOP) control retinal ganglion cell death after axonal injury. *Mol. Neurodegener.* **2017**, *12*, 71. [CrossRef] [PubMed]
4. Lieven, C.J.; Hoegger, M.J.; Schlieve, C.R.; Levin, L.A. Retinal ganglion cell axotomy induces an increase in intracellular superoxide anion. *Investig. Ophthalmol. Vis. Sci.* **2006**, *47*, 1477–1485. [CrossRef]
5. Kanamori, A.; Catrinescu, M.M.; Kanamori, N.; Mears, K.A.; Beaubien, R.; Levin, L.A. Superoxide is an associated signal for apoptosis in axonal injury. *Brain* **2010**, *133*, 2612–2625. [CrossRef]
6. Geiger, L.K.; Kortuem, K.R.; Alexejun, C.; Levin, L.A. Reduced redox state allows prolonged survival of axotomized neonatal retinal ganglion cells. *Neuroscience* **2002**, *109*, 635–642. [CrossRef]
7. Swanson, K.I.; Schlieve, C.R.; Lieven, C.J.; Levin, L.A. Neuroprotective effect of sulfhydryl reduction in a rat optic nerve crush model. *Investig. Ophthalmol. Vis. Sci.* **2005**, *46*, 3737–3741. [CrossRef]
8. Schlieve, C.R.; Tam, A.; Nilsson, B.L.; Lieven, C.J.; Raines, R.T.; Levin, L.A. Synthesis and characterization of a novel class of reducing agents that are highly neuroprotective for retinal ganglion cells. *Exp. Eye Res.* **2006**, *83*, 1252–1259. [CrossRef]
9. Niemuth, N.J.; Thompson, A.F.; Crowe, M.E.; Lieven, C.J.; Levin, L.A. Intracellular disulfide reduction by phosphine-borane complexes: Mechanism of action for neuroprotection. *Neurochem. Int.* **2016**, *99*, 24–32. [CrossRef]
10. Dmitrenko, O.; Thorpe, C.; Bach, R.D. Mechanism of S_N2 disulfide bond cleavage by phosphorus nucleophiles. Implications for biochemical disulfide reducing agents. *J. Org. Chem.* **2007**, *72*, 8298–8307. [CrossRef]
11. Lana-Peixoto, M.A.; Talim, N. Neuromyelitis optica spectrum disorder and anti-MOG syndromes. *Biomedicines* **2019**, *7*, 42. [CrossRef] [PubMed]
12. Kuhle, J.; Lindberg, R.L.; Regeniter, A.; Mehling, M.; Hoffmann, F.; Reindl, M.; Berger, T.; Radue, E.W.; Leppert, D.; Kappos, L. Antimyelin antibodies in clinically isolated syndromes correlate with inflammation in MRI and CSF. *J. Neurol.* **2007**, *254*, 160–168. [CrossRef] [PubMed]
13. Reindl, M.; Linington, C.; Brehm, U.; Egg, R.; Dilitz, E.; Deisenhammer, F.; Poewe, W.; Berger, T. Antibodies against the myelin oligodendrocyte glycoprotein and the myelin basic protein in multiple sclerosis and other neurological diseases: A comparative study. *Brain* **1999**, *122*, 2047–2056. [CrossRef]
14. Karni, A.; Bakimer-Kleiner, R.; Abramsky, O.; Ben-Nun, A. Elevated levels of antibody to myelin oligodendrocyte glycoprotein is not specific for patients with multiple sclerosis. *Arch. Neurol.* **1999**, *56*, 311–315. [CrossRef]
15. Markovic, M.; Trajkovic, V.; Drulovic, J.; Mesaros, S.; Stojsavljevic, N.; Dujmovic, I.; Stojkovic, M.M. Antibodies against myelin oligodendrocyte glycoprotein in the cerebrospinal fluid of multiple sclerosis patients. *J. Neurol. Sci.* **2003**, *211*, 67–73. [CrossRef]
16. Lolli, F.; Rovero, P.; Chelli, M.; Papini, A.M. Antibodies against glycosylated native MOG are elevated in patients with multiple sclerosis. *Neurology* **2005**, *65*, 781–782. [CrossRef]
17. Egg, R.; Reindl, M.; Deisenhammer, F.; Linington, C.; Berger, T. Anti-MOG and anti-MBP antibody subclasses in multiple sclerosis. *Mult. Scler. J.* **2001**, *7*, 285–289. [CrossRef]
18. Tajfirouz, D.A.; Bhatti, M.T.; Chen, J.J. Clinical characteristics and treatment of MOG-IgG–associated optic neuritis. *Curr. Neurol. Neurosci. Rep.* **2019**, *19*, 100. [CrossRef]
19. Jarius, S.; Ruprecht, K.; Kleiter, I.; Borisow, N.; Asgari, N.; Pitarokoili, K.; Pache, F.; Stich, O.; Beume, L.-A.; Hümmert, M.W. MOG-IgG in NMO and related disorders: A multicenter study of 50 patients. Part 2: Epidemiology, clinical presentation, radiological and laboratory features, treatment responses, and long-term outcome. *J. Neuroinflammation* **2016**, *13*, 1–45. [CrossRef]
20. Ramanathan, S.; Mohammad, S.; Tantsis, E.; Nguyen, T.K.; Merheb, V.; Fung, V.S.; White, O.B.; Broadley, S.; Lechner-Scott, J.; Vucic, S. Clinical course, therapeutic responses and outcomes in relapsing MOG antibody-associated demyelination. *J. Neurol. Neurosurg. Psychiatry* **2018**, *89*, 127–137. [CrossRef]
21. Breithaupt, C.; Schubart, A.; Zander, H.; Skerra, A.; Huber, R.; Linington, C.; Jacob, U. Structural insights into the antigenicity of myelin oligodendrocyte glycoprotein. *Proc. Natl. Acad. Sci. USA* **2003**, *100*, 9446–9451. [CrossRef] [PubMed]
22. Johns, T.G.; Bernard, C.C. The structure and function of myelin oligodendrocyte glycoprotein. *J. Neurochem.* **1999**, *72*, 1–9. [CrossRef] [PubMed]
23. Pham-Dinh, D.; Mattei, M.-G.; Nussbaum, J.-L.; Roussel, G.; Pontarotti, P.; Roeckel, N.; Mather, I.H.; Artzt, K.; Lindahl, K.F.; Dautigny, A. Myelin/oligodendrocyte glycoprotein is a member of a subset of the immunoglobulin superfamily encoded within the major histocompatibility complex. *Proc. Natl. Acad. Sci. USA* **1993**, *90*, 7990–7994. [CrossRef] [PubMed]
24. Hilton, A.A.; Slavin, A.J.; Hilton, D.J.; Bernard, C.C. Characterization of cDNA and genomic clones encoding human myelin oligodendrocyte glycoprotein. *J. Neurochem.* **1995**, *65*, 309–318. [CrossRef] [PubMed]

25. Clements, C.S.; Reid, H.H.; Beddoe, T.; Tynan, F.E.; Perugini, M.A.; Johns, T.G.; Bernard, C.C.; Rossjohn, J. The crystal structure of myelin oligodendrocyte glycoprotein, a key autoantigen in multiple sclerosis. *Proc. Natl. Acad. Sci. USA* **2003**, *100*, 11059–11064. [CrossRef]
26. Gonzalez-Perez, O. The ventricular-subventricular zone: A source of oligodendrocytes in the adult brain. *Front. Media SA* **2014**, *8*, 137.
27. Gardinier, M.; Amiguet, P.; Linington, C.; Matthieu, J.M. Myelin/oligodendrocyte glycoprotein is a unique member of the immunoglobulin superfamily. *J. Neurosci. Res.* **1992**, *33*, 177–187. [CrossRef]
28. Kroepfl, J.F.; Viise, L.R.; Charron, A.J.; Linington, C.; Gardinier, M.V. Investigation of myelin/oligodendrocyte glycoprotein membrane topology. *J. Neurochem.* **1996**, *67*, 2219–2222. [CrossRef]
29. Sun, X.; Qiu, W.; Wang, J.; Wang, S.; Wang, Y.; Zhong, X.; Liu, C.; Cui, C.; Hong, H.; Yang, H.; et al. Myelin oligodendrocyte glycoprotein-associated disorders are associated with HLA subtypes in a Chinese paediatric-onset cohort. *J. Neurol. Neurosurg. Psychiatry* **2020**, *91*, 733–739. [CrossRef]
30. Slavin, A.; Johns, T.; Orian, J.; Bernard, C. Regulation of myelin oligodendrocyte glycoprotein in different species throughout development. *Dev. Neurosci.* **1997**, *19*, 69–78. [CrossRef]
31. Linnington, C.; Webb, M.; Woodhams, P.L. A novel myelin-associated glycoprotein defined by a mouse monoclonal antibody. *J. Neuroimmunol.* **1984**, *6*, 387–396. [CrossRef] [PubMed]
32. Linington, C.; Berger, T.; Perry, L.; Weerth, S.; Hinze-Selch, D.; Zhang, Y.; Lu, H.C.; Lassmann, H.; Wekerle, H. T cells specific for the myelin oligodendrocyte glycoprotein mediate an unusual autoimmune inflammatory response in the central nervous system. *Eur. J. Immunol.* **1993**, *23*, 1364–1372. [CrossRef] [PubMed]
33. Slavin, A.; Ewing, C.; Liu, J.; Ichikawa, M.; Slavin, J.; Bernard, C.C. Induction of a multiple sclerosis-like disease in mice with an immunodominant epitope of myelin oligodendrocyte glycoprotein. *Autoimmunity* **1998**, *28*, 109–120. [CrossRef]
34. Amor, S.; Groome, N.; Linington, C.; Morris, M.M.; Dornmair, K.; Gardinier, M.V.; Matthieu, J.-M.; Baker, D. Identification of epitopes of myelin oligodendrocyte glycoprotein for the induction of experimental allergic encephalomyelitis in SJL and Biozzi AB/H mice. *J. Immunol.* **1994**, *153*, 4349–4356. [CrossRef]
35. Brehm, U.; Piddlesden, S.J.; Gardinier, M.V.; Linington, C. Epitope specificity of demyelinating monoclonal autoantibodies directed against the human myelin oligodendrocyte glycoprotein (MOG). *J. Neuroimmunol.* **1999**, *97*, 9–15. [CrossRef]
36. Almasieh, M.; Lieven, C.J.; Levin, L.A.; Di Polo, A. A cell-permeable phosphine-borane complex delays retinal ganglion cell death after axonal injury through activation of the pro-survival extracellular signal-regulated kinases 1/2 pathway. *J. Neurochem.* **2011**, *118*, 1075–1086. [CrossRef]
37. Chong, W.L.; Chupradit, K.; Chin, S.P.; Khoo, M.M.; Khor, S.M.; Tayapiwatana, C.; Nimmanpipug, P.; Thongkum, W.; Lee, V.S. Protein-protein interactions: Insight from molecular dynamics simulations and nanoparticle tracking analysis. *Molecules* **2021**, *26*, 5696. [CrossRef]
38. Hanafiah, A.; Abd Aziz, S.N.; Nesran, Z.N.; Wezen, X.C.; Ahmad, M.F. Molecular investigation of antimicrobial peptides against Helicobacter pylori proteins using a peptide-protein docking approach. *Heliyon* **2024**, *10*, e28128. [CrossRef]
39. Li, J.; Monk, K.R. Healthy attachments: Cell adhesion molecules collectively control myelin integrity. *J. Cell Biol.* **2019**, *218*, 2824–2825. [CrossRef]
40. Elazar, N.; Vainshtein, A.; Rechav, K.; Tsoory, M.; Eshed-Eisenbach, Y.; Peles, E. Coordinated internodal and paranodal adhesion controls accurate myelination by oligodendrocytes. *J. Cell Biol.* **2019**, *218*, 2887–2895. [CrossRef]
41. Spiegel, I.; Adamsky, K.; Eshed, Y.; Milo, R.; Sabanay, H.; Sarig-Nadir, O.; Horresh, I.; Scherer, S.S.; Rasband, M.N.; Peles, E. A central role for Necl4 (SynCAM4) in Schwann cell-axon interaction and myelination. *Nat. Neurosci.* **2007**, *10*, 861–869. [CrossRef] [PubMed]
42. Kim, G.H.; Kim, J.E.; Rhie, S.J.; Yoon, S. The Role of Oxidative Stress in Neurodegenerative Diseases. *Exp. Neurobiol.* **2015**, *24*, 325–340. [CrossRef] [PubMed]
43. Naik, A.K.; Tandan, S.K.; Dudhgaonkar, S.P.; Jadhav, S.H.; Kataria, M.; Prakash, V.R.; Kumar, D. Role of oxidative stress in pathophysiology of peripheral neuropathy and modulation by N-acetyl-L-cysteine in rats. *Eur. J. Pain* **2006**, *10*, 573–579. [CrossRef] [PubMed]
44. Mallet, M.L.; Hadjivassiliou, M.; Sarrigiannis, P.G.; Zis, P. The Role of Oxidative Stress in Peripheral Neuropathy. *J. Mol. Neurosci.* **2020**, *70*, 1009–1017. [CrossRef]
45. von Büdingen, H.-C.; Mei, F.; Greenfield, A.; Jahn, S.; Shen, Y.-A.A.; Reid, H.H.; McKemy, D.D.; Chan, J.R. The myelin oligodendrocyte glycoprotein directly binds nerve growth factor to modulate central axon circuitry. *J. Cell Biol.* **2015**, *210*, 891–898. [CrossRef]
46. Artim, S.C.; Kiyatkin, A.; Lemmon, M.A. Comparison of tyrosine kinase domain properties for the neurotrophin receptors TrkA and TrkB. *Biochem. J.* **2020**, *477*, 4053–4070. [CrossRef]
47. de Rosbo, N.K.; Honegger, P.; Lassmann, H.; Matthieu, J.M. Demyelination induced in aggregating brain cell cultures by a monoclonal antibody against myelin/oligodendrocyte glycoprotein. *J. Neurochem.* **1990**, *55*, 583–587. [CrossRef]
48. Piddlesden, S.; Lassmann, H.; Zimprich, F.; Morgan, B.; Linington, C. The demyelinating potential of antibodies to myelin oligodendrocyte glycoprotein is related to their ability to fix complement. *Am. J. Pathol.* **1993**, *143*, 555. [CrossRef]
49. Rus, H.; Cudrici, C.; Niculescu, F.; Shin, M.L. Complement activation in autoimmune demyelination: Dual role in neuroinflammation and neuroprotection. *J. Neuroimmunol.* **2006**, *180*, 9–16. [CrossRef]

50. Stevens, B.; Allen, N.J.; Vazquez, L.E.; Howell, G.R.; Christopherson, K.S.; Nouri, N.; Micheva, K.D.; Mehalow, A.K.; Huberman, A.D.; Stafford, B.; et al. The classical complement cascade mediates CNS synapse elimination. *Cell* **2007**, *131*, 1164–1178. [CrossRef]
51. Elazar, N.; Vainshtein, A.; Golan, N.; Vijayaragavan, B.; Schaeren-Wiemers, N.; Eshed-Eisenbach, Y.; Peles, E. Axoglial Adhesion by Cadm4 Regulates CNS Myelination. *Neuron* **2019**, *101*, 224–231.e5. [CrossRef] [PubMed]
52. Serguera, C.; Stimmer, L.; Fovet, C.M.; Horellou, P.; Contreras, V.; Tchitchek, N.; Massonneau, J.; Leroy, C.; Perrin, A.; Flament, J.; et al. Anti-MOG autoantibodies pathogenicity in children and macaques demyelinating diseases. *J. Neuroinflamm.* **2019**, *16*, 244. [CrossRef] [PubMed]
53. Roos, K.; Wu, C.; Damm, W.; Reboul, M.; Stevenson, J.M.; Lu, C.; Dahlgren, M.K.; Mondal, S.; Chen, W.; Wang, L. OPLS3e: Extending force field coverage for drug-like small molecules. *J. Chem. Theory Comput.* **2019**, *15*, 1863–1874. [CrossRef] [PubMed]
54. Laskowski, R.A.; Watson, J.D.; Thornton, J.M. ProFunc: A server for predicting protein function from 3D structure. *Nucleic Acids Res.* **2005**, *33*, W89–W93. [CrossRef] [PubMed]
55. *Maestro Schrödinger*; Version 11.2; Schrödinger LLC: New York, NY, USA, 2021.
56. Agostino, M.; Mancera, R.L.; Ramsland, P.A.; Fernández-Recio, J. Optimization of protein–protein docking for predicting Fc–protein interactions. *J. Mol. Recognit.* **2016**, *29*, 555–568. [CrossRef]
57. Rosell, M.; Rodríguez-Lumbreras, L.A.; Romero-Durana, M.; Jiménez-García, B.; Díaz, L.; Fernández-Recio, J. Integrative modeling of protein-protein interactions with pyDock for the new docking challenges. *Proteins Struct. Funct. Bioinform.* **2020**, *88*, 999–1008. [CrossRef]
58. Ongaro, A.; Oselladore, E.; Memo, M.; Ribaudo, G.; Gianoncelli, A. Insight into the LFA-1/SARS-CoV-2 Orf7a complex by protein–protein docking, molecular dynamics, and MM-GBSA calculations. *J. Chem. Inf. Model.* **2021**, *61*, 2780–2787. [CrossRef]
59. Durham, J.; Zhang, J.; Humphreys, I.R.; Pei, J.; Cong, Q. Recent advances in predicting and modeling protein–protein interactions. *Trends Biochem. Sci.* **2023**, *48*, 527–538. [CrossRef]
60. Saikia, S.; Bordoloi, M. Molecular docking: Challenges, advances and its use in drug discovery perspective. *Curr. Drug Targets* **2019**, *20*, 501–521. [CrossRef]
61. Singh, S.; Srivastava, H.K.; Kishor, G.; Singh, H.; Agrawal, P.; Raghava, G.P. Evaluation of protein-ligand docking methods on peptide-ligand complexes for docking small ligands to peptides. *bioRxiv* **2017**, 212514. [CrossRef]
62. Batra, S.; Pancholi, P.; Roy, M.; Kaushik, S.; Jyoti, A.; Verma, K.; Srivastava, V.K. Exploring insights of syntaxin superfamily proteins from Entamoeba histolytica: A prospective simulation, protein-protein interaction, and docking study. *J. Mol. Recognit.* **2021**, *34*, e2886. [CrossRef] [PubMed]
63. Li, Q.; Wang, H.; Yang, W.L.; Yang, J.K. An approach combining deep learning and molecule docking for drug discovery of cathepsin L. *Expert Opin. Drug Discov.* **2023**, *18*, 347–356. [CrossRef] [PubMed]
64. Al-Khafaji, K.; Al-Duhaidahawi, D.; Taskin Tok, T. Using integrated computational approaches to identify safe and rapid treatment for SARS-CoV-2. *J. Biomol. Struct. Dyn.* **2021**, *39*, 3387–3395. [CrossRef] [PubMed]
65. Toledo Warshaviak, D.; Golan, G.; Borrelli, K.W.; Zhu, K.; Kalid, O. Structure-based virtual screening approach for discovery of covalently bound ligands. *J. Chem. Inf. Model.* **2014**, *54*, 1941–1950. [CrossRef]
66. Zhu, K.; Borrelli, K.W.; Greenwood, J.R.; Day, T.; Abel, R.; Farid, R.S.; Harder, E. Docking covalent inhibitors: A parameter free approach to pose prediction and scoring. *J. Chem. Inf. Model.* **2014**, *54*, 1932–1940. [CrossRef]
67. Kmiecik, S.; Kouza, M.; Badaczewska-Dawid, A.E.; Kloczkowski, A.; Kolinski, A. Modeling of Protein Structural Flexibility and Large-Scale Dynamics: Coarse-Grained Simulations and Elastic Network Models. *Int. J. Mol. Sci.* **2018**, *19*, 3496. [CrossRef]
68. Wang, E.; Sun, H.; Wang, J.; Wang, Z.; Liu, H.; Zhang, J.Z.; Hou, T. End-point binding free energy calculation with MM/PBSA and MM/GBSA: Strategies and applications in drug design. *Chem. Rev.* **2019**, *119*, 9478–9508. [CrossRef]
69. Friesner, R.A.; Murphy, R.B.; Repasky, M.P.; Frye, L.L.; Greenwood, J.R.; Halgren, T.A.; Sanschagrin, P.C.; Mainz, D.T. Extra precision glide: Docking and scoring incorporating a model of hydrophobic enclosure for protein– ligand complexes. *J. Med. Chem.* **2006**, *49*, 6. [CrossRef]
70. Velázquez-Libera, J.L.; Durán-Verdugo, F.; Valdés-Jiménez, A.; Núñez-Vivanco, G.; Caballero, J. LigRMSD: A web server for automatic structure matching and RMSD calculations among identical and similar compounds in protein-ligand docking. *Bioinformatics* **2020**, *36*, 2912–2914. [CrossRef]
71. Ahmad, S.; Singh, S.; Srivastava, M.R.; Shukla, S.; Rai, S.; Shamsi, A.S. Molecular Docking Simplified: Literature Review. *Adv. Med. Dent. Health Sci.* **2021**, *4*, 37–44. [CrossRef]

pharmaceuticals

Article

Glutathione Modulates Hydrogen Sulfide Release and the Ocular Hypotensive Action of Diallyl Polysulfide Compounds

Susmit Mhatre [1,2,3], Rai Anjali [1,4,5], Pulkit Sahai [1], John Auden [1], Somnath Singh [1], Ya Fatou Njie Mbye [6], Sunny E. Ohia [6] and Catherine A. Opere [1,*]

1 Department of Pharmacy Sciences, School of Pharmacy and Health Professionals, Creighton University, Omaha, NE 68178, USA; susmitmhatre@creighton.edu or mhatre2@wisc.edu (S.M.); anjalirai@creighton.edu or rai27@wisc.edu (R.A.); pulkitsahai@creighton.edu (P.S.); johnauden@creighton.edu (J.A.); somnathsingh@creighton.edu (S.S.)
2 Cell and Molecular Biology Graduate Program, University of Wisconsin-Madison, Madison, WI 53706, USA
3 Department of Neuroscience, University of Wisconsin-Madison, Madison, WI 53706, USA
4 Comparative Biomedical Sciences Graduate Program, University of Wisconsin-Madison, Madison, WI 53706, USA
5 School of Veterinary Medicine, University of Wisconsin-Madison, Madison, WI 53706, USA
6 Department of Pharmaceutical Sciences, College of Pharmacy and Health Sciences, Texas Southern University, Houston, TX 77004, USA; yafatou.mbye@tsu.edu (Y.F.N.M.); sunny.ohia@tsu.edu (S.E.O.)
* Correspondence: copere@creighton.edu; Tel.: +1-402-280-3726

Abstract: Background: Hydrogen sulfide (H_2S) is an endogenous transmitter with the potential to regulate aqueous humor dynamics and protect retinal neurons from degeneration. The aim of the present study was two-fold: (a) to evaluate the release of H_2S from two polysulfides, diallyl disulfide (DADS), and diallyl trisulfide (DATS); and (b) to investigate their ocular hypotensive actions in normotensive male and female rabbits in the presence and absence of GSH. Materials and Methods: H_2S was quantified hourly for up to 6 h using a H_2S-Biosensor (World Precision Instruments, Sarasota, Fl). Intraocular pressure (IOP) was assessed in normotensive New Zealand Albino rabbits using a pneumotonometer (model 30 classic; Reichert Ophthalmic Instruments, Depew, NY, USA). Results: In the presence of GSH, there was an increase in the in vitro release of H_2S produced by DADS and DATS. Both DADS and DATS also caused a dose-dependent reduction in IOP in male and female rabbits, in both treated and untreated eyes. For instance, in male animals, the presence of GSH (3% and 5%) significantly ($p < 0.05$, $n = 5$) enhanced the ocular hypotensive action of DADS (2%) and DATS (2%) from 14.02 ± 2.89% to 18.67 ± 5.6% and from 16.22 ± 3.48 to 23.62 ± 5.79%, respectively. Conclusions: GSH enhanced both H_2S release and ocular hypotensive action of the polysulfides in a manner that was dependent on the number of sulfur atoms present in each polysulfide. Furthermore, female animals were less sensitive to the IOP-lowering action of the polysulfides, when compared to their male counterparts.

Keywords: hydrogen sulfide; glaucoma; intraocular pressure; glutathione; diallyl disulfide; diallyl trisulfide

1. Introduction

Hydrogen sulfide (H_2S), an inflammable toxic gas with application in organic chemistry as base material for elemental sulfur, has found applications in pharmacology after its discovery as an endogenous gaseous transmitter in various biological systems [1,2]. H_2S has been reported to mediate multiple critical processes such as neurotransmission [2–4], cytoprotection [5–8], neuroprotection [6,9], smooth muscle relaxation [10,11], vasorelaxation [10,12–17], anti-inflammation [18–23], and cellular respiration [24,25], and it can serve as a free radical scavenger [2,26], amongst others.

H_2S is endogenously produced in mammalian ocular tissues based on reports of its presence in tissues of the anterior and posterior segments [4]. Furthermore, the enzymes responsible for H_2S biosynthesis, cystathionine-β-synthase (CBS), cystathionine-γ-lyase (CSE), and 3-mercaptopyruvate sulfurtransferase (3MST), along with cysteine aminotransferase (CAT), have been localized in different ocular tissues [4,27–29]. CBS insufficiency has been associated with ocular conditions such as lens dislocation, retina degeneration, retinal detachment, and acute glaucoma [30], suggesting a pivotal role for H_2S in the pathophysiology of eye diseases. Compounds that release H_2S in biological media such as sodium hydrosulfide (NaHS) have been reported to exert pharmacological actions in ocular tissues. For example, H_2S-releasing compounds have been reported to relax iris smooth muscle via K^+-ATP channel-dependent mechanisms [4], decrease the electrically evoked norepinephrine release [4] and norepinephrine levels in mammalian anterior uvea [4], and increase aqueous humor outflow facility in the anterior segment of the eye [4]. Additionally, H_2S attenuates excitatory amino acid neurotransmission in retinal neurons [3], suggesting a neuroprotective role for this gas in the posterior segment of the eye. The ability to measure H_2S in biological tissues involves the use of several methodologies, such as colorimetric assays for methylene blue formation, sulfide-ion-selective or -polarographic electrodes, high-performance liquid chromatography analysis with fluorescent derivatization of sulfides, and gas chromatography with flame photometric or sulfur chemiluminescence detection [31].

We have evidence that the inorganic compound, NaHS; a slow-releasing synthetic H_2S donor, GYY4137; the H_2S-latanoprost hybrid compound, ACS67; and the precursor for the endogenous production of H_2S, L-cysteine, can reduce IOP in male New Zealand Albino normotensive rabbits [4]. However, the effect of the garlic-derived, bioactive diallyl compounds diallyl disulfide (DADS) and diallyl trisulfide (DATS) on IOP have not been described. Both DADS and DATS are bioactive compounds generated via the decomposition of allicin, the oil-soluble organosulfur derived from garlic [32]. These compounds are structurally composed of disulfide and trisulfide groups in between two allyl groups, respectively (Figure 1), and are reported to release H_2S by nucleophilic substitution at the sulfur atom via their interaction with GSH [33]. Moreover, both DADS and DATS are reported to exhibit potent anticancer [32,34,35], cardioprotection [33], and anti-inflammatory [36] properties. Indeed, it has been postulated that production of H_2S from the polysulfides such as DADS and DATS account for the beneficial actions of garlic [33]. In ocular tissues, we have evidence that DATS mitigated both time-dependent and oxidative stress-induced cataractogenesis via an increase in both GSH content and superoxide dismutase activity in isolated bovine lenses [37,38]. Therefore, the aim of the present study was two-fold: (a) to evaluate the in vitro release of H_2S from DADS and DATS in the presence and absence of GSH and (b) to elucidate the ocular hypotensive action of these two diallyl compounds on IOP in the presence and absence of GSH in normotensive male and female New Zealand Albino rabbits.

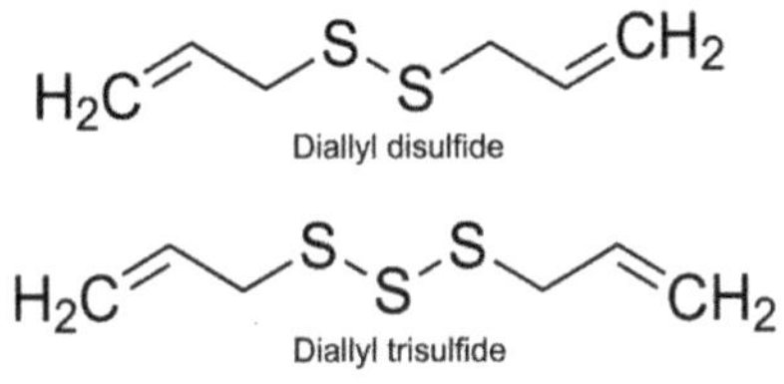

Figure 1. Chemical structure of diallyl disulfide (DADS) and diallyl trisulfide (DATS) with two allyl groups on each end, with either two or three sulfur atoms with disulfide bonds.

2. Results

2.1. In Vitro Release

Whereas DADS (2%) did not release any H_2S in solution for 6 h, DATS (2%) exhibited an immediate release of H_2S, with a small peak of 0.2 ± 0.004 nmoles/mL observed after 5 h. The presence of GSH (0.5–5%) stimulated an instantaneous production of H_2S from DADS (2%) that achieved a maximum release of 15.26 ± 1.29 nmoles ($p < 0.001$; $n = 3$) at the GSH (3%) concentration (Figure 2, left panel). Furthermore, DATS (2%) released H_2S in a concentration-dependent manner, achieving a maximum release of 23.79 ± 1.03 nmoles ($p < 0.001$; $n = 3$) at the GSH (5%) concentration (Figure 2, right panel). Following the burst release, DADS (2%) in the presence of GSH (3%) exhibited a steady decline in H_2S release in a time-dependent manner, while DATS (2%) in the presence of GSH (5%) sustained the release of H_2S for over a period of 6 h (Figure 3).

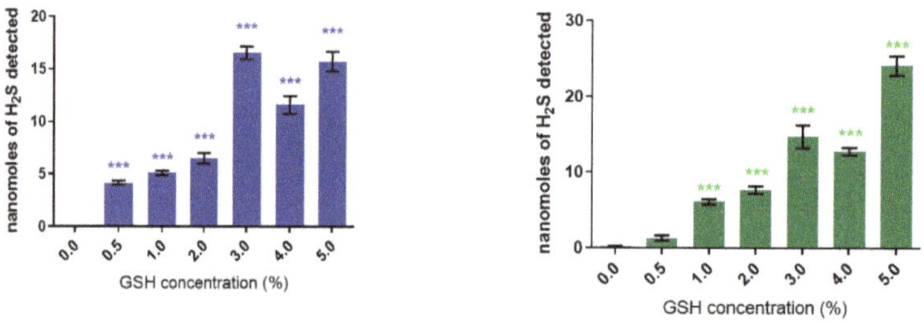

Figure 2. Burst release of hydrogen sulfide (H_2S) from diallyl disulfide (DADS, **left panel**) and diallyl trisulfide (DATS, **right panel**) in the presence of glutathione (GSH), in vitro. Vertical bars represent means ± SEM; each data point represents 3 observations. One-way ANOVA *** $p < 0.001$ significantly different from control.

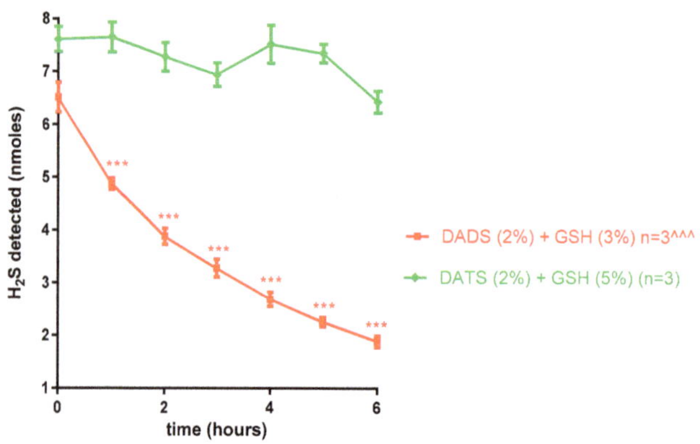

Figure 3. Time-dependent release of hydrogen sulfide (H_2S) from diallyl disulfide (DADS) and diallyl trisulfide (DATS) in the presence of glutathione (GSH), in vitro. Vertical bars represent means ± SEM; each data point represents 3 observations. Two-way ANOVA ^^^ $p < 0.001$, significantly different from DATS/GSH; one-way ANOVA *** $p < 0.0001$, significantly different from H_2S at time zero.

2.2. Effect of Polysulfides on Intraocular Pressure

Both of the diallyl polysulfides, DADS and DATS, exerted an ocular hypotensive effect on male New Zealand Albino rabbits. DADS (2%) achieved a maximum ($p > 0.05$; $n = 5$) reduction in IOP of 14.02 ± 2.89% after 4 h in the treated eye, and 6.51 ± 1.45% after 30 min in the contralateral eye (Figure 4 and Table 1). Whereas glutathione (GSH) did not significantly alter IOP pretreatment of animals with GSH (3%) significantly ($p < 0.05$, $n = 5$) enhanced the maximum IOP-lowering action of DADS (2%) to 18.67 ± 5.60% (Figure 4) after 1 h in the treated eye, while the contralateral eye exhibited a reduction of 15.69 ± 2.24% in 1 h as well (Table 1). Whereas topical administration of the vehicle (DMSO (20%)/saline) and normal saline, which served as negative controls, did not elicit a significant effect on IOP in all normotensive New Zealand Albino rabbits, the IOP-lowering action of DADS (2%) was less than that of timolol (0.25%; used as a positive control), which elicited a maximal effect of 27.33 ± 2.46% ($p < 0.001$; $n = 5$) in 3 h in the treated eye and 18.86 ± 6.59% ($p > 0.05$; $n = 5$) (Figure 4).

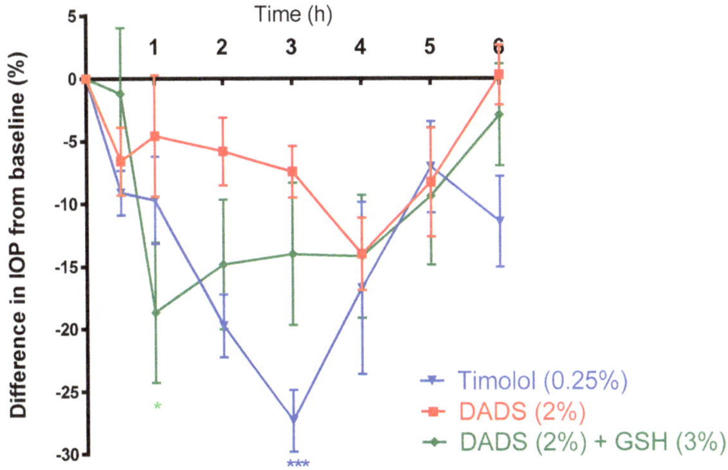

Figure 4. Effect of diallyl disulfide (DADS) on intraocular pressure (IOP) in the treated eye in the presence and absence of glutathione ethyl ester (GSH) in male normotensive conscious New Zealand Albino rabbits in vivo. Vertical bars represent means ± SEM of data obtained from five rabbits. * $p < 0.05$; *** $p < 0.001$, significantly different from baseline IOP.

Table 1. Effect of DADS on IOP in the contralateral eyes of normotensive male New Zealand Albino rabbits.

Treatment	Contralateral Eye		
	Peak Activity (h)	Duration of Action (h)	Maximum % IOP Reduction Mean ± SEM
DADS (2%)	0.5	2	6.51 ± 1.45
DADS (2%) + GSH (3%)	1	6	15.69 ± 2.24
Timolol gel drops (0.25%)	3	6	18.86 ± 6.59

Effects of DADS (2%) in the absence and presence of GSH (3%) were not significantly different from baseline IOP ($p > 0.05$). Similarly, the effect of timolol, the positive control, was not statistically significant from baseline IOP ($p > 0.05$). Each IOP reading represent means ± SEM of data obtained from five rabbits. DADS, diallyl disulfide; GSH, glutathione ethyl ester; IOP, intraocular pressure.

It was interesting to note that DATS (0.1%, 1%, and 2%) elicited a reduction in IOP in a concentration-dependent manner. For instance, in male animals, DATS (0.1%; 1% and 2%) achieved a maximum reduction in IOP of 11.03 ± 2.57%, 13.08 ± 2.01%, and 16.22 ± 3.48 ($p > 0.05$; $n = 5$) in 0.5, 2, and 1 h, respectively, in the treated eye (Table 2 and Figure 5). In the contralateral eye, DATS (0.1%; 1% and 2%) achieved a maximum reduction in IOP by 11.70 ± 4.85%, 8.59 ± 6.78, and 9.38 ± 4.17% ($p > 0.05$; $n = 5$) in 1, 5, and 1 h, respectively (Table 2).

Table 2. Effect of DATS on IOP in the treated eye normotensive male New Zealand Albino rabbits.

Treatment	Treated Eye			Contralateral Eye		
	Peak Activity (h)	Duration of Action (h)	Maximum % IOP Reduction Mean ± SEM	Peak Activity (h)	Duration of Action (h)	Maximum % IOP Reduction Mean ± SEM
DATS (0.1%)	0.5	2	11.03 ± 2.57	1	4	11.70 ± 4.85
DATS (1%)	2	8	13.08 ± 2.01	5	8	8.59 ± 6.78
DATS (2%)	1	6	16.22 ± 3.48	1	6	9.38 ± 4.17
DATS (2%) + GSH (5%)	2	5	23.62 ± 5.79 *	3	6	21.14 ± 7.42 *
Timolol (0.25%)	3	6	27.33 ± 2.46 ***	3	6	18.86 ± 6.59

Each IOP reading represent means ± SEM of data obtained from five rabbits. * $p < 0.05$ means that the effect of DATS (2%) in the presence of GSH (5%) was significantly different from baseline IOP in the treated and the contralateral eye. *** $p > 0.001$ means that the effect of timolol was significantly different from baseline IOP in the treated eye. DATS, diallyl trisulfide; GSH, glutathione; IOP, intraocular pressure.

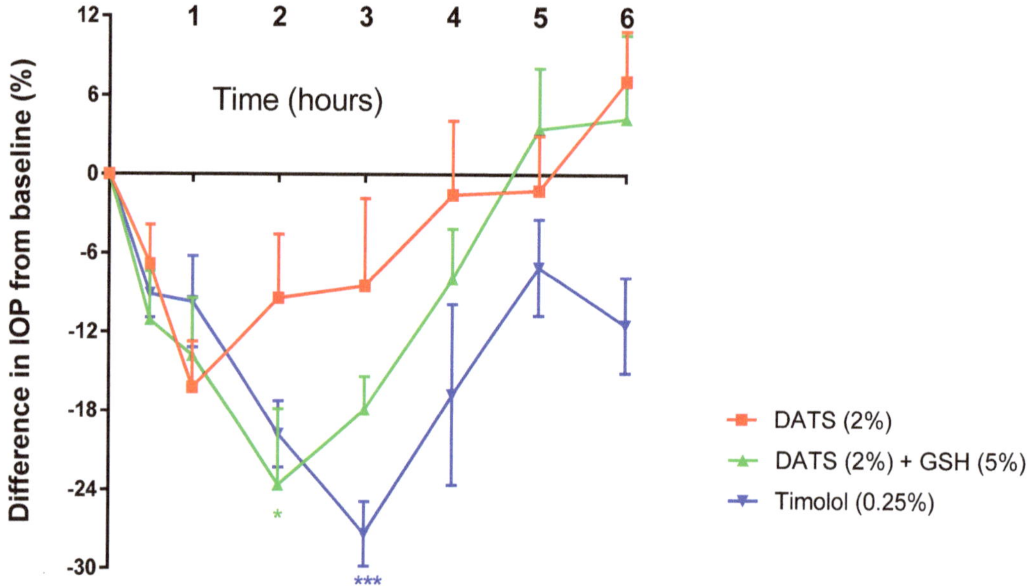

Figure 5. Effect of diallyl trisulfide (DATS) on intraocular pressure (IOP) in the treated eye in the presence and absence of glutathione ethyl ester (GSH) in male, normotensive conscious New Zealand Albino rabbits in vivo. Vertical bars represent means ± SEM of data obtained from five rabbits. * $p < 0.05$; *** $p < 0.001$, significantly different from baseline IOP.

The presence of GSH-EE (5%) enhanced the IOP-lowering effect of DATS (2%), achieving a significant ($p < 0.05$; $n = 5$) maximum reduction of 23.62 ± 5.79% and 21.14 ± 7.42% in the treated and contralateral eyes in 2 h, respectively (Figure 5 and Table 2). It was interesting to note that this decrease in the IOP was either less than (for the treated eye) or comparable (for the contralateral eye) to that elicited by timolol (0.25%) in male animals (Figure 5 and Table 2).

To determine the role of gender in the IOP-lowering effect of polysulfides, the effect of DATS was evaluated in female normotensive animals. DATS (2%) elicited a lower IOP-lowering action in females (10.26 ± 2.06%; $p > 0.05$), compared to 16.22 ± 3.48% ($p > 0.05$) in males after 1 h in the treated eye. The contralateral effect for DATS (2%) was insignificant and equivalent in males (9.38 ± 4.17%) and females (8.05 ± 3.29 (Figure 6 and Table 3a,b). The presence of GSH (5%) enhanced the maximum IOP-lowering action of DATS (2%) to 18.14 ± 2.08% ($p < 0.01$; $n = 5$) in 3 h in females, compared to 23.62 ± 5.79 ($p < 0.05$) in 2 h in males in the treated eye (Figure 6 and Table 3a,b). The contralateral effect was higher in males (21.14 ± 7.42%; $p < 0.05$ in 3 h) compared to that in females (11.14 ± 2.92%; $p > 0.05$ in 4 h). Moreover, timolol elicited a lower ocular hypotensive effect in females as well (17.78 ± 1.68; $p > 0.05$ in 2 h) compared to that in males (27.33 ± 2.46; $p < 0.001$ in 3 h). The contralateral effect of timolol was not significant in both males and females (Figure 6 and Table 3a,b).

Table 3. (**a**) Effect of DATS on IOP in the treated eye in normotensive male and female New Zealand Albino rabbits. (**b**) Effect of DATS on IOP in the contralateral eye in normotensive male and female New Zealand Albino rabbits.

a

Treatment	Male, Treated Eye			Female, Treated Eye		
	Peak Activity (h)	Duration of Action (h)	Maximum % IOP Reduction Mean ± SEM	Peak Activity (h)	Duration of Action (h)	Maximum % IOP Reduction Mean ± SEM
DATS (2%)	1	6	16.22 ± 3.48	1	6	10.26 ± 2.06
DATS (2%) + GSH (5%)	2	5	23.62 ± 5.79 *	3	7	18.14 ± 2.08 **
Timolol (0.25%)	3	6	27.33 ± 2.46 ***	2	8	17.78 ± 1.68

b

Treatment	Male, Contralateral Eye			Female, Contralateral Eye		
	Peak Activity (h)	Duration of Action (h)	Maximum % IOP Reduction Mean ± SEM	Peak Activity (h)	Duration of Action (h)	Maximum % IOP Reduction Mean ± SEM
DATS (2%)	1	6	9.38 ± 4.17	3	6	8.05 ± 3.29
DATS (2%) + GSH (5%)	3	6	21.14 ± 7.42 *	4	7	11.14 ± 2.92
Timolol (0.25%)	3	6	18.86 ± 6.59	2	8	13.17 ± 2%

Each IOP reading represent means ± SEM of data obtained from five rabbits. (a) * $p < 0.05$; ** $p < 0.01$ mean that the effect of DATS (2%) in the presence of GSH (5%) was significantly different from baseline IOP in the treated eyes. *** $p > 0.001$ means that the effect of timolol was significantly different from baseline IOP in the treated eye. (b) * $p < 0.05$ means that the effect of DATS (2%) in the presence of GSH (5%) was significantly different from baseline IOP in the contralateral, untreated eye. DATS, diallyl trisulfide; GSH, glutathione; IOP, intraocular pressure.

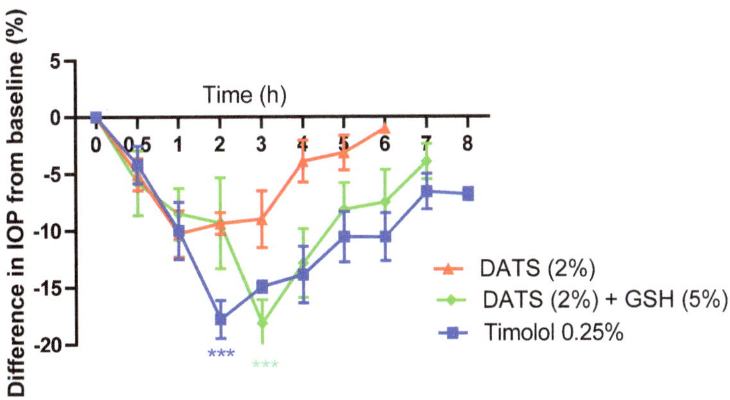

Figure 6. Effect of diallyl trisulfide (DATS) on intraocular pressure (IOP) in the treated eye in the presence and absence of glutathione ethyl ester (GSH) in female normotensive conscious New Zealand Albino rabbits in vivo. Vertical bars represent means ± SEM of data obtained from five rabbits. *** $p < 0.001$, significantly different from baseline IOP.

3. Discussion

In the past few decades, H_2S has evolved from an industrial toxicant to a gaseous transmitter that plays significant pathophysiological roles in multiple biological systems [39]. The potential therapeutic applications of this gas have been described for several diseases, including cardiovascular, inflammation, cancer, Parkinson's disease, Alzheimer's disease, etc. [39,40]. In ocular tissues, several investigators have localized the enzymes responsible for the biosynthesis of H_2S in the retina, iris ciliary bodies, lens, etc. [4,27–29]. In the anterior segment of the eye, H_2S modulates sympathetic neurotransmission [4] and relaxes smooth muscle [4] in mammalian isolated iris ciliary bodies in vitro. In the posterior segment, H_2S-releasing compounds have also been reported to relax posterior ciliary arteries and to inhibit amino acid neurotransmission in the neural retina [3,4]. We have further demonstrated that multiple H_2S-producing compounds, NaHS, GYY4137, and ACS67, can reduce IOP in normotensive male New Zealand Albino rabbits [4]. However, the effect of the diallyl compounds DADS and DATS in ocular tissues has not been investigated. DADS and DATS are bioactive polysulfide compounds generated from the decomposition of allicin and are believed to contribute to beneficial roles of garlic via the production of H_2S. Therefore, this study investigated the in vitro release of H_2S from DADS and DATS and their ocular hypotensive action in normotensive male and female New Zealand Albino rabbits.

It was interesting to note that the DADS did not release any detectable H_2S in aqueous solution, while DATS exhibited a small release of the gas over a period of 6 h. The cellular release of H_2S from the polysulfides is reported to be dependent on the presence of GSH. Therefore, in vitro, the release of H_2S was further determined in the presence of GSH, where the presence of exogenous GSH (0.5 to 5%) triggered a burst release of the H_2S from both polysulfides. The maximum burst production of H_2S from DADS was observed after GSH (3%), while that of DATS was concentration-dependent, achieving a maximum production at the highest concentration tested, GSH (5%). Moreover, DATS exhibited a higher and consistent rate of H_2S release compared to that of DATS for the 6-h duration of the study. In corroboration, DADS has been reported to exhibit less burst release and a slower rate of release compared to that of DATS in PBS [39,41]. Benavides et al. (2007) proposed that both DATS and DADS release H_2S via an α-carbon nucleophilic substitution pathway which requires an interaction with GSH [33]. However, Liang et al. (2015) contended that DADS does not significantly produce H_2S via the α-carbon nucleophilic reaction with GSH but rather interacts with GSH via thiol–disulfide exchange to form products that do not release H_2S. These researchers further proposed that the presence of DATS impurities in commercially available DADS could account for the H_2S production observed from DADS [41]. It also is conceivable that, irrespective of mechanism, DADS, which contains two sulfur atoms, can potentially produce less H_2S compared to DATS, which has three sulfur atoms and can produce more H_2S per molecule of each polysulfide. In preliminary in vitro studies, we observed that exogenous glutathione (0.5–5%) released a small, insignificant amount of H_2S in solution. Furthermore, these concentrations of GSH did not affect IOP but enhanced the maximum IOP-lowering actions of the polysulfides. Since the purpose of this study was to determine the effect of GSH-mediated H_2S release on IOP, subsequent in vivo experiments were conducted using GSH quantities that exhibited the highest H_2S release for each polysulfide (3% GSH for DADS and 5% for DATS).

To determine if the release of H_2S from the polysulfides correlates with their IOP-lowering ability, the effect of topically administered DADS and DATS was determined in normotensive male and female New Zealand Albino rabbits. GSH has been localized in mammalian ocular tissues such as the lens, cornea, retina, aqueous humor, etc., where it plays an integral role in the maintenance of ocular homeostasis and defensive protective mechanisms against oxidative stress in ocular tissues [42,43]. Therefore, endogenous GSH was leveraged to trigger the in vivo release of H_2S via undergo nucleophilic substitution reaction. Although DADS (2%) did not produce any detectable H_2S, it exhibited a minor reduction in IOP ($14.02 \pm 2.89\%$; $p > 0.05$; $n = 5$) that could be attributed to the presence of endogenous GSH in the anterior uvea of the eye. Interestingly, DATS (2%), which released a small quantity of H_2S in vitro, also elicited a minor reduction in IOP (16.22 ± 3.48 ($p > 0.05$; $n = 5$). Taken together, these observations suggest that endogenous GSH does not trigger H_2S production from the polysulfides in quantities sufficient to significantly reduce IOP. However, the magnitude of IOP reduction appears to be proportional to the number of sulfur atoms in each polysulfide.

Congruent to the in vitro H_2S-release studies, the presence of exogenous GSH enhanced the IOP-lowering action of both polysulfides in the treated and contralateral eyes of male normotensive animals. In addition to the production of H_2S from DATS via the α-carbon nucleophilic substitution pathway [33], in mammalian cells, DATS has been proposed to generate endogenous H_2S via other mechanisms, including interaction with other thiol–SH proteins, upregulation of the biosynthetic enzyme CSE, and via an oxidoreductase reaction with catalase enzyme [39]. Indeed, these mechanisms may contribute to the higher rate of H_2S production from DATS. It is conceivable that, in the presence of GSH, H_2S produced in the anterior chamber easily diffuses through cell membranes to gain central access and distributes to the contralateral eye [44], where it elicits its hypotensive effects. Consistent with the in vitro release studies, the IOP-lowering effect of DATS/GSH was higher than that of DADS/GSH in both the treated and contralateral eyes. In corroboration, Kim et al. (2014) reported that the antioxidant capacity of the polysulfides to activate nuclear factor erythroid 2-related factor 2 (Nrf2), a transcription factor that regulates the cellular defense against toxic and oxidative insults, in human gastric epithelial cells was directly proportional to the number of sulfur atoms in each polysulfide (DATS > DADS > DAS) [45]. Taken together, IOP reduction correlates with H_2S production, both in the absence and presence of GSH.

To determine the role of gender in the IOP-lowering action of the polysulfides, the effect of DATS (2%) in males was compared to that in females in normotensive animals. In general, female animals were more resistant to the IOP-lowering action of DATS and timolol when compared to male animals. To the best of our knowledge, no gender-related responses to the IOP-lowering action of H_2S-releasing compounds have been reported in the literature for rabbits. Wang et al. (2013) conducted studies to assess IOP over a 24-h period in normotensive New Zealand rabbits and found no significant gender differences in IOP between males and females [46]. However, exogenous progesterone was associated with an increase in IOP in normotensive rabbits, implying that elevation in progesterone associated with the luteal phase of the reproductive cycle could disrupt IOP [47,48]. Since the reproductive cycle in the female animals in this study was unknown, it is conceivable that variations associated with the reproductive cycle could contribute to the apparent resistance to the IOP-lowering actions of the drugs tested.

4. Materials and Methods

4.1. In Vitro Hydrogen Sulfide Release Studies

To quantify the release of H_2S from DADS and DATS, stock solutions (25 mg/250 μL) were prepared in DMSO and GSH (0.5–5%) in argonated/deionized water. Aliquots of GSH (0.5–5%) were added to each diallyl compound (2%; 10 μL), and the volume was stocked up to 100 μL using argonated water, as indicated on Table 4. The amount of H_2S was quantified amperometrically using a H_2S-Biosensor ion-selective electrode and TBR4100 Free Radical Analyzer, both from World Precision Instruments (WPI), (Sarasota, FL, USA), using a protocol described elsewhere [49].

Table 4. Preparations of diallyl compounds with GSH for assessment of in vitro release of H_2S.

Target GSH %	10% Donor Solution (μL)	GSH 5% (μL)	Argonated Water (μL)	Total (μL)
0%	10	0	90	100
0.5%	10	5	85	100
1%	10	10	80	100
2%	10	20	70	100
3%	10	30	60	100
4%	10	40	50	100
5%	10	50	40	100

4.2. Intraocular Pressure Studies in Normotensive Rabbits

Animal protocols were approved by the Institutional Animal Care and Use Committee (Protocol #1165, 2021). Animal studies were conducted in adherence to the Association for Research in Vision and Ophthalmology's statement for the Use of Animals in Ophthalmic and Vision Research.

Male and female normotensive New Zealand Albino rabbits (weighing about 2 kg) were purchased from Charles River Laboratories, Wilmington, MA, USA. In addition to conditioning to 12-h light–dark cycles, animals were acclimatized for 7 days and divided into groups of five animals/group (n = 5). On the day of the experiment, two baseline IOP measurements were taken following topical application of proparacaine 0.5% (local anesthetic) and averaged to serve as the baseline reading for comparison to IOP readings following the treatment. IOP readings were taken using a Pneumatonometer (model 30 classic; Reichert Ophthalmic Instruments, Depew, NY, USA). At 0 h, 50 µL of the test compounds (DATS and DATS) was applied topically to one eye of each animal, while the contralateral eye received the same quantity of vehicle (saline). IOP readings were obtained at 30 min, 1 h, and 2 h and then every hour until the IOP returned to baseline. When used, GSH-EE was topically administered 10 min prior to the topical application of H_2S-producing compounds. GSH-EE was used in the in vivo studies instead of reduced GSH, which was used in in vitro studies to increase access of the intracellular antioxidant into the anterior chamber of the eye. Normal saline and vehicle (DMSO/saline/Tween 20) were used as negative controls, while timolol was used as a positive control. Animal eyes were monitored for signs of ocular discomfort, such as tearing, redness, conjunctival swelling, squinting, and eyelid irritation, that could occur due to the administration of experimental compounds. At the conclusion of the study, animals were painlessly euthanized, as approved by IACUC, using intravenous injection of Euthanasia solution (1 mL/10 lb weight; 1 mL contains 390 mg pentobarbital sodium and 50 mg phenytoin sodium, Patterson Veterinary, Patterson Vet Supply, Inc., manufactured for: Virbac AH, Inc, Fort Worth, TX, USA).

4.3. Data Analysis

Results were expressed as change in IOP (mm of Hg) and/or percentage inhibition of IOP. Except where indicated otherwise, values given are arithmetic means ± S.E.M. Significance of differences between control and agent-treated preparations was evaluated using analysis of variance (ANOVA), followed by Tukey's post-test. Differences with p-values < 0.05 were accepted as statistically significant.

5. Conclusions

Although DADS did not release H_2S in solution, while DATS released minute quantities of H_2S in a simple aqueous environment, both polysulfides elicited a burst release of the gas in the presence of GSH. The presence of GSH elicited an increased release of the gas from DATS, compared to DADS. The IOP-lowering action of the polysulfides was directly dependent on the number of sulfur atoms in each polysulfide and was enhanced by the presence of GSH. Moreover, female animals were less sensitive to the IOP-lowering effects of the polysulfide DATS. The possible pharmacological mechanisms by which the polysulfides decrease IOP in normotensive male and female New Zealand Albino rabbits are shown in Figure 7.

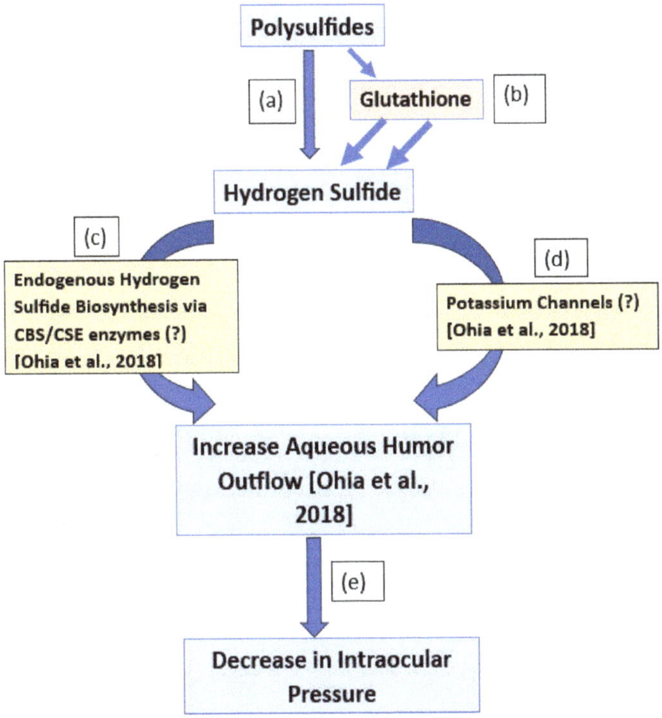

Figure 7. Schematic presentation of the of the possible pharmacological mechanisms by which polysulfides decrease intraocular pressure in normotensive male and female New Zealand Albino rabbits. The production of hydrogen sulfide from polysulfides (**a**) is enhanced in the presence of glutathione (**b**). Hydrogen sulfide elicits an increase in aqueous humor outflow (**e**) via CBS/CSE biosynthetic pathways (**c**) and/or potassium channels (**d**) to reduce intraocular pressure (**e**) [4].

Author Contributions: Conceptualization, C.A.O., Y.F.N.M. and S.E.O.; methodology, C.A.O., Y.F.N.M. and S.E.O.; investigation, S.M., R.A., J.A., P.S., S.S. and C.A.O.; writing—original draft preparation, S.M.; writing—review and editing, C.A.O., S.E.O. and Y.F.N.M.; supervision, C.A.O.; project administration, C.A.O.; funding acquisition, Y.F.N.M. and C.A.O. All authors have read and agreed to the published version of the manuscript.

Funding: This research was funded by NIH NEI, grant number 1R15EY032274-01.

Institutional Review Board Statement: The animal study protocol was approved by the Institutional Animal Care and Use Committee (IACUC) Institutional Review Board of Creighton University, Omaha, NE, Protocol number #1165, August 2021) for studies involving animals.

Informed Consent Statement: Not applicable.

Data Availability Statement: The original contributions presented in the study are included in the article, further inquiries can be directed to the corresponding author.

Conflicts of Interest: The authors declare no conflicts of interest.

References

1. Kimura, H. Physiological role of hydrogen sulfide and polysulfide in the central nervous system. *Neuro-Chem. Int.* **2013**, *63*, 492–497. [CrossRef] [PubMed]
2. Abe, K.; Kimura, H. The possible role of hydrogen sulfide as an endogenous neuromodulator. *J. Neurosci.* **1996**, *16*, 1066–1071. [CrossRef] [PubMed]

3. Bankhele, P.; Salvi, A.; Jamil, J.; Njie-Mbye, F.; Ohia, S.; Opere, C.A. Comparative Effects of Hydrogen Sul-fide-Releasing Compounds on [^{3}H]D-Aspartate Release from Bovine Isolated Retinae. *Neurochem. Res.* **2018**, *43*, 692–701. [CrossRef] [PubMed]
4. Ohia, S.E.; Robinson, J.; Mitchell, L.; Ngele, K.K.; Heruye, S.; Opere, C.A.; Njie-Mbye, Y.F. Regulation of Aqueous Humor Dynamics by Hydrogen Sulfide: Potential Role in Glaucoma Pharmacotherapy. *J. Ocul. Pharmacol. Ther.* **2018**, *34*, 61–69. [CrossRef] [PubMed] [PubMed Central]
5. Kimura, H.; Shibuya, N.; Kimura, Y. Hydrogen sulfide is a signaling molecule and a cytoprotectant. *Antioxid. Redox Signal.* **2012**, *17*, 45–57. [CrossRef] [PubMed] [PubMed Central]
6. Kimura, Y.; Kimura, H. Hydrogen sulfide protects neurons from oxidative stress. *FASEB J.* **2004**, *18*, 165–167. [CrossRef]
7. Elrod, J.W.; Calvert, J.W.; Morrison, J.; Doeller, J.E.; Kraus, D.W.; Tao, L.; Jiao, X.; Scalia, R.; Kiss, L.; Szabo, C.; et al. Hydrogen sulfide attenuates myo-cardial ischemia-reperfusion injury by preservation of mitochondrial function. *Proc. Natl. Acad. Sci. USA* **2007**, *104*, 15560–15565. [CrossRef]
8. King, A.L.; Polhemus, D.J.; Bhushan, S.; Otsuka, H.; Kondo, K.; Nicholson, C.K.; Bradley, J.M.; Islam, K.N.; Calvert, J.W.; Tao, Y.X.; et al. Hydrogen sulfide cytopro-tective signaling is endothelial nitric oxide synthase-nitric oxide dependent. *Proc. Natl. Acad. Sci. USA* **2014**, *111*, 3182–3187. [CrossRef]
9. Whiteman, M.; Armstrong, J.S.; Chu, S.H.; Jia-Ling, S.; Wong, B.S.; Cheung, N.S.; Halliwell, B.; Moore, P.K. The novel neuromodulator hydrogen sulfide: An endogenous peroxynitrite 'scavenger'? *J. Neurochem.* **2004**, *90*, 765–768. [CrossRef] [PubMed]
10. Hosoki, R.; Matsuki, N.; Kimura, H. The possible role of hydrogen sulfide as an endogenous smooth muscle relaxant in synergy with nitric oxide. *Biochem. Biophys. Res. Commun.* **1997**, *237*, 527–531. [CrossRef]
11. Tang, G.; Wu, L.; Liang, W.; Wang, R. Direct Stimulation of KATP Channels by Exogenous and Endogenous Hydrogen Sulfide in Vascular Smooth Muscle Cells. *Mol. Pharmacol.* **2005**, *68*, 1757–1764. [CrossRef] [PubMed]
12. Cheng, Y.; Ndisang, J.F.; Tang, G.; Cao, K.; Wang, R. Hydrogen sulfide-induced relaxation of resistance mes-enteric artery beds of rats. *AJP-Heart Circ. Physiol.* **2004**, *287*, H2316–H2323. [CrossRef]
13. Mustafa, A.K.; Sikka, G.; Gazi, S.K.; Steppan, J.; Jung, S.M.; Bhunia, A.K.; Barodka, V.M.; Gazi, F.K.; Barrow, R.K.; Wang, R.; et al. Hydrogen sulfide as endotheli-um-derived hyperpolarizing factor sulfhydrates potassium channels. *Circ. Res.* **2011**, *109*, 1259–1268. [CrossRef] [PubMed]
14. Wang, R. Two's company, three's a crowd: Can H_2S be the third endogenous gaseous transmitter? *FASEB J.* **2002**, *16*, 1792–1798. [CrossRef]
15. Yang, G.; Wu, L.; Jiang, B.; Yang, W.; Qi, J.; Cao, K.; Meng, Q.; Mustafa, A.K.; Mu, W.; Zhang, S.; et al. H_2S as a Physiologic Vasorelaxant: Hypertension in Mice with Deletion of Cystathionine γ-Lyase. *Science* **2008**, *322*, 587–590. [CrossRef] [PubMed]
16. Zhao, W.; Zhang, J.; Lu, Y.; Wang, R. The vasorelaxant effect of H2S as a novel endogenous gaseous KATP channel opener. *EMBO J.* **2001**, *20*, 6008–6016. [CrossRef]
17. Zhao, W.; Wang, R. H2S-induced vasorelaxation and underlying cellular and molecular mechanisms. *AJP-Heart Circ. Physiol.* **2002**, *283*, H474–H480. [CrossRef]
18. Krishnan, N.; Fu, C.; Pappin, D.J.; Tonks, N.K. H2S-Induced Sulfhydration of the Phosphatase PTP1B and Its Role in the Endoplasmic Reticulum Stress Response. *Sci. Signal.* **2011**, *4*, ra86. [CrossRef]
19. Li, L.; Bhatia, M.; Zhu, Y.C.; Ramnath, R.D.; Wang, Z.J.; Anuar, F.B.M.; Whiteman, M.; Salto-Tellez, M.; Moore, P.K. Hydrogen sulfide is a novel mediator of lipopolysaccharide-induced inflammation in the mouse. *FASEB J.* **2005**, *19*, 1196–1198. [CrossRef]
20. Mustafa, A.K.; Gadalla, M.M.; Sen, N.; Kim, S.; Mu, W.; Gazi, S.K.; Barrow, R.K.; Yang, G.; Wang, R.; Snyder, S.H. H2S Signals Through Protein S-Sulfhydration. *Sci. Signal.* **2009**, *2*, ra72. [CrossRef]
21. Oh, G.S.; Pae, H.O.; Lee, B.S.; Kim, B.N.; Kim, J.M.; Kim, H.R.; Jeon, S.B.; Jeon, W.K.; Chae, H.J.; Chung, H.T. Hydrogen sulfide inhibits nitric oxide pro-duction and nuclear factor-κB via heme oxygenase-1 expression in RAW264.7 macrophages stimulated with lipopolysaccharide. *Free Radic. Biol. Med.* **2006**, *41*, 106–119. [CrossRef] [PubMed]
22. Sen, N.; Paul, B.D.; Gadalla, M.M.; Mustafa, A.K.; Sen, T.; Xu, R.; Kim, S.; Snyder, S.H. Hydrogen sulfide-linked sulfhydration of NF-κB mediates its antiapoptotic actions. *Mol. Cell* **2012**, *45*, 13–24. [CrossRef] [PubMed]
23. Zanardo, R.C.O.; Brancaleone, V.; Distrutti, E.; Fiorucci, S.; Cirino, G.; Wallace, J.L. Hydrogen sulfide is an en-dogenous modulator of leukocyte-mediated inflammation. *FASEB J.* **2006**, *20*, 2118–2120. [CrossRef]
24. Szabo, C.; Coletta, C.; Chao, C.; Módis, K.; Szczesny, B.; Papapetropoulos, A.; Hellmich, M.R. Tumor-derived hydrogen sulfide, produced by cystathionine-β-synthase, stimulates bioenergetics, cell proliferation, and angio-genesis in colon cancer. *Proc. Natl. Acad. Sci. USA* **2013**, *110*, 12474–12479. [CrossRef]
25. Yong, R.; Searcy, D.G. Sulfide oxidation coupled to ATP synthesis in chicken liver mitochondria. *Com-parative Biochemistry and Physiology Part B: Biochem. Mol. Biol.* **2001**, *129*, 129–137. [CrossRef]
26. Zhang, Z.; Jin, S.; Teng, X.; Duan, X.; Chen, Y.; Wu, Y. Hydrogen sulfide attenuates cardiac injury in takotsubo cardiomyopathy by alleviating oxidative stress. *Nitric Oxide.* **2017**, *67*, 10–25. [CrossRef] [PubMed]
27. De, L.G.; Ruggeri, P.; Macaione, S. Cystathionase activity in rat tissues during development. *Ital. J. Biochem.* **1974**, *23*, 371–379.
28. Persa, C.; Osmotherly, K.; Chao-Wei, C.K.; Moon, S.; Lou, M.F. The distribution of cystathionine beta-synthase (CBS) in the ey: Implication of the presence of a trans-sulfuration pathway for oxidative stress defense. *Exp. Eye Res.* **2006**, *83*, 817–823. [CrossRef]
29. Pong, W.W.; Stouracova, R.; Frank, N.; Kraus, J.P.; Eldred, W.D. Comparative localization of cystathionine beta-synthase and cystathionine gamma-lyase in retina: Differences between amphibians and mannals. *J. Comp. Neurol.* **2007**, *505*, 158–165. [CrossRef]

30. Kraus, J.P.; Kozich, V. Cystathionine B synthase and its deficiency. In *Homocysteine in Health and Disease*; Carmel, R., Jacobsen, D.W., Eds.; Cambridge University Press: Cambridge, UK, 2001; pp. 223–243.
31. Koike, S.; Ogasawara, Y. Analysis and characterization of sulfane sulfur. *Anal. Biochem.* **2024**, *687*, 115458. [CrossRef] [PubMed]
32. Subramanian, S.; Nandagopal, G.; Nordin, S.A.; Joseph, N. Prevailing knowledge on the bioavailability and biological activities of sulphur compounds from alliums: A potential drug candidate. *Molecules* **2020**, *25*, 4111. [CrossRef]
33. Benavides, G.A.; Squadrito, G.L.; Mills, R.W.; Patel, H.D.; Isbell, T.S.; Patel, R.P.; Darley-Usmar, V.M.; Doeller, J.E.; Kraus, D.W. Hydrogen sulfide mediates the vasoactivity of garlic. *Proc. Natl. Acad. Sci. USA* **2007**, *104*, 17977–17982. [CrossRef] [PubMed]
34. Mitra, S.; Das, R.; Emran, T.B.; Labib, R.K.; Islam, F.; Sharma, R.; Ahmad, I.; Nainu, F.; Chidambaram, K.; Alhumaydhi, F.A.; et al. Diallyl Disulfide: A Bioac-tive Garlic Compound with Anticancer Potential. *Front. Pharmacol.* **2022**, *13*, 943967. [CrossRef] [PubMed]
35. Puccinelli, M.T.; Stan, S.D. Dietary Bioactive Diallyl Trisulfide in Cancer Prevention and Treatment. *Int. J. Mol. Sci.* **2017**, *18*, 1645. [CrossRef] [PubMed]
36. Zhang, H.; Shang, C.; Tian, Z.; Amin, H.K.; Kassab, R.B.; Abdel Moneim, A.E.; Zhang, Y. Diallyl Disulfide Suppresses Inflammatory and Oxidative Machineries following Carra-geenan Injection-Induced Paw Edema in Mice. *Mediators Inflamm.* **2020**, *2020*, 8508906. [CrossRef] [PubMed]
37. Heruye, S.H.; Mbye, Y.F.; Ohia, S.E.; Opere, C.A. Protective Action of Hydrogen Sulfide-Releasing Compounds against Oxidative Stress-Induced Cataract Formation in Cultured Bovine Lenses. *Curr. Eye Res.* **2021**, *19*, 1–7. [CrossRef]
38. Heruye, S.H.; Maffofou Nkenyi, L.N.; Singh, N.U.; Yalzadeh, D.; Ngele, K.K.; Njie-Mbye, Y.F.; Ohia, S.E.; Opere, C.A. Current Trends in the Pharmacotherapy of Cataracts. *Pharmaceuticals* **2020**, *13*, 15. [CrossRef]
39. Szabo, C.; Papapetropoulos, A. International Union of Basic and Clinical Pharmacology. CII: Pharmacological Modulation of H2S Levels: H2S Donors and H2S Biosynthesis Inhibitors. *Pharmacol. Rev.* **2017**, *69*, 497–564. [CrossRef] [PubMed] [PubMed Central]
40. Sharif, A.H.; Iqbal, M.; Manhoosh, B.; Gholampoor, N.; Ma, D.; Marwah, M.; Sanchez-Aranguren, L. Hydrogen Sulphide-Based Therapeutics for Neurological Conditions: Perspectives and Challenges. *Neurochem. Res.* **2023**, *48*, 1981–1996. [CrossRef] [PubMed] [PubMed Central]
41. Liang, D.; Wu, H.; Wong, M.W.; Huang, D. Diallyl Trisulfide Is a Fast H_2S Donor, but Diallyl Disulfide Is a Slow One: The Reaction Pathways and Intermediates of Glutathione with Polysulfides. *Org. Lett.* **2015**, *17*, 4196–4199. [CrossRef]
42. Megaw, J.M. Gluthathione and ocular photobiology. *Curr. Eye Res.* **1984**, *3*, 83–87. [CrossRef] [PubMed]
43. Ganea, E.; Harding, J.J. Glutathione-Related Enzymes and the Eye. *Curr. Eye Res.* **2006**, *31*, 1–11. [CrossRef]
44. Han, Y.; Shang, Q.; Yao, J.; Ji, Y. Hydrogen sulfide: A gaseous signaling molecule modulates tissue homeo-stasis: Implications in ophthalmic diseases. *Cell Death Dis.* **2019**, *10*, 1–12. [CrossRef] [PubMed]
45. Kim, S.; Lee, H.G.; Park, S.A.; Kundu, J.K.; Keum, Y.S.; Cha, Y.N.; Na, H.K.; Surh, Y.J. Keap1 Cysteine 288 as a Potential Target for Diallyl Trisulfide-Induced Nrf2 Activation. *PLoS ONE* **2014**, *9*, e85984. [CrossRef]
46. Wang, X.; Dong, J.; Wu, Q. Twenty-four-hour measurement of IOP in rabbits using rebound tonometer. *Vet. Ophthalmol.* **2013**, *16*, 423–428. [CrossRef] [PubMed]
47. Green, K.; Cheeks, L.; Slagle, T.; Phillips, C.I. Interaction Between Progesterone and Mifepristone on Intraocular Pressure in Rabbits. *Curr. Eye Res.* **1989**, *8*, 317–320. [CrossRef] [PubMed]
48. Ofri, R.; Shore, L.S.; Kass, P.H.; Horowitz, I.H. The effect of elevated progesterone levels on intraocular pressure in lions (Panthera leo). *Res. Vet. Sci.* **1999**, *67*, 121–123. [CrossRef]
49. Mhatre, S.; Rai, A.; Ali, H.; Patil, A.; Singh, N.; Verma, R.; Auden, J.; Chandler, C.; Dash, A.; Opere, C.; et al. Comparison of Colorimetric, Spectroscopic and Electrochemical Techniques for Quantification of Hydrogen Sulfide. *BioTechniques* **2023**, *76*, 71–80. Available online: https://www.future-science.com/doi/full/10.2144/btn-2023-0075 (accessed on 10 September 2024). [CrossRef]

pharmaceuticals

MDPI

Article

The Identification of New Pharmacological Targets for the Treatment of Glaucoma: A Network Pharmacology Approach

Erika Giuffrida [1,†], Chiara Bianca Maria Platania [1,2,†], Francesca Lazzara [1], Federica Conti [1], Nicoletta Marcantonio [1], Filippo Drago [1,2] and Claudio Bucolo [1,2,*]

1 Department of Biomedical and Biotechnological Sciences, School of Medicine, University of Catania, 95125 Catania, Italy; erika.giuffri@gmail.com (E.G.); chiara.platania@unict.it (C.B.M.P.); francesca.lazzara@unict.it (F.L.); federica.conti@unict.it (F.C.); nicolettamarcantonio61@gmail.com (N.M.); fdrago@unict.it (F.D.)

2 Center for Research in Ocular Pharmacology-CERFO, University of Catania, 95125 Catania, Italy

* Correspondence: claudio.bucolo@unict.it

† These authors equally contributed to the study.

Abstract: Background: Glaucoma is a progressive optic neuropathy characterized by the neurodegeneration and death of retinal ganglion cells (RGCs), leading to blindness. Current glaucoma interventions reduce intraocular pressure but do not address retinal neurodegeneration. In this effort, to identify new pharmacological targets for glaucoma management, we employed a network pharmacology approach. **Methods:** We first retrieved transcriptomic data from GEO, an NCBI database, and carried out GEO2R (an interactive web tool aimed at comparing two or more groups of samples in a GEO dataset). The GEO2R statistical analysis aimed at identifying the top differentially expressed genes (DEGs) and used these as input of STRING (Search Tool for the Retrieval of Interacting Genes/Proteins) app within Cytoscape software, which builds networks of proteins starting from input DEGs. Analyses of centrality metrics using Cytoscape were carried out to identify nodes (genes or proteins) involved in network stability. We also employed the web-server software MIRNET 2.0 to build miRNA–target interaction networks for a re-analysis of the GSE105269 dataset, which reports analyses of microRNA expressions. **Results:** The pharmacological targets, identified in silico through analyses of the centrality metrics carried out with Cytoscape, were rescored based on correlations with entries in the PubMed and clinicaltrials.gov databases. When there was no match (82 out of 135 identified central nodes, in 8 analyzed networks), targets were considered "potential innovative" targets for the treatment of glaucoma, after further validation studies. **Conclusions:** Several druggable targets, such as GPCRs (e.g., 5-hydroxytryptamine 5A (5-HT5A) and adenosine A_{2B} receptors) and enzymes (e.g., lactate dehydrogenase A or monoamine oxidase B), were found to be rescored as "potential innovative" pharmacological targets for glaucoma treatment.

Keywords: network pharmacology; computational systems biology; pharmacological targets; neuroinflammation; glaucoma

1. Introduction

Glaucoma is a progressive optic neuropathy and one of the leading causes of irreversible blindness worldwide. In 2020, 3.6 million people, aged 50 years or older, irreversibly lost their vision due to glaucoma, which accounted for a total of 11% of global blindness among adults of this age [1]. Furthermore, an increased burden of care in glaucoma management is estimated due to the ageing of populations worldwide, especially in the industrialized world [2].

The death of retinal ganglion cells (RGCs), optic disc cupping, and alterations of the peripheral and central visual fields represent the hallmarks of this disease [3,4]. Although glaucoma etiology has not yet been fully elucidated, the loss of RGCs, along with a damage of the optic nerve head (ONH), is linked to multifactorial pathogenic mechanisms such as axonal transport blockades, the deprivation of neurotrophic factors, mitochondrial dysfunction, excitotoxic damage, oxidative stress, and reactive gliosis [5]. Although several risk factors for glaucoma have been identified, such as aging, ethnic origin, and family history, as well as polygenic associations, high intraocular pressure (IOP, >21 mmHg) remains the main and the only modifiable risk factor of glaucoma [6,7]. The increase of IOP is caused by an imbalance between the production and the drainage of the aqueous humor (AH). Up to now, approved topical ocular hypotensive drugs belong to the following therapeutic classes: prostaglandin derivatives, beta-blockers, carbonic anhydrase inhibitors, sympathomimetics, miotics, and the recently approved rho-kinase inhibitors [8]. Prostaglandin analogues and beta-blockers represent first- and second-line treatments, respectively, followed by carbonic anhydrase inhibitors and alpha-2 adrenergic agonists [9]. In the case of an inadequate response to these drugs, ocular surgery, such as trabeculectomy, is recommended. It should be mentioned that normotensive glaucoma patients may take advantage of IOP-lowering drugs, as reported in a clinical trial where a decreased odds ratio for disease progression was registered in patients with glaucoma without hypertension [10]. However, about 40% of patients do not respond to IOP-lowering drugs, and in such patients the disease still slowly progresses to sight loss [11]. Therefore, new therapeutic approaches, able to preserve the retinal visual system and to directly protect RGCs, are needed [12].

In this regard, network pharmacology can be used to identify, in silico, new pharmacological targets for the treatment of complex and multifactorial diseases, including Alzheimer's disease [13], COVID-19 infections [14], and diabetic retinopathy [15], and for the management of cancer metastasis [16]. Through the application of this method, a disease can be represented as a network, in which the nodes (genes, proteins, molecules, and pathways), establishing multiple interactions (edges), provide a holistic representation of the pathology [17]. Moreover, the integration of multi-layer omics data would help to unveil the pathogenic mechanisms of predicted new diseases [17,18].

Our in silico study exploited the network pharmacology approach, which borrows methods from the in silico systems biology approach, also defined as the reverse engineering approach [19]. Specifically, we hereby re-analyzed experimental data already published or deposited in databases to build a network with new nodes and connections through the enrichment of knowledge approaches [15]. A network centrality analysis then helped with the identification of the nodes essential for network stability [20,21], which can be considered predicted pharmacological targets [15] for the management of a given disease. The following steps were applied, according to our previously published work [15]: first, we carried out a re-analysis of transcriptomic data retrieved from the Gene Expression Omnibus (GEO) database [22] of expression datasets associated with tissues from glaucoma patients and control subjects of experimental in vivo models of glaucoma. The datasets were directly analyzed using the GEO2R statistical tool, which was employed to identify the genes that were significantly ($p < 0.05$) differentially expressed (DEGs) (i.e., glaucoma vs. healthy control) [23,24]. Thereafter, we applied the following reverse engineering approach, giving, as input, the DEGs from the GEO2R analysis to the STRING (Search Tool for the Retrieval of Interacting Genes/Proteins) app embedded within the Cytoscape software v. 3.7.0 [25] (https://apps.cytoscape.org/apps/stringapp, accessed on 1 March 2022). The STRING app ensures enrichment of the information it receives from enrichment analyses of gene ontology, KEGG, Reactome, and Wiki Pathways. Finally, the predicted key genes, identified after a centrality analysis of the networks, were rescored as predicted new or not new, on the basis of the lack or presence of a matching with entries on the PubMed and clinicaltrials.gov databases, respectively.

The network pharmacology re-analysis evidenced connections with pathways already experimentally associated with glaucoma pathogenesis, e.g., inflammatory pathways and pathways involved in the regulation of the immune system and fibrosis [26,27]. Furthermore, our study evidenced that several G protein-coupled receptors (GPCRs), ion channels, or enzymes could be involved in the pathological mechanisms of glaucoma. In particular, metabotropic serotoninergic and adenosine receptors, such as the 5-hydroxytryptamine

5A (5-HT5A) and adenosine A_{2B} receptors, were identified as innovative (not yet investigated) drug targets for glaucoma treatment. Moreover, different ligand–gated ion channels belonging to purinergic (e.g., the purinergic ionotropic receptor P2X3) and glutamatergic systems (e.g., the glutamate ionotropic receptor AMPA type subunit 1), might be predicted new pharmacological targets for glaucoma treatment, since no in vitro, in vivo, or clinical studies have been linked to these targets yet. Finally, a series of enzymes, such as lactate dehydrogenase A or monoamine oxidase B, were also identified as predicted new targets for the management of glaucoma.

2. Results

The search in the GEO database for GEO datasets associated with glaucoma (search query "glaucoma") resulted in six GEO datasets (Table 1) reporting transcriptomic data associated with preclinical in vivo models of glaucoma and postmortem isolated human eye tissues. These retrieved GEO datasets were automatically linked to the GEO2R analysis tool.

Table 1. Gene Expression Omnibus datasets regarding preclinical (in vivo) models of glaucoma and postmortem isolated tissues from donor glaucoma subjects. POAG = primary open-angle glaucoma; XFG = exfoliation glaucoma; OHT = ocular hypertension; IOP = intraocular pressure.

GEO Dataset Code	Reference	Sample Group 1	Sample Group 2
GSE133563	[28]	Optic nerve crush tissue (retina) of 8-week-old Lister Hooded rats	Sham-operated tissue (retina) of 8-week-old Lister Hooded rats
GSE105269	[29]	Aqueous humor of POAG or XFG patients	Aqueous humor of control patients
GSE27276	[30]	Trabecular meshwork cell culture, whose cells were isolated from POAG patients	Trabecular meshwork cell culture, whose cells were isolated from control subjects
GSE4316	[31]	POAG human trabecular meshwork tissues	Human trabecular meshwork tissues of control donors
GSE45570		Optic nerve head of POAG or OHT patients	Optic nerve head of control subjects
GSE3554	[32]	Tissues (retina) of 8-month-old DBA/2J mice of (with an elevated IOP)	Tissues (retina) of DBA/2J 3-month-old mice (with a normal IOP)

2.1. *GSE133563: Analysis of Retinal Transcriptomic Data 2 Weeks after Optic Nerve Crush*

GSE133563 is a repository of transcriptomes of the retinas of Lister Hooded rats (aged 8 weeks) 14 days after an optic nerve crush (ONC) compared to sham-operated animals, i.e., the control (CTRL). Since the ONC caused a strong reduction in the number of RGCs, this transcriptome analysis provided information about adaptive modifications in response to the loss of RGCs [28]. In this regard, these authors found that the complement cascade and Notch signaling pathways were the main dysregulated pathways in RGCs after an ONC injury. Moreover, oxidative stress and KIT receptor pathways were also highlighted as enriched pathways. In line with these findings, our re-analysis evidenced the involvement of different genes related to oxidative stress and inflammatory pathways in ONC-induced damages on retinal tissues. Starting from the 250 DEGs, retrieved through a GEO2R analysis of GSE133563, we have built an enriched network, as shown in Figure 1. From a centrality analysis of this network and from rescoring, 12 out of 23 identified genes were predicted as new pharmacological targets for the treatment of glaucoma, since no entries, related to literature studies or clinical trials, were retrieved: GRIA1 (glutamate ionotropic receptor AMPA type subunit 1), GRIK2 (glutamate ionotropic receptor kainate type subunit 2), TRPC3 (transient receptor potential cation channel subfamily C member 3), TRPC7 (transient receptor potential cation channel subfamily C member 7), HYPK (huntingtin interacting protein K), FIBCD1 (fibrinogen C domain containing 1), GJC2 (gap junction protein gamma 2), FZD6 (frizzled class receptor 6), GABRR2 (gamma-aminobutyric acid type A receptor subunit rho2), ANGPTL6 (angiopoietin like 6), IL25 (interleukin 25), and F5 (coagulation factor V).

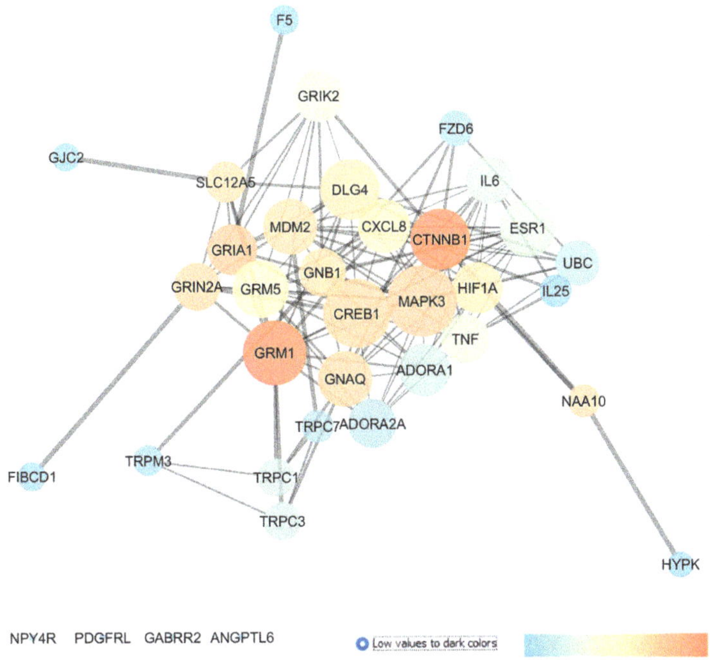

Figure 1. Dysregulated genes in retina of Lister Hooded rats 2 weeks after an optic nerve crush (ONC) injury. Network was analyzed with Cytoscape. Nodes are represented on the basis of closeness centrality values (node dimension) and betweenness centrality values, as shown by the legend of this figure (temperature color scale blue < red). Edge thickness is proportional to edge betweenness values.

Moreover, this network analysis evidenced 11 genes already experimentally linked to glaucoma pathogenesis. Among these, the glutamate metabotropic receptors mGluR1 and mGluR5, along with the adenosine A_1 and A_{2A} receptors, showed elevated centrality values in the network. Other interesting druggable targets were two transient receptor potential cation channels, encoded by TRPC1, found upregulated in glaucoma lamina cribrosa cells [33] and TRPM3. Additionally, DLG4 (discs large MAGUK scaffold protein 4) and several proinflammatory chemokines and cytokines already known to have a role in glaucoma, e.g., C-X-C motif chemokine ligand 8 (CXCL8), tumor necrosis factor (TNF), and interleukin-6 (IL-6), were other central nodes in the network. Specifically, IL-6 was found to be upregulated in the tears of POAG patients [34] and in the aqueous humor of angle-closure glaucoma (ACG) patients, respectively [35]. Moreover, a clinical study evaluated the effect of latanoprost (IOP-lowering drug) on IL-6 levels in the tear films of recruited subjects (NCT04957329). Finally, HIF1A (hypoxia inducible factor 1 subunit alpha), a transcription factor activated in tissues under hypoxic conditions and a regulator of several downstream pathways, i.e., of the vascular endothelial growth factor (VEGF), displayed high closeness and betweenness centrality values in the network. Interestingly, HIF-1α was found to be upregulated in tissues from glaucoma patients [36] and in experimental animal models of glaucoma, such as in mice [37] and rats [38].

2.2. GSE105269: Analysis of Differentially Expressed miRNAs in Aqueous Humor of Exfoliation Glaucoma and Primary Open-Angle Glaucoma Patients

GSE105269 is a collection of microRNAs (miRNAs) expression profile in AH of patients with exfoliation glaucoma (XFG), primary open-angle glaucoma (POAG) and cataract controls (CTRLs). miRNAs are small single-stranded noncoding RNAs that regulate gene expression at post-transcriptional level, through inhibition of the translation of messenger RNAs (mRNAs) or by promoting mRNA degradation [39]. As suggested by Drewry et al., 2018 [29], that deposited the GSE105269, miRNAs in AH could not only have key roles in disease onset progression but also in AH outflow dysregulation. Therefore, we exploited these data to identify new genes, that are also targets of differentially expressed (DE) miRNAs in GSE105269. At first, a GEO2R analysis was carried out to identify the most significant DE miRNAs in each group comparison, i.e., POAG vs. CTRL and XFG vs. CTRL. After this, the top 30 DE miRNAs of the GEO2R analysis (listed in the Supplementary Materials, List S1, S2) were set as an input of the MIRNET web server to create a miRNA–target gene network that was then exported and analyzed with the Cytoscape environment.

2.2.1. GSE105269 AH from Primary Open-Angle Glaucoma Patients

After miRNA profiling, Drewry et al. (2018) [29] identified three significant DE miRNAs in the POAG group compared to the CTRL group: hsa-miR-125b-5p, hsa-miR-302d-3p, and hsa-miR-451a. Furthermore, by evaluating the target genes of the above-mentioned DE miRNAs, genes related to the AKT and Bcl-2-regulated apoptosis signaling pathways were identified as gene targets of these miRNAs [29]. Our GEO2R analysis was carried out by comparing 12 AH samples of the POAG group with the 11 samples of the CTRL group. This analysis evidenced that hsa-miR-517c-3p, hsa-miR-339-5p, and hsa-miR-375 were the most significant DE miRNAs, and only hsa-miR-375 was previously found to be upregulated in the AH of patients with normal-tension glaucoma (NTG) compared to control subjects [40]. After the GEO2R analysis, the top 30 DE miRNAs were set as an input of the MIRNET web server, and a network was generated on the basis of miRNAs and their target genes. In the MIRNET network, shown in Figure 2A, the miRNAs (nodes) with the highest degree of centrality were hsa-let-7b-5p, hsa-mir-484, and hsa-mir-21-5p. For these miRNAs, no hits were retrieved on Pubmed. Furthermore, the above-mentioned hsa-mir-375, hsa-mir-125b-5p (one of the DE identified in Drewry et al. (2018) [29], and hsa-mir-302b-3p displayed high centrality values in our network. Meanwhile, the genes showing the highest degree and betweenness centrality values were HNF4A (hepatocyte nuclear factor 4 alpha), APP (amyloid beta precursor protein), and SUMO2 (small ubiquitin-like modifier 2).

An enrichment analysis of the Kyoto encyclopedia of genes and genomes (KEGG), aimed at linking genomic data with higher-order functional information [41], was carried out with MIRNET, which evidenced the following signaling pathways known to be implicated in this disease: the "Neurotrophin signaling pathway" (linked nodes: GRB2, YWHAZ, YWHAB, AKT1, and BAD); the "Insulin signaling pathway" (linked nodes: PYGB, FLOT1, and GRB2); the "p53 signaling pathway"; "Apoptosis"; and "Axon guidance". Pathways such as "Cholinergic synapse" were enriched by genes encoding CHRM3 and CHRNA6. Moreover, "Dopaminergic synapse" (linked nodes: GRI2A, GRIN2B, and GRIA4) and "Long-term potentiation" could be also implicated in this disease.

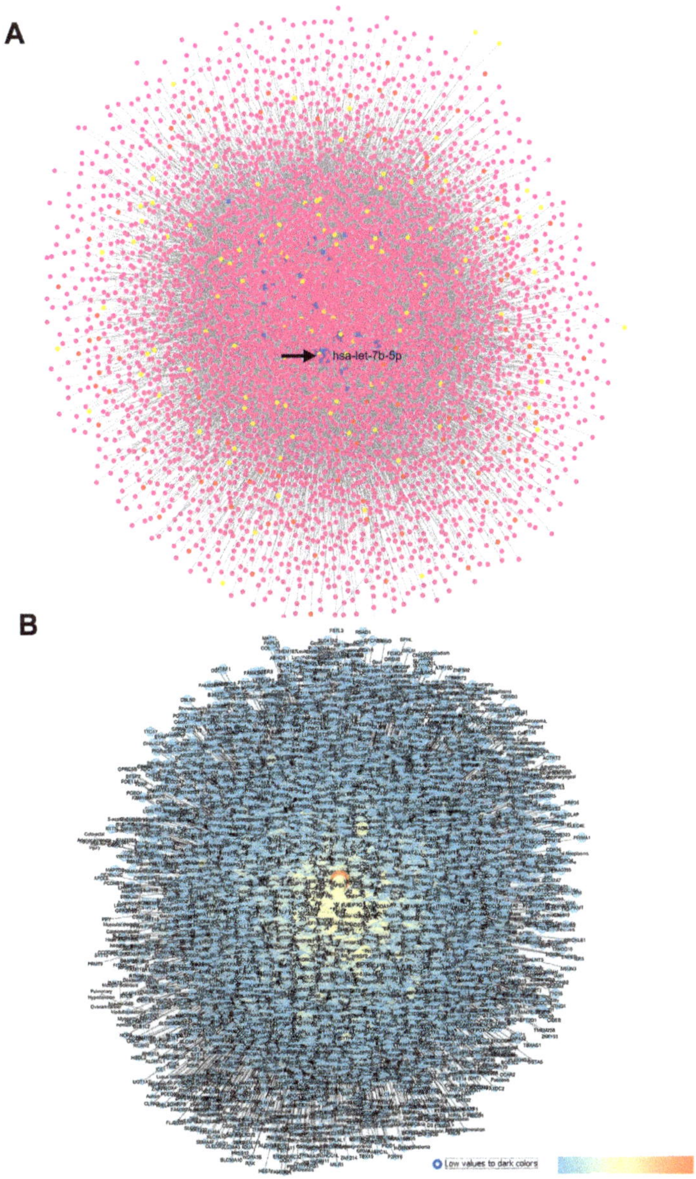

Figure 2. GSE105269 network: primary open-angle vs. CTRL cataract patients. (**A**) Output of MIRNET network within the MIRNET environment. Magenta nodes correspond to genes, blue nodes to miRNAs, yellow nodes to compounds (predicted to be able to modulate miRNA expressions), and red nodes to diseases associated to miRNAs. The arrow indicates the has-let-7b-5p miRNA, i.e., the most central node of this network. (**B**) For the MIRNET-generated network represented within Cytoscape, centrality parameters are mapped as follows: closeness centrality is proportional to node dimension; betweenness centrality is represented with a temperature color scale (blue to red for increasing values, as shown by the legend of this figure); and edge betweenness is proportional to edge thickness.

Then, the MIRNET network, providing miRNA–target genes interactions, was exported in Cytoscape and represented according to centrality metric parameters (Figure 2B).

Overall, after this network analysis, we found that 8 genes, out of 17, encode new pharmacological targets, which have not yet been associated with glaucoma and were, thereby, rescored as innovative targets: ADORA2B (the adenosine A_{2B} receptor); HTR1D (the 5-hydroxytryptamine 1D receptor, 5-HT1D); P2RX4 (the purinergic receptor P2X4); HSPA8 (heat shock protein family A (Hsp70) member 8); TNRC6A (the trinucleotide repeat-containing adaptor 6A); NUFIP2 (nuclear FMR1-interacting protein 2); SRRM2 (serine/arginine repetitive matrix 2, required for pre-mRNA splicing as a component of the spliceosome); and SUMO2 (small ubiquitin-like modifier 2).

Within the miRNA–gene network, built on the basis of a re-analysis of GSE105269 POAG vs. CTRL patients, ELAVL1 (embryonic lethal abnormal vision-like protein 1) was the most central gene node of this network (Table 2). ELAVL1 encodes an RNA-binding protein which is able to stabilize the mRNAs, with an AU-rich element, of several genes, such as growth factors and inflammatory cytokines. ELAVL1 has already been proven to be involved in glaucoma pathogenesis [42,43]. The second most central node was APP (the amyloid beta precursor protein), implicated in patients with Alzheimer's disease (AD) and also in glaucomatous neurodegeneration [44,45]. Moreover, EGR1 (early growth response 1), IGF1R (the insulin-like growth factor 1 receptor), THBS1 (thrombospondin 1), CDKN1A (the cyclin-dependent kinase inhibitor 1A), P2RY2 (the purinergic receptor P2Y2), ADRA2A (the alpha-2A adrenergic receptor and α2A adrenoceptor), and ADRB2 (the beta-2 adrenergic receptor) were other genes displaying elevated closeness centrality values in the network. EGR1 was found to be significantly upregulated in C57BL/6 mice, following an axonal injury through RNA-sequencing technology [46]. The insulin-like growth factor 1 receptor (IGF1R) was found to be phosphorylated in rat RGCs shortly after an episcleral vein cauterization (EVC), identifying this activation as a compensatory response to EVC stress [47]. Finally, a missense variant of the THBS1 gene was associated with POAG, as reported in a study that carried whole-genome sequencing (WGS) in a large cohort of POAG subjects with an elevated IOP [48]. With respect to the cyclin-dependent kinase inhibitor 1A, this encoding gene was associated with an increased risk of POAG, as reported in a meta-analysis by Springelkamp et al. (2017) [49]. The metabotropic purinergic receptor P2Y2 has already been investigated as a modulator of IOPs in vivo [50,51]. With respect to adrenergic receptors, the alpha-2A and beta-2 adrenoceptors are the target of two IOP-lowering drug classes, i.e., alpha-2adrenergic-receptor agonists and beta-blockers. In this regard, an interventional study (NCT00279253) assessed the effects of the intravenous administration of clonidine (an alpha-2 adrenergic agonist) on the ocular blood flow and IOP of healthy subjects; however, no report on this study has been published yet. Moreover, several studies regarding the association of ADRB2 genetic polymorphisms with POAG were retrieved from the PubMed database [52–54]. Specifically, ADRB2 polymorphisms were associated with glaucoma in an Indian population [52], and in Japanese patients [54] and were also found to correlate to increased therapeutical response to β-blockers eye drops, used for the treatment of ocular hypertension [55].

Table 2. Central genes in the network built fromGSE105269 re-analysis (POAG vs. CTRL). Closeness centrality depends on the sum of shortest paths between all nodes. The stress centrality of a node measures the number of shortest paths passing through that node, while degree represents the number of nodes it is in contact with.

Node	Closeness Centrality	Stress	Degree
Hsa-let-7b-5p	0.62	2,455,969,486	4489
Elavl1	0.51	123,020,754	1244
Myc	0.49	18,006,672	512
Parp1	0.49	10,943,968	122

2.2.2. GSE105269 AH of Exfoliation Glaucoma

XFG is the most common secondary form of POAG, a complication of the exfoliation syndrome (XFS), characterized by the accumulation of fibrillar extracellular material in the anterior segment of the eye, leading to the obstruction of AH outflow pathways and an increase in the IOP [56]. Drewry et al. (2018) [29] reported five miRNAs with significantly different expressions in an XFG group compared to a CTRL group: hsa-miR-122-5p, hsa-miR-3144-3p, hsa-miR-320a, hsa-miR-320e, and hsa-miR-630. Moreover, these authors underlined BCL2L2, IGF1R, and RAC1 as the genes targeted by most of the identified DE miRNAs in the XFG group. Our GEO2R analysis revealed that the most significant DE miRNAs from samples of the XFG and CTRL groups were hsa-miR-548ar-5p, hsa-miR-612, hsa-miR-122-5p, and hsa-miR-320a. The hsa-miR-122-5p and hsa-miR-320a were validated by Drewry et al. (2018), that deposited the GSE105269 dataset [29]. Furthermore, hsa-miR-122-5p was found to be significantly upregulated in later stages of the disease for the XFG group compared to earlier phases. In particular, this miRNA is linked to the TGF–β1 pathway (involved in fibrosis and extracellular matrix remodeling) [57]. On the other hand, hsa-miR-320a was found to be upregulated in the AH of XFG patients in comparison with normal subjects [58]. From our re-analysis of the network built with the MIRNET web server (Figure 3), the node with the highest centrality was hsa-let-7a-5p, which was upregulated in the XFG group compared to the CTRL group, according to the GEO2R analysis. Others top nodes in the MIRNET network were hsa-mir-122-5p and hsa-mir-320a, followed by APP, which was also found in the network built on the basis of a POAG vs. CTRL comparison. Additionally, hsa-mir-128-3p, hsa-mir-603 miRNAs, and SUMO2, not previously associated with this disease, were the most central nodes of the XFG vs. CTRL network. A KEGG enrichment analysis with MIRNET (https://www.mirnet.ca/ accessed on 1 March 2022) evidenced that the "Neurotrophin signaling pathway" (linked nodes: BDNF, YWHAQ, and NTRK3); the "p53 signaling pathway"; (linked nodes: TP53, CDK4, CDKN2A, CDKN1A, BAX, and CASP3); and the "Insulin signaling pathway" (linked nodes: MTOR and AKT3) were the most significant signaling pathways of the network. Therefore, this enrichment evidenced pathways that are common to the two different networks representing GSE105269, XFG vs. CTRL, and POAG vs. CTRL. Other interesting significant signaling pathways were the "TGF-beta signaling pathway" (linked nodes: SMAD3 and TGFBR1), the "Wnt signaling pathway", and the "MAPK signaling pathway".

Moreover, a rescoring based on entries in PubMed and clinicaltrials.gov evidenced new predicted pharmacological targets for the treatment of XFG: GRK5 (the G protein-coupled receptor kinase 5), GRID1 (the glutamate ionotropic receptor delta type subunit 1), NFKB2 (nuclear factor kappa B subunit 2), PIK3R1 (phosphoinositide-3-kinase regulatory subunit 1), VHL (the von Hippel Lindau tumor suppressor), PKM (pyruvate kinase M1/2), TRPC4 (the transient receptor potential cation channel subfamily C member 4), P2RX3 (the purinergic receptor P2X3), and IGF2R (the insulin-like growth factor 2 receptor). Interestingly, SUMO2, which was identified in the POAG vs. CTRL network, also displayed high closeness centrality values in the XFG vs. CTRL network.

Moreover, 13 genes, already known to be implicated in this disease, were central, according to our network analysis, including the small ubiquitin-like modifier 1 (SUMO1) and IGF1R, which are target genes of the miRNA hsa-let-7a-5p. SUMO1 has been identified as a gene involved in POAG pathogenesis [59], as well as IGF1R [29]. Our analysis also identified the androgen receptor (AR) as a central node, which has been found to be upregulated in glaucomatous ONH astrocytes both in vivo and in vitro [60]. The genes largely described to be involved in glaucoma, which were identified in the XFG vs. CTRL network, include the following: HIF1A (node linked to the enriched "mTOR signaling pathway"); SOD2, which is a target of the central node hsa-let-7a-5p; and TGFBR2, belonging to the "TGF-beta signaling pathway" [61]. Other examples of "non innovative targets" were the nuclear factor NFKB1 and BAX (nodes linked to the "Neurotrophin signaling pathway", the "p53 signaling pathway", and "Apoptosis" enriched pathways). Another central node in the XFG vs. CTRL network was VEGFC, which encodes the vascular endothelial growth factor C that was found to be expressed in the trabecular meshwork (TM) and Schlemm's canal of the endothelial cells of patients with NTG and POAG [62]. Additionally, Hase et al. (2021) showed that the VEGFC expression is induced in TM cells by hypoxia, thereby evidencing the link between VEGFC and HIF1A [62].

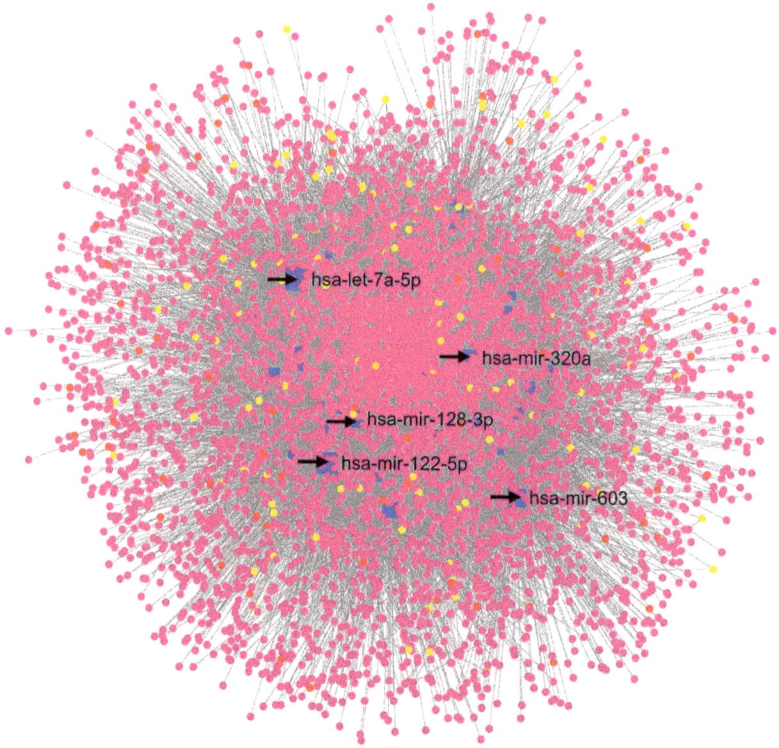

Figure 3. GSE105269 MIRNET network, XFG vs. CTRL. The MIRNET web server provided miRNA–gene, miRNA–disease, and miRNA–small molecule interactions, depicted in this network. In this network, has-let-7a-5p, hsa-miR-122-5p, hsa-miR-320a, hsa-miR-128-3p, and hsa-miR-603 are the nodes with the highest centrality values. Magenta nodes correspond to genes, blue nodes to miRNAs, yellow nodes to compounds (predicted to be able to modulate miRNA expressions), and red nodes to diseases associated with miRNAs.

2.3. GSE27276: Analysis of Differentially Expressed Gene Profiles in Cultured Human Trabecular Meshwork Tissues from Primary Open-Angle Glaucoma Patients Compared to Cells Isolated from Control Healthy Subjects

GSE27276 is a repository containing the gene expression profile of trabecular meshwork (TM) cells isolated from POAG patients compared to cells isolated from the TM of healthy subjects (CTRLs) [30]. Liu et al., by comparing the transcriptome of POAG patients with control subjects, found a list of significant DEGs in POAG, which were mainly associated with endocytic and exosome pathways [30]. With respect to the article by Liu et al. 2013 [30], our re-analysis identified CALML3 and LCN2, included by these authors in the list of top 30 DEGs. The top 250 DEGs retrieved with our GEO2R analysis (POAG vs. CTRL) were used as an input of the STRING app within the Cytoscape environment, and the network in Figure 4 was built.

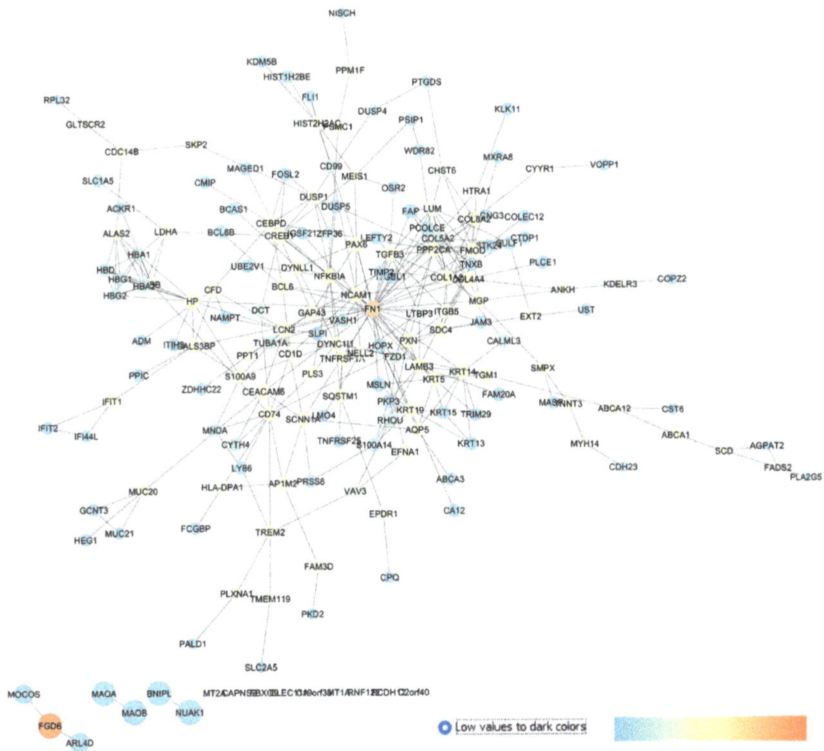

Figure 4. GSE27276 network describing the differences between the expression pattern of cultured trabecular meshwork cells in isolated tissues from POAG patients and healthy subjects. The network has been built with STRING app of Cytoscape, and the centrality parameters are mapped as follows: closeness centrality is proportional to node dimension, betweenness centrality is represented with a temperature color scale, and edge betweenness is proportional to edge thickness.

An analysis of the centrality metrics of this network identified a series of drug targets that can be classified as predicted new pharmacological targets for the treatment of glaucoma, since no entries on PubMed or clinicaltrials.gov were found: NUAK1 (NUAK family kinase 1), MAOB (monoamine oxidase B), HTRA1 (HtrA serine peptidase 1), LDHA (lactate dehydrogenase A), TNFRSF1A (the TNF receptor superfamily member 1A), FZD1 (frizzled class receptor 1), and MTNR1A (the metabotropic melatonin 1A receptor).

Among the in silico identified new drug targets, i.e., genes or proteins not previously linked to glaucoma, we identified the following: SCNN1A (sodium channel epithelial 1 subunit alpha); CACNG3 (calcium voltage-gated channel auxiliary subunit gamma 3); PKD2 (polycystin 2, the transient receptor potential cation channel); SDC4 (syndecan 4, transmembrane heparan sulfate proteoglycan); and HP (haptoglobin).

Our analysis identified several drug targets already linked to glaucoma, strengthening the validity of the reverse engineering approach, hereby used. Fibronectin 1 (FN1) had the highest closeness and betweenness centrality values in our network and was found to be upregulated in immortalized and primary human TM cells through a stimulation of the $\alpha v\beta 3$ integrin pathway and contributes to increased TM resistance in glaucoma [63]. As mentioned above, LCN2 (i.e., lipocalin 2) was between the top 30 DEGs identified by Liu et al. (2013), who deposited the GSE objective of this study [30]. LCN2 was also a central gene in our network and encodes for an iron-trafficking protein which plays key roles in innate immunity, apoptosis, and cell proliferation. Moreover, the involvement of lipocalin 2 was confirmed in two in vivo models of glaucoma, spontaneous glaucomatous DBA/2J mice [64], and in a model of mice with an optic nerve crush (OCN) injury [65], respectively. Within the high centrality nodes, GAP43 (the growth-associated protein 43) was found to be involved in retinal ganglion cell death in an experimental chronic glaucomatous injury [66].

DUSP1 also showed elevated betweenness centrality parameters in our network. This gene encodes a nuclear phosphatase (with dual specificity for tyrosine and threonine) and was found to regulate mitogen-activated protein kinase p38 in a retinal ischemic preconditioning [67]. Among the high centrality nodes in our network, AQP5 (aquaporin 5) was recently identified, through exome sequencing, as a new candidate gene for familial cases of POAG [68].

In summary, from an analysis of GSE27276, we found 15 genes with no hits on PubMed and clinicaltrials.gov and 10 genes already known to be involved in glaucoma.

2.4. GSE4316: Analysis of Differentially Expressed Gene Profiles in Trabecylar Meshwork Tissues of Primary Open-Angle Glaucoma Patients

This GEO dataset includes the gene expression profiles of the TM tissues of POAG patients [31]. Liton et al. 2006 [31], comparing the gene expression profiles of CTRL and POAG TM tissues (isolated postmortem tissues from patients), found that the top deregulated genes were coding proteins involved in inflammation such as: chemokine (C-X-C motif) ligand 6, chemokine (C-C motif) ligand 5, immune-associated nucleotides, the interleukin 1 receptor type II, transthyretin, haptoglobin, and myelin basic proteins. In addition, Liton et al. 2006 identified the following as upregulated genes in POAG TM tissues: prokineticin 2, the G protein-coupled receptor 146, the regulator of G-protein signaling 1, the neuropeptide Y receptor Y2, and the adenosine A3 receptor [31].

According to the methods reported in our study, we carried out a GEO2R statistical analysis, retrieving 250 DEGs in GSE4316, which were the input of the STRING app of Cytoscape, which was used to build the interaction network shown in Figure 5.

Our study highlights several genes and druggable targets which gave no result in the search done on PubMed or on clinicaltrials.gov database; therefore, these may be innovative targets for the treatment of glaucoma: HPDL (4-hydroxyphenylpyruvate dioxygenase like); QDPR (quinoid dihydropteridine reductase); GPR112 (the orphan receptor, G protein-coupled receptor G4); FMO5 (flavin-containing dimethylaniline monooxygenase 5); CYP2C8 (cytochrome P450 monooxygenase); B3GAT2 (beta-1,3-glucuronyltransferase 2); FUCA1 (alpha-L-fucosidase 1); ADAMTS13 (ADAM metallopeptidase); RIC1 (the RIC1 homolog, RAB6A GEF complex partner 1); HTR2C (the 5-hydroxytryptamine receptor 2C); GAD2 (glutamate decarboxylase 2); and HIPK3 (homeodomain-interacting protein kinase 3).

The role of complement system in IOP elevations and then in glaucoma was recently investigated [69]. Interestingly, we found two members of the complement system central to the network describing GSE4316, CD28 and the TNF receptor superfamily member 9, TNFRSF9, but no entries about the involvement of these genes in glaucoma were found. Moreover, ITGB3, which encodes integrin beta-3 (β3), showed an elevated betweenness centrality value in our network, and its role in POAG has not yet been investigated; however, integrin signaling pathways may have a role in the pathogenesis of glaucoma and in the increase in TM resistance to the AH outflow [70].

Our re-analysis of GSE4316 identified several genes already experimentally associated with glaucoma. For example, BCAN (brevican), which is a member of the lectican family of chondroitin sulphate proteoglycans specifically expressed in the central nervous system and known to be released from reactive astrocytes [71], and it showed elevated centrality values in our network. BCAN upregulation was found in rats subjected to ONC injury [72] and in the optic nerve of rats exposed to ischemia/reperfusion (I/Rdamage [73]. Furthermore, TNF, i.e., the tumor necrosis factor, showed elevated betweenness centrality values in the network, and this proinflammatory cytokine has already been proven to have a role in POAG pathophysiology [74].

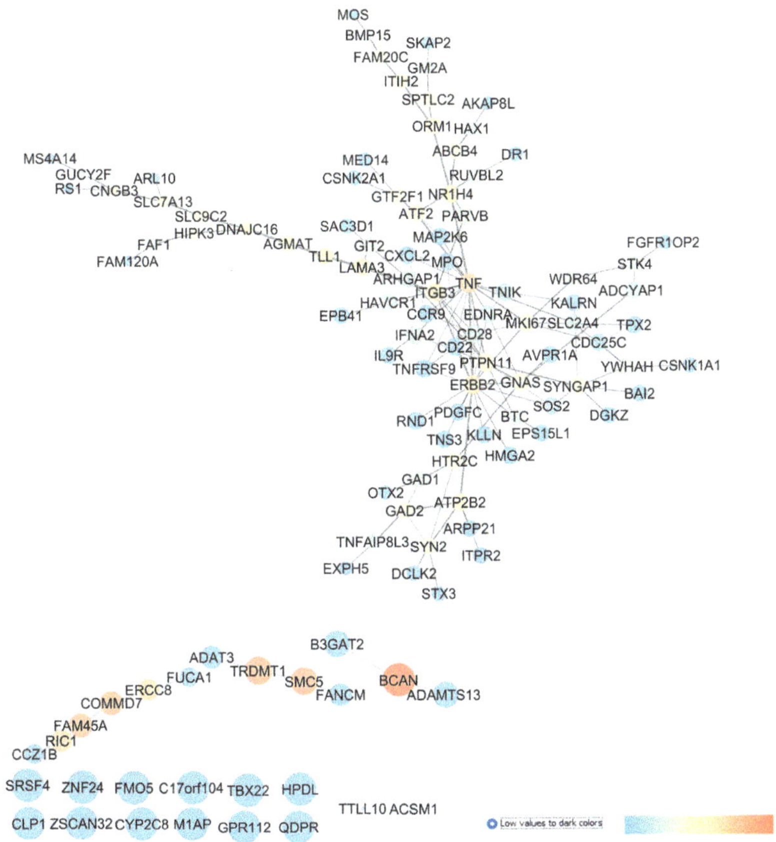

Figure 5. GSE4316 re-analysis and network building, comparing gene expression profile of TM control tissues vs. tissues isolated from POAG patients. In this network, analyzed in Cytoscape, nodes are represented on the basis of betweenness centrality values (color scale blue < red) and closeness centrality values (node dimension), and edge thickness is proportional to edge betweenness values.

Moreover, our network centrality analysis identified, as a central node, the gene EDNRA, which encodes the GPCR endothelin receptor type A (ET-A), whose activation triggers vasoconstriction. In the study conducted by Howell et al., in 2014 [75], ET-A was expressed in the ONH vascular endothelial cells of DBA/2J mice, which spontaneously developed with aging glaucoma signs, resembling features in patients. Particularly, the endothelin receptor antagonist bosentan, protected RGCs from neurodegeneration and death [75]. Recently, it has been shown that the retinas of EDNRA-conditioned KO mice in vascular mural cells were protected by endothelin-1 administration, confirming the role of ET-A in glaucomatous neurovascular dysfunctions [76]. Furthermore, a phase I clinical trial was carried out on POAG subjects treated with bosentan (500 mg o.s. and an 8-day treatment) to study the effects of endothelin antagonist treatment in ocular blood flow (NCT00701597). No study results have been published about this trial, but bosentan is currently under evaluation in a randomized controlled trial (NCT02377271) including patients with nonarteritic anterior ischemic optic neuropathy (NAAION) [77].

Finally, YWHAH (tyrosine 3-monooxygenase/tryptophan 5-monooxygenase activation protein eta) was also found to be central in the network, as shown in Figure 5, and it has already been investigated in glaucoma [78].

2.5. *GSE45570: Analysis of Differentially Expressed Gene Profiles in Optic Nerve Head of Primary Open-Angle Glaucoma Patients and Ocular Hypertensive Patients without Glaucoma*

GSE45570 includes transcriptomic data from the optic nerve head (ONH) of three experimental groups: patients with primary open-angle glaucoma (POAG), subjects with ocular hypertension (OHT) without glaucoma, and healthy control donors (CTRLs). The aim of our analysis was to discriminate the dysregulated genes involved in IOP elevation (OHT conditions) from genes differentially expressed in the ONH of POAG subjects. In the first comparison, 6 sample groups of ONH from OHT patients and 6 sample groups of ONH from control donors were analyzed with GEO2R. Then, the network in Figure 6 was built, setting as input of the app STRING of Cytoscape the first 250 found DEGs. An analysis of the centrality metrics of the network was carried out, followed by match of genes with PubMed and clinicaltrials.gov entries, which evidenced that 11 genes have not already been associated with glaucoma, meaning that these genes have a high potential to be innovative pharmacological targets for the management of glaucoma: ADGRA3 (the adhesion G protein-coupled receptor A3), NPY2R (the neuropeptide Y receptor Y2), UNC79 (encoding for an auxiliary subunit of the NALCN sodium channel), BMPR1B (the bone morphogenetic protein receptor type 1B), OR2B3 (the olfactory receptor family 2 subfamily B member 3), THRB (the thyroid hormone receptor beta), RXRG (retinoid x receptor gamma), LBR (the lamin B receptor), ADIPOQ (adipokine), BBOX1 (gamma-butyrobetaine hydroxylase 1), and TRHDE (the thyrotropin-releasing hormone-degrading enzyme).

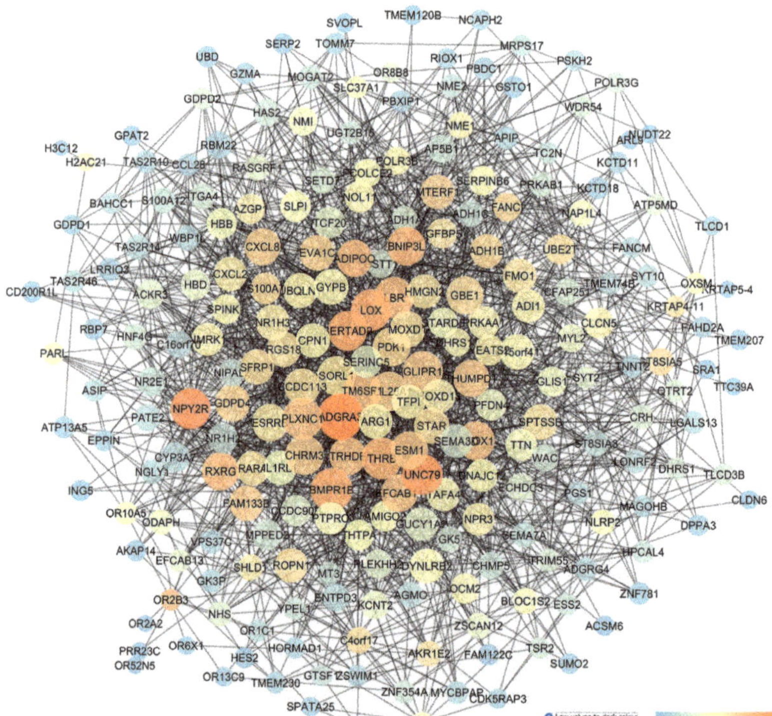

Figure 6. GSE45570 network representing a comparison of the optic nerve head expression profiles of patients with ocular hypertension (without glaucoma) vs. CTRL subjects. This network was analyzed with Cytoscape: the node size is proportional to closeness centrality, the color node represents betweenness values ranging from blue to red (blue to red for increasing values), and edge thickness values are proportional to those of edge betweenness.

Furthermore, the network describing GSE45570 (OHT vs. CTRL) identified other genes with high closeness or betweenness centrality values; however, these were already found to be involved in glaucoma. Among these, LOX (lysyl oxidase) showed high network centrality parameters; this gene encodes a copper-dependent enzyme that regulates the oxidative deamination of the lysine residues of elastin and collagen. LOX activity has been found to increase trabecular meshwork stiffness, thereby increasing the outflow of the aqueous humor and then IOP [79]. Our network analysis identified the gene CHRM3 (the cholinergic muscarinic receptor 3, M_3R) which has high betweenness or centrality values, and this receptor has already been studied as an intriguing pharmacological target for glaucoma treatment. In fact, the acetylcholinesterase inhibitor, huperzine A, is able to decrease IOP in New Zealand rabbits, through the indirect activation of M_3R [80].

The re-analysis of GSE45570, comparing the following samples, POAG (6 samples) vs. CTRL (6 samples), was conducted, which led to the network shown in Figure 7. The network centrality analysis evidenced the following five central nodes as the new predicted genes involved in glaucoma, that can be potentially further validated for glaucoma treatment, given that no entries were retrieved about these genes on PubMed and clinicaltrials.gov: HTR5A (5-HT5A), FSTL5 (follistatin-like 5), EXOSC3, (exosome component 3), FBXL2 (the f-box and leucine-rich repeat protein 2), and CLIC2 (the chloride intracellular channel 2).

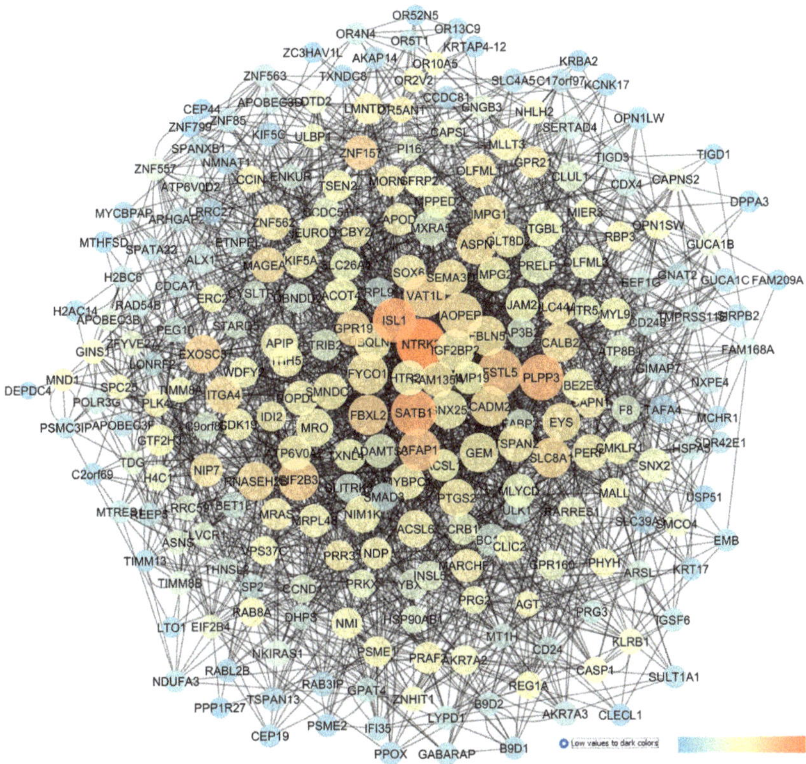

Figure 7. GSE45570 network describing a comparison of the optic nerve head expression profiles of patients with POAG vs. CTRL subjects. This network was analyzed with Cytoscape: the node size is proportional to closeness centrality, the color node represents betweenness values ranging from blue (low betweenness values) to red (high betweenness values), and edge thickness values are proportional to those of edge betweenness.

Other central nodes in this network have already been associated with glaucoma, such as NTRK2 (the neurotrophic receptor tyrosine kinase 2, alias TRKB), which showed the highest centrality values inside the network. TRKB is activated by the brain-derived neurotrophic factor (BDNF), and the BDNF–TRKB signaling pathway has already been proven to stimulate the survival of RGCs in different models of retinal injuries [81,82]. PLPP3 (phospholipid phosphatase 3) showed high centrality metric parameters, and it was found to be significantly overexpressed in the optic nerve of glaucoma patients [83]. With respect to AFAP1 (the actin filament-associated protein 1), implicated in the cross-linking of actin filaments, this gene has been associated with elevated IOP values or to increase POAG risk factor, as evidenced by several gene-wide association studies (GWASs) [84,85]. Among the nodes with moderate centrality values, the HTR2A gene (the 5-HT2A receptor) has already been linked to glaucoma. Particularly, a treatment using 5-HT2A agonists decreased the IOP in Cynomolgus monkeys [86]. Furthermore, cabergoline (a dopaminergic and serotoninergic agonist) decreased the IOP in a mouse model of steroid-induced ocular hypertension [87]. However, up to now, the exact mechanism of the serotonergic system in modulation of IOP levels has not been fully elucidated.

Finally, with respect to GSE45570, the GEO2R analysis was carried out comparing 6 samples from POAG patients with 6 samples of OHT patients. This comparison was carried out in order to highlight the DEGs putatively involved in ocular hypertension progression to POAG, or to at least differentiate between these two conditions. The top 250 DEGs were set as input of the STRING app within Cytoscape, and the enriched network (Figure 8) was built.

The centrality analysis of the network highlighted (high centrality values) 8 genes as new in silico identified dysregulated genes in OHT and POAG, with no entries in the literature or in clinical trials: LBR (the laminin B receptor, already identified in the OHT vs. CTRL network); FSTL5 (follistatin-like 5, already identified in the POAG vs. CTRL network); IL3 (interleukin 3); BTK (bruton tyrosine kinase); MSR1 (the macrophage scavenger receptor 1); ARHGEF6 (the Rac/Cdc42 guanine nucleotide exchange factor 6); PAK1 (serine/threonine p21-activating kinases); and MIS18BP1 (the MIS18-binding protein 1).

Other nodes central in this network have been also identified in the re-analysis of a comparison of POAG vs. OHT in GSE45570, and these nodes have already been linked to glaucomatous condition. Specifically, we again identified the gene HTR2A (serotonergic 5-HT2A receptor) [86,87]. Furthermore, CXCL12 (C-X-C motif chemokine ligand 12) was the most central node in the network, and it was found that through an activation of the CXCR3 receptor, this chemokine promotes caspase activation, leading to apoptosis, and IOP elevation in a rat model of ocular hypertension [88]. Another chemokine, the node CXCR4 (C-X-C motif chemokine receptor 4), was found to be central in the network (Figure 8), and its expression influenced the trabecular meshwork morphology of POAG patients [88–90]. ITPR1 (the inositol 1,4,5-trisphosphate receptor, type 1) also displayed elevated closeness centrality values in the network, and it was found to be dysregulated in the AH of POAG patients [91]. Finally, CDC7 (cell division cycle 7) was identified as a central node in our analysis, and polymorphisms of CDC7/TGFBR3 have been associated with glaucoma [92], along with visual field progression and a reduction in the thickness of the retinal nerve fiber layer of glaucoma patients [92–94].

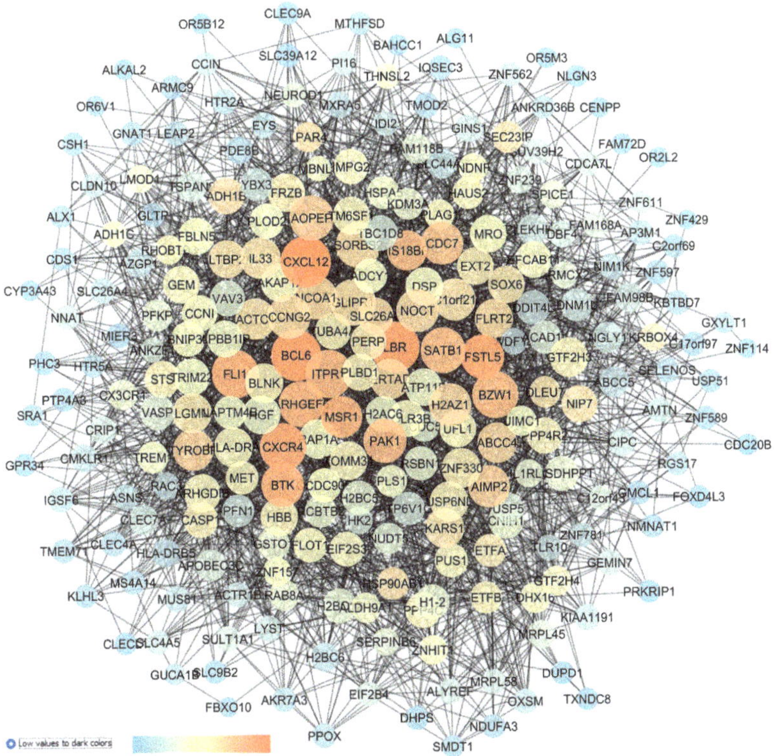

Figure 8. GSE45570 network representing the differences of the optic nerve head expression profiles of patients with POAG vs. patients with ocular hypertension (without glaucoma). This network was analyzed with Cytoscape: the node size is proportional to closeness centrality; the color node represents betweenness values, ranging from a blue to red of temperature color scale (shown in figure), with increasing values going from blue to red; and the edge thickness values are proportional to those of edge betweenness.

2.6. *GSE3554: Analysis of DBA/2J Mice Retinal Gene Expressions*

The GSE3554 reports the expression profiles of the retinas of 3-month- and 8-month-old DBA/2J mice [32]. These DBA/2J mice spontaneously developed glaucoma, and at 8 months, the IOP significantly increase, and the mice also display retinal dysfunction and an altered cell morphology [95]. Steele et al., 2006 [32], validated, with a quantitative RT-PCR the microarray results for ceruloplasmin, chitinase 3-like 1, lipocalin 2, the complement component 1q, chemokine ligand 12, and the interferon-induced transmembrane protein 1 [32]. We were also able to identify, with our informatic analysis, lipocalin 2 (GSE27276) and chemokine ligand 12 (GSE45570: POAG vs. OHT). We applied to GSE3554 the bioinformatic protocol hereby presented, but we were not able to identify any central node in the network, because we did not found any centrality or topological parameter able to discriminate nodes in the network.

2.7. *Integration with Preclinical and Clinical Data*

After the network analysis of the different expression datasets and identification of the central genes in the network, we matched results with entries from the PubMed (https://pubmed.ncbi.nlm.nih.gov/ accessed on 1 October 2022) and/or clinicaltrials.gov (https://clinicaltrials.gov/ accessed on 1 October 2022) databases. The genes, without hits related to literature studies or clinical trials, can be considered as new predicted pharmacological targets for the treatment of glaucoma and would need an in vitro and in vivo validation. Overall, we identified 135 genes over the six analyzed GEO datasets. A total of 82 out of 135 had not match within PubMed and clinicaltrials.gov database. Figure 9 shows the whole set of genes identified after an analysis and rescoring of the GSE dataset networks, built de novo through our in silico approach. It is worth noting that genes encoding for GPCRs and ligand-gated ionic channels and enzymes (easily druggable targets) were in the subset of "new pharmacological targets". Clinical studies were identified only for three genes out of 135 identified central nodes: IL6, ADRA2A, and EDNRA. This result would highlight the current attrition in the clinical development of new drugs or interventions for glaucoma management.

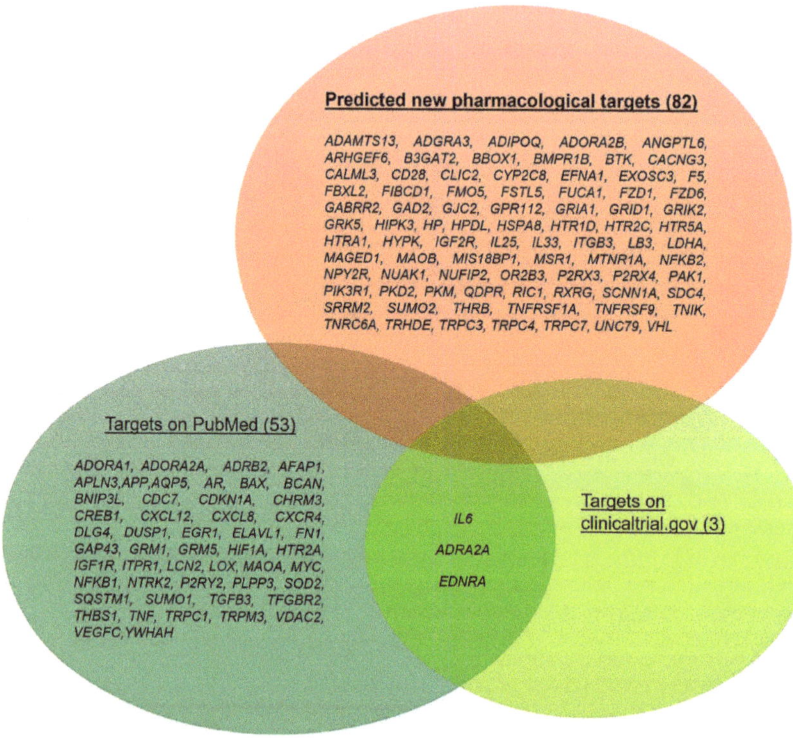

Figure 9. Set of dysregulated genes identified after our re-analyses of the GSE133563, GSE105269, GSE27276, GSE4316, GSE45570, and GSE3554 datasets. Genes that were already evaluated in clinical trials or preclinical studies are distinguished from those without a match in the PubMed and clinicaltrials.gov databases, i.e., predicted new pharmacological targets. The number of genes is shown in brackets.

3. Discussion

In this work, six GEO datasets were analyzed. These datasets report transcriptomic analyses of tissues primarily implicated in the etiopathogenesis of glaucoma (trabecular meshwork, optic nerve head, whole retinal samples, retinal ganglion cells, and aqueous humor) and isolated from in vivo models of diseases or isolated from patients. A bioinformatic analysis of these datasets was carried out, using a system biology or system pharmacology approach, with the aim of identifying genes and miRNAs that may have a role in glaucoma pathogenesis. As was already found in a previous study published by our group [15], this approach is unbiased, and a main strength of this methodology is the rescoring of nodes central in the networks through a match with the PubMed and clinicaltrials.gov databases. In fact, we were able to identify genes that were already found to have an experimental link to glaucoma. Specifically, genes have been linked, through a gene ontology analysis or directly through a network analysis, to inflammatory, immune system, oxidative stress, and fibrotic pathways, which have already been associated with mechanisms involved in glaucoma [26,27,96–98].

For instance, several inflammatory cytokines and chemokines, i.e., IL-6, TNF, CXCL12, and CXCL8, showed elevated centrality values in various networks, evidencing the detrimental role of neuroinflammation in glaucoma [99]. Several studies have found an overexpression of these pro-inflammatory cytokines in retinal glaucomatous tissues and AH samples [100,101]. Moreover, elevated levels of the pro-inflammatory cytokine CXCL8 were detected in the AH of POAG, XFG, and neovascular glaucoma (NVG) subjects and were also associated with an elevated preoperative IOP or visual field defects in the eyes of patients [100]. Furthermore, an upregulation of TNF-α was found in the AH of POAG and XFG eyes [102]. Interestingly, TNF-α and the TNF-α receptor 1 were found to be upregulated in retinal glial cells and RGCs postmortem in the eyes of POAG donors [103]. Accordingly, different genes related to the TNF-α signaling pathway, such as TNFRSF1A and TNFRSF9, were enriched in our networks, strengthening the involvement of immunopathological mechanisms in glaucomatous retinas [74].Additionally, several components related to the complement cascade, such as CD28 and TNFRSF9, were identified in our in silico analysis, in accordance with evidence from other already published works [104,105]. Our study also highlighted the involvement of the NFKB1 and NFKB2 genes, that encode transcription factors regulating innate immune system-related responses. In this regard, Reinehr et al., in 2018 [106], found an increase in NFkB in S100 retinas (autoimmune glaucoma model). Furthermore, we identified the gene encoding HIF-1α as a crucial gene in our network analyses, and several experimental reports, along with one on the etiopathogenic hypothesis of the onset and progression of glaucoma (i.e., perfusion deficits at the optic nerve head), corroborated the detrimental role of this transcription factor in glaucoma pathogenesis [36–38]. Interestingly, HIF-1α was also found to be upregulated in an inherited glaucoma model of DBA/2J mice and, along with its target gene VEGF, was suggested to participate to in the breakdown of the blood–retinal barrier, thus promoting the loss of RGCs [107]. In this regard, our in silico analysis also identified VEGFC, which was found to be upregulated in the trabecular meshwork of patients with POAG and neovascular glaucoma [62]. Specifically, the VEGF pathway is already known to promote neovascularization in different retinal neurodegenerative diseases, as well as in glaucoma [108]. Additionally, our analysis identified different glutamate receptors in the de novo built and analyzed networks, confirming the experimental finding regarding the role of glutamate excitotoxicity as a detrimental factor in glaucoma neurodegeneration. In detail, GRM1, encoding for mGlur1, was identified as a central node in the network describing GSE133563. The signaling of mGluR1 has already been described in the degeneration of RGCs by Liberatore et al., 2017 [109], where the excitotoxic damage of RGCs, induced by a toxic dose of monosodium glutamate in mice, was reversed by treating the animals with an mGluR1-negative allosteric modulator [109]. Furthermore, mGluR1 also has a role in the activation of Müller cell gliosis [110,111]

With respect to apoptotic events that lead to the programmed death of RGCs, our in silico analysis identified several genes central in the networks, such as BAX (identified as "GSE105269 XFG vs. CTRL" in the network analyses) and the "p53 signaling pathway" (enriched in the MIRNET GSE105269 networks). Furthermore, there is strong evidence to support the claim that apoptosis is linked to neurotrophin deprivation [112]. Indeed, our in silico analysis identified the genes involved in neurotrophin signaling, such as NTRK2, i.e., the tyrosine kinase receptor of BDNF is central gene in the network describing the GSE45570 network "POAG vs. CTRL".

Our bioinformatic analysis identified the gene ELAVL1 as a central node in the networks, and this chaperonine is a key regulator of gene expressions at the post-transcriptional level, which was recently found to be fundamental for the survival and protection of

RGCs [113]. In the work by Smedowski et al., (2018) [42], ELAVL1 was detected in the retinas and optic nerves of glaucomatous rats, with an altered nucleus and cytoplasmic localization. Additionally, after the bioinformatic re-analysis of GS105269, we identified, as central node, APP. These results strengthen the validity of our in silico method, since there are several pieces of evidence supporting the common pathogenesis of glaucoma and Alzheimer's diseases [44,45,114].

Furthermore, our bioinformatic analysis was able to identify several druggable targets for glaucoma treatment, e.g., the A_1 and A_{2A} adenosine receptors, along with the 5-hydroxytryptamine 2A and purinergic P2Y2 receptors, which have already been investigated in other studies. Indeed, adenosine receptors (identified in the GSE133563 network analysis) have already been hypothesized to be useful for the treatment of glaucoma, due to their recognized role in controlling the IOP [115]. More specifically, a blockade of the microglial adenosine A2A receptor proved to be protective against a retinal dysfunction triggered by an elevation of the IOP [116], while an activation of the adenosine A1 receptor demonstrated to reduce the IOP by increasing the outflow facility in mice [117]. Additionally, caffeine, a non-selective adenosine receptor antagonist with a higher affinity for the adenosine A_1 and A_{2A} receptors, showed retinal protective effects in a mouse model of I/R injuries, which resembled glaucomatous neurodegeneration [118]. Furthermore, the serotoninergic system and sigma receptor have already been investigated for their role in modulating the IOP and neurodegeneration [86,87,119]. Additionally, the purinergic receptor P2Y2, identified by the GSE105269 network analysis, was previously found to be a potential dysregulated receptor in glaucoma. Specifically, P2RY2 receptor agonists were found to increase the IOP [50]. Furthermore, Fonseca et al., in 2017 [51], found an overexpression of P2RY2 in DBA/2J mice, strengthening the involvement of this receptor in the onset and progression of glaucoma [51].

Rescoring on the basis of matching nodes in the built networks with entries in the PubMed and clinicaltrials.gov databases was necessary in order to identify the genes, and then related proteins, that can be considered innovative targets for the treatment of glaucoma. For instance, 82 out of 135 identified targets have not yet been experimentally associated with glaucoma. For instance, we characterized these in silico identified nodes in networks as predicted new pharmacological targets for the treatment of glaucoma. A main limitation of our study is related to the in silico approach, which is merely a predictive approach, given that these targets need experimental validation. One strength of the reverse engineering approach hereby used is related to the methods of in silico system biology approaches, namely, these are successful in their description of the pathogenic mechanisms of complex multifactorial diseases, facilitating the R&D process of drug development [120]. Several GPCRs were classified, through rescoring, as predicted new pharmacological targets: the adenosine A_{2B} receptor, the 5-hydroxytryptamine receptor 1D, the 5-hydroxytryptamine receptor 2C, the 5-hydroxytryptamine receptor 5A, the melatonin receptor 1A, and the olfactory receptor family 2 subfamily B member 3. Moreover, among the innovative druggable targets, we found the following: the purinergic ionotropic receptors P2X3 and P2X4; the chloride intracellular channel 2 (CLIC2); and the cation channels TRPC3, TRPC4, and TRPC7. Interestingly, the role of the purinergic system in glaucoma has already been described for the P2X7 receptor [121,122]. Moreover, different subunits of glutamatergic receptors, e.g., GRIA1 and GRIK2, encoding the glutamate ionotropic receptor AMPA type subunit 1 and the glutamate ionotropic receptor kainate type subunit 2, respectively, identified as central nodes in the GSE133563 network, have not yet been associated with glaucoma pathogenesis. We were also able to identify, as intriguing and innovative targets, genes encoding for enzymes: CYP2C8, LDHA, MAOB, and B3GAT2.

4. Materials and Methods

4.1. GEO Dataset

The Gene Expression Omnibus Dataset repository (GEO dataset https://www.ncbi.nlm.nih.gov/gds accessed on 1 March 2022) [22] was the database from which the datasets listed in Table 1 were retrieved, by the means of the keyword "glaucoma". DEGs ($p < 0.05$) for each sample of the GEO datasets (e.g., glaucoma vs. healthy control) were identified using the GEO2R statistical analysis (https://www.ncbi.nlm.nih.gov/geo/info/geo2r.html accessed on 1 March 2022), applying the Benjamini and Hochberg (false discovery rate) method for the adjustment of p-values using standard parameters in GEO2R [23,123,124]. In each GEO dataset comparison, approximately 250 DEGs with a p-value of <0.05 were then selected as an input of the STRING app within Cytoscape.

4.2. Network Definition and Descriptor Computations

STRING (© STRING CONSORTIUM 2018), applied with Cytoscape v. 3.7.0 (2001–2017 Cytoscape Consortium), was employed to carry out an enrichment of information and to build gene networks (https://apps.cytoscape.org/apps/stringapp, accessed on 1 March 2022) [25]. After the entry of a list of genes, setting a confidence score cut-off (0.40) and the maximal number of additional interactors as 20, the STRING app, by using a wide dataset of direct (physical) as well as indirect (functional) association [125], provided the annotated Cytoscape networks based on the predicted integration of protein–protein interactions. The enrichment data were loaded. A network analysis was carried out through Cytoscape, and all networks were treated as unweighted and undirected graphs. The adjacency matrix A is the mathematical representation of the network:

$$A_{ij} = \begin{cases} 1 & if\ the\ residue\ i\ and\ j\ are\ in\ contact \\ 0 & otherwise \end{cases}$$

On the basis of the definition of matrix A, a series of network descriptors, representing the topological role of nodes (genes) in the networks, could be calculated. Node degree k_i is the network primary descriptor: it defines the number of nodes with which every node is in contact. The shortest path between two nodes is determined as the minimum number of links connecting the two nodes. Based on the shortest paths, two centrality metrics, i.e., closeness centrality and betweenness centrality, could be defined.

Closeness centrality depends on the sum of the minimal distance of one vertex (V) to all other vertices, considering the length of the shortest paths connecting them. In particular, the closeness centrality (close) of the i-th node can be mathematically calculated as follows:

$$close(i) = \frac{1}{\sum_{u\epsilon V, u\neq i} sp_{ui}}$$

Betweenness centrality quantifies the times in which one vertex acts as a bridge between two other vertices. Specifically, given a set of vertices (V), the mathematical representation of the betweenness (betw) centrality of the i-th node is follows:

$$betw(i) = \sum_{v\epsilon V, v\neq i}\ \sum_{u\epsilon V, u\epsilon i} \frac{\sigma_{uv}(i)}{\sigma_{uv}}$$

where σ_{uv} is the total number of shortest paths linking nodes u and v, and $\sigma_{uv}\ (i)$ is the number of shortest paths linking the two nodes and passing through the node i.

An analysis of centrality metrics is necessary for the identification of genes that would play a key role in the stability and modularity of a network [126]. Specifically, nodes were represented with a size proportional to closeness centrality values. Betweenness values were plotted by assigning a node a color according to a temperature color scale, which changed from blue to red according to the increase values of the node's betweenness values. Moreover, edge betweenness was set as proportional as to edge thickness, and the prefuse force layout of the networks was represented on the basis of edge betweenness. Furthermore, in the analysis, the bioinformatic tool MIRNET (https://www.mirnet.ca/ accessed on 1 March 2022) was employed [127].

Finally, after identifying the nodes that were central in the built networks, we matched the nodes with entries in PubMed and clinicaltrials.gov. This rescoring approach was necessary to discriminating nodes (genes or miRNA) that have already been studied in preclinical models of this disease or in glaucoma subjects recruited in clinical trials. Specifically, to match the in silico identified targets with experimental studies (preclinical and/or clinical), the query "target name AND glaucoma" was used to mine data in the PubMed and clinicaltrials.gov databases.

5. Conclusions

In conclusion, we identified, through an in silico approach, pharmacological targets that can be considered predicted and new for the potential treatment or management of glaucoma, due to their lack of a match with published data (PubMed) or clinical trials (clinicaltrials.gov). One step forward would be to identify the potential ligands, agomirs, and antagomirs, along with transcription factors or suppressors, of these 82 predicted new targets, i.e., through access to enrichment approaches and related databases: STITCH (http://stitch.embl.de/) [128], DIANA tools (https://diana.e-ce.uth.gr/home) [129], and GeneCards.org (https://www.genecards.org/) [130]. This is the first theoretical step toward the validation of pharmacological targets that can reveal novel treatments of glaucoma, and the most time-consuming and expensive steps are experimental validation of these targets through preclinical and clinical studies. Since the assumed network pharmacology approach, which borrows in silico system biology methods, is unbiased, it can facilitate an R&D process analyzing a complex multifactorial disease through a holistic perspective instead of a reductionist approach. It is worth noting that most of the identified "predicted new" targets are GPCRs, channels, and enzymes; therefore, drug repurposing programs could further facilitate clinical research in glaucoma, hopefully with a focus on neuroprotective strategies.

Supplementary Materials: The following supporting information can be downloaded at: https://www.mdpi.com/article/10.3390/ph17101333/s1, List S1: List of the differentially expressed miRNAs input on MIRNET for the analysis of GSE105269 POAG vs. CTRL; List S2: List of the differentially expressed miRNAs input on MIRNET for the analysis of GSE105269 XFG vs. CTRL.

Author Contributions: E.G. and N.M. carried out the analyses. C.B.M.P. designed, interpreted, and conceived the study. C.B.M.P., E.G. and F.L. wrote the manuscript. C.B.M.P., F.L., F.C., F.D. and C.B. reviewed and edited the manuscript. C.B. conceived the study and provided funding and hardware and software resources. All authors have read and agreed to the published version of the manuscript.

Funding: This research was funded by PRIN 2020FR7TCL, PIAno inCEntivi RIcerca (PIACERI) Ateneo 2020/2022 (NanoRET).

Institutional Review Board Statement: Not applicable.

Informed Consent Statement: Not applicable.

Data Availability Statement: All data generated or analyzed in this study are included in this published article.

Acknowledgments: E.G. was supported by PON Ricerca e Innovazione D.M. 1061/21-Dottorati di ricerca, from the Italian Ministry of University (MUR). C.B.M.P. was supported by PON Ricerca e Innovazione D.M. 1062/21–Contratti di ricerca, from the Italian Ministry of University (MUR), contract #: 08-I-17629-2. F.L. was supported by Missione 4 "Istruzione e Ricerca"-"Dalla Ricerca all'impresa"-NextGenerationUE-"ANTHEM: AdvaNced Technologies for Human-centrEd Medicine", Spoke 4, PNC0000003.

Conflicts of Interest: The authors declare that this research was conducted in the absence of any commercial or financial relationships that could be construed as a potential conflict of interest.

Abbreviations

ACG	Angle-closure glaucoma
AD	Alzheimer's disease
AH	Aqueous humour
CTRL	Control
DE	Differentially expressed
DEGs	Differentially expressed genes
ECM	Extracellular matrix
EVC	Episcleral vein cauterization
GPCRs	G protein-coupled receptors
HTM	Human trabecular meshwork
IOP	Intraocular pressure
I/R	Ischemia/reperfusion
KEGG	Kyoto encyclopedia of genes and genomes
LGN	Lateral geniculate nucleus
NF-kB	Nuclear factor kappa-B
NTG	Normal-tension glaucoma
NVG	Neovascular glaucoma
OHT	Ocular hypertension
ONC	Optic nerve crush
ONH	Optic nerve head
POAG	Primary open-angle glaucoma
RGCs	Retinal ganglion cells
TM	Trabecular meshwork
VEGF	Vascular endothelial growth factor
WGS	Whole-genome sequencing
XFG	Exfoliation glaucoma
XFS	Exfoliation syndrome

References

1. GBD 2019 Blindness and Vision Impairment Collaborators; Vision Loss Expert Group of the Global Burden of Disease Study. Causes of blindness and vision impairment in 2020 and trends over 30 years, and prevalence of avoidable blindness in relation to VISION 2020: The Right to Sight: An analysis for the Global Burden of Disease Study. *Lancet Glob. Health* **2021**, *9*, e144–e160. [CrossRef] [PubMed]
2. Mohan, N.; Chakrabarti, A.; Nazm, N.; Mehta, R.; Edward, D.P. Newer advances in medical management of glaucoma. *Indian J. Ophthalmol.* **2022**, *70*, 1920–1930. [CrossRef] [PubMed]
3. Weinreb, R.N.; Aung, T.; Medeiros, F.A. The pathophysiology and treatment of glaucoma: A review. *JAMA* **2014**, *311*, 1901–1911. [CrossRef] [PubMed]
4. Chou, T.-H.; Musada, G.; Romano, G.; Bolton, E.; Porciatti, V. Anesthetic Preconditioning as Endogenous Neuroprotection in Glaucoma. *Int. J. Mol. Sci.* **2018**, *19*, 237. [CrossRef]
5. Almasieh, M.; Wilson, A.M.; Morquette, B.; Vargas, J.L.C.; Di Polo, A. The molecular basis of retinal ganglion cell death in glaucoma. *Prog. Retin. Eye Res.* **2012**, *31*, 152–181. [CrossRef]
6. Youngblood, H.; Schoenlein, P.V.; Pasquale, L.R.; Stamer, W.D.; Liu, Y. Estrogen dysregulation, intraocular pressure, and glaucoma risk. *Exp. Eye Res.* **2023**, *237*, 109725. [CrossRef]
7. Liu, Y.; Allingham, R.R. Major review: Molecular genetics of primary open-angle glaucoma. *Exp. Eye Res.* **2017**, *160*, 62–84. [CrossRef] [PubMed]
8. Casson, R.J. Medical therapy for glaucoma: A review. *Clin. Exp. Ophthalmol.* **2022**, *50*, 198–212. [CrossRef]
9. Jayaram, H.; Kolko, M.; Friedman, D.S.; Gazzard, G. Glaucoma: Now and beyond. *Lancet* **2023**, *402*, 1788–1801. [CrossRef]
10. Jeong, J.H.; Park, K.H.; Jeoung, J.W.; Kim, D.M. Preperimetric normal tension glaucoma study: Long-term clinical course and effect of therapeutic lowering of intraocular pressure. *Acta Ophthalmol.* **2014**, *92*, e185–e193. [CrossRef]
11. Tribble, J.R.; Hui, F.; Quintero, H.; El Hajji, S.; Bell, K.; Di Polo, A.; Williams, P.A. Neuroprotection in glaucoma: Mechanisms beyond intraocular pressure lowering. *Mol. Asp. Med.* **2023**, *92*, 101193. [CrossRef] [PubMed]
12. Sharif, N.A. Therapeutic Drugs and Devices for Tackling Ocular Hypertension and Glaucoma, and Need for Neuroprotection and Cytoprotective Therapies. *Front. Pharmacol.* **2021**, *12*, 729249. [CrossRef] [PubMed]
13. Yeh, S.-J.; Chung, M.-H.; Chen, B.-S. Investigating Pathogenetic Mechanisms of Alzheimer's Disease by Systems Biology Approaches for Drug Discovery. *Int. J. Mol. Sci.* **2021**, *22*, 11280. [CrossRef] [PubMed]

14. Auwul, M.R.; Rahman, M.R.; Gov, E.; Shahjaman, M.; Moni, M.A. Bioinformatics and machine learning approach identifies potential drug targets and pathways in COVID-19. *Brief. Bioinform.* **2021**, *22*, bbab120. [CrossRef] [PubMed]
15. Platania, C.B.M.; Leggio, G.M.; Drago, F.; Salomone, S.; Bucolo, C. Computational systems biology approach to identify novel pharmacological targets for diabetic retinopathy. *Biochem. Pharmacol.* **2018**, *158*, 13–26. [CrossRef]
16. Peng, H.; Tan, H.; Zhao, W.; Jin, G.; Sharma, S.; Xing, F.; Watabe, K.; Zhou, X. Computational systems biology in cancer brain metastasis. *Front. Biosci. (Sch. Ed.)* **2016**, *8*, 169–186. [CrossRef]
17. Lazzara, F.; Conti, F.; Giuffrida, E.; Eandi, C.M.; Drago, F.; Platania, C.B.M.; Bucolo, C. Integrating network pharmacology: The next-generation approach in ocular drug discovery. *Curr. Opin. Pharmacol.* **2024**, *74*, 102425. [CrossRef]
18. Yan, J.; Risacher, S.L.; Shen, L.; Saykin, A.J. Network approaches to systems biology analysis of complex disease: Integrative methods for multi-omics data. *Brief. Bioinform.* **2018**, *19*, 1370–1381. [CrossRef]
19. Marbach, D.; Schaffter, T.; Mattiussi, C.; Floreano, D. Generating realistic in silico gene networks for performance assessment of reverse engineering methods. *J. Comput. Biol.* **2009**, *16*, 229–239. [CrossRef]
20. Hu, G.; Di Paola, L.; Pullara, F.; Liang, Z.; Nookaew, I. Network Proteomics: From Protein Structure to Protein-Protein Interaction. *Biomed. Res. Int.* **2017**, *2017*, 8929613. [CrossRef]
21. Platania, C.B.M.; Di Paola, L.; Leggio, G.M.; Romano, G.L.; Drago, F.; Salomone, S.; Bucolo, C. Molecular features of interaction between VEGFA and anti-angiogenic drugs used in retinal diseases: A computational approach. *Front. Pharmacol.* **2015**, *6*, 248. [CrossRef] [PubMed]
22. Clough, E.; Barrett, T.; Wilhite, S.E.; Ledoux, P.; Evangelista, C.; Kim, I.F.; Tomashevsky, M.; Marshall, K.A.; Phillippy, K.H.; Sherman, P.M.; et al. NCBI GEO: Archive for gene expression and epigenomics data sets: 23-year update. *Nucleic Acids Res.* **2024**, *52*, D138–D144. [CrossRef]
23. Barrett, T.; Wilhite, S.E.; Ledoux, P.; Evangelista, C.; Kim, I.F.; Tomashevsky, M.; Marshall, K.A.; Phillippy, K.H.; Sherman, P.M.; Holko, M.; et al. NCBI GEO: Archive for functional genomics data sets--update. *Nucleic Acids Res.* **2013**, *41*, D991–D995. [CrossRef]
24. Sun, P.; Feng, S.; Yu, H.; Wang, X.; Fang, Y. Two hub genes of bipolar disorder, a bioinformatics study based on the GEO database. *IBRO Neurosci. Rep.* **2024**, *17*, 122–130. [CrossRef]
25. Doncheva, N.T.; Morris, J.H.; Gorodkin, J.; Jensen, L.J. Cytoscape StringApp: Network Analysis and Visualization of Proteomics Data. *J. Proteome Res.* **2019**, *18*, 623–632. [CrossRef] [PubMed]
26. Geiduschek, E.K.; McDowell, C.M. The Fibro-Inflammatory Response in the Glaucomatous Optic Nerve Head. *Int. J. Mol. Sci.* **2023**, *24*, 13240. [CrossRef]
27. Mavlyutov, T.A.; Myrah, J.J.; Chauhan, A.K.; Liu, Y.; McDowell, C.M. Fibronectin extra domain A (FN-EDA) causes glaucomatous trabecular meshwork, retina, and optic nerve damage in mice. *Cell Biosci.* **2022**, *12*, 72. [CrossRef] [PubMed]
28. Victorino, P.H.; Marra, C.; Iacobas, D.A.; Iacobas, S.; Spray, D.C.; Linden, R.; Adesse, D.; Petrs-Silva, H. Retinal Genomic Fabric Remodeling after Optic Nerve Injury. *Genes* **2021**, *12*, 403. [CrossRef]
29. Drewry, M.D.; Challa, P.; Kuchtey, J.G.; Navarro, I.; Helwa, I.; Hu, Y.; Mu, H.; Stamer, W.D.; Kuchtey, R.W.; Liu, Y. Differentially expressed microRNAs in the aqueous humor of patients with exfoliation glaucoma or primary open-angle glaucoma. *Hum. Mol. Genet.* **2018**, *27*, 1263–1275. [CrossRef]
30. Liu, Y.; Allingham, R.R.; Qin, X.; Layfield, D.; Dellinger, A.E.; Gibson, J.; Wheeler, J.; Ashley-Koch, A.E.; Stamer, W.D.; Hauser, M.A. Gene expression profile in human trabecular meshwork from patients with primary open-angle glaucoma. *Investig. Ophthalmol. Vis. Sci.* **2013**, *54*, 6382–6389. [CrossRef]
31. Liton, P.B.; Luna, C.; Challa, P.; Epstein, D.L.; Gonzalez, P. Genome-wide expression profile of human trabecular meshwork cultured cells, nonglaucomatous and primary open angle glaucoma tissue. *Mol. Vis.* **2006**, *12*, 774–790. [PubMed]
32. Steele, M.R.; Inman, D.M.; Calkins, D.J.; Horner, P.J.; Vetter, M.L. Microarray analysis of retinal gene expression in the DBA/2J model of glaucoma. *Investig. Ophthalmol. Vis. Sci.* **2006**, *47*, 977–985. [CrossRef] [PubMed]
33. Irnaten, M.; O'Malley, G.; Clark, A.F.; O'Brien, C.J. Transient receptor potential channels TRPC1/TRPC6 regulate lamina cribrosa cell extracellular matrix gene transcription and proliferation. *Exp. Eye Res.* **2020**, *193*, 107980. [CrossRef] [PubMed]
34. Benitez-Del-Castillo, J.; Cantu-Dibildox, J.; Sanz-González, S.M.; Zanón-Moreno, V.; Pinazo-Duran, M.D. Cytokine expression in tears of patients with glaucoma or dry eye disease: A prospective, observational cohort study. *Eur. J. Ophthalmol.* **2019**, *29*, 437–443. [CrossRef] [PubMed]
35. Chen, H.; Zheng, G.; Chen, H.; Li, L.; Xu, Z.; Xu, L. Evaluations of aqueous humor protein markers in different types of glaucoma. *Medicine* **2022**, *101*, e31048. [CrossRef]
36. Tezel, G.; Wax, M.B. Hypoxia-inducible factor 1alpha in the glaucomatous retina and optic nerve head. *Arch. Ophthalmol.* **2004**, *122*, 1348–1356. [CrossRef]
37. Nsiah, N.Y.; Morgan, A.B.; Donkor, N.; Inman, D.M. Long-term HIF-1α stabilization reduces respiration, promotes mitophagy, and results in retinal cell death. *Sci. Rep.* **2023**, *13*, 20541. [CrossRef]
38. Ergorul, C.; Ray, A.; Huang, W.; Wang, D.Y.; Ben, Y.; Cantuti-Castelvetri, I.; Grosskreutz, C.L. Hypoxia inducible factor-1α (HIF-1α) and some HIF-1 target genes are elevated in experimental glaucoma. *J. Mol. Neurosci.* **2010**, *42*, 183–191. [CrossRef]
39. de Sousa, M.C.; Gjorgjieva, M.; Dolicka, D.; Sobolewski, C.; Foti, M. Deciphering miRNAs' Action through miRNA Editing. *Int. J. Mol. Sci.* **2019**, *20*, 6249. [CrossRef]

40. Seong, H.; Cho, H.-K.; Kee, C.; Song, D.H.; Cho, M.-C.; Kang, S.S. Profiles of microRNA in aqueous humor of normal tension glaucoma patients using RNA sequencing. *Sci. Rep.* **2021**, *11*, 19024. [CrossRef]
41. Kanehisa, M.; Goto, S. KEGG: Kyoto encyclopedia of genes and genomes. *Nucleic Acids Res.* **2000**, *28*, 27–30. [CrossRef] [PubMed]
42. Smedowski, A.; Liu, X.; Podracka, L.; Akhtar, S.; Trzeciecka, A.; Pietrucha-Dutczak, M.; Lewin-Kowalik, J.; Urtti, A.; Ruponen, M.; Kaarniranta, K.; et al. Increased intraocular pressure alters the cellular distribution of HuR protein in retinal ganglion cells—A possible sign of endogenous neuroprotection failure. *Biochim. Biophys. Acta Mol. Basis Dis.* **2018**, *1864*, 296–306. [CrossRef] [PubMed]
43. Mochizuki, H.; Murphy, C.J.; Brandt, J.D.; Kiuchi, Y.; Russell, P. Altered stability of mRNAs associated with glaucoma progression in human trabecular meshwork cells following oxidative stress. *Investig. Ophthalmol. Vis. Sci.* **2012**, *53*, 1734–1741. [CrossRef] [PubMed]
44. Guo, L.; Salt, T.E.; Luong, V.; Wood, N.; Cheung, W.; Maass, A.; Ferrari, G.; Russo-Marie, F.; Sillito, A.M.; Cheetham, M.E.; et al. Targeting amyloid-beta in glaucoma treatment. *Proc. Natl. Acad. Sci. USA* **2007**, *104*, 13444–13449. [CrossRef]
45. Goldblum, D.; Kipfer-Kauer, A.; Sarra, G.-M.; Wolf, S.; Frueh, B.E. Distribution of amyloid precursor protein and amyloid-beta immunoreactivity in DBA/2J glaucomatous mouse retinas. *Investig. Ophthalmol. Vis. Sci.* **2007**, *48*, 5085–5090. [CrossRef]
46. Yasuda, M.; Tanaka, Y.; Ryu, M.; Tsuda, S.; Nakazawa, T. RNA sequence reveals mouse retinal transcriptome changes early after axonal injury. *PLoS ONE* **2014**, *9*, e93258. [CrossRef]
47. Kanamori, A.; Nakamura, M.; Nakanishi, Y.; Nagai, A.; Mukuno, H.; Yamada, Y.; Negi, A. Akt is activated via insulin/IGF-1 receptor in rat retina with episcleral vein cauterization. *Brain Res.* **2004**, *1022*, 195–204. [CrossRef] [PubMed]
48. Wirtz, M.K.; Sykes, R.; Samples, J.; Edmunds, B.; Choi, D.; Keene, D.R.; Tufa, S.F.; Sun, Y.Y.; Keller, K.E. Identification of Missense Extracellular Matrix Gene Variants in a Large Glaucoma Pedigree and Investigation of the N700S Thrombospondin-1 Variant in Normal and Glaucomatous Trabecular Meshwork Cells. *Curr. Eye Res.* **2022**, *47*, 79–90. [CrossRef]
49. Springelkamp, H.; Iglesias, A.I.; Mishra, A.; Höhn, R.; Wojciechowski, R.; Khawaja, A.P.; Nag, A.; Wang, Y.X.; Wang, J.J.; Cuellar-Partida, G.; et al. New insights into the genetics of primary open-angle glaucoma based on meta-analyses of intraocular pressure and optic disc characteristics. *Hum. Mol. Genet.* **2017**, *26*, 438–453. [CrossRef]
50. AMartin-Gil, A.; de Lara, M.J.P.; Crooke, A.; Santano, C.; Peral, A.; Pintor, J. Silencing of P2Y(2) receptors reduces intraocular pressure in New Zealand rabbits. *Br. J. Pharmacol.* **2012**, *165*, 1163–1172. [CrossRef]
51. Fonseca, B.; Martínez-Águila, A.; de Lara, M.J.P.; Miras-Portugal, M.T.; Gómez-Villafuertes, R.; Pintor, J. Changes in P2Y Purinergic Receptor Expression in the Ciliary Body in a Murine Model of Glaucoma. *Front. Pharmacol.* **2017**, *8*, 719. [CrossRef] [PubMed]
52. Gowtham, L.; Halder, N.; Singh, S.B.; Angmo, D.; Jayasundar, R.; Dada, T.; Velpandian, T. Evaluation of ADRB2 and OATP2A1 genetic polymorphisms in Indian patients with primary open-angle glaucoma. *Pharmacogenet. Genom.* **2024**, *34*, 20–24. [CrossRef]
53. Sidjanin, D.J.; McCarty, C.A.; Patchett, R.; Smith, E.; Wilke, R.A. Pharmacogenetics of ophthalmic topical beta-blockers. *Pers. Med.* **2008**, *5*, 377–385. [CrossRef]
54. Inagaki, Y.; Mashima, Y.; Fuse, N.; Funayama, T.; Ohtake, Y.; Yasuda, N.; Murakami, A.; Hotta, Y.; Fukuchi, T.; Tsubota, K. Polymorphism of beta-adrenergic receptors and susceptibility to open-angle glaucoma. *Mol. Vis.* **2006**, *12*, 673–680. [PubMed]
55. McCarty, C.A.; Burmester, J.K.; Mukesh, B.N.; Patchett, R.B.; Wilke, R.A. Intraocular pressure response to topical beta-blockers associated with an ADRB2 single-nucleotide polymorphism. *Arch. Ophthalmol.* **2008**, *126*, 959–963. [CrossRef] [PubMed]
56. Aboobakar, I.F.; Allingham, R.R. Genetics of exfoliation syndrome and glaucoma. *Int. Ophthalmol. Clin.* **2014**, *54*, 43–56. [CrossRef] [PubMed]
57. Rao, A.; Chakraborty, M.; Roy, A.; Sahay, P.; Pradhan, A.; Raj, N. Differential miRNA Expression: Signature for Glaucoma in Pseudoexfoliation. *Clin. Ophthalmol.* **2020**, *14*, 3025–3038. [CrossRef]
58. Cho, H.-K.; Seong, H.; Kee, C.; Song, D.H.; Kim, S.J.; Seo, S.W.; Kang, S.S. MicroRNA profiles in aqueous humor between pseudoexfoliation glaucoma and normal tension glaucoma patients in a Korean population. *Sci. Rep.* **2022**, *12*, 6217. [CrossRef]
59. Colak, D.; Morales, J.; Bosley, T.M.; Al-Bakheet, A.; AlYounes, B.; Kaya, N.; Abu-Amero, K.K. Genome-wide expression profiling of patients with primary open angle glaucoma. *Investig. Ophthalmol. Vis. Sci.* **2012**, *53*, 5899–5904. [CrossRef]
60. Agapova, O.A.; Kaufman, P.L.; Hernandez, M.R. Androgen receptor and NFkB expression in human normal and glaucomatous optic nerve head astrocytes in vitro and in experimental glaucoma. *Exp. Eye Res.* **2006**, *82*, 1053–1059. [CrossRef]
61. Hernandez, H.; Medina-Ortiz, W.E.; Luan, T.; Clark, A.F.; McDowell, C.M. Crosstalk Between Transforming Growth Factor Beta-2 and Toll-Like Receptor 4 in the Trabecular Meshwork. *Investig. Ophthalmol. Vis. Sci.* **2017**, *58*, 1811–1823. [CrossRef] [PubMed]
62. Hase, K.; Kase, S.; Kanda, A.; Shinmei, Y.; Noda, K.; Ishida, S. Expression of Vascular Endothelial Growth Factor-C in the Trabecular Meshwork of Patients with Neovascular Glaucoma and Primary Open-Angle Glaucoma. *J. Clin. Med.* **2021**, *10*, 2977. [CrossRef] [PubMed]
63. Filla, M.S.; Faralli, J.A.; Desikan, H.; Peotter, J.L.; Wannow, A.C.; Peters, D.M. Activation of $\alpha v\beta 3$ Integrin Alters Fibronectin Fibril Formation in Human Trabecular Meshwork Cells in a ROCK-Independent Manner. *Investig. Ophthalmol. Vis. Sci.* **2019**, *60*, 3897–3913. [CrossRef] [PubMed]
64. Yoneshige, A.; Hagiyama, M.; Takashima, Y.; Ueno, S.; Inoue, T.; Kimura, R.; Koriyama, Y.; Ito, A. Elevated Hydrostatic Pressure Causes Retinal Degeneration through Upregulating Lipocalin-2. *Front. Cell Dev. Biol.* **2021**, *9*, 664327. [CrossRef]
65. Ueno, S.; Yoneshige, A.; Koriyama, Y.; Hagiyama, M.; Shimomura, Y.; Ito, A. Early Gene Expression Profile in Retinal Ganglion Cell Layer After Optic Nerve Crush in Mice. *Investig. Ophthalmol. Vis. Sci.* **2018**, *59*, 370–380. [CrossRef]

66. Huang, C.; Cen, L.-P.; Liu, L.; Leaver, S.G.; Harvey, A.R.; Cui, Q.; Pang, C.P.; Zhang, M. Adeno-associated virus-mediated expression of growth-associated protein-43 aggravates retinal ganglion cell death in experimental chronic glaucomatous injury. *Mol. Vis.* **2013**, *19*, 1422–1432.
67. Dreixler, J.C.; Bratton, A.; Du, E.; Shaikh, A.R.; Savoie, B.; Alexander, M.; Marcet, M.M.; Roth, S. Mitogen-activated protein kinase phosphatase-1 (MKP-1) in retinal ischemic preconditioning. *Exp. Eye Res.* **2011**, *93*, 340–349. [CrossRef] [PubMed]
68. Narta, K.; Teltumbade, M.R.; Vishal, M.; Sadaf, S.; Faruq, M.; Jama, H.; Waseem, N.; Rao, A.; Sen, A.; Ray, K.; et al. Whole Exome Sequencing Reveals Novel Candidate Genes in Familial Forms of Glaucomatous Neurodegeneration. *Genes* **2023**, *14*, 495. [CrossRef]
69. Reinehr, S.; Wulf, J.; Theile, J.; Schulte, K.K.; Peters, M.; Fuchshofer, R.; Dick, H.B.; Joachim, S.C. In a novel autoimmune and high-pressure glaucoma model a complex immune response is induced. *Front. Immunol.* **2024**, *15*, 1296178. [CrossRef]
70. Diskin, S.; Cao, Z.; Leffler, H.; Panjwani, N. The role of integrin glycosylation in galectin-8-mediated trabecular meshwork cell adhesion and spreading. *Glycobiology* **2009**, *19*, 29–37. [CrossRef]
71. Beggah, A.T.; Dours-Zimmermann, M.T.; Barras, F.M.; Brosius, A.; Zimmermann, D.R.; Zurn, A.D. Lesion-induced differential expression and cell association of Neurocan, Brevican, Versican V1 and V2 in the mouse dorsal root entry zone. *Neuroscience* **2005**, *133*, 749–762. [CrossRef] [PubMed]
72. Pearson, C.S.; Solano, A.G.; Tilve, S.M.; Mencio, C.P.; Martin, K.R.; Geller, H.M. Spatiotemporal distribution of chondroitin sulfate proteoglycans after optic nerve injury in rodents. *Exp. Eye Res.* **2020**, *190*, 107859. [CrossRef]
73. Reinhard, J.; Renner, M.; Wiemann, S.; Shakoor, D.A.; Stute, G.; Dick, H.B.; Faissner, A.; Joachim, S.C. Ischemic injury leads to extracellular matrix alterations in retina and optic nerve. *Sci. Rep.* **2017**, *7*, 43470. [CrossRef]
74. Samelska, K.; Zaleska-Żmijewska, A.; Bałan, B.; Grąbczewski, A.; Szaflik, J.P.; Kubiak, A.J.; Skopiński, P. Immunological and molecular basics of the primary open angle glaucoma pathomechanism. *Cent. Eur. J. Immunol.* **2021**, *46*, 111–117. [CrossRef]
75. Howell, G.R.; MacNicoll, K.H.; Braine, C.E.; Soto, I.; Macalinao, D.G.; Sousa, G.L.; John, S.W.M. Combinatorial targeting of early pathways profoundly inhibits neurodegeneration in a mouse model of glaucoma. *Neurobiol. Dis.* **2014**, *71*, 44–52. [CrossRef]
76. Marola, O.J.; Howell, G.R.; Libby, R.T. Vascular derived endothelin receptor A controls endothelin-induced retinal ganglion cell death. *Cell Death Discov.* **2022**, *8*, 207. [CrossRef]
77. Chiquet, C.; Vignal, C.; Gohier, P.; Heron, E.; Thuret, G.; Rougier, M.B.; Lehmann, A.; Flet, L.; Quesada, J.-L.; Roustit, M.; et al. Treatment of nonarteritic anterior ischemic optic neuropathy with an endothelin antagonist: ENDOTHELION (ENDOTHELin antagonist receptor in Ischemic Optic Neuropathy)-a multicentre randomised controlled trial protocol. *Trials* **2022**, *23*, 916. [CrossRef]
78. Ivanov, D.; Dvoriantchikova, G.; Nathanson, L.; McKinnon, S.J.; Shestopalov, V.I. Microarray analysis of gene expression in adult retinal ganglion cells. *FEBS Lett.* **2006**, *580*, 331–335. [CrossRef] [PubMed]
79. Rahman, N.; O'Neill, E.; Irnaten, M.; Wallace, D.; O'Brien, C. Corneal Stiffness and Collagen Cross-Linking Proteins in Glaucoma: Potential for Novel Therapeutic Strategy. *J. Ocul. Pharmacol. Ther.* **2020**, *36*, 582–594. [CrossRef]
80. Yu, P.; Dong, W.-P.; Tang, Y.-B.; Chen, H.-Z.; Cui, Y.-Y.; Bian, X.-L. Huperzine A lowers intraocular pressure via the M3 mAChR and provides retinal neuroprotection via the M1 mAChR: A promising agent for the treatment of glaucoma. *Ann. Transl. Med.* **2021**, *9*, 332. [CrossRef]
81. Cheng, L.; Sapieha, P.; Kittlerova, P.; Hauswirth, W.W.; Di Polo, A. TrkB gene transfer protects retinal ganglion cells from axotomy-induced death in vivo. *J. Neurosci.* **2002**, *22*, 3977–3986. [CrossRef] [PubMed]
82. Kimura, A.; Namekata, K.; Guo, X.; Harada, C.; Harada, T. Neuroprotection, Growth Factors and BDNF-TrkB Signalling in Retinal Degeneration. *Int. J. Mol. Sci.* **2016**, *17*, 1584. [CrossRef]
83. Milbeck, S.M.; Bhattacharya, S.K. Alteration in Lysophospholipids and Converting Enzymes in Glaucomatous Optic Nerves. *Investig. Ophthalmol. Vis. Sci.* **2020**, *61*, 60. [CrossRef]
84. Zukerman, R.; Harris, A.; Vercellin, A.V.; Siesky, B.; Pasquale, L.R.; Ciulla, T.A. Molecular Genetics of Glaucoma: Subtype and Ethnicity Considerations. *Genes* **2020**, *12*, 55. [CrossRef]
85. Gharahkhani, P.; Burdon, K.P.; Fogarty, R.; Sharma, S.; Hewitt, A.W.; Martin, S.; Law, M.H.; Cremin, K.; Bailey, J.N.C.; Loomis, S.J.; et al. Common variants near ABCA1, AFAP1 and GMDS confer risk of primary open-angle glaucoma. *Nat. Genet.* **2014**, *46*, 1120–1125. [CrossRef] [PubMed]
86. Sharif, N.A. Serotonin-2 receptor agonists as novel ocular hypotensive agents and their cellular and molecular mechanisms of action. *Curr. Drug Targets* **2010**, *11*, 978–993. [CrossRef]
87. Platania, C.B.M.; Leggio, G.M.; Drago, F.; Salomone, S.; Bucolo, C. Regulation of intraocular pressure in mice: Structural analysis of dopaminergic and serotonergic systems in response to cabergoline. *Biochem. Pharmacol.* **2013**, *86*, 1347–1356. [CrossRef]
88. Denoyer, A.; Godefroy, D.; Célérier, I.; Frugier, J.; Degardin, J.; Harrison, J.K.; Brignole-Baudouin, F.; Picaud, S.; Baleux, F.; Sahel, J.A.; et al. CXCR3 antagonism of SDF-1(5-67) restores trabecular function and prevents retinal neurodegeneration in a rat model of ocular hypertension. *PLoS ONE* **2012**, *7*, e37873. [CrossRef]
89. Yu, N.; Zhang, Z.; Chen, P.; Zhong, Y.; Cai, X.; Hu, H.; Yang, Y.; Zhang, J.; Li, K.; Ge, J.; et al. Tetramethylpyrazine (TMP), an Active Ingredient of Chinese Herb Medicine Chuanxiong, Attenuates the Degeneration of Trabecular Meshwork through SDF-1/CXCR4 Axis. *PLoS ONE* **2015**, *10*, e0133055. [CrossRef]
90. Zhou, Y.; Xia, X.; Yang, E.; Wang, Y.; Marra, K.G.; Ethier, C.R.; Schuman, J.S.; Du, Y. Adipose-derived stem cells integrate into trabecular meshwork with glaucoma treatment potential. *FASEB J.* **2020**, *34*, 7160–7177. [CrossRef]

91. Kaur, I.; Kaur, J.; Sooraj, K.; Goswami, S.; Saxena, R.; Chauhan, V.S.; Sihota, R. Comparative evaluation of the aqueous humor proteome of primary angle closure and primary open angle glaucomas and age-related cataract eyes. *Int. Ophthalmol.* **2019**, *39*, 69–104. [CrossRef]
92. Ramdas, W.D.; van Koolwijk, L.M.; Lemij, H.G.; Pasutto, F.; Cree, A.J.; Thorleifsson, G.; Janssen, S.F.; Jacoline, T.B.; Amin, N.; Rivadeneira, F.; et al. Common genetic variants associated with open-angle glaucoma. *Hum. Mol. Genet.* **2011**, *20*, 2464–2471. [CrossRef]
93. Chai, X.; Low, K.Y.; Tham, Y.C.; Chee, M.L.; Thakur, S.; Zhang, L.; Tan, N.Y.; Khor, C.C.; Aung, T.; Wong, T.Y.; et al. Association of Glaucoma Risk Genes with Retinal Nerve Fiber Layer in a Multi-ethnic Asian Population: The Singapore Epidemiology of Eye Diseases Study. *Investig. Ophthalmol. Vis. Sci.* **2020**, *61*, 37. [CrossRef]
94. Trikha, S.; Saffari, E.; Nongpiur, M.; Baskaran, M.; Ho, H.; Li, Z.; Tan, P.-Y.; Allen, J.; Khor, C.-C.; Perera, S.A.; et al. A Genetic Variant in TGFBR3-CDC7 Is Associated with Visual Field Progression in Primary Open-Angle Glaucoma Patients from Singapore. *Ophthalmology* **2015**, *122*, 2416–2422. [CrossRef]
95. Lazzara, F.; Amato, R.; Platania, C.B.M.; Conti, F.; Chou, T.-H.; Porciatti, V.; Drago, F.; Bucolo, C. 1α,25-dihydroxyvitamin D3 protects retinal ganglion cells in glaucomatous mice. *J. Neuroinflamm.* **2021**, *18*, 206. [CrossRef]
96. Quaranta, L.; Bruttini, C.; Micheletti, E.; Konstas, A.G.P.; Michelessi, M.; Oddone, F.; Katsanos, A.; Sbardella, D.; De Angelis, G.; Riva, I. Glaucoma and neuroinflammation: An overview. *Surv. Ophthalmol.* **2021**, *66*, 693–713. [CrossRef]
97. Fernández-Albarral, J.A.; Ramírez, A.I.; de Hoz, R.; Matamoros, J.A.; Salobrar-García, E.; Elvira-Hurtado, L.; López-Cuenca, I.; Sánchez-Puebla, L.; Salazar, J.J.; Ramírez, J.M. Glaucoma: From pathogenic mechanisms to retinal glial cell response to damage. *Front. Cell Neurosci.* **2024**, *18*, 1354569. [CrossRef]
98. Naguib, S.; Backstrom, J.R.; Gil, M.; Calkins, D.J.; Rex, T.S. Retinal oxidative stress activates the NRF2/ARE pathway: An early endogenous protective response to ocular hypertension. *Redox Biol.* **2021**, *42*, 101883. [CrossRef]
99. Baudouin, C.; Kolko, M.; Melik-Parsadaniantz, S.; Messmer, E.M. Inflammation in Glaucoma: From the back to the front of the eye, and beyond. *Prog. Retin. Eye Res.* **2021**, *83*, 100916. [CrossRef]
100. Chono, I.; Miyazaki, D.; Miyake, H.; Komatsu, N.; Ehara, F.; Nagase, D.; Kawamoto, Y.; Shimizu, Y.; Ideta, R.; Inoue, Y. High interleukin-8 level in aqueous humor is associated with poor prognosis in eyes with open angle glaucoma and neovascular glaucoma. *Sci. Rep.* **2018**, *8*, 14533. [CrossRef]
101. Gramlich, O.W.; Beck, S.; von Thun Und Hohenstein-Blaul, N.; Boehm, N.; Ziegler, A.; Vetter, J.M.; Pfeiffer, N.; Grus, F.H. Enhanced insight into the autoimmune component of glaucoma: IgG autoantibody accumulation and pro-inflammatory conditions in human glaucomatous retina. *PLoS ONE* **2013**, *8*, e57557. [CrossRef] [PubMed]
102. Khalef, N.; Labib, H.; Helmy, H.; El Hamid, M.A.; Moemen, L.; Fahmy, I. Levels of cytokines in the aqueous humor of eyes with primary open angle glaucoma, pseudoexfoliation glaucoma and cataract. *Electron. Physician* **2017**, *9*, 3833–3837. [CrossRef]
103. Tezel, G.; Li, L.Y.; Patil, R.V.; Wax, M.B. TNF-alpha and TNF-alpha receptor-1 in the retina of normal and glaucomatous eyes. *Investig. Ophthalmol. Vis. Sci.* **2001**, *42*, 1787–1794.
104. Williams, P.A.; Tribble, J.R.; Pepper, K.W.; Cross, S.D.; Morgan, B.P.; Morgan, J.E.; John, S.W.M.; Howell, G.R. Inhibition of the classical pathway of the complement cascade prevents early dendritic and synaptic degeneration in glaucoma. *Mol. Neurodegener.* **2016**, *11*, 26. [CrossRef] [PubMed]
105. Reinehr, S.; Reinhard, J.; Gandej, M.; Kuehn, S.; Noristani, R.; Faissner, A.; Dick, H.B.; Joachim, S.C. Simultaneous Complement Response via Lectin Pathway in Retina and Optic Nerve in an Experimental Autoimmune Glaucoma Model. *Front. Cell Neurosci.* **2016**, *10*, 140. [CrossRef]
106. Reinehr, S.; Reinhard, J.; Gandej, M.; Gottschalk, I.; Stute, G.; Faissner, A.; Dick, H.B.; Joachim, S.C. S100B immunization triggers NFκB and complement activation in an autoimmune glaucoma model. *Sci. Rep.* **2018**, *8*, 9821. [CrossRef]
107. Amato, R.; Cammalleri, M.; Melecchi, A.; Bagnoli, P.; Porciatti, V. Natural History of Glaucoma Progression in the DBA/2J Model: Early Contribution of Müller Cell Gliosis. *Cells* **2023**, *12*, 1272. [CrossRef]
108. Wang, X.; Wang, M.; Liu, H.; Mercieca, K.; Prinz, J.; Feng, Y.; Prokosch, V. The Association between Vascular Abnormalities and Glaucoma-What Comes First? *Int. J. Mol. Sci.* **2023**, *24*, 13211. [CrossRef]
109. Liberatore, F.; Bucci, D.; Mascio, G.; Madonna, M.; Di Pietro, P.; Beneventano, M.; Puliti, A.M.; Battaglia, G.; Bruno, V.; Nicoletti, F.; et al. Permissive role for mGlu1 metabotropic glutamate receptors in excitotoxic retinal degeneration. *Neuroscience* **2017**, *363*, 142–149. [CrossRef]
110. Ji, M.; Miao, Y.; Dong, L.-D.; Chen, J.; Mo, X.-F.; Jiang, S.-X.; Sun, X.-H.; Yang, X.-L.; Wang, Z. Group I mGluR-mediated inhibition of Kir channels contributes to retinal Müller cell gliosis in a rat chronic ocular hypertension model. *J. Neurosci.* **2012**, *32*, 12744–12755. [CrossRef]
111. Miao, Y.; Zhao, G.-L.; Cheng, S.; Wang, Z.; Yang, X.-L. Activation of retinal glial cells contributes to the degeneration of ganglion cells in experimental glaucoma. *Prog. Retin. Eye Res.* **2023**, *93*, 101169. [CrossRef] [PubMed]
112. Lambuk, L.; Lazaldin, M.A.M.; Ahmad, S.; Iezhitsa, I.; Agarwal, R.; Uskoković, V.; Mohamud, R. Brain-Derived Neurotrophic Factor-Mediated Neuroprotection in Glaucoma: A Review of Current State of the Art. *Front. Pharmacol.* **2022**, *13*, 875662. [CrossRef] [PubMed]
113. Pacwa, A.; Machowicz, J.; Akhtar, S.; Rodak, P.; Liu, X.; Pietrucha-Dutczak, M.; Lewin-Kowalik, J.; Amadio, M.; Smedowski, A. Deficiency of the RNA-binding protein ELAVL1/HuR leads to the failure of endogenous and exogenous neuroprotection of retinal ganglion cells. *Front. Cell Neurosci.* **2023**, *17*, 1131356. [CrossRef]

114. Burgaletto, C.; Platania, C.B.M.; Di Benedetto, G.; Munafò, A.; Giurdanella, G.; Federico, C.; Caltabiano, R.; Saccone, S.; Conti, F.; Bernardini, R.; et al. Targeting the miRNA-155/TNFSF10 network restrains inflammatory response in the retina in a mouse model of Alzheimer's disease. *Cell Death Dis.* **2021**, *12*, 905. [CrossRef] [PubMed]
115. Agarwal, R.; Agarwal, P. Newer targets for modulation of intraocular pressure: Focus on adenosine receptor signaling pathways. *Expert. Opin. Ther. Targets* **2014**, *18*, 527–539. [CrossRef]
116. Aires, I.D.; Boia, R.; Rodrigues-Neves, A.C.; Madeira, M.H.; Marques, C.; Ambrósio, A.F.; Santiago, A.R. Blockade of microglial adenosine A2A receptor suppresses elevated pressure-induced inflammation, oxidative stress, and cell death in retinal cells. *Glia* **2019**, *67*, 896–914. [CrossRef]
117. Li, G.; Torrejon, K.Y.; Unser, A.M.; Ahmed, F.; Navarro, I.D.; Baumgartner, R.A.; Albers, D.S.; Stamer, W.D. Trabodenoson, an Adenosine Mimetic With A1 Receptor Selectivity Lowers Intraocular Pressure by Increasing Conventional Outflow Facility in Mice. *Investig. Ophthalmol. Vis. Sci.* **2018**, *59*, 383–392. [CrossRef]
118. Conti, F.; Lazzara, F.; Romano, G.L.; Platania, C.B.M.; Drago, F.; Bucolo, C. Caffeine Protects Against Retinal Inflammation. *Front. Pharmacol.* **2021**, *12*, 824885. [CrossRef]
119. Bucolo, C.; Campana, G.; Di Toro, R.; Cacciaguerra, S.; Spampinato, S. Sigma1 recognition sites in rabbit iris-ciliary body: Topical sigma1-site agonists lower intraocular pressure. *J. Pharmacol. Exp. Ther.* **1999**, *289*, 1362–1369.
120. Hopkins, A.L. Network pharmacology: The next paradigm in drug discovery. *Nat. Chem. Biol.* **2008**, *4*, 682–690. [CrossRef]
121. Platania, C.B.M.; Drago, F.; Bucolo, C. The P2X7 receptor as a new pharmacological target for retinal diseases. *Biochem. Pharmacol.* **2022**, *198*, 114942. [CrossRef] [PubMed]
122. Romano, G.L.; Amato, R.; Lazzara, F.; Porciatti, V.; Chou, T.-H.; Drago, F.; Bucolo, C. P2X7 receptor antagonism preserves retinal ganglion cells in glaucomatous mice. *Biochem. Pharmacol.* **2020**, *180*, 114199. [CrossRef] [PubMed]
123. Smyth, G.K. Linear models and empirical bayes methods for assessing differential expression in microarray experiments. *Stat. Appl. Genet. Mol. Biol.* **2004**, *3*, 3. [CrossRef] [PubMed]
124. Benjamini, Y.; Hochberg, Y. Controlling the False Discovery Rate: A Practical and Powerful Approach to Multiple. *J. R. Stat. Society. Ser. B (Methodol.)* **1995**, *57*, 289–300. [CrossRef]
125. Szklarczyk, D.; Franceschini, A.; Wyder, S.; Forslund, K.; Heller, D.; Huerta-Cepas, J.; Simonovic, M.; Roth, A.; Santos, A.; Tsafou, K.P.; et al. STRING v10: Protein-protein interaction networks, integrated over the tree of life. *Nucleic Acids Res.* **2015**, *43*, D447–D452. [CrossRef]
126. Koschützki, D.; Schreiber, F. Centrality analysis methods for biological networks and their application to gene regulatory networks. *Gene Regul. Syst. Biol.* **2008**, *2*, 193–201. [CrossRef]
127. Chang, L.; Zhou, G.; Soufan, O.; Xia, J. miRNet 2.0: Network-based visual analytics for miRNA functional analysis and systems biology. *Nucleic Acids Res.* **2020**, *48*, W244–W251. [CrossRef]
128. Kuhn, M.; Szklarczyk, D.; Pletscher-Frankild, S.; Blicher, T.H.; von Mering, C.; Jensen, L.J.; Bork, P. STITCH 4: Integration of protein-chemical interactions with user data. *Nucleic Acids Res.* **2014**, *42*, D401–D407. [CrossRef] [PubMed]
129. Vlachos, I.S.; Kostoulas, N.; Vergoulis, T.; Georgakilas, G.; Reczko, M.; Maragkakis, M.; Paraskevopoulou, M.D.; Prionidis, K.; Dalamagas, T.; Hatzigeorgiou, A.G. DIANA miRPath v.2.0: Investigating the combinatorial effect of microRNAs in pathways. *Nucleic Acids Res.* **2012**, *40*, W498–W504. [CrossRef]
130. Safran, M.; Dalah, I.; Alexander, J.; Rosen, N.; Stein, T.I.; Shmoish, M.; Nativ, N.; Bahir, I.; Doniger, T.; Krug, H.; et al. GeneCards Version 3: The human gene integrator. *Database* **2010**, *2010*, baq020. [CrossRef]

pharmaceuticals MDPI

Article

Neuroprotective Actions of Hydrogen Sulfide-Releasing Compounds in Isolated Bovine Retinae

Leah Bush [1], Jenaye Robinson [1], Anthonia Okolie [1], Fatima Muili [1], Catherine A. Opere [2], Matthew Whiteman [3], Sunny E. Ohia [1] and Ya Fatou Njie Mbye [1,*]

1 Department of Pharmaceutical Sciences, College of Pharmacy and Health Sciences, Texas Southern University, Houston, TX 77004, USA; lemwjb@gmail.com (L.B.); jenayerobinson@yahoo.com (J.R.); a.okolie9552@student.tsu.edu (A.O.); f.muili1867@student.tsu.edu (F.M.); sunny.ohia@tsu.edu (S.E.O.)

2 Department of Pharmacy Sciences, School of Pharmacy and Health Professions, Creighton University, Omaha, NE 68178, USA; catherineopere@creighton.edu

3 University of Exeter Medical School, St. Luke's Campus, Magdalen Road, Exeter EX1 2LU, UK; m.whiteman@exeter.ac.uk

* Correspondence: yafatou.mbye@tsu.edu

Abstract: Background: We have evidence that hydrogen sulfide (H_2S)-releasing compounds can reduce intraocular pressure in normotensive and glaucomatous rabbits by increasing the aqueous humor (AH) outflow through the trabecular meshwork. Since H_2S has been reported to possess neuroprotective actions, the prevention of retinal ganglion cell loss is an important strategy in the pharmacotherapy of glaucoma. Consequently, the present study aimed to investigate the neuroprotective actions of H_2S-releasing compounds against hydrogen peroxide (H_2O_2)-induced oxidative stress in an isolated bovine retina. **Materials and Methods:** The isolated neural retinae were pretreated with a substrate for H_2S biosynthesis called L-cysteine, with the fast H_2S-releasing compound sodium hydrosulfide, and with a mitochondrial-targeting H_2S-releasing compound, AP123, for thirty minutes before a 30-min oxidative insult with H_2O_2 (100 μM). Lipid peroxidation was assessed via an enzyme immunoassay by measuring the stable oxidative stress marker, 8-epi PGF2α (8-isoprostane), levels in the retinal tissues. To determine the role of endogenous H_2S, studies were performed using the following biosynthesis enzyme inhibitors: aminooxyacetic acid (AOAA, 30 μM); a cystathione-β-synthase/cystathionine-γ-lyase (CBS/CSE) inhibitor, α–ketobutyric acid (KBA, 1 mM); and a 3-mercaptopyruvate-s-sulfurtransferase (3-MST) inhibitor, in the absence and presence of H_2S-releasing compounds. **Results:** Exposure of the isolated retinas to H_2O_2 produced a time-dependent (10–40 min) and concentration-dependent (30–300 μM) increase in the 8-isoprostane levels when compared to the untreated tissues. L-cysteine (10 nM–1 μM) and NaHS (30–100 μM) significantly ($p < 0.001$; n = 12) prevented H_2O_2-induced oxidative damage in a concentration-dependent manner. Furthermore, AP123 (100 nM–1 μM) attenuated oxidative H_2O_2 damage resulted in an approximated 60% reduction in 8-isoprostane levels compared to the tissues treated with H_2O_2 alone. While AOAA (30 μM) and KBA (1 mM) did not affect the L-cysteine evoked attenuation of H_2O_2-induced oxidative stress, KBA reversed the antioxidant responses caused by AP123. **Conclusions:** In conclusion, various forms of H_2S-releasing compounds and the substrate, L-cysteine, can prevent H_2O_2-induced lipid peroxidation in an isolated bovine retina.

Keywords: hydrogen sulfide; retina; oxidative stress; lipid peroxidation

1. Introduction

Lipid peroxidation-induced damage has been implicated in the pathogenesis of a variety of neurodegenerative diseases [1,2]. Evidence from the literature indicates that oxidative stress (OS) also plays a major role in ocular pathologies such as cataracts, age-related macular degeneration (ARMD), glaucoma, and diabetic retinopathy (DR) [3–6]. Under normal physiological conditions, the presence of several intrinsic antioxidant enzymes in the eye prevents OS formation as a consequence of normal cellular metabolism [2,7,8]. Hydrogen peroxide (H_2O_2) is a biologically derived oxidant intermediate which acts with other peroxide-induced free radicals to inflict damage on ocular tissues by changing cellular structure and function [9]. Because of its high content of polyunsaturated fatty acids, the retina is highly susceptible to damage via free radical oxidation [2,10]. Products of the oxygen-derived free radical pathways, including peroxides and long-chain polyunsaturated fatty acid (LCPUFA) metabolites such as isoprostanes, contribute to oxidative reactions in the eye and are important in the pathophysiology of most ocular diseases [2,11–14]. Indeed, isoprostanes are involved in the pharmacological actions of H_2O_2 in ocular tissues and have been shown to regulate sympathetic and excitatory amino acid neurotransmission in the eye [2,15]. The use of pharmacological agents to prevent OS could be an effective strategy in protecting ocular tissues such as the retina from oxidative damage.

Hydrogen sulfide (H_2S) is a colorless gas that has long been recognized as an environmental pollutant and a highly toxic gas [16]. Since the discovery of its basal production in mammalian tissues about two decades ago, H_2S has assumed the role of a gaseous mediator with biological importance in the central and peripheral nervous systems [17–21]. This novel gaseous transmitter, H_2S, is endogenously produced in mammalian tissues from the amino acids L-cysteine and homocysteine in the presence of two pyridoxal-5-phosphate-dependant enzymes, cystathionine b-synthase (CBS), and cystathionine γ-lyase (CSE). 3-mercaptopyruvate sulfurtransferase (3-MST), has been reported to be involved in the calcium-dependent production of H_2S [22–24]. In recent years, d-amino oxidase (DAO) has been identified as the fourth enzyme involved in the biosynthesis of H_2S, as it is responsible for the conversion of d-cysteine into H_2S through the 3-MST/CAT pathway [25,26]. As research into the biological activities of H_2S evolved, the sources of H_2S expanded from the creation of fast- and slow-releasing H_2S compounds such as sodium hydrosulfide (NaHS) and GYY4137, respectively, to the development of H_2S donors targeted at the mitochondrial sources of the gas such as AP123 and AP39 [27–29]. Indeed, H_2S has been reported to be involved in several physiological and pathophysiological processes such as learning and memory, cell survival, inflammation, modulation of synaptic activities in the central nervous system, and maintenance of vascular tone in the cardiovascular system [20,30–34]. Because of the localization of the enzymes responsible for the biosynthesis of H_2S in cells and tissues, the endogenously produced gas has been shown to mediate biological processes [22,35,36]. In the eye, H_2S is present in the retina [37] and has been shown to exert pharmacological actions such as the inhibition of excitatory amino acid neurotransmission [38] and increases in cyclic AMP formation [39,40]. The presence of H_2S in the retina and its transsulfuration pathways indicates a role for this gas in several physiological processes that affect cellular signaling and redox homeostasis [41].

As a retinal neurodegenerative disease, current strategies for the treatment of glaucoma include using agents that can provide neuroprotection for the ganglion cell layer [42]. The gaseous neurotransmitter, H_2S, has been reported to play a role in several neurodegenerative diseases such as Parkinson's disease, Alzheimer's disease, Amyotrophic Lateral Sclerosis, and Down Syndrome [43]. There is evidence that H_2S can prevent ischemia/reperfusion injury and promote retinal glial cell survival indicating a possible role for this gas in glaucoma-related damage [44]. While there is evidence that H_2S is produced in the eye and can protect the retina from light- and NMDA-induced neurodegenerative damage [22,45], the effect of this gas on LCPUFA metabolites and lipid peroxidation in the retina is unknown. The aim of the present study was, therefore, to investigate the pharmacological actions of H_2S derived from several sources (a substrate for enzymatic H_2S biosynthesis, L-cysteine; a fast-releasing H_2S compound, NaHS; and a mitochondrial-targeting H_2S compound, AP123) against oxidative damage induced by H_2O_2 in an isolated bovine neural retina.

2. Results

2.1. Effect of H_2O_2 on 8-Isoprostane Production in the Retina

Isoprostanes are stable compounds whose production increases with exposure to OS, thus making them dependable markers of oxidative damage in both in vivo and in vitro animal models [46,47]. In a series of experiments, we studied the effects of time and different concentrations of H_2O_2 on the production of 8-isoprostane in an isolated bovine neural retina. As illustrated in Figure 1, 10–40 min insults with H_2O_2 (30–300 μM) elicited a concentration- and time-dependent increase in the 8-isoprostane levels (Figure 1). There was a time-dependent increase in the basal 8-isoprostane concentrations, reaching a maximum at 40 min. The effects elicited by H_2O_2 (100 μM) also showed a time-dependent increase reaching a maximum at 40 min. As a result of the data obtained in the studies described in Figure 1, all subsequent experiments assessing the pharmacological actions of H_2S-producing compounds against H_2O_2-induced damage on isolated bovine retinae were performed in the presence of H_2O_2 (100 μM) for 30 min.

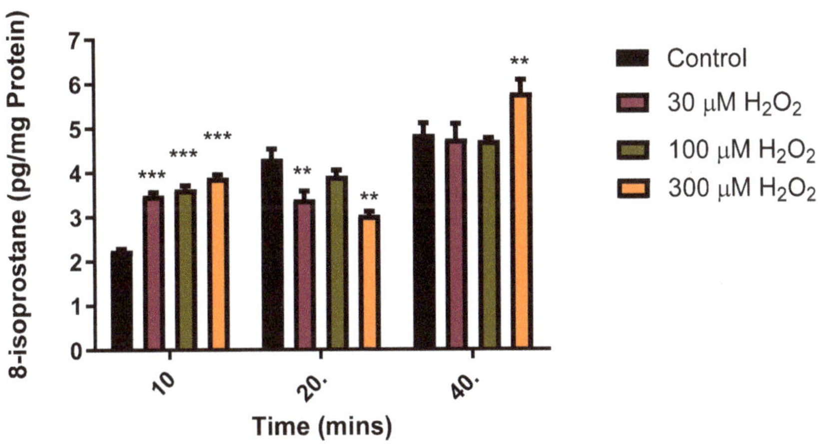

Figure 1. Time- and concentration-dependent effect of H_2O_2 on 8-isoprostane production in the bovine retinae. Each value represents the mean ± SEM for n = 12; *** $p < 0.001$ is significantly different from the control and ** $p < 0.01$ is significantly different from the control.

2.2. Effects of H_2S-Releasing Compounds on Lipid Peroxidation in the Bovine Retina

Although evidence in the literature supports a protective role for H_2S against OS in neurons in vivo [46,48], the neuroprotective action of this gas in ocular tissues under conditions of OS induced by H_2O_2 in vitro warrants investigation. To assess the pharmacological role of endogenously produced H_2S in the presence of the substrate, L-cysteine, isolated retinal tissues were pretreated with L-cysteine (10 nM–1 μM) for 30 min before an insult with H_2O_2 (100 μM). L-cysteine (10 nM) significantly ($p < 0.001$; n = 12) reversed the H_2O_2-induced increase in the 8-isoprostane levels, producing a 43% reduction in 8-isoprostane generation compared to the control tissues in the presence of H_2O_2. (Figure 2). In contrast, higher concentrations of L-cysteine (100 nM–1 μM) do not affect H_2O_2-induced damage.

We then examined the effects of a fast-releasing H_2S compound, NaHS, on 8-isoprostane production in the retina under conditions of H_2O_2-induced OS. Treatment of the tissues with NaHS (10–30 μM) for 10 min before the administration of H_2O_2 (100 μM) for 30 min significantly ($p < 0.001$; n = 12) attenuated 8-isoprostane production, whereas a high concentration of NaHS (100 μM) significantly ($p < 0.001$; n = 12) augmented the retinal 8-isoprostane levels, approximately 12% over the control (H_2O_2 treated tissues) (Figure 3).

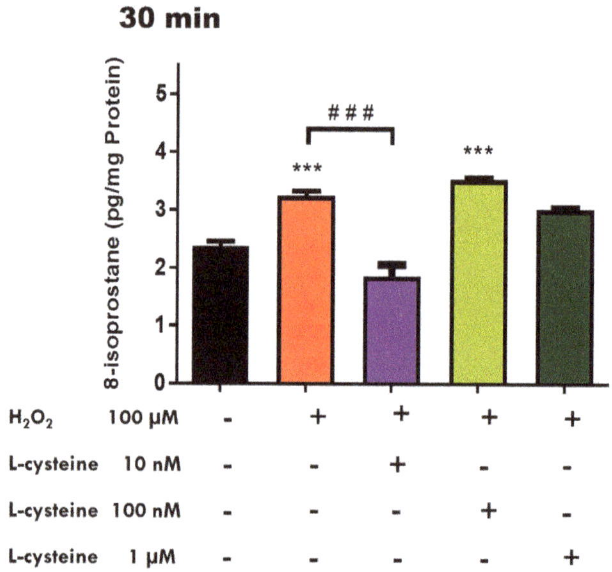

Figure 2. Concentration-dependent effect of L-cysteine on H_2O_2-induced oxidative stress in the bovine retinae. Each value represents the mean ± SEM for n = 12; *** $p < 0.001$ is significantly different from the control and ### $p < 0.001$ is significantly different from the H_2O_2 treated tissues.

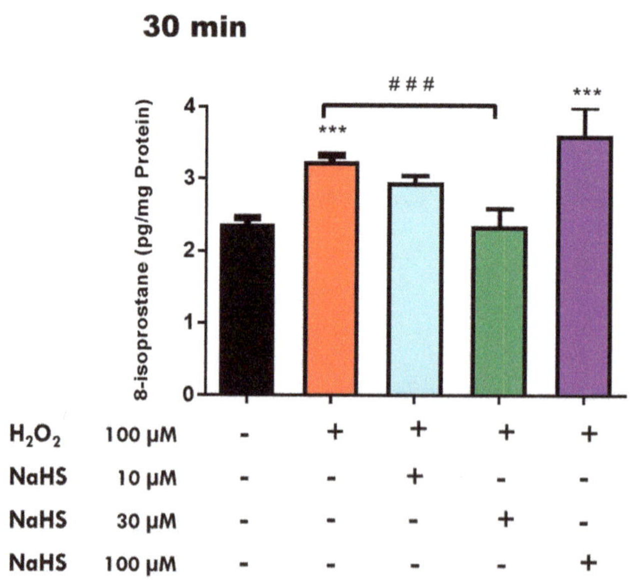

Figure 3. Concentration-dependent Effect of NaHS on H_2O_2–induced oxidative stress in the bovine retinae. Each value represents the mean ± SEM for n = 12; *** $p < 0.001$ is significantly different from the control and ### $p < 0.001$ is significantly different from the H_2O_2 treated tissues.

To examine the pharmacological action of the novel mitochondrial-targeting H_2S-releasing compound, AP123, retinal tissues were exposed to varying concentrations of AP123 (10 nM–1 µM) before exposure to H_2O_2 (100 µM). After 30 min of treatment, AP123 (10 nM–1 µM) attenuated the H_2O_2-induced increase in 8-isoprostane production and even reduced the 8-isoprostane levels in the bovine retinae by 60% at higher concentrations (100 nM–1 µM) compared to the control tissues in the presence of H_2O_2. A maximal inhibitory effect on isoprostane production was observed with 100 nM of AP123 (Figure 4).

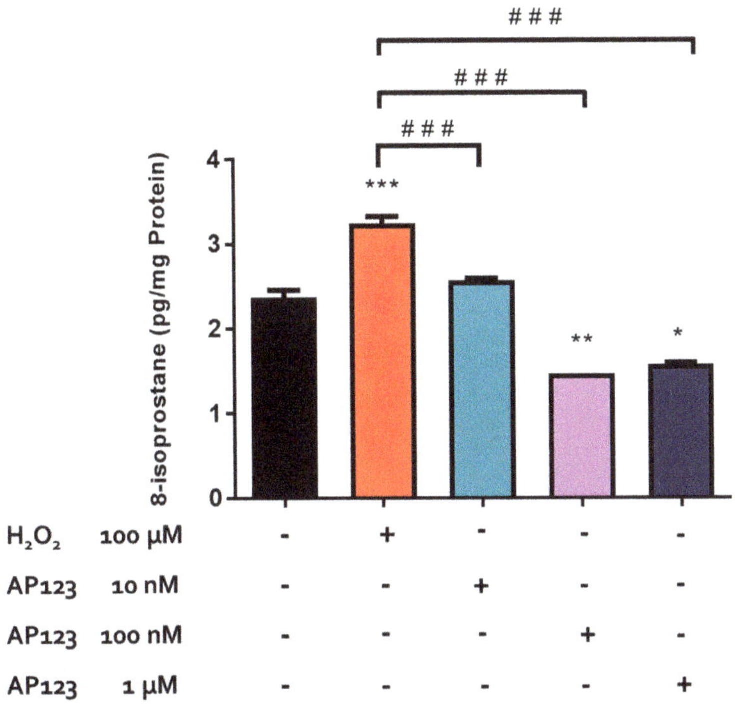

Figure 4. Concentration-dependent effect of AP123 on H_2O_2-induced oxidative stress in the bovine retinae. Each value represents the mean ± SEM for n = 12; *** $p < 0.001$ is significantly different from the control, ** $p < 0.01$ is significantly different from the control, * $p < 0.05$ is significantly different from the control, and ### $p < 0.001$ is significantly different from the H_2O_2 treated tissues.

2.3. *Effect of Inhibitors of CBS/CSE and 3-MST on the Antioxidant Actions of H_2S-Producing Compounds*

Endogenously produced H_2S has been shown to protect retinal photoreceptors against light-induced oxidative damage [22]. In the present study, we investigated the role of endogenously produced H_2S in the neuroprotective action of L-cysteine against H_2O_2-induced lipid peroxidation in isolated bovine neural retina. In a series of experiments, we tested the effect of an inhibitor of both CBS and CSE, amino-oxyacetic acid [AOAA, (30 µM)], and an inhibitor of 3-MST, ketobutyric acid [49] (KBA, 1 mM), on the pharmacological effect elicited by L-cysteine. Both AOAA (30 µM) and KBA (1 mM) did not affect the basal 8-isoprostane levels and had no significant effect on the reversal of H_2O_2-induced 8-isoprostane production caused by L-cysteine (100 µM) (Figure 5). The evidence in the literature suggests that AP123, the mitochondrial H_2S donor, can exert antioxidant effects in the microvascular endothelial cells [27]. In the present study, we evaluated the role of endogenous H_2S production through the 3-MST/H_2S pathway in the antioxidant activity of AP123 in an isolated bovine retina under the conditions of H_2O_2-induced OS. On its own, KBA (1 mM) did not affect basal 8-isoprostane production. In the presence of KBA (1 mM), the inhibitory actions of AP123 on 8-isoprostane generation were significantly ($p < 0.001$; n = 12) reversed (Figure 6).

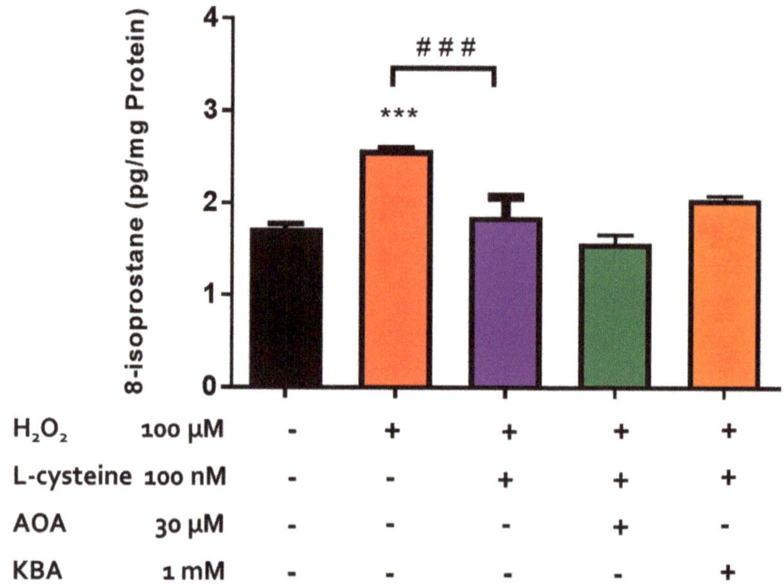

Figure 5. Role of the 3-MST/CAT pathway in AP123-mediated neuroprotection. Each value represents the mean ± SEM for n = 12; *** $p < 0.001$ is significantly different from the control and ### $p < 0.001$ is significantly different from the H_2O_2 treated tissues.

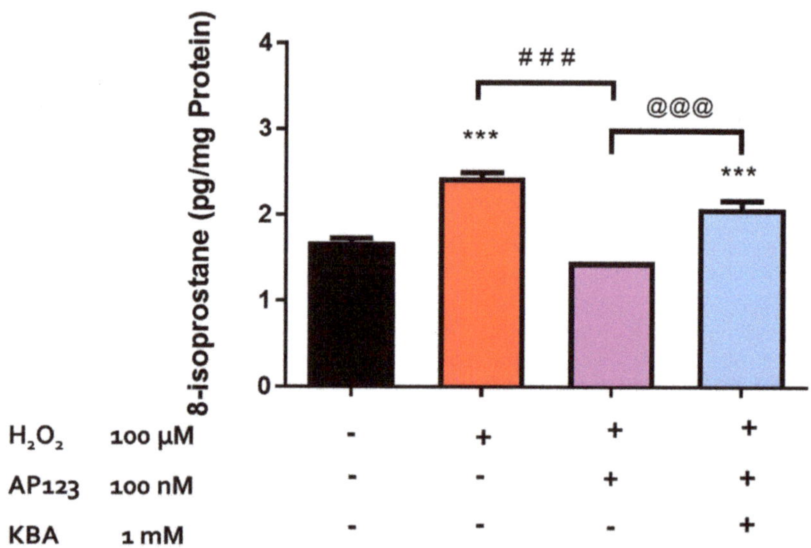

Figure 6. Role of CBS/CSE and the 3-MST/CAT pathway in L-cysteine-mediated neuroprotection. Each value represents the mean ± SEM for n = 12; *** $p < 0.001$ is significantly different from the control, ### $p < 0.001$ is significantly different from the H_2O_2 treated tissues, and @@@ $p < 0.001$ is significantly different from the AP123 + H_2O_2 treated tissues.

3. Discussion

ROS are formed during normal cellular metabolism, and there is evidence that they play an important, yet complex role in biological systems [1,2]. In excess, ROS can react with

DNA, proteins, and lipids to form pharmacologically active metabolites that can perpetuate pathophysiological conditions [1,2]. Isoprostanes are chemically stable, prostaglandin-like lipid peroxidation products that are endogenously formed from oxidative damage to LCPUFAs [50]. These stable molecules are utilized as markers of lipid peroxidation in mammalian tissues, as their levels increase with oxidative damage [51,52]. While the evidence available supports the use of several markers to validate oxidative stress in biological samples [53,54], we selected 8-isoprostanes because of their direct relevance to the measurement of oxidative stress in retinal tissues [55–57]. In the present study, we found that various concentrations of H_2O_2 elicited a time-dependent increase in 8-isoprostane production in isolated bovine retinal tissues. The ability of exogenously administered H_2O_2 to cause a measurable increase in 8-isoprostane production supports the fact that lipid peroxidation occurs in response to this peroxide. The exposure of the tissues to increasing concentrations of H_2O_2 (30–300 μM) caused a time- and concentration-related increase in H_2O_2-induced lipid peroxidation, reaching a maximum at 300 μM. For this reason, an insult of 30 min duration with a submaximal concentration of H_2O_2 (100 μM) was selected as oxidative stimuli for all the experiments involving OS. It is pertinent to note that the concentration of H_2O_2 employed for inducing oxidative stress in retinal cells in the present study (100 μM) is much lower than that reported by other investigators such as Cui et al. [58] (200–800 μM), Hu et al. [59] (300–400 μM), and Wang et al. [60] (300 μM).

The biological actions of H_2S have been documented in various mammalian tissues and systems [61]. This gaseous neurotransmitter is involved in a variety of pathophysiological processes such as inflammation, memory and learning, and in the regulation of blood pressure [16,62]. In the cardiovascular system, H_2S has been shown to play a vital role in the maintenance of vascular smooth muscle tone, and in the central nervous system, this gas has been found to act as a neurotransmitter at synapses [63–65]. Cytoprotection is one of the many roles of this gas, as H_2S has been reported to exert a neuroprotective action on neurons [48]. In 2011, Biermann and co-workers demonstrated that the exposure of animals to inhalational H_2S decreased the apoptosis of retinal ganglion cells in a rat model of ischemia/reperfusion injury, suggesting a role for this gas as a neuroprotectant [66]. Because of the inherent challenges involved in the use of H_2S gas in biological studies, the kinetic profiles of the release of this gas from some organic and inorganic compounds were extensively studied [28,67]. For instance, NaHS was reported to release H_2S gas rapidly in biological media, whereas GYY4137 is a slow-releasing H_2S compound [28]. There is evidence that H_2S-releasing compounds such as NaHS can inhibit both light-induced degeneration and NMDA-induced oxidative damage in the retina, supporting a neuroprotective role for H_2S [22,41]. H_2S-releasing compounds have also been reported to inhibit excitatory amino acid neurotransmission (a marker of excitotoxicity) in an isolated bovine retina [38,68,69] and in H_2O_2-induced OS and cataract formation in cultured bovine lenses [70].

In the present study, the amino acid substrate for H_2S production, L-cysteine, the fast-releasing H_2S compound, NaHS, and the mitochondrial-targeted H_2S compound, AP123, all protected the neural retina against H_2O_2-induced lipid peroxidation. Pretreatment with low concentrations of L-cysteine for 30 min attenuated lipid peroxidation whereas higher concentrations of this amino acid had no such action. The amino acid L-cysteine acts as a substrate for the enzymatic production of H_2S and has been found to play a developmental and cytoprotective role in the CNS [71,72]. L-cysteine has been reported to promote the proliferation and differentiation of neural stem cells via the H_2S/CBS pathway [72] and can protect against cellular damage in brain tissue [71]. Indeed, L-cysteine was found to upregulate the expression and activity of CBS and reduce brain edema in rats following a subarachnoid hemorrhage [71]. Taken together, our data suggest that the endogenous production of H_2S from the substrate L-cysteine can protect the neural retina against H_2O_2-induced oxidative damage.

As observed with L-cysteine, pretreatment of tissues with low concentrations of NaHS for 30 min prevented H_2O_2-induced lipid peroxidation whereas a higher concentration of this fast-releasing H_2S compound had no effect on oxidative damage. Interestingly, the pretreatment of tissues with a higher concentration of NaHS significantly increased ($p < 0.001$; n = 12) the levels of this lipid metabolite by 44% when compared with the control tissues. Both NaHS and N-acetyl-cysteine (NAC), a derivative of L-cysteine, have both been shown to exert concentration-related pharmacological actions that can be either protective or deleterious in biological systems [62,73,74]. Depending on the timing and duration of the H_2S exposure prior to an oxidative insult as well as the concentration of the H_2S-releasing compound utilized, the pharmacological actions of this gas can vary from cytoprotective and anti-inflammatory [62,71,74] to apoptotic and pro-inflammatory [62, 73,74]. NaHS, an inorganic sulfide salt that instantaneously releases H_2S in an aqueous solution, has been found to protect tissues from oxidative damage [45,66,75]. In a mouse model for diabetes, NaHS (100 μmol/kg) attenuated the production of ROS via the nicotine adenine dinucleotide phosphate (NADPH) oxidase and restored endothelial function in the aorta of diabetic mice [45]. The cortical neurons were protected against glutamate-induced excitotoxicity by NaHS (100 μM) via an observed increase in the activity of glutamate cysteine ligase, an enzyme involved in the production of the antioxidant GSH, and the augmentation of the intracellular GSH levels [75]. It is tempting to speculate that the observed protective effects of NaHS in the present study could be attributed to the stimulatory effect of NaHS on glutamate cysteine ligase and GSH production; however, further studies are needed to confirm this theory.

In the present study, the treatment of tissues with L-cysteine (10 nM) elicited a 43% reduction in the 8-isoprostane levels in the presence of H_2O_2, making it the most potent compound tested. The high potency of L-cysteine compared to the other H_2S-releasing compounds has been observed in its pharmacological actions in ocular tissues. In a 2017 study, L-cysteine (100 nM) was found to increase the aqueous humor outflow facility by 150%, whereas a much higher dose of NaHS (10 μM) was required to exert a similar effect [76]. Similarly, Ohia and colleagues observed that L-cysteine elicited the most potent relaxation of carbachol-induced tone in porcine isolated irides when compared to the H_2S-releasing compounds such as NaHS and Na_2S [77].

AP123, a novel mitochondrial-targeting H_2S-releasing compound, has been found to play a cytoprotective role in the cardiovascular system [27]. Indeed, AP123 (10 nM to 3 μM) was reported to stabilize mitochondrial membrane potential and significantly reduce the generation of ROS in hyperglycemic vascular endothelial cells [27]. In the present study, we found that the neuroprotective actions of AP123 in bovine retinae were observed in a similar concentration range as those demonstrated by Gero et al. [27]. Interestingly, AP123 not only prevented the H_2O_2-induced lipid peroxidation but it also elicited a further decrease in the basal isoprostane levels at higher concentrations. Through its actions as an electron donor in the electron transport chain, H_2S has been found to stabilize membrane potential, thereby inhibiting the production of mitochondrial ROS in the vascular endothelium [78]. It is tempting to speculate that the observed marked effects of AP123 on lipid peroxidation in the present study may be related to its actions in H_2S biosynthesis at the mitochondrial level.

In another series of experiments, we studied the role of endogenously produced H_2S in the pharmacological action of L-cysteine against lipid peroxidation. The inhibitors of the enzymes responsible for the biosynthesis of H_2S, (CBS, CSE, and 3-MST) were ineffective at diminishing the inhibitory effect elicited by L-cysteine on isoprostane production. Taken together, our data suggest that the protective actions of this amino acid in an isolated bovine retina are independent of its de novo enzymatic conversion to H_2S. It is important to note that L-cysteine is also the rate-limiting agent in the production of glutathione (GSH), which possesses a cytoprotective function in various cell types due to its ROS-scavenging capabilities [8,79]. It may well be that the high potency displayed by L-cysteine in our studies could be related to its additional ability to regulate the production of GSH, a potent antioxidant.

3-MST is an enzyme that is constitutively expressed in mitochondria, and the production of H_2S through this pathway is neuroprotective in the brain and retina [22,24]. In the present study, inhibition of 3-MST with KBA abolished the pharmacological actions of AP123, suggesting that the endogenous biosynthesis of H_2S via the 3-MST pathway is involved in the pharmacological action of this H_2S-releasing compound in an isolated bovine retina. Since AP123 is a mitochondrial-targeting H_2S-releasing compound and 3-MST is localized to the mitochondria, it appears that the direct delivery of H_2S to the mitochondria may account, at least in part, for the ability of KBA to effectively prevent the H_2O_2-induced lipid peroxidation. In summary, the data obtained using AP123 highlight a role for the mitochondria in serving as an important source of H_2S in the observed protective action exhibited by some H_2S-releasing compounds.

H_2S-releasing compounds have been shown to reduce intraocular pressure in normotensive rabbits [80] and to increase the outflow facility in porcine ocular anterior segments, ex vivo [76], suggesting a potential role for this gas in glaucoma pharmacotherapy [81]. The current observation that H_2S-releasing compounds can prevent oxidative damage in the retina supports a dual role for this gas in altering aqueous humor dynamics and serving as a neuroprotectant for the eye.

4. Materials and Methods

4.1. Chemicals

The H_2O_2, L-cysteine, sodium hydrosulfide (NaHS), amino-oxyacetic acid (AOAA), and α-ketobutyric acid (KBA) were purchased from Sigma Chemical (St. Louis, MO, USA). GYY4137, the 8-Isoprostane enzyme-linked immunoassay kit, and the protein determination assay kit were purchased from Cayman Chemical (Ann Harbor, MI, USA). AP123 was a gift from Dr. Matthew Whiteman's laboratory. All test agents were freshly prepared immediately before use in the series of experiments. Stock solutions of AOAA were prepared in 70% ethanol. All other stock solutions were prepared in deionized water.

4.2. Tissue Preparation

Studies were performed using bovine eyes procured from a local slaughterhouse and transported to the laboratory in an ice bath. A cut was made along the equator of each eye, and the vitreous humor and lens were delicately dissected out. The neural retina was isolated by a gentle dislocation from the posterior segment of the eye and immediately immersed in a warm, oxygenated Krebs buffer solution (pH 7.4). After the isolation of the neural retina, the tissues were cut into four pieces and randomly assigned to treatment groups. The control groups received neither H_2O_2 nor H_2S-releasing compounds/enzyme inhibitors. The time elapsed between the animal sacrifice and tissue preparation in the present study was less than 24 h; a protocol consistent with observations made by Murali et al. [82] who reported that human retinal tissues remained resilient to cell death, with a substantial number of cells remaining viable for up to five days postmortem.

4.3. 8-Isoprostane ELISA Assay

The methodology used for the extraction of 8-isoprostane was essentially the same as described by our laboratory and other investigators [83–86] with some modifications. The isolated bovine retinae were equilibrated in an oxygenated Krebs solution at 37 °C for 20 min. Tissues were then transferred and incubated in a Krebs solution in the presence and absence of H_2O_2 for 10 to 40 min. To determine the pharmacological effect of H_2S-releasing compounds on H_2O_2-induced oxidative stress, the retinal tissues were treated with test compounds for 10–30 min before being subjected to a 30 min exposure to H_2O_2. To determine the role of enzymes in the biosynthesis of H_2S, the retinae were treated with the CBS and CSE inhibitor (AOAA (30 μM)) and the noncompetitive 3-MST inhibitor, α-ketobutyric acid, (KBA (1 mM) to determine the involvement of the enzymes responsible for H_2S biosynthesis in AP123- and L-cysteine-mediated actions. After incubation, tissues were homogenized in a 0.1 M phosphate buffer (pH 7.4) containing 1 mM of EDTA and 0.005% BHT (1 mL/100 mg tissue) and centrifuged at 3000 rpm for ten minutes at 5 °C. The supernatant was collected and purified using potassium hydroxide (15% *w*/*v*). The 8-isoprotane was extracted from purified samples using solid phase extraction cartridges and an ethyl acetate–methanol (99:1) mixture [87]. The 8-isoprotane was concentrated into a pellet by evaporating the ethyl acetate–methanol solution under N_2 gas. The enzyme immunoassay buffer was used to re-suspend the 8-isoprotane prior to plating. Protein content was determined from the unpurified supernatant using a Cayman protein determination kit.

4.4. Data Analysis

The results were expressed as 8-isoprotane concentrations per milligram of soluble protein (pg/mg protein). Except where indicated, the values given are means ± S.E.M. The significance of differences between the values obtained in the control and drug-treated preparations were evaluated using one-way ANOVA followed by Tukey's post hoc analysis. The time- and concentration-dependent effects were determined using two-way ANOVA. The differences in p values < 0.05 were accepted as statistically significant.

5. Conclusions

In conclusion, the substrate for the biosynthesis of H_2S, L-cysteine, and H_2S-releasing compounds, NaHS and AP123, can prevent H_2O_2-induced OS in an isolated bovine neural retina in a concentration-dependent manner. The pharmacological effect of L-cysteine was independent of the de novo intramural biosynthesis of H_2S in the retina. On the other hand, the pharmacological action of the mitochondrial-targeted H_2S-releasing compound, AP123, was mediated by the mechanisms that involve the 3-MST pathway for the biosynthesis of this gas in an isolated bovine retina. Taken together, our data support a neuroprotective role for H_2S in the retina, a gas that has been reported to play a role in the regulation of aqueous humor dynamics in mammalian eyes.

Author Contributions: Conceptualization, L.B., S.E.O. and Y.F.N.M.; data curation, L.B., S.E.O. and Y.F.N.M.; formal analysis, L.B., S.E.O. and Y.F.N.M.; funding acquisition, C.A.O. and Y.F.N.M.; investigation, L.B., J.R., S.E.O. and Y.F.N.M.; methodology, L.B., S.E.O. and Y.F.N.M.; project administration, S.E.O. and Y.F.N.M.; resources, M.W. and Y.F.N.M.; supervision, C.A.O., M.W., S.E.O. and Y.F.N.M.; validation, L.B. and S.E.O.; visualization, L.B. and A.O.; writing—original draft, L.B.; writing—review & editing, L.B., J.R., A.O., F.M., C.A.O., S.E.O. and Y.F.N.M. All authors have read and agreed to the published version of the manuscript.

Funding: This project was supported in part by Grant Number 1R15EY022215-01 from the National Institute of Health at the National Eye Institute.

Institutional Review Board Statement: Not applicable.

Informed Consent Statement: Not applicable.

Data Availability Statement: The data are presented in the main text.

Acknowledgments: The authors would like to thank Matthew Whiteman for providing the H_2S-releasing compound AP123 for use in these studies.

Conflicts of Interest: The authors declare that they have no known competing financial interests.

References

1. Kim, G.H.; Kim, J.E.; Rhie, S.J.; Yoon, S. The Role of Oxidative Stress in Neurodegenerative Diseases. *Exp. Neurobiol.* **2015**, *24*, 325–340. [CrossRef] [PubMed]
2. Njie-Mbye, Y.F.; Kulkarni-Chitnis, M.; Opere, C.A.; Barrett, A.; Ohia, S.E. Lipid peroxidation: Pathophysiological and pharmacological implications in the eye. *Front. Physiol.* **2013**, *4*, 366. [CrossRef]
3. Al-Shabrawey, M.; Smith, S. Prediction of diabetic retinopathy: Role of oxidative stress and relevance of apoptotic biomarkers. *EPMA J.* **2010**, *1*, 56–72. [CrossRef]
4. Berthoud, V.M.; Beyer, E.C. Oxidative stress, lens gap junctions, and cataracts. *Antioxid. Redox Signal.* **2009**, *11*, 339–353. [CrossRef]
5. Sacca, S.; Bolognesi, C.; Battistella, A.; Bagnis, A.; Izzotti, A. Gene–environment interactions in ocular diseases. *Mutat. Res. Mol. Mech. Mutagen.* **2009**, *667*, 98–117. [CrossRef] [PubMed]
6. Hollyfield, J.G. Age-related macular degeneration: The molecular link between oxidative damage, tissue-specific inflammation and outer retinal disease: The Proctor lecture. *Investig. Opthalmology Vis. Sci.* **2010**, *51*, 1276–1281. [CrossRef] [PubMed]

7. De La Paz, M.A.; Zhang, J.; Fridovich, I. Antioxidant enzymes of the human retina: Effect of age on enzyme activity of macula and periphery. *Curr. Eye Res.* **1996**, *15*, 273–278. [CrossRef]
8. Shih, A.Y.; Erb, H.; Sun, X.; Toda, S.; Kalivas, P.W.; Murphy, T.H. Cystine/glutamate exchange modulates glutathione supply for neuroprotection from oxidative stress and cell proliferation. *J. Neurosci.* **2006**, *26*, 10514–10523. [CrossRef] [PubMed]
9. Wielgus, A.R.; Sarna, T. Ascorbate enhances photogeneration of hydrogen peroxide mediated by the iris melanin. *Photochem. Photobiol.* **2008**, *84*, 683–691. [CrossRef] [PubMed]
10. Nishimura, Y.; Hara, H.; Kondo, M.; Hong, S.; Matsugi, T. Oxidative Stress in Retinal Diseases. *Oxidative Med. Cell. Longev.* **2017**, *2017*, 4076518. [CrossRef]
11. Greene, E.L.; Paller, M.S. Xanthine oxidase produces O^{2-}. in posthypoxic injury of renal epithelial cells. *Am. J. Physiol. Physiol.* **1992**, *263 Pt 2*, F251–F255. [CrossRef]
12. SanGiovanni, J.P.; Chew, E.Y. The role of omega-3 long-chain polyunsaturated fatty acids in health and disease of the retina. *Prog. Retin. Eye Res.* **2005**, *24*, 87–138. [CrossRef] [PubMed]
13. Shichi, H. Cataract formation and prevention. *Expert Opin. Investig. Drugs* **2004**, *13*, 691–701. [CrossRef] [PubMed]
14. van Reyk, D.M.; Gillies, M.C.; Davies, M.J. The retina: Oxidative stress and diabetes. *Redox Rep.* **2003**, *8*, 187–192. [CrossRef] [PubMed]
15. Ohia, S.E.; Opere, C.A.; LeDay, A.M. Pharmacological consequences of oxidative stress in ocular tissues. *Mutat. Res. Mol. Mech. Mutagen.* **2005**, *579*, 22–36. [CrossRef]
16. Łowicka, E.; Bełtowski, J. Hydrogen sulfide (H_2S)—The third gas of interest for pharmacologists. *Pharmacol. Rep.* **2007**, *59*, 4–24.
17. Goodwin, L.R.; Francom, D.; Dieken, F.P.; Taylor, J.D.; Warenycia, M.W.; Reiffenstein, R.J.; Dowling, G. Determination of sulfide in brain tissue by gas dialysis/ion chromatography: Postmortem studies and two case reports. *J. Anal. Toxicol.* **1989**, *13*, 105–109. [CrossRef]
18. Lefer, D.J. A new gaseous signaling molecule emerges: Cardioprotective role of hydrogen sulfide. *Proc. Natl. Acad. Sci. USA* **2007**, *104*, 17907–17908. [CrossRef]
19. Savage, J.; Gould, D. Determination of sulfide in brain tissue and rumen fluid by ion-interaction reversed-phase high-performance liquid chromatography. *J. Chromatogr.* **1990**, *526*, 540–545. [CrossRef]
20. Szabó, C. Hydrogen sulphide and its therapeutic potential. *Nat. Rev. Drug Discov.* **2007**, *6*, 917–935. [CrossRef]
21. Tan, B.H.; Wong, P.T.-H.; Bian, J.-S. Hydrogen sulfide: A novel signaling molecule in the central nervous system. *Neurochem. Int.* **2010**, *56*, 3–10. [CrossRef] [PubMed]
22. Mikami, Y.; Shibuya, N.; Kimura, Y.; Nagahara, N.; Yamada, M.; Kimura, H. Hydrogen sulfide protects the retina from light-induced degeneration by the modulation of Ca^{2+} influx. *J. Biol. Chem.* **2011**, *286*, 39379–39386. [CrossRef] [PubMed]
23. Shibuya, N.; Mikami, Y.; Kimura, Y.; Nagahara, N.; Kimura, H. Vascular endothelium expresses 3-mercaptopyruvate sulfurtransferase and produces hydrogen sulfide. *J. Biochem.* **2009**, *146*, 623–626. [CrossRef] [PubMed]
24. Shibuya, N.; Tanaka, M.; Yoshida, M.; Ogasawara, Y.; Togawa, T.; Ishii, K.; Kimura, H. 3-Mercaptopyruvate sulfurtransferase produces hydrogen sulfide and bound sulfane sulfur in the brain. *Antioxid. Redox Signal.* **2009**, *11*, 703–714. [CrossRef] [PubMed]
25. Shibuya, N.; Kimura, H. Production of hydrogen sulfide from d-cysteine and its therapeutic potential. *Front. Endocrinol.* **2013**, *4*, 57041. [CrossRef]
26. Shibuya, N.; Koike, S.; Tanaka, M.; Ishigami-Yuasa, M.; Kimura, Y.; Ogasawara, Y.; Fukui, K.; Nagahara, N.; Kimura, H. A novel pathway for the production of hydrogen sulfide from D-cysteine in mammalian cells. *Nat. Commun.* **2013**, *4*, 1366. [CrossRef]
27. Gerő, D.; Torregrossa, R.; Perry, A.; Waters, A.; Le-Trionnaire, S.; Whatmore, J.L.; Wood, M.; Whiteman, M. The novel mitochondria-targeted hydrogen sulfide (H_2S) donors AP123 and AP39 protect against hyperglycemic injury in microvascular endothelial cells in vitro. *Pharmacol. Res.* **2016**, *113 Pt A*, 186–198. [CrossRef]
28. Li, L.; Whiteman, M.; Guan, Y.Y.; Neo, K.L.; Cheng, Y.; Lee, S.W.; Zhao, Y.; Baskar, R.; Tan, C.H.; Moore, P.K. Characterization of a novel, water-soluble hydrogen sulfide-releasing molecule (GYY4137): New insights into the biology of hydrogen sulfide. *Circulation* **2008**, *117*, 2351–2360. [CrossRef]
29. Zhao, Y.; Biggs, T.D.; Xian, M. Hydrogen sulfide (H_2S) releasing agents: Chemistry and biological applications. *Chem. Commun.* **2014**, *50*, 11788–11805. [CrossRef]
30. Abe, K.; Kimura, H. The possible role of hydrogen sulfide as an endogenous neuromodulator. *J. Neurosci.* **1996**, *16*, 1066–1071. [CrossRef]
31. Cheng, Y.; Ndisang, J.F.; Tang, G.; Cao, K.; Wang, R. Hydrogen sulfide-induced relaxation of resistance mesenteric artery beds of rats. *Am. J. Physiol. Heart Circ. Physiol.* **2004**, *287*, H2316–H2323. [CrossRef] [PubMed]
32. Hosoki, R.; Matsuki, N.; Kimura, H. The possible role of hydrogen sulfide as an endogenous smooth muscle relaxant in synergy with nitric oxide. *Biochem. Biophys. Res. Commun.* **1997**, *237*, 527–531. [CrossRef]
33. Kimura, H. Hydrogen sulfide as a neuromodulator. *Mol. Neurobiol.* **2002**, *26*, 013–020. [CrossRef]
34. Qu, K.; Lee, S.W.; Bian, J.S.; Low, C.-M.; Wong, P.T.-H. Hydrogen sulfide: Neurochemistry and neurobiology. *Neurochem. Int.* **2008**, *52*, 155–165. [CrossRef] [PubMed]
35. Kimura, Y.; Mikami, Y.; Osumi, K.; Tsugane, M.; Oka, J.; Kimura, H. Polysulfides are possible H_2S-derived signaling molecules in rat brain. *FASEB J.* **2013**, *27*, 2451–2457. [CrossRef]
36. Zhang, W.; Xu, C.; Yang, G.; Wu, L.; Wang, R. Interaction of H_2S with Calcium Permeable Channels and Transporters. *Oxid. Med. Cell Longev.* **2015**, *2015*, 323269. [CrossRef] [PubMed]

37. Kulkarni, M.; Njie-Mbye, Y.F.; Okpobiri, I.; Zhao, M.; Opere, C.A.; Ohia, S.E. Endogenous production of hydrogen sulfide in isolated bovine eye. *Neurochem. Res.* **2011**, *36*, 1540–1545. [CrossRef] [PubMed]
38. Bankhele, P.; Salvi, A.; Jamil, J.; Njie-Mbye, F.; Ohia, S.; Opere, C.A. Comparative Effects of Hydrogen Sulfide-Releasing Compounds on [3H]D-Aspartate Release from Bovine Isolated Retinae. *Neurochem. Res.* **2018**, *43*, 692–701. [CrossRef] [PubMed]
39. Njie-Mbye, Y.F.; Bongmba, O.Y.N.; Onyema, C.C.; Chitnis, A.; Kulkarni, M.; Opere, C.A.; LeDay, A.M.; Ohia, S.E. Effect of hydrogen sulfide on cyclic AMP production in isolated bovine and porcine neural retinae. *Neurochem. Res.* **2010**, *35*, 487–494. [CrossRef]
40. Njie-Mbye, Y.F.; Kulkarni, M.; Opere, C.A.; Ohia, S.E. Mechanism of action of hydrogen sulfide on cyclic AMP formation in rat retinal pigment epithelial cells. *Exp. Eye Res.* **2012**, *98*, 16–22. [CrossRef]
41. Cornwell, A.; Badiei, A. The role of hydrogen sulfide in the retina. *Exp. Eye Res.* **2023**, *234*, 109568. [CrossRef] [PubMed]
42. Tribble, J.R.; Hui, F.; Quintero, H.; El Hajji, S.; Bell, K.; Di Polo, A.; Williams, P.A. Neuroprotection in glaucoma: Mechanisms beyond intraocular pressure lowering. *Mol. Asp. Med.* **2023**, *92*, 101193. [CrossRef] [PubMed]
43. Tripathi, S.J.; Chakraborty, S.; Miller, E.; Pieper, A.A.; Paul, B.D. Hydrogen sulfide signalling in neurodegenerative diseases. *Br. J. Pharmacol.* **2023**. [CrossRef] [PubMed]
44. Liu, H.; Perumal, N.; Manicam, C.; Mercieca, K.; Prokosch, V. Proteomics reveals the potential protective mechanism of hydrogen sulfide on retinal ganglion cells in an ischemia/reperfusion injury animal model. *Pharmaceuticals* **2020**, *13*, 213. [CrossRef]
45. Sakamoto, K.; Suzuki, Y.; Kurauchi, Y.; Mori, A.; Nakahara, T.; Ishii, K. Hydrogen sulfide attenuates NMDA-induced neuronal injury via its anti-oxidative activity in the rat retina. *Exp. Eye Res.* **2014**, *120*, 90–96. [CrossRef] [PubMed]
46. Delanty, N.; Reilly, M.P.; Pratico, D.; Lawson, J.A.; McCarthy, J.F.; Wood, A.E.; Ohnishi, S.T.; Fitzgerald, D.J.; FitzGerald, G.A. 8-epi PGF2 alpha generation during coronary reperfusion. A potential quantitative marker of oxidant stress in vivo. *Circulation* **1997**, *95*, 2492–2499. [CrossRef]
47. Gopaul, N.K.; Anggard, E.E.; Mallet, A.I.; Betteridge, D.J.; Wolff, S.P.; Nourooz-Zadeh, J. Plasma 8-epi-PGF2 alpha levels are elevated in individuals with non-insulin dependent diabetes mellitus. *FEBS Lett.* **1995**, *368*, 225–229. [CrossRef] [PubMed]
48. Kimura, Y.; Kimura, H. Hydrogen sulfide protects neurons from oxidative stress. *FASEB J.* **2004**, *18*, 1165–1167. [CrossRef]
49. Baskin, S.I.; Porter, D.W.; Rockwood, G.A.; Romano, J.J.A.; Patel, H.C.; Kiser, R.C.; Cook, C.M.; Ternay, J.A.L. In vitro andin vivo comparison of sulfur donors as antidotes to acute cyanide intoxication. *J. Appl. Toxicol.* **1999**, *19*, 173–183. [CrossRef]
50. Milatovic, D.; Montine, T.J.; Aschner, M. Measurement of isoprostanes as markers of oxidative stress. *Methods Mol Biol.* **2011**, *758*, 195–204.
51. Kadiiska, M.B.; Gladen, B.C.; Baird, D.; Germolec, D.; Graham, L.B.; Parker, C.E.; Nyska, A.; Wachsman, J.T.; Ames, B.N.; Basu, S.; et al. Biomarkers of oxidative stress study II: Are oxidation products of lipids, proteins, and DNA markers of CCl4 poisoning? *Free Radic. Biol. Med.* **2005**, *38*, 698–710. [CrossRef] [PubMed]
52. Labuschagne, C.F.; Broek, N.J.F.v.D.; Postma, P.; Berger, R.; Brenkman, A.B. A protocol for quantifying lipid peroxidation in cellular systems by F2-isoprostane analysis. *PLoS ONE* **2013**, *8*, e80935. [CrossRef] [PubMed]
53. Frijhoff, J.; Winyard, P.G.; Zarkovic, N.; Davies, S.S.; Stocker, R.; Cheng, D.; Knight, A.R.; Taylor, E.L.; Oettrich, J.; Ruskovska, T.; et al. Clinical relevance of biomarkers of oxidative stress. *Antioxid. Redox Signal.* **2015**, *23*, 1144–1170. [CrossRef] [PubMed]
54. Czerska, M.; Mikołajewska, K.; Zieliński, M.; Gromadzińska, J.; Wąsowicz, W. Today's oxidative stress markers. *Med. Pr.* **2015**, *66*, 393–405. [CrossRef] [PubMed]
55. Nourooz-Zadeh, J.; Pereira, P. F2 isoprostanes, potential specific markers of oxidative damage in human retina. *Ophthalmic Res.* **2000**, *32*, 133–137. [CrossRef] [PubMed]
56. Cervellati, F.; Cervellati, C.; Romani, A.; Cremonini, E.; Sticozzi, C.; Belmonte, G.; Pessina, F.; Valacchi, G. Hypoxia induces cell damage via oxidative stress in retinal epithelial cells. *Free Radic. Res.* **2014**, *48*, 303–312. [CrossRef] [PubMed]
57. Wong, L.L.; Pye, Q.N.; Chen, L.; Seal, S.; McGinnis, J.F. Defining the catalytic activity of nanoceria in the P23H-1 rat, a photoreceptor degeneration model. *PLoS ONE* **2015**, *10*, e0121977. [CrossRef] [PubMed]
58. Cui, Y.; Xu, N.; Xu, W.; Xu, G. Mesenchymal stem cells attenuate hydrogen peroxide-induced oxidative stress and enhance neuroprotective effects in retinal ganglion cells. *Vitr. Cell. Dev. Biol.-Anim.* **2017**, *53*, 328–335. [CrossRef] [PubMed]
59. Hu, L.; Guo, J.; Zhou, L.; Zhu, S.; Wang, C.; Liu, J.; Hu, S.; Yang, M.; Lin, C. Hydrogen sulfide protects retinal pigment epithelial cells from oxidative stress-induced apoptosis and affects autophagy. *Oxid. Med. Cell. Longev.* **2020**, *2020*, 8868564. [CrossRef]
60. Wang, Y.; Wang, J.; Zhang, X.; Feng, Y.; Yuan, Y. Neuroprotective effects of idebenone on hydrogen peroxide-induced oxidative damage in retinal ganglion cells-5. *Int. Ophthalmol.* **2023**, *43*, 3831–3839. [CrossRef]
61. Wallace, J.L.; Wang, R. Hydrogen sulfide-based therapeutics: Exploiting a unique but ubiquitous gasotransmitter. *Nat. Rev. Drug Discov.* **2015**, *14*, 329–345. [CrossRef] [PubMed]
62. Wu, D.; Hu, Q.; Zhu, Y. Therapeutic application of hydrogen sulfide donors: The potential and challenges. *Front. Med.* **2015**, *10*, 18–27. [CrossRef] [PubMed]
63. Van Goor, H.; van den Born, J.C.; Hillebrands, J.-L.; Joles, J.A. Hydrogen sulfide in hypertension. *Curr. Opin. Nephrol. Hypertens.* **2016**, *25*, 107–113. [CrossRef] [PubMed]
64. Sattar, M.; Ahmad, A.; Rathore, H.; Khan, S.; Lazhari, M.; Afzal, S.; Hashmi, F.; Abdullah, N.; Johns, E. A critical review of pharmacological significance of Hydrogen Sulfide in hypertension. *Indian J. Pharmacol.* **2015**, *47*, 243–247. [CrossRef]
65. Kida, K.; Ichinose, F. Hydrogen Sulfide and Neuroinflammation. *Handb. Exp. Pharmacol.* **2015**, *230*, 181–189. [PubMed]

66. Biermann, J.; Lagrèze, W.A.; Schallner, N.; Schwer, C.I.; Goebel, U. Inhalative preconditioning with hydrogen sulfide attenuated apoptosis after retinal ischemia/reperfusion injury. *Mol. Vis.* **2011**, *17*, 1275–1286.
67. Powell, C.R.; Dillon, K.M.; Matson, J.B. A review of hydrogen sulfide (H_2S) donors: Chemistry and potential therapeutic applications. *Biochem. Pharmacol.* **2018**, *149*, 110–123. [CrossRef] [PubMed]
68. Ohia, S.E.; Opere, C.A.; Awe, S.O.; Adams, L.; Sharif, N.A. Human, bovine, and rabbit retinal glutamate-induced [3H]D-aspartate release: Role in excitotoxicity. *Neurochem. Res.* **2000**, *25*, 853–860. [CrossRef] [PubMed]
69. Opere, C.A.; Heruye, S.; Njie-Mbye, Y.-F.; Ohia, S.E.; Sharif, N.A. Regulation of excitatory amino acid transmission in the retina: Studies on neuroprotection. *J. Ocul. Pharmacol. Ther.* **2018**, *34*, 107–118. [CrossRef]
70. Heruye, S.H.; Mbye, Y.F.; Ohia, S.E.; Opere, C.A. Protective action of hydrogen sulfide-releasing compounds against oxidative stress-induced cataract formation in cultured bovine lenses. *Curr. Eye Res.* **2022**, *47*, 239–245. [CrossRef]
71. Li, T.; Wang, L.; Hu, Q.; Liu, S.; Bai, X.; Xie, Y.; Zhang, T.; Bo, S.; Gao, X.; Wu, S.; et al. Neuroprotective Roles of l-Cysteine in Attenuating Early Brain Injury and Improving Synaptic Density via the CBS/H_2S Pathway Following Subarachnoid Hemorrhage in Rats. *Front. Neurol.* **2017**, *8*, 176. [CrossRef] [PubMed]
72. Wang, Z.; Liu, D.-X.; Wang, F.-W.; Zhang, Q.; Du, Z.-X.; Zhan, J.-M.; Yuan, Q.-H.; Ling, E.-A.; Hao, A.-J. L-Cysteine promotes the proliferation and differentiation of neural stem cells via the CBS/H(2)S pathway. *Neuroscience* **2013**, *237*, 106–117. [CrossRef] [PubMed]
73. Qanungo, S.; Wang, M.; Nieminen, A.L. N-Acetyl-L-cysteine enhances apoptosis through inhibition of nuclear factor-kappaB in hypoxic murine embryonic fibroblasts. *J. Biol. Chem.* **2004**, *279*, 50455–50464. [CrossRef] [PubMed]
74. Whiteman, M.; Li, L.; Rose, P.; Tan, C.-H.; Parkinson, D.B.; Moore, P.K. The effect of hydrogen sulfide donors on lipopolysaccharide-induced formation of inflammatory mediators in macrophages. *Antioxid. Redox Signal.* **2010**, *12*, 1147–1154. [CrossRef] [PubMed]
75. Kimura, Y.; Goto, Y.-I.; Kimura, H. Hydrogen sulfide increases glutathione production and suppresses oxidative stress in mitochondria. *Antioxid. Redox Signal.* **2010**, *12*, 1–13. [CrossRef] [PubMed]
76. Robinson, J.; Okoro, E.; Ezuedu, C.; Bush, L.; Opere, C.A.; Ohia, S.E.; Njie-Mbye, Y.F. Effects of Hydrogen Sulfide-Releasing Compounds on Aqueous Humor Outflow Facility in Porcine Ocular Anterior Segments, Ex Vivo. *J. Ocul. Pharmacol. Ther.* **2017**, *33*, 91–97. [CrossRef]
77. Ohia, S.E.; Opere, C.A.; Monjok, E.M.; Kouamou, G.; Leday, A.M.; Njie-Mbye, Y.F. Role of hydrogen sulfide production in inhibitory action of L-cysteine on isolated porcine irides. *Curr. Eye Res.* **2010**, *35*, 402–407. [CrossRef]
78. Suzuki, K.; Olah, G.; Modis, K.; Coletta, C.; Kulp, G.; Gerö, D.; Szoleczky, P.; Chang, T.; Zhou, Z.; Wu, L.; et al. Hydrogen sulfide replacement therapy protects the vascular endothelium in hyperglycemia by preserving mitochondrial function. *Proc. Natl. Acad. Sci. USA* **2011**, *108*, 13829–13834. [CrossRef]
79. Franklin, C.C.; Backos, D.S.; Mohar, I.; White, C.C.; Forman, H.J.; Kavanagh, T.J. Structure, function, and post-translational regulation of the catalytic and modifier subunits of glutamate cysteine ligase. *Mol. Asp. Med.* **2008**, *30*, 86–98. [CrossRef]
80. Salvi, A.; Bankhele, P.; Jamil, J.; Chitnis, M.K.; Njie-Mbye, Y.F.; Ohia, S.E.; Opere, C.A. Effect of hydrogen sulfide donors on intraocular pressure in rabbits. *J. Ocul. Pharmacol. Ther.* **2016**, *32*, 371–375. [CrossRef] [PubMed]
81. Ohia, S.E.; Robinson, J.; Mitchell, L.; Ngele, K.K.; Heruye, S.; Opere, C.A.; Njie-Mbye, Y.F. Regulation of aqueous humor dynamics by hydrogen sulfide: Potential role in glaucoma pharmacotherapy. *J. Ocul. Pharmacol. Ther.* **2018**, *34*, 61–69. [CrossRef] [PubMed]
82. Murali, A.; Ramlogan-Steel, C.A.; Steel, J.C.; Layton, C.J. Characterisation and validation of the 8-fold quadrant dissected human retinal explant culture model for pre-clinical toxicology investigation. *Toxicol. Vitr.* **2020**, *63*, 104716. [CrossRef]
83. Kulkarni, P.; Payne, S. Eicosanoids in bovine retinal microcirculation. *J. Ocul. Pharmacol. Ther.* **1997**, *13*, 139–149. [CrossRef] [PubMed]
84. LeDay, A.M.; Kulkarni, K.H.; Opere, C.A.; Ohia, S.E. Arachidonic acid metabolites and peroxide-induced inhibition of [3H]D-aspartate release from bovine isolated retinae. *Curr. Eye Res.* **2004**, *28*, 367–372. [CrossRef] [PubMed]
85. Matsuda, K.; Ohnishi, K.; Misaka, E.; Yamazaki, M. Decrease of urinary prostaglandin E2 and prostaglandin F2 alpha excretion by nonsteroidal anti-inflammatory drugs in rats. Relationship to anti-inflammatory activity. *Biochem. Pharmacol.* **1983**, *32*, 1347–1352. [CrossRef] [PubMed]
86. Zhan, G.-L.; Ohia, S.; Camras, C.; Ohia, E.; Wang, Y. Superior cervical ganglionectomy-induced lowering of intraocular pressure in rabbits: Role of prostaglandins and neuropeptide Y. *Gen. Pharmacol. Vasc. Syst.* **1999**, *32*, 189–194. [CrossRef] [PubMed]
87. Nucci, C.; Gasperi, V.; Tartaglione, R.; Cerulli, A.; Terrinoni, A.; Bari, M.; De Simone, C.; Agro, A.F.; Morrone, L.A.; Corasaniti, M.T.; et al. Involvement of the endocannabinoid system in retinal damage after high intraocular pressure-induced ischemia in rats. *Investig. Opthalmology Vis. Sci.* **2007**, *48*, 2997–3004. [CrossRef]

pharmaceuticals

MDPI

Article

Roles of Prostaglandins and Hydrogen Sulfide in an Outflow Model of the Porcine Ocular Anterior Segment Ex Vivo

Jenaye Robinson [1], Leah Bush [1], Anthonia Okolie [1], Fatima Muili [1], Sunny Ohia [1], Catherine Opere [2] and Ya Fatou Njie Mbye [1,*]

[1] Department of Pharmaceutical Sciences, College of Pharmacy and Health Sciences, Texas Southern University, Houston, TX 77004, USA; jenayerobinson@yahoo.com (J.R.); lemwjb@gmail.com (L.B.); a.okolie9552@student.tsu.edu (A.O.); f.muili1867@student.tsu.edu (F.M.); sunny.ohia@tsu.edu (S.O.)

[2] Department of Pharmacy Sciences, School of Pharmacy and Health Professions, Creighton University, Omaha, NE 68178, USA; catherineopere@creighton.edu

* Correspondence: yafatou.mbye@tsu.edu

Abstract: Background: Hydrogen sulfide (H_2S)-releasing compounds can reduce intraocular pressure in normotensive rabbits by increasing aqueous humor (AH) outflow through the trabecular meshwork. In the present study, we investigated the contribution of endogenous H_2S and the role of intramurally generated prostaglandins in the observed increase in AH outflow facility in an ex vivo porcine ocular anterior segment model. Material and Methods: Porcine ocular anterior segment explants were perfused with Dulbecco's Modified Eagle's Medium maintained at 37 °C and gassed with 5% CO_2 and 95% air under an elevated pressure of 15 mmHg for four hours. Perfusates from the anterior segment explants were collected and immediately assayed for their H_2S and prostaglandin E_2 content. Results: Elevating perfusion pressure from 7.35 to 15 mm Hg significantly ($p < 0.001$) increased H_2S concentration in the perfusate from 0.4 ± 0.1 to 67.6 ± 3.6 nM/μg protein. In the presence of an inhibitor of cystathionine ß-synthase/cystathionine γ-lyase, aminooxyacetic acid (AOAA, 30 μM), or an inhibitor of 3-mercaptopyruvate sulfurtransferase, α-ketobutyric acid (KBA, 1 mM), the effects of elevated pressure on H_2S levels in the perfusate was significant ($p < 0.001$). Furthermore, flurbiprofen (30 μM) and indomethacin (10 μM) attenuated the elevated pressure-induced increase in H_2S levels in the perfusate. Interestingly, elevating perfusion pressure had no significant effect on PGE_2 concentrations in the perfusate. While the inhibition of H_2S biosynthesis by AOAA or KBA did not affect PGE_2 levels in perfusate, flurbiprofen (30 μM) caused a significant ($p < 0.05$) decrease in the concentration of PGE_2 under conditions of elevated perfusion pressure. Conclusions: We conclude that the elevated perfusion pressure-induced increase in H_2S concentrations depends upon the endogenous biosynthesis of H_2S and intramurally produced prostaglandins in the porcine anterior segment explants. While the concentration of PGE_2 in the perfusate under elevated perfusion pressure was unaffected by pretreatment with inhibitors of H_2S biosynthesis, it was reduced in the presence of an inhibitor of cyclooxygenase.

Keywords: aqueous humor outflow; hydrogen sulfide; prostaglandins; trabecular meshwork; perfusion pressure

1. Introduction

For centuries, hydrogen sulfide (H_2S) was known as a toxic gaseous compound until recently when new evidence emerged to support its biological role in mammalian organs and tissues [1–3]. Indeed, H_2S has been reported to have important physiological and pharmacological actions similar to those of other gaseous transmitters such as nitric oxide and carbon monoxide [4]. In the cytosol, H_2S biosynthesis occurs through its substrate, L-cysteine, in a reaction that is catalyzed by the pyridoxal-5-phosphate (P5P)-dependent enzymes, cystathionine β-synthase (CBS) and cystathionine γ-lyase (CSE) [1,5].

In addition to the cytosol, L-cysteine in the mitochondria can produce H_2S in the presence of cysteine aminotransferase (CAT) and 3-mercaptopyruvate sulfur transferase (3-MST) in a reaction that is dependent upon calcium [6,7]. Recently, a fourth enzyme, D-amino acid oxidase (DAO), has been shown to catalyze the formation of H_2S from D-cysteine in mammalian tissues [8]. In the eye, CBS, CSE, 3-MST, and DAO are involved in the biosynthesis of H_2S and are highly expressed in tissues such as the iris, ciliary body, and retina [9,10]. It is well established that intraocular pressure (IOP) is maintained by a delicate balance between aqueous humor production in the ciliary body and its drainage via two outflow pathways. Approximately 75% of aqueous humor outflow occurs through the trabecular meshwork (TM) and Schlemm's canal [11]. There is evidence that a dysfunction of the TM is one of the primary contributory factors that leads to elevated IOP and subsequent retinal damage and vision loss [11]. Compounds that lower IOP exert their pharmacological effects through various mechanisms, one of which is by modulating TM outflow facility [12]. Interestingly, H_2S (using H_2S-releasing compounds as gas donors) has been reported to lower IOP and increase trabecular meshwork outflow facility in experimental animal models [13,14]. In 2017, Huang and coworkers reported that endogenous levels of H_2S and the expression of enzymes responsible for the biosynthesis of this gas (CBS, CSE, and 3MST) are significantly reduced in retinal ganglion cells in a rat model of ocular hypertension [15]. This finding suggests a possible involvement of an endogenous pathway leading to the biosynthesis of H_2S in the regulation of IOP and maintenance of aqueous humor dynamics.

Prostaglandins are pro-inflammatory mediators that have been reported to be present at high concentrations in the aqueous humor of glaucomatous humans and experimental animal models [16,17]. Evidence from studies performed in our laboratory demonstrates that endogenous prostaglandins are involved in the pharmacological actions of H_2S-releasing compounds in the eye [13,14,18,19]. Taken together, evidence from our laboratory and the published work of other investigators support a potential role for prostaglandins in the pharmacological actions of H_2S on intraocular pressure. The demonstration of a possible interaction between these modulators in the TM outflow pathway may prove useful in understanding their role in aqueous humor dynamics. In the present study, we sought to determine whether a relationship exists between endogenous concentrations of H_2S and prostaglandins in an ex vivo porcine anterior segment explant model designed to study TM outflow facility. Consequently, the aim of the present study was two-fold: (1) to determine the role of the endogenous pathway for the biosynthesis of H_2S in a porcine ex vivo model of TM outflow facility under conditions of elevated perfusion pressure and (2) to investigate the interaction between endogenous prostaglandins and H_2S in the pathway leading to the outflow of aqueous humor under conditions of elevated perfusion pressure.

2. Results

2.1. Effect of Elevated Perfusion Pressure on Hydrogen Sulfide Levels in TM Outflow Pathway

To assess the effects of elevated perfusion pressure on endogenous H_2S production in porcine anterior segment explants, an experiment was designed to measure TM outflow facility with the perfusion pressure raised from 7.35 mm Hg to 15 mm Hg. The baseline concentration of H_2S in the perfusate at 7.35 mm Hg was 0.41 ± 0.1 nM/µg protein, which increased significantly ($p < 0.001$) after the perfusion pressure was elevated to 15 mm Hg, yielding a concentration of 67.6 ± 3.7 nM/µg protein ($p < 0.001$) (Figure 1). We found the basal concentration of H_2S in the porcine aqueous humor (used as a reference point) to be 1.5 ± 0.2 nM/µg protein. In a second set of experiments, we evaluated whether the perfusion pressure-induced increase in H_2S concentrations involved enzymes responsible for the biosynthesis of H_2S. Anterior segment explants were perfused with an inhibitor of CBS/CSE (AOAA, 30–100 µM; [20]) or an inhibitor of 3-MST (KBA, 1 mM; [21]) for four hours before the measurement of concentrations of H_2S in the perfusate. In the presence of AOAA and KBA, the elevated perfusion pressure-induced increase in H_2S levels was significantly ($p < 0.001$) attenuated (Figure 2). At a perfusion pressure of 15 mm Hg, the

H_2S concentration in the perfusate (67.6 ± 3.7 nM/μg protein) was significantly ($p < 0.001$) lowered to 5.7 ± 0.3 nM/μg protein and 9.0 ± 0.7 nM/μg protein following treatment with AOAA (30 μM and 100 μM, respectively). In the presence of KBA (1 mM), H_2S levels were significantly ($p < 0.001$) lowered to 4.9 ± 0.8 nM/μg protein in the perfusate when compared to untreated controls (67.6 ± 3.7 nM/μg protein) (Figure 3).

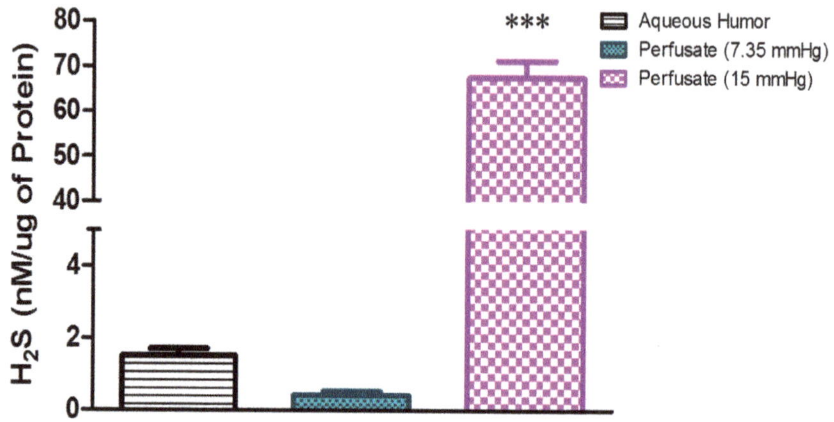

Figure 1. Effect of different perfusion pressures on hydrogen sulfide (H_2S) concentrations in the perfusate of porcine anterior segment explant trabecular meshwork outflow facility, ex vivo. Untreated porcine anterior segment explants were mounted in perfusion culture chambers perfused with DMEM for four hours at 7.35 mm Hg (normal) or 15 mm Hg (elevated). Porcine aqueous humor H_2S levels were measured as a reference point for comparison. Results are expressed as mean ± standard error. N = 12. *** significant difference between 7.35 and 15 mm Hg perfusate $p < 0.001$.

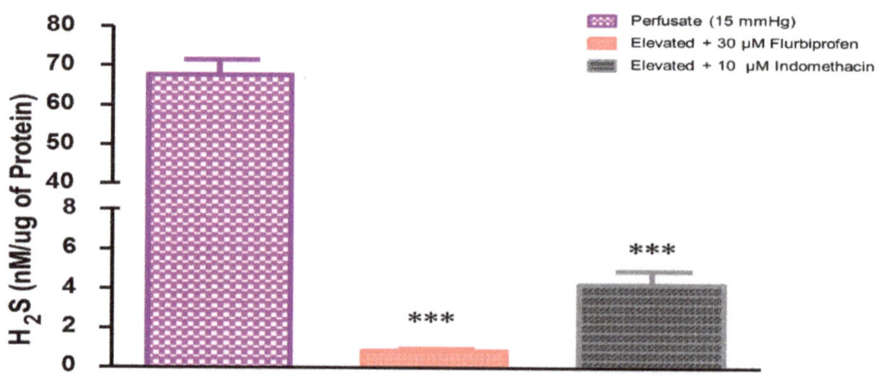

Figure 2. Effects of inhibitors of cyclooxygenase on hydrogen sulfide (H_2S) concentrations in porcine anterior segment explant trabecular meshwork outflow facility, ex vivo. Control (at elevated perfusion pressure of 15 mm Hg) and in the presence of flurbiprofen (10 μM) or indomethacin (10 μM) for four hours. Results are expressed as mean ± standard error. N = 12. *** significant difference between flurbiprofen- and indomethacin-treated eyes and 15 mm Hg perfusate $p < 0.001$.

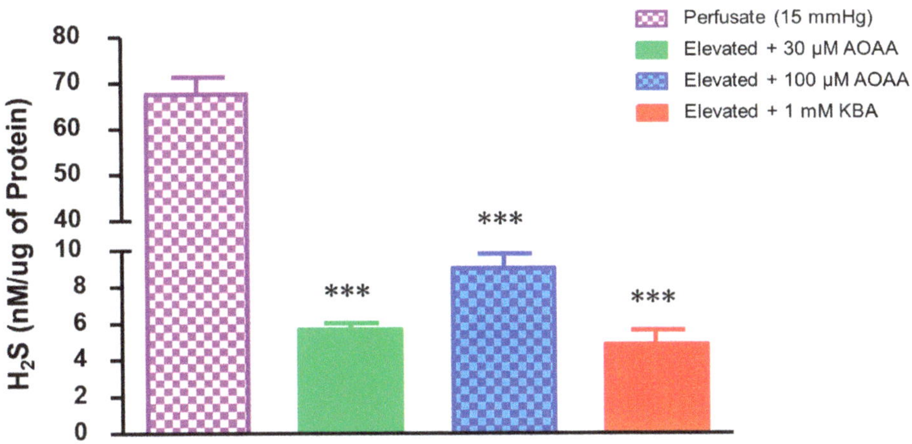

Figure 3. Effects of inhibitors of enzymes responsible for the biosynthesis of hydrogen sulfide (H_2S) on H_2S concentrations in porcine anterior segment explant trabecular meshwork outflow facility, ex vivo. Control (at elevated perfusion pressure of 15 mm Hg) and in the presence of Aminooxyacetic acid (AOAA) (30–100 μM) or α-Ketobutyric acid (KBA) (1 mM) for four hours. Results are expressed as mean ± standard error. N = 12. *** significant difference AOAA- or KBA-treated eyes and 15 mm Hg perfusate $p < 0.001$.

2.2. Role of Prostaglandins in Elevated Perfusion Pressure-Induced Increase in Hydrogen Sulfide Levels in TM Outflow Pathways

Evidence from our laboratory and that reported by other investigators shows that H_2S can stimulate the biosynthesis of prostaglandins and that prostaglandins may play a modulatory role on the pharmacological actions of H_2S [13,18–23]. In the present study, we evaluated the effect of intramurally generated prostaglandins on the endogenous levels of H_2S under conditions of elevated perfusion pressure in a porcine anterior segment explant model designed to measure TM outflow facility, ex vivo. Pretreatment with the non-selective COX inhibitor flurbiprofen (30 μM) or indomethacin (10 μM) significantly ($p < 0.001$) lowered the levels of endogenous H_2S (Figure 3). In the presence of flurbiprofen or indomethacin, the elevated H_2S levels due to increased perfusion pressure were 0.8 ± 0.1 and 4.2 ± 0.7 nM/μg protein, respectively (Figure 3).

2.3. Role of Hydrogen Sulfide in Elevated Perfusion Pressure on Prostaglandin Levels in TM Outflow Pathways

In another set of experiments, we investigated whether elevated perfusion pressure-induced increase in H_2S concentrations involved pathways leading to the biosynthesis of prostanoids such as prostaglandin E_2 (PGE_2). As shown in Figure 4, the concentration of PGE_2 in the aqueous humor was 27.67 ± 1.59 pg/μg protein. Furthermore, there was no significant difference ($p > 0.05$) in PGE_2 levels in the perfusate at normal and elevated perfusion pressure. The inhibition of enzymes responsible for the biosynthesis of endogenous H_2S (CBS/CSE) by AOAA (30 μM) or 3-MST by KBA (1 mM) did not affect PGE_2 levels. However, the non-specific COX inhibitor, flurbiprofen, significantly ($p < 0.05$) lowered PGE_2 levels to 0.18 ± 0.01 pg/μg protein when compared to the untreated 15 mm Hg perfusate (0.26 ± 0.03 pg/μg protein).

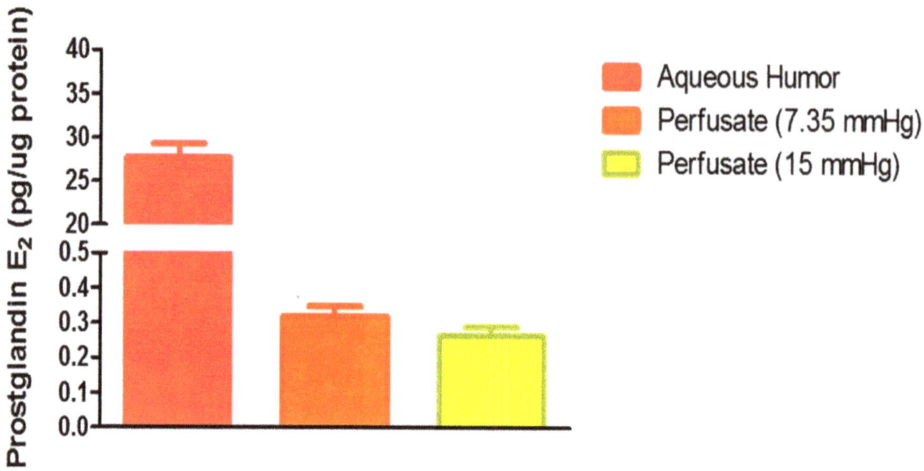

Figure 4. Effect of different perfusion pressures on prostaglandin E_2 (PGE_2) concentrations in the perfusate of porcine anterior segment explant trabecular meshwork outflow facility, ex vivo. Untreated porcine anterior segments were mounted in perfusion culture chambers perfused with DMEM for four hours at 7.35 mm Hg (normal) or 15 mm Hg (elevated). Porcine aqueous humor PGE_2 levels were measured as a reference point for comparison. Results are expressed as mean ± standard error. N = 6.

3. Discussion

The presence of an endogenous pathway for the biosynthesis of H_2S in ocular tissues indicates a potential physiological role for H_2S in the eye, especially in the maintenance of IOP and aqueous humor dynamics [6,24]. There is evidence that H_2S-releasing compounds can lower IOP in normal and glaucomatous animal models and that these ocular hypotensive actions are mediated by the endogenous biosynthesis of H_2S [13]. We also have evidence that under normal perfusion pressure, H_2S (administered as NaHS) and a substrate for the biosynthesis of this gas (L-cysteine) can increase aqueous humor outflow facility in porcine anterior segment explants, ex vivo, a response mediated at least in part by endogenously produced H_2S, K_{ATP} channels and adenylyl cyclase [14]. The concept of doubling perfusion pressures as a marker of elevated intraocular pressures has been extensively reported by Acott and his coworkers [25–27]. Consequently, we used 15 mmHg to reflect the doubling of our normal value of 7.35 mm Hg. The rationale for studying trabecular meshwork (TM) outflow facility under conditions of elevated pressure is designed to mimic the pathophysiology of glaucoma. In the present study, we observed that there is a significant increase in H_2S concentrations in the perfusate of porcine anterior segment explants TM outflow facility model following an increase in pressure from 7.35 mm Hg to 15 mm Hg. This observed increase in endogenous H_2S levels may play a vital role in regulating TM outflow facility and subsequently IOP. Our findings are similar to those reported by Huang et al. [15], which showed that there was a three-fold increase in the expression of the 3-MST enzyme in the rat retina of an animal model of glaucoma, indicating a possible increase in the endogenous biosynthesis of H_2S under conditions of elevated IOP [15]. Although there are no studies reported in the literature about changes in concentrations of H_2S in perfusate from an anterior segment explant model designed for measurement of TM outflow facility, there is evidence of a significant increase in nitric oxide levels in perfusate of human anterior segments explants following changes in perfusion pressure from 10 mm Hg to 25 mm Hg [28]. Interestingly, in a previous study, we found that nitric oxide donors can increase H_2S levels in isolated ocular tissues [29]. It is, therefore, tempting to speculate that the increased concentrations of H_2S observed in response to elevated perfusion pressure in the porcine anterior segment explants may be secondary to an increase in nitric oxide in the TM outflow channels [28]. In summary, our results demonstrate that the concentration of H_2S in the perfusate of porcine ocular anterior segment explants is dependent upon perfusion pressure. The enzymatic pathway for the biosynthesis of H_2S from L-cysteine by CBS, CSE, and 3-MST is well documented [6,7,30]. However, it is pertinent to note that the endogenous production of H_2S can also occur independently of these enzymes from processes such as glutathione reduction or the reduction in sulfur and polysulfides [31]. Based on evidence from previous studies, it was of interest to determine if an enzymatic conversion of L-cysteine to H_2S was involved in the perfusion pressure-induced increase in H_2S levels. We found that the CBS/CSE inhibitor, AOAA, and the 3-MST inhibitor, KBA, abolished the increase in H_2S levels due to the increased perfusion pressure. These observations strongly suggest that the perfusion pressure-induced increase in the endogenous biosynthesis of H_2S accounts, at least in part, for the observed augmented gas concentrations in the perfusate.

To elucidate a possible mechanism of perfusion pressure-induced increase in H_2S levels in the perfusate of porcine anterior segment explants, we considered the potential involvement of prostaglandins as mediators of this response. In the present study, we found that the inhibition of prostaglandin biosynthesis by flurbiprofen or indomethacin elicited a significant ($p < 0.001$) attenuation of H_2S levels induced by the increased perfusion pressure. Based on this observation, it appears that the increased level of H_2S in the perfusate in response to elevated perfusion pressure in the anterior segment explants is due, at least in part, to the biosynthesis of prostaglandins. While there is ample evidence in the literature of prostaglandins being involved in the pharmacological actions of H_2S [13,22,23,32], this is the first report of these eicosanoids being involved in the pathway leading to the production of H_2S. Be that as it may, the present observation that changes in the biosynthesis of prostaglandins can alter the concentrations of H_2S in the perfusate of porcine anterior segment explants is interesting and merits further investigation.

Prostaglandin analogs are widely used as IOP-lowering compounds in experimental animals and as drugs for the treatment of glaucoma [33,34]. Although prostaglandins lower IOP by increasing outflow through the unconventional (uveoscleral) pathway, there is evidence that suggests that they may also mediate the outflow of aqueous humor via the TM outflow channels [33,35]. In the present study, we sought to determine the relationship between changes in the perfusion pressure of porcine anterior segment explants and PGE_2 levels in the perfusate. We now report that increasing perfusion pressure from 7.35 mm Hg to 15 mm Hg had no significant effect on PGE_2 concentrations in the perfusate. A study by Maihoefner and colleagues reported a decrease in PGE_2 levels in the aqueous humor of patients with primary open-angle glaucoma (POAG) and complete loss of COX-2 expression in the ciliary body when compared to healthy eyes [36]. These observations led these authors to conclude that the loss of COX-2 expression in the ciliary body and decreased PGE_2 concentrations in the aqueous humor could be linked to the incidence of POAG [36]. In contrast, in the present study, we found that increasing ocular perfusion pressure did not affect the concentrations of PGE_2 in the perfusate indicating that no such relationship exists between perfusion pressure and eicosanoid levels in the TM outflow channels in the porcine anterior segment explants, ex vivo.

There is evidence of the ability of H_2S to regulate prostaglandin levels in mammalian tissues and organs. For instance, while Na_2S and L-cysteine can inhibit PGE_2 production in the rat hypothalamus, it was found to stimulate the expression of prostaglandin metabolizing enzymes in human uterine tissue [37,38]. In the present study, we investigated the possibility that increasing perfusion pressure and subsequent increases in H_2S concentrations in the perfusate of porcine anterior segment explants can alter PGE_2 levels in this system. There is evidence that endogenously formed H_2S can regulate COX-2 and PGE_2 expression in an acute lung injury mouse model [39]. Surprisingly, we observed that inhibition of enzymes responsible for the biosynthesis of H_2S by AOAA or KBA did not affect PGE_2 levels. As a positive control, we found that the cyclooxygenase inhibitor, flurbiprofen, significantly lowered the levels of PGE_2 in the perfusate. A possible explanation for the lack of effect of inhibitors of the biosynthesis of H_2S on PGE_2 levels in the perfusate could be because the synthesis of new gas may not have a direct action on the concentration of already formed PGE_2 detected in the perfusate. Indeed, the lack of an expected change in the concentration of PGE_2 in the perfusate under conditions of elevated perfusion in the anterior segment explants appears to support our explanation. Another plausible explanation could be that the concentrations of newly synthesized gas may be too small to affect the biosynthetic pathway, leading to the biosynthesis of PGE_2 in the explants.

A possible practical application of the findings of the present study is in understanding the role of H_2S (both exogenous and endogenous) in the regulation of outflow facility in the TM. Data from our previous work demonstrate the ability of H_2S-releasing compounds to increase outflow facility, a response that was dependent on endogenous biosynthesis of the gas, adenylyl cyclase activity, and K^+-ATP channels [14]. In the present study, we found that the elevation of perfusion pressure increased the production of endogenous H_2S, suggesting a role for this gas in the regulation of aqueous humor dynamics during normal and increased intraocular pressure observed during glaucoma.

In summary, the effect of changes in perfusion pressure on eicosanoid levels in the absence and/or presence of H_2S in the perfusate of porcine anterior segment explant model of TM outflow channels is of interest and should be thoroughly investigated.

4. Materials and Methods

4.1. Chemicals

The substrate for H_2S synthesis (L-cysteine), CBS/CSE inhibitor (aminooxyacetic acid (AOAA)), 3-MST inhibitor (alpha-ketobutyric acid (KBA), and indomethacin were procured from Sigma-Aldrich (St. Louis, MO, USA). Flurbiprofen was purchased from Cayman Chemical (Ann Arbor, MI, USA). All test agents were freshly prepared immediately before use on the day of the study. Stock solutions of flurbiprofen were dissolved in 70% ethanol. All other test compounds were dissolved in distilled water. Fresh porcine aqueous humor was collected for the measurement of basal H_2S concentrations. The number of explants used for each experiment ranged from 8 to 12. Parallel measurements were made for control explants and those exposed to drugs/chemicals.

4.2. Porcine Anterior Segment Perfusion Model, Ex Vivo

Fresh porcine eyes obtained from Vision Tech (Dallas, TX, USA) were transported to the laboratory in saline on ice within 24 h following enucleation. The methodology de-scribed by us for setting up the perfused porcine ocular anterior segment organ culture model was employed for these experiments [14,40]. Briefly, porcine ocular anterior segment explants were carefully dissected by removing the iris and ciliary muscle and leaving only the trabecular meshwork, cornea, and small (5 mm) ring of sclera. Explants were then mounted onto perfusion culture chambers perfused with DMEM (supplemented with penicillin and streptomycin) using a constant perfusion headset at a height of 10 or 20 cm for a constant pressure of 7.35 and 15 mm Hg, respectively [41]. Explants were treated with enzyme inhibitors of the biosynthesis of H2S (AOAA, 30–100 μM or KBA, 1 mM) or inhibitors of cyclooxygenase (flurbiprofen, 10 μM or indomethacin, 10 μM) for four hours. At the end of the study, outflow was clamped off, and the perfusate was carefully collected from the anterior segment using a syringe. (Figure 5).

The isolated anterior segment of the eye includes the cornea, sclera, and the trabecular meshwork.

DMEM—Dulbecco's Modified Eagle's Medium; AOAA—aminooxyacetic acid; KBA—α-ketobutyric acid; COX inhibitors—flurbiprofen, and indomethacin.

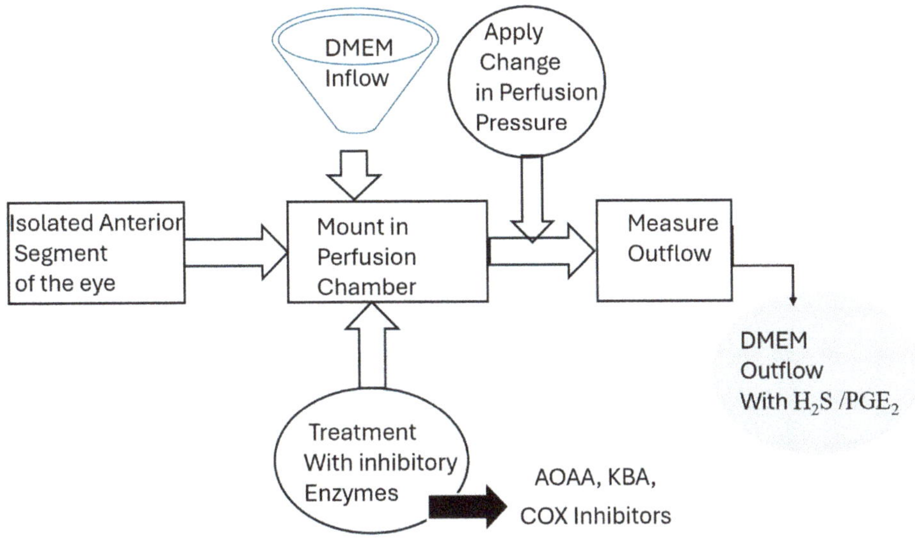

Figure 5. Experimental design.

4.3. Measurement of H_2S and Prostaglandin E_2 Concentrations

The concentration of H_2S in the media perfusing the explants in the chamber (i.e., perfusate) was measured by the Methylene Blue assay method as described by several investigators [29,42,43] and prostaglandin E_2 (PGE_2) was measured using the Enzyme Immunoassay (EIA) kit following the manufacturer's instructions (Cayman Chemical, Ann Arbor, MI, USA). After four hours, perfusates were collected, combined with 1% EDTA, 1% zinc acetate, and borate buffer (pH 11). The mixture was incubated in a 37 °C water bath for 30 min. Thereafter, N,N dimethyl p-phenylendiamine (20 mM in 7.2 M HCl) and $FeCl_3$ (300 mM) were added and incubated at 37 °C with shaking for 10 min. The resulting solution absorbance was measured at 670 nm with a spectrophotometer (Bio Tek Microplate Instrumentation, U.S. Biotek Laboratories, Shoreline, WA, USA). The concentrations of H_2S in perfusate was calculated based on a standard curve of sodium hydrosulfide (NaHS, 0–750 μM). H_2S concentrations were expressed as nM/μg protein.

In assays to measure PGE_2 concentrations, the perfusate was pretreated with indomethacin to prevent the further biosynthesis of eicosanoids during the assay period. PGE_2 levels were measured using a Prostaglandin E_2 Enzyme Immunoassay (EIA) kit following the manufacturer's instructions (Cayman Chemical, Ann Arbor, MI, USA). PGE_2 levels are expressed as pg/μg of protein.

4.4. Protein Measurement

Protein content in all samples from H_2S and PGE_2 assays was measured using the Bradford Method (Cayman Chemical, Ann Arbor, MI, USA) [44].

4.5. Statistics

Data are represented as the mean ± standard error and were analyzed using a GraphPad software (10.2.3 (403)). Increase in outflow facility compared to vehicle was considered statistically significant if $p < 0.05$ as analyzed by Student's *t*-test or one-way ANOVA with Tukey's post-test. Changes in endogenous levels of H_2S or PGE_2 was deemed statistically significant if $p < 0.05$ using Student's *t*-test.

5. Conclusions

We conclude that in the perfused porcine ocular anterior segment model of TM outflow facility, the elevation of perfusion pressure elicits an increase in H_2S levels but not that of PGE_2. Furthermore, the perfusion pressure-induced changes in H_2S levels are dependent upon the biosynthesis of endogenous H_2S and prostaglandins. Further studies are needed to elucidate the mechanism/s that underlie the observed changes in the concentrations of H_2S and prostaglandins in the perfusate of anterior segment explants under elevated perfusion pressure. It remains to be determined whether these observations ex vivo have any correlation with the outflow of aqueous humor from TM pathways during the regulation of aqueous humor dynamics and IOP in vivo.

Author Contributions: Conceptualization, J.R., S.O. and Y.F.N.M.; data curation, J.R., S.O. and Y.F.N.M.; formal analysis, J.R., C.O., S.O. and Y.F.N.M.; funding acquisition, C.O. and Y.F.N.M.; investigation, J.R., L.B., A.O., F.M., C.O., S.O. and Y.F.N.M.; methodology, J.R. and Y.F.N.M.; project administration, S.O. and Y.F.N.M.; resources, C.O. and Y.F.N.M.; software, Y.F.N.M.; supervision, S.O. and Y.F.N.M.; validation, J.R. and Y.F.N.M.; visualization, Y.F.N.M.; writing—original draft, J.R.; writing—review and editing, J.R., L.B., A.O., F.M., C.O., S.O. and Y.F.N.M. All authors have read and agreed to the published version of the manuscript.

Funding: This work was supported by the National Institutes of Health, National Eye Institute Grants 1R15EY032274-01 and 1RI5EY022215-01.

Institutional Review Board Statement: Not applicable.

Informed Consent Statement: Not applicable.

Data Availability Statement: The data have been presented in the main text.

Conflicts of Interest: The authors declare that they have no known competing financial interests.

References

1. Kimura, H. Hydrogen sulfide: Its production, release and functions. *Amino Acids* **2011**, *41*, 113–121. [CrossRef] [PubMed]
2. Kimura, H. Hydrogen Sulfide (H_2S) and Polysulfide (H_2S_n) Signaling: The First 25 Years. *Biomolecules* **2021**, *11*, 896. [CrossRef] [PubMed]
3. Tan, B.H.; Wong, P.T.; Bian, J.S. Hydrogen sulfide: A novel signaling molecule in the central nervous system. *Neurochem. Int.* **2010**, *56*, 3–10. [CrossRef] [PubMed]
4. Li, L.; Rossoni, G.; Sparatore, A.; Lee, L.C.; Del Soldato, P.; Moore, P.K. Anti-inflammatory and gastrointestinal effects of a novel diclofenac derivative. *Free Radic. Biol. Med.* **2007**, *42*, 706–719. [CrossRef] [PubMed]
5. Kimura, H. Hydrogen sulfide as a neuromodulator. *Mol. Neurobiol.* **2002**, *26*, 13–19. [CrossRef]
6. Mikami, Y.; Shibuya, N.; Kimura, Y.; Nagahara, N.; Yamada, M.; Kimura, H. Hydrogen sulfide protects the retina from light-induced degeneration by the modulation of Ca^{2+} influx. *J. Biol. Chem.* **2011**, *286*, 39379–39386. [CrossRef]
7. Shibuya, N.; Tanaka, M.; Yoshida, M.; Ogasawara, Y.; Togawa, T.; Ishii, K.; Kimura, H. 3-Mercaptopyruvate sulfurtransferase produces hydrogen sulfide and bound sulfane sulfur in the brain. *Antioxid. Redox Signal.* **2009**, *11*, 703–714. [CrossRef]
8. Mikami, Y.; Shibuya, N.; Ogasawara, Y.; Kimura, H. Hydrogen sulfide is produced by cystathionine γ-lyase at the steady-state low intracellular Ca^{2+} concentrations. *Biochem. Biophys. Res. Commun.* **2013**, *431*, 131–135. [CrossRef]
9. Beard, M.E.; Davies, T.; Holloway, M.; Holtzman, E. Peroxisomes in pigment epithelium and Müller cells of amphibian retina possess D-amino acid oxidase as well as catalase. *Exp. Eye Res.* **1988**, *47*, 795–806. [CrossRef]
10. Koga, R.; Miyoshi, Y.; Sakaue, H.; Hamase, K.; Konno, R. Mouse d-Amino-Acid Oxidase: Distribution and Physiological Substrates. *Front. Mol. Biosci.* **2017**, *4*, 82. [CrossRef]
11. Buffault, J.; Labbé, A.; Hamard, P.; Brignole-Baudouin, F.; Baudouin, C. The trabecular meshwork: Structure, function and clinical implications. A review of the literature. *J. Fr. D'ophtalmol.* **2020**, *43*, e217–e230. [CrossRef] [PubMed]
12. Stamer, W.D.; Acott, T.S. Current understanding of conventional outflow dysfunction in glaucoma. *Curr. Opin. Ophthalmol.* **2012**, *23*, 135–143. [CrossRef]
13. Salvi, A.; Bankhele, P.; Jamil, J.M.; Kulkarni-Chitnis, M.; Njie-Mbye, Y.F.; Ohia, S.E.; Opere, C.A. Pharmacological Actions of Hydrogen Sulfide Donors on Sympathetic Neurotransmission in the Bovine Anterior Uvea, In Vitro. *Neurochem. Res.* **2016**, *41*, 1020–1028. [CrossRef]
14. Robinson, J.; Okoro, E.; Ezuedu, C.; Bush, L.; Opere, C.A.; Ohia, S.E.; Njie-Mbye, Y.F. Effects of Hydrogen Sulfide-Releasing Compounds on Aqueous Humor Outflow Facility in Porcine Ocular Anterior Segments, Ex Vivo. *J. Ocul. Pharmacol. Ther. Off. J. Assoc. Ocul. Pharmacol. Ther.* **2017**, *33*, 91–97. [CrossRef] [PubMed]

15. Huang, S.; Huang, P.; Liu, X.; Lin, Z.; Wang, J.; Xu, S.; Guo, L.; Leung, C.K.; Zhong, Y. Relevant variations and neuroprotective effect of hydrogen sulfide in a rat glaucoma model. *Neuroscience* **2017**, *341*, 27–41. [CrossRef]
16. Freedman, J.; Goddard, D. Elevated levels of transforming growth factor beta and prostaglandin E2 in aqueous humor from patients undergoing filtration surgery for glaucoma. *Can. J. Ophthalmol.* **2008**, *43*, 370.
17. Mermoud, A.; Baerveldt, G.; Minckler, D.S.; Rao, N.A. Prostaglandines E2 et F2-alpha au cours du glaucome uvéitique chez le rat Lewis [Prostaglandins E2 and F2-alpha in uveitic glaucoma in the Lewis rat]. *Klin. Monatsblatter Augenheilkd.* **1995**, *206*, 409–412. [CrossRef] [PubMed]
18. Monjok, E.M.; Kulkarni, K.H.; Kouamou, G.; McKoy, M.; Opere, C.A.; Bongmba, O.N.; Njie, Y.F.; Ohia, S.E. Inhibitory action of hydrogen sulfide on muscarinic receptor-induced contraction of isolated porcine irides. *Exp. Eye Res.* **2008**, *87*, 612–616. [CrossRef]
19. Ohia, S.E.; Opere, C.A.; Monjok, E.M.; Kouamou, G.; Leday, A.M.; Njie-Mbye, Y.F. Role of hydrogen sulfide production in inhibitory action of L-cysteine on isolated porcine irides. *Curr. Eye Res.* **2010**, *35*, 402–407. [CrossRef]
20. Asimakopoulou, A.; Panopoulos, P.; Chasapis, C.T.; Coletta, C.; Zhou, Z.; Cirino, G.; Giannis, A.; Szabo, C.; Spyroulias, G.A.; Papapetropoulos, A. Selectivity of commonly used pharmacological inhibitors for cystathionine β synthase (CBS) and cystathionine γ lyase (CSE). *Br. J. Pharmacol.* **2013**, *169*, 922–932. [CrossRef]
21. Porter, D.W.; Baskin, S.I. The effect of three alpha-keto acids on 3-mercaptopyruvate sulfurtransferase activity. *J. Biochem. Toxicol.* **1996**, *11*, 45–50. [CrossRef]
22. Chitnis, M.K.; Njie-Mbye, Y.F.; Opere, C.A.; Wood, M.E.; Whiteman, M.; Ohia, S.E. Pharmacological actions of the slow-release hydrogen sulfide donor GYY4137 on phenylephrine-induced tone in isolated bovine ciliary artery. *Exp. Eye Res.* **2013**, *116*, 350–354. [CrossRef] [PubMed]
23. Kulkarni-Chitnis, M.; Njie-Mbye, Y.F.; Mitchell, L.; Robinson, J.; Whiteman, M.; Wood, M.E.; Opere, C.A.; Ohia, S.E. Inhibitory action of novel hydrogen sulfide donors on bovine isolated posterior ciliary arteries. *Exp. Eye Res.* **2015**, *134*, 73–79. [CrossRef]
24. Pong, W.W.; Stouracova, R.; Frank, N.; Kraus, J.P.; Eldred, W.D. Comparative localization of cystathionine beta-synthase and cystathionine gamma-lyase in retina: Differences between amphibians and mammals. *J. Comp. Neurol.* **2007**, *505*, 158–165. [CrossRef] [PubMed]
25. Acott, T.S.; Kelley, M.J.; Keller, K.E.; Vranka, J.A.; Abu-Hassan, D.W.; Li, X.; Aga, M.; Bradley, J.M. Intraocular pressure homeostasis: Maintaining balance in a high-pressure environment. *J. Ocul. Pharmacol. Ther.* **2014**, *30*, 94–101. [CrossRef]
26. Vranka, J.A.; Acott, T.S. Pressure-induced expression changes in segmental flow regions of the human trabecular meshwork. *Exp. Eye Res.* **2017**, *158*, 67–72. [CrossRef]
27. Vranka, J.A.; Staverosky, J.A.; Raghunathan, V.; Acott, T.S. Elevated pressure influences relative distribution of segmental regions of the trabecular meshwork. *Exp. Eye Res.* **2020**, *190*, 107888. [CrossRef]
28. Schneemann, A.; Leusink-Muis, A.; van den Berg, T.; Hoyng, P.F.; Kamphuis, W. Elevation of nitric oxide production in human trabecular meshwork by increased pressure. *Graefes Arch. Clin. Exp.* **2003**, *241*, 321–326. [CrossRef]
29. Kulkarni, M.; Njie-Mbye, Y.F.; Okpobiri, I.; Zhao, M.; Opere, C.A.; Ohia, S.E. Endogenous production of hydrogen sulfide in isolated bovine eye. *Neurochem. Res.* **2011**, *36*, 1540–1545. [CrossRef]
30. Abe, K.; Kimura, H. The possible role of hydrogen sulfide as an endogenous neuromodulator. *J. Neurosci. Off. J. Soc. Neurosci.* **1996**, *16*, 1066–1071. [CrossRef]
31. Kolluru, G.K.; Shen, X.; Bir, S.C.; Kevil, C.G. Hydrogen sulfide chemical biology: Pathophysiological roles and detection. *Nitric Oxide Biol. Chem.* **2013**, *35*, 5–20. [CrossRef] [PubMed]
32. Huang, K.; Wang, Z.; Gu, Y.; Hu, Y.; Ji, Z.; Wang, S.; Lin, Z.; Li, X.; Xie, Z.; Pan, S. Glibenclamide Is Comparable to Target Temperature Management in Improving Survival and Neurological Outcome After Asphyxial Cardiac Arrest in Rats. *J. Am. Heart Assoc.* **2016**, *5*, e003465. [CrossRef] [PubMed]
33. Winkler, N.S.; Fautsch, M.P. Effects of prostaglandin analogues on aqueous humor outflow pathways. *J. Ocul. Pharmacol. Ther.* **2014**, *30*, 102–109. [CrossRef]
34. Toris, C.B.; Gabelt, B.T.; Kaufman, P.L. Update on the mechanism of action of topical prostaglandins for intraocular pressure reduction. *Surv. Ophthalmol.* **2008**, *53* (Suppl. S1), S107–S120. [CrossRef] [PubMed]
35. Bahler, C.K.; Howell, K.G.; Hann, C.R.; Fautsch, M.P.; Johnson, D.H. Prostaglandins increase trabecular meshwork outflow facility in cultured human anterior segments. *Am. J. Ophthalmol.* **2008**, *145*, 114–119. [CrossRef]
36. Maihöfner, C.; Schlötzer-Schrehardt, U.; Gühring, H.; Zeilhofer, H.U.; Naumann, G.O.; Pahl, A.; Mardin, C.; Tamm, E.R.; Brune, K. Expression of cyclooxygenase-1 and -2 in normal and glaucomatous human eyes. *Investig. Ophthalmol. Vis. Sci.* **2001**, *42*, 2616–2624.
37. Kwiatkoski, M.; Soriano, R.N.; Araujo, R.M.; Azevedo, L.U.; Batalhao, M.E.; Francescato, H.D.; Coimbra, T.M.; Carnio, E.C.; Branco, L.G. Hydrogen sulfide inhibits preoptic prostaglandin E2 production during endotoxemia. *Exp. Neurol.* **2013**, *240*, 88–95. [CrossRef]
38. Sun, Q.; Chen, Z.; He, P.; Li, Y.; Ding, X.; Huang, Y.; Gu, H.; Ni, X. Reduced Expression of Hydrogen Sulfide-Generating Enzymes Down-Regulates 15-Hydroxyprostaglandin Dehydrogenase in Chorion during Term and Preterm Labor. *Am. J. Pathol.* **2018**, *188*, 63–71. [CrossRef]

39. Ang, S.F.; Sio, S.W.; Moochhala, S.M.; MacAry, P.A.; Bhatia, M. Hydrogen sulfide upregulates cyclooxygenase-2 and prostaglandin E metabolite in sepsis-evoked acute lung injury via transient receptor potential vanilloid type 1 channel activation. *J. Immunol.* **2011**, *187*, 4778–4787. [CrossRef]
40. Njie, Y.F.; Qiao, Z.; Xiao, Z.; Wang, W.; Song, Z.H. N-arachidonylethanolamide-induced increase in aqueous humor outflow facility. *Investig. Ophthalmol. Vis. Sci.* **2008**, *49*, 4528–4534. [CrossRef]
41. Yang, Y.F.; Sun, Y.Y.; Acott, T.S.; Keller, K.E. Effects of induction and inhibition of matrix cross-linking on remodeling of the aqueous outflow resistance by ocular trabecular meshwork cells. *Sci. Rep.* **2016**, *6*, 30505. [CrossRef] [PubMed]
42. Smith, H.M.; Pluth, M.D. Advances and Opportunities in H_2S Measurement in Chemical Biology. *JACS Au* **2023**, *3*, 2677–2691. [CrossRef] [PubMed]
43. Zou, S.; Shimizu, T.; Shimizu, S.; Higashi, Y.; Nakamura, K.; Ono, H.; Aratake, T.; Saito, M. Possible role of hydrogen sulfide as an endogenous relaxation factor in the rat bladder and prostate. *Neurourol. Urodyn.* **2018**, *37*, 2519–2526. [CrossRef] [PubMed]
44. Kruger, N.J. The Bradford method for protein quantitation. *Methods Mol. Biol.* **1994**, *32*, 9–15. [PubMed]

pharmaceuticals

Article

Assessment of Brain-Derived Neurotrophic Factor on Retinal Structure and Visual Function in Rodent Models of Optic Nerve Crush

Takazumi Taniguchi [1], Najam A. Sharif [2,3,4,5,6,7,8,9,10], Takashi Ota [1], Rafal A. Farjo [11] and Rebecca Rausch [11,*]

1 Ophthalmology Innovation Center, Santen Pharmaceutical Co., Ltd., Nara 630-0101, Japan
2 Ophthalmology Innovation Center, Santen Inc., Emeryville, CA 94608, USA
3 Institute of Ophthalmology, University College London, London WC1E 6BT, UK
4 Imperial College of Science and Technology, St. Mary's Campus, London SW7 2AZ, UK
5 Eye-APC Duke-NUS Medical School, Singapore 169857, Singapore
6 Department of Pharmacy Sciences, Creighton University, Omaha, NE 68178, USA
7 Department of Pharmaceutical Sciences, College of Pharmacy and Health Sciences, Texas Southern University, Houston, TX 77004, USA
8 Department of Pharmacology and Neuroscience, University of North Texas Health Sciences Center, Fort Worth, TX 76107, USA
9 Singapore Eye Research Institute, Singapore 169856, Singapore
10 Global Research and Development, Nanoscope Therapeutics Inc., Dallas, TX 75207, USA
11 EyeCRO LLC, Oklahoma City, OK 73105, USA
* Correspondence: rrausch@eyecro.com

Abstract: The effects of brain-derived neurotrophic factor (BDNF) on retinal ganglion cell (RGC) survival and visual function were assessed in rat and mouse models of optic nerve (ON) crush. ONs were crushed on Day 1, followed by intravitreal injections of a vehicle or BDNF on Days 1 and 8. The spatial frequency threshold was measured using optokinetic tracking on Days 7 and 14. On Day 15, ganglion cell complex (GCC) thickness was quantified using optical coherence tomography. Furthermore, all eyes were enucleated for immunohistochemical analysis of the surviving RGC somas and axons. BDNF significantly reduced the RGC soma in mice and increased GCC thickness in intact eyes, with apparent axonal swelling in both species. It displayed significantly greater RGC soma survival in eyes with ON injury, with moderately thicker axonal bundles in both species and a thicker GCC in rats. Visual function was significantly reduced in all ON-crushed animals, regardless of BDNF treatment. Thus, we obtained a comprehensive analysis of the structural and functional impact of BDNF in intact and ON-crushed eyes in two rodent models. Our results provide a foundation for further BDNF evaluation and the design of preclinical studies on neuroprotectants using BDNF as a reference positive control.

Keywords: brain-derived neurotrophic factor; optic nerve crush; retinal ganglion cells; glaucoma; neuroprotection

1. Introduction

Glaucoma, a major optic neuropathy, is the leading cause of irreversible vision loss and blindness worldwide [1,2]. This complex neurodegenerative disease affects the retinal ganglion cell (RGC) somas and axonal projections that comprise the optic nerve [1,3]. Glaucomatous optic neuropathy (GON) is characterized by damage to the optic nerve head, with progressive cupping and distinctive progressive changes in the visual field. Elevated intraocular pressure (IOP) is a major risk factor for the development and progression of glaucoma, and lowering the IOP pharmaceutically or surgically is the only known treatment option for this disease [1,3]. However, further visual field deterioration continues despite good IOP management in some cases. This suggests that the presence of IOP-independent mechanisms, such as neurotrophic factor deprivation, oxidative stress, ocular blood flow disturbance, and inflammation, contributes to the development and progression of glaucoma [3–6]. Therefore, new treatment options that directly protect RGCs and the optic nerve in an IOP-independent manner are needed to prevent blindness. Many neuroprotective strategies using small-molecule drugs in addition to cell and gene therapies with various modes of action have been proposed. However, there remain no approved treatment options to date [3–6].

Animal models that mirror or closely resemble human disease mechanisms are critical to elucidating disease etiology and molecular pathobiology. These models enable the progression of new therapies into clinical development [7]. To obtain meaningful preclinical data on candidate neuroprotective agents for glaucoma during drug discovery and development, the establishment of clinically relevant animal models with non-IOP-related factors associated with glaucomatous damage to the RGCs and/or the optic nerve is critical. One of the hypothesized pathogeneses of GON is the local damage of lamina cribrosa, disrupting axoplasmic transport within the optic nerve fibers [1,3]. Therefore, animal models in which forceps or a surgical clip are used to crush the optic nerve in a controlled manner, thereby causing specific RGC soma and optic nerve degeneration in an IOP-independent manner, have been extensively used to evaluate candidate neuroprotective agents in preclinical settings in glaucoma research. Additionally, these models have been used to study the pathophysiological events involved in RGC damage caused by the loss of retrograde transport of trophic factors [8–15]. Thus, optic nerve crush injuries in rats and mice have become popular animal models with several advantages. Optic nerve injury caused by mechanical crush in these species is relatively easy to establish and highly reproducible. In addition, rat and mouse eyes have several similarities to the human eye, although the non-human primate eye has the closest anatomical structure to that of humans [13,14]. Furthermore, because rodents are much less expensive to purchase and maintain than non-human primates, larger numbers can be used to obtain statistically meaningful data [13]. Many potential agents with neuroprotective effects have been identified from various studies using these optic nerve injury models [8–11,15].

Neurotrophic factors, particularly brain-derived neurotrophic factor (BDNF), exert a robust neuroprotective effect on RGC somas histologically analyzed in rodent optic nerve crush models [10,15–17]. This factor confers neuroprotective effects directly via the Tropomyosin receptor kinase B (TrkB) expressed in RGCs and/or indirectly via the TrkB expressed in glial cells. BDNF is locally produced by RGCs and other retinal cells, but the retrograde transport of BDNF from the brain to the retina is crucial [18]. In addition to the interruption of this retrograde transport in the early stages of glaucoma, lower levels of BDNF were detected in the serum and ocular fluid of patients with glaucoma compared to people without the disease, supporting the theory that neurotrophic deprivation is a potential mechanism of GON [18]. Although the precise factors that contribute to glaucoma are still debated, the neurotrophin deprivation theory represents one of the most predominant proposed contributors to this disease [4–6].

BDNF could be a useful agent for various purposes, such as elucidating the mechanism underlying RGC damage following optic nerve crush. In addition, it can be used as a comparator/reference active compound during the development of neuroprotective candidates. On the other hand, there is a lack of assessment of BDNF on retinal thickness using clinically useful optical coherence tomography (OCT) and on visual function measurements. These are important indices for evaluating the structure and function of candidate drugs in optic nerve crush models. Furthermore, although an understanding of the impact of BDNF in intact eyes is needed to correctly interpret the effects of this trophic factor in crushed eyes, these parameters have not yet been fully investigated.

In this study, we simultaneously assessed the effects of intravitreal injection of BDNF on RGC soma survival, retinal thickness, and visual function in naïve and crushed optic nerve eyes of rats and mice, as illustrated in Figure 1. Additionally, data were compared between the two species.

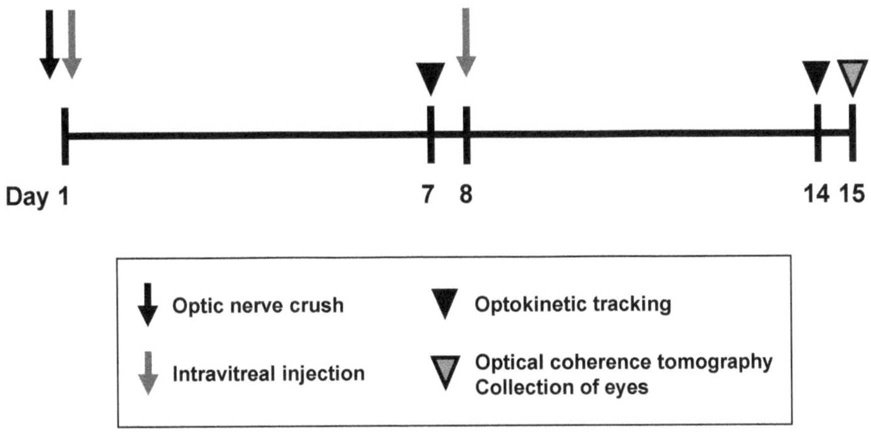

Figure 1. An outline of optic nerve crush experiments in rats and mice. The optic nerve was crushed just before the intravitreal injection of the vehicle, or BDNF, on Day 1. A second intravitreal injection was performed on Day 8. Optokinetic tracking was performed to assess visual function on Days 7 and 14. Optical coherence tomography was performed to quantify the ganglion cell complex (GCC) thickness on Day 15, followed by animal euthanasia, after which eyeballs were enucleated for flatmount immunohistochemistry to label retinal ganglion cells (RGCs) with RBPMS and TUJ1.

2. Results

2.1. Retinal Ganglion Cell Survival

Rat retinal flatmounts were immunolabeled with RNA-binding protein with multiple splicing (RBPMS) antibodies to visualize RGC somas following terminal collection two weeks post-crush. Optic nerve crush resulted in a significant loss of RGC somas in the vehicle-treated group ($p < 0.001$, Aspin-Welch t-test). No statistically significant difference in RGC soma survival was observed between uncrushed vehicle-treated eyes and uncrushed BDNF-treated eyes, although there was a slight trend toward a reduction in the BDNF group. Crushed eyes injected with BDNF exhibited significantly greater RGC survival than the vehicle-injected controls (Figure 2a,b). All retinas were co-labeled with beta-tubulin III (anti-TUJ1), which labels the RGC axon in addition to the soma, to qualitatively assess RGC axon degeneration proximal to the site of the optic nerve crush injury. Uncrushed nerves showed typical RGC axon integrity regardless of BDNF injection. However, crushed nerves treated with the vehicle displayed characteristic signs of axonal degeneration, such as thinning, beading, and fragmentation. The axon bundles were moderately thicker in eyes administered with BDNF than in crushed eyes receiving the vehicle injection (Figure 2c). These findings directly correlate with the significant protection observed in RGC soma counts.

Optic nerve crush significantly induced the loss of RGC soma in the vehicle-treated group in mouse experiments ($p < 0.001$, Aspin-Welch t-test). The RGC soma number was significantly reduced in BDNF-injected, uncrushed eyes. In crushed eyes, the protective effect of BDNF was small but exerted statistically significant RGC soma protection compared to vehicle-injected controls (Figure 3a,b). Uncrushed nerves labeled with anti-TUJ1 corresponded with typical RGC axon integrity, regardless of BDNF injection. However, crushed nerves with vehicle injection displayed signs of axonal degeneration. In the group receiving BDNF, the axon bundles were moderately thicker than those in the group of crushed optic nerves that received the vehicle injection (Figure 3c). These mouse-based findings of immunostaining with anti-TUJ1 were similar to those observed in rat experiments.

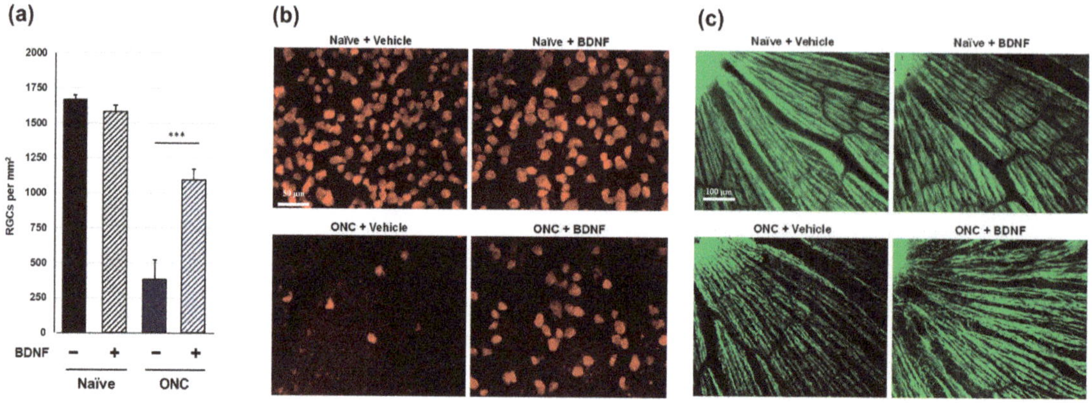

Figure 2. Effects of BDNF on retinal ganglion cell survival in rats. RGC soma count (**a**), as measured in uncrushed and crushed eyes administered a vehicle or BDNF injection. Vehicle, closed column; BDNF, diagonal column. Each value represents the mean ± standard error of the mean (SEM) for n = 8 (**a**). *** $p < 0.001$, as determined using Student's t-test. Representative images from retinal flatmounts stained with anti-RBPMS to label surviving RGCs in peripheral regions (**b**) and with anti-TUJ1 to label RGC axons close to the optic nerve head (**c**) in rats. No difference in RGC soma count was observed between the uncrushed eyes, regardless of treatment. Crushed eyes receiving BDNF displayed significantly greater RGC survival compared to those injected with the vehicle. Axonal density and integrity were decreased in the crushed eyes. However, axon bundles appeared moderately thicker in BDNF-injected eyes compared to vehicle-injected controls. Note: The bright spot in the top left represents the optic nerve head (**c**).

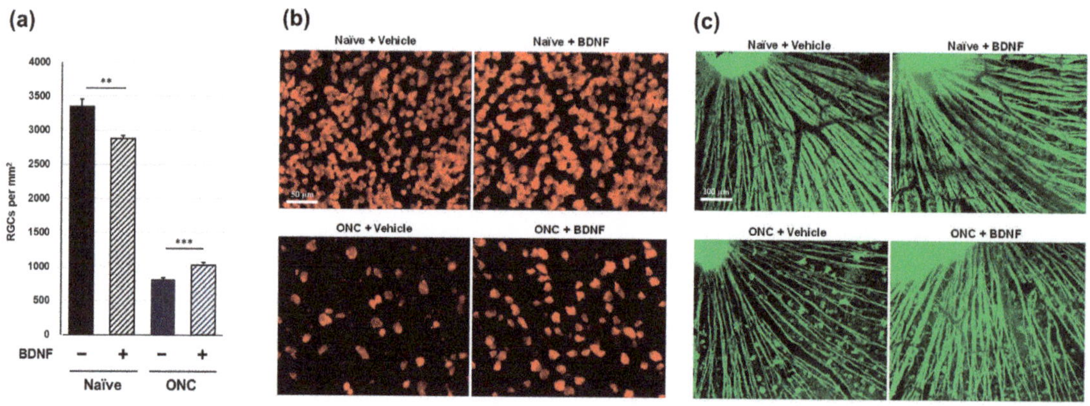

Figure 3. Effects of BDNF on retinal ganglion cell survival in mice. RGC soma number, as measured in uncrushed and crushed eyes administered a vehicle or BDNF injection. Vehicle, closed column; BDNF, diagonal column. Each value represents the mean ± SEM for n = 8 (**a**). ** $p < 0.01$, *** $p < 0.001$, as determined using Student's t-test. Representative images from retinal flatmounts stained with anti-RBPMS to label surviving RGCs in peripheral regions (**b**) and with anti-TUJ1 to label RGC axons close to the optic nerve head (**c**) in mice. BDNF slightly decreased RGC soma survival in uncrushed eyes compared to vehicle-injected eyes. Crushed eyes receiving BDNF displayed greater RGC survival compared to those injected with the vehicle. Axonal density and integrity were decreased in crushed eyes. However, axon bundles appeared thicker in eyes administered BDNF compared to vehicle-injected controls. Note: The bright spot in the top left represents the optic nerve head (**c**).

2.2. Ganglion Cell Complex Thickness

In vivo OCT was performed immediately prior to euthanasia to examine the effect of BDNF administration after optic nerve crush on the retinal structure in rats. The ganglion cell complex (GCC), measured from the retinal nerve fiber layer (RNFL) to the inner plexiform layer and comprising RGC somas, dendrites, and axons, was quantified using circular scans acquired from the peripapillary region of the retina. Optic nerve crush caused a significant reduction in GCC thickness in the vehicle-treated group ($p < 0.01$, Student's *t*-test). Administration of BDNF significantly increased GCC thickness in uncrushed eyes, likely due to axonal swelling in the RNFL, though this was not directly confirmed (Figure 4). This result was less apparent in crushed eyes, although GCC thickness was significantly greater in the BDNF injection group than in the vehicle group (Figure 4).

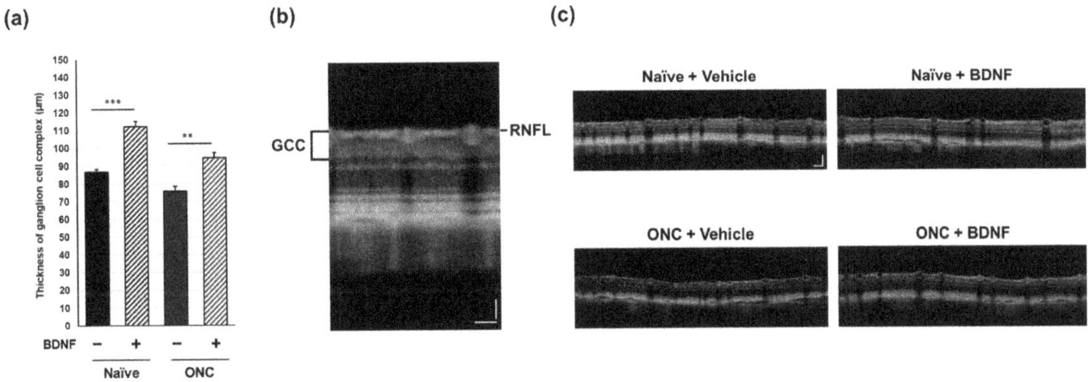

Figure 4. Effects of BDNF on ganglion cell complex (GCC) thickness in rats. GCC thickness (**a**), as measured in uncrushed and crushed eyes administered a vehicle or BDNF injection. Vehicle, closed column; BDNF, diagonal column. Each value represents the mean ± standard error of the mean (SEM) for n = 4–8 (**a**). ** $p < 0.01$, *** $p < 0.001$, as determined using Student's *t*-test. Representative peripapillary circle scans from optical coherence tomography measurements in rats (**b**). The GCC thickness, which consists of the retinal nerve fiber layer (RNFL), ganglion cell layer (GCL), and inner plexiform layer (IPL), was measured. ONC reduced GCC thickness in vehicle-injected eyes. Increased thickness was evident in the RNFL of BDNF-injected eyes without optic nerve crush. A slight RNFL increase was also qualitatively observed in BDNF-injected crushed eyes (**c**). Scale bar = 50 mm.

In the mouse experiments, optic nerve crush caused a statistically significant decrease in GCC thickness in the vehicle-treated group ($p < 0.001$, Student's *t*-test). Administration of BDNF significantly increased the GCC thickness in uncrushed eyes, similar to that observed in the rat experiments (Figure 5). In the crushed eyes, GCC thickness in the BDNF injection group was slightly thicker than that in the vehicle-injected group, but the difference was not statistically significant (Figure 5).

2.3. Optokinetic Tracking

Spatial frequency thresholds (SFT) were analyzed on Days 7 and 14 in the rats and mice to assess visual function following an optic nerve crush injury and BDNF treatment. There was no significant difference in SFT between the uncrushed groups at either time point for either species (Figures 6 and 7). A significant reduction in SFT was observed in animals with crushed optic nerves, regardless of BDNF treatment, although there was a small trend of functional recovery in the group that received BDNF in rats (Figures 6 and 7).

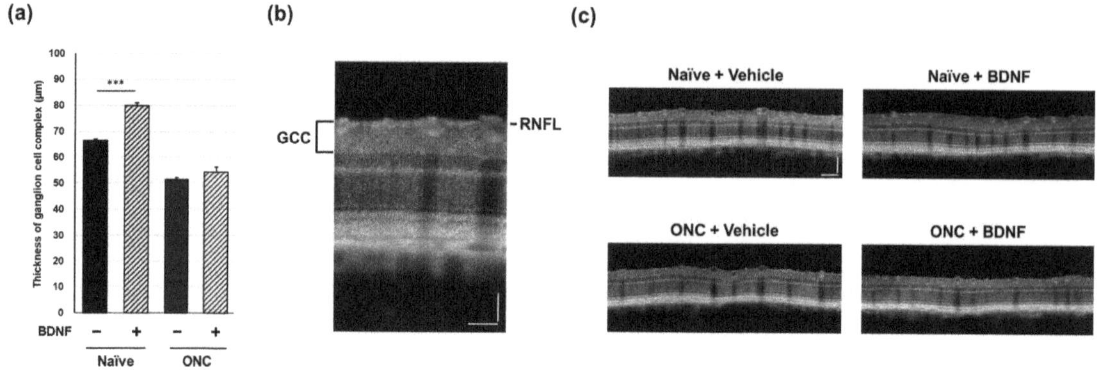

Figure 5. Effects of BDNF on ganglion cell complex (GCC) thickness in mice. GCC thickness (**a**), as measured in uncrushed and crushed eyes administered a vehicle or BDNF injection. Vehicle, closed column; BDNF, diagonal column. Each value represents the mean ± SEM for n = 7–8 (**a**). *** $p < 0.001$, as determined using Student's *t*-test. Representative peripapillary circle scans from optical coherence tomography measurements in mice (**b**). The GCC thickness was measured. ONC reduced GCC thickness in vehicle-injected eyes. Increased thickness was only evident in the RNFL of uncrushed eyes administered BDNF (**c**). Scale bar = 100 mm.

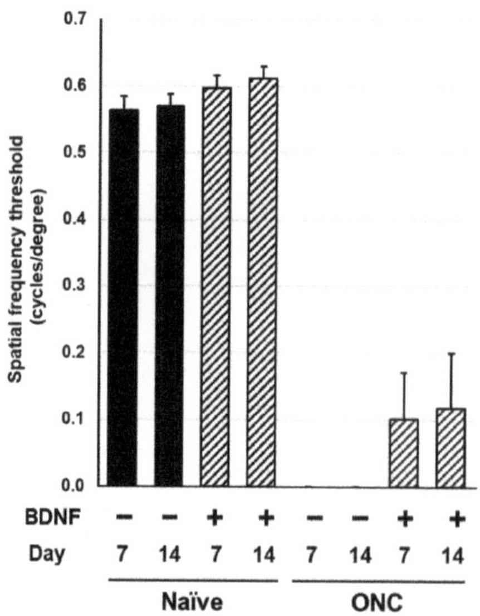

Figure 6. Effects of BDNF on optokinetic tracking in rats. Spatial frequency threshold, as measured in uncrushed and crushed eyes administered a vehicle or BDNF injection. Vehicle, closed column; BDNF, diagonal column. Each value represents the mean ± standard error of the mean (SEM) for n = 8–10.

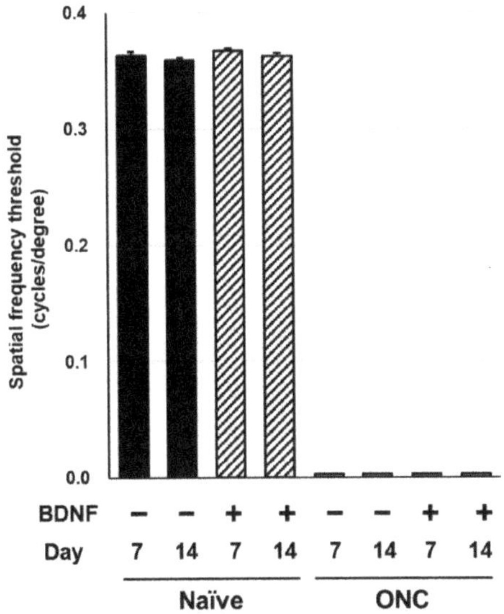

Figure 7. Effects of BDNF on optokinetic tracking in mice. Spatial frequency threshold, as measured in uncrushed and crushed eyes administered a vehicle or BDNF injection. Vehicle, closed column; BDNF, diagonal column. Each value represents the mean ± SEM for n = 5–8.

3. Discussion

We simultaneously assessed the effects of intravitreally administered BDNF on RGC survival and visual function in rat and mouse eyes with intact and crushed optic nerves, providing neuroprotective effect profiles for this trophic factor in a relevant disease model. The key findings of the present study are as follows: (1) RGC soma survival is a robust index for detecting the neuroprotective effects of BDNF in the optic nerve crush injury model of GON; (2) BDNF administration causes RGC soma loss (only in mice) and increased GCC thickness in intact eyes (both species), indicating that the evaluation of BDNF in uncrushed eyes is required to properly assess the effects of BDNF in crushed eyes; and (3) the neuroprotective effects of BDNF were similar between rats and mice, with a few exceptions.

Consistent with previous studies [15–17], intravitreal injection of BDNF afforded some protection of the RGC soma from optic nerve injury in rats and mice (Figures 2a and 3a). The two-time injection of BDNF (10 μg/eye in rats and 2 μg/eye in mice) demonstrated that its neuroprotective efficacy was stronger in rats than in mice. In both species, the partial protective effects of BDNF on RGC death may be due to the relatively short half-life of this trophic factor [18]. On the other hand, BDNF significantly induced RGC soma loss in uncrushed mouse eyes (Figure 3a). Thus, a lower BDNF dose may exert stronger neuroprotective effects in mice. BDNF binds to two receptors (the p75 neurotrophin receptor and the TrkB receptor), and stimulation of the p75 receptor by BDNF activates cell death signaling [19,20]. The mechanism underlying the RGC soma loss observed in the intact eyes of mice is unclear. However, stimulation of the p75 receptor may be a contributing factor.

OCT has been widely used in clinical settings to analyze retinal structures and help assess the diagnosis and progression of GON [21]. Therefore, OCT analysis in animal models is valuable for translating pre-clinical data to human subjects. Intriguingly, the intravitreal injection of BDNF increased GCC thickness in the intact eyes of both rats and mice (Figures 4a and 5a). Although the causative mechanism of this potentially negative impact of BDNF on the retina is unclear, this is likely due to axonal swelling in the RNFL (Figures 4c and 5c). However, this is not unprecedented, as a previous study also indicated that BDNF induces axonal swelling in the injured optic nerve following a lens injury [22]. In rats, GCC thickness in BDNF-injected eyes with crushed optic nerves was greater than that in the corresponding vehicle control group (Figure 4a). This effect seems to be reflected in the presumed direct axonal swelling effect of BDNF, which remains unclear.

Optic nerve crush completely ablated visual function, as assessed through SFT in rats and mice (Figures 6 and 7). The intraretinal axons labeled with the anti-TUJ1 antibody (Figures 2c and 3c) were not completely damaged, suggesting that this functional index of the SFT is more sensitive than the structural indices assessed in this study. Under the current experimental conditions, BDNF did not provide significant functional protection in either rodent species (Figures 6 and 7). We crushed the optic nerve using self-closing forceps to induce mechanical injury. It is important to note that the level of optic nerve damage can vary depending on the crush technique, such as the magnitude of the applied crush force and the duration of the crush injury using forceps. Although our technique produced consistent and reproducible damage, the extent of the injury may have been too severe to allow for any positive effects of BDNF on the visual function of the animals. Interestingly, uncrushed eyes injected with BDNF did not exhibit changes to visual function (Figures 6 and 7), despite GCC swelling in both species and the loss of RGC somas observed in mice. The reason behind the apparent lack of correlation between functional and structural indexes remains unclear, but intrinsic complementary mechanisms might have been activated in these uncrushed eyes.

Our findings suggest that RGC soma survival is a more robust index for detecting the protective effects of BDNF in rodent optic nerve crush injury models compared to other structural and functional indices. However, the latter may be altered under different experimental conditions, such as the degree of optic nerve injury, administered BDNF dose, frequency of BDNF injections, and timing of evaluation following optic nerve crush. Further, extensive studies are required to confirm these hypotheses. To the best of our knowledge, only a few studies have investigated the effects of BDNF on intact rodent eyes (e.g., the effect on the TrkB receptor and GFAP expressions) [23,24]. The present study revealed the effects of BDNF on RGC soma survival, GCC thickness, and visual function in intact rodent eyes.

Overall, there were data similarities between rats and mice in the present study. Therefore, we could use either species for optic nerve crush experiments, depending on the purpose. Rats may be preferable in studies focusing on the pharmacokinetics and tolerability of candidate neuroprotective drugs, owing to their larger eyeballs, whereas the greater availability of mice makes them the ideal species for studies using genetically modified animals. Furthermore, the surgical procedure differs slightly between the two species. Rats require the use of larger and stronger self-closing forceps, more time spent exposing the nerve, and additional pain relief post-surgery. While nerve exposure is performed more quickly in mice, there is a greater chance of unintentional blood vessel laceration inhibiting proper visualization for a successful crush. Mice tend to tolerate the procedure better than rats, requiring less pain reliving intervention post-surgery. If properly crushed, both species produce highly reproducible results.

In summary, we obtained fundamental data on the effects of intravitreally injected BDNF in intact eyes and those with crushed optic nerves in rats and mice. As for the comparative standpoint between intact and crushed eyes, the impacts of BDNF were different depending on the indexes, as described above. A series of these baseline data will be useful to further analyze the effects of BDNF and its relevant receptor agonists, such as TrkB receptor agonists, on optic nerve injury in rodents.

4. Methods and Materials

4.1. Animals

Female Brown Norway rats (10–13 weeks old) were purchased from Charles River Laboratories. Female C57Bl/6J mice (6–9 weeks old) were purchased from Jackson Laboratories. All animal studies were reviewed and approved by the Institutional Animal

Care and Use Committee of EyeCRO (Ethics number #2021-07-16-001), and all procedures adhered to the ARVO Statement for the Use of Animals in Ophthalmic and Vision Research, state guidelines, and local regulations. Animals were housed in groups of 3–5 in cages and kept on ventilated shelves under standard animal care conditions (humidity: 30–70%; temperature: 20–26 °C; 12 h light/dark cycle). In addition, they had ad libitum access to water and food (LabDiet PicoLab Rodent Diet 20; #5053).

4.2. Anesthesia

The animals were sedated with ketamine (rats, 60 mg/kg; mice, 85 mg/kg) and xylazine (rats, 9 mg/kg; mice, 14 mg/kg), which were administered intraperitoneally. Sedation was verified based on the lack of response to a toe pinch before any invasive procedure. Breathing and heart rate were monitored throughout the procedure until consciousness was regained.

4.3. Optic Nerve Crush

A small incision was made in the temporal conjunctiva of fully sedated animals, taking care to avoid the underlying musculature and vasculature. Forceps were then used to retract the conjunctiva to expose the posterior globe and allow access to and visualization of the optic nerve. To proceed with the optic nerve crush surgery, self-closing forceps were used to grasp and pinch the optic nerve approximately 1–3 mm from the globe, and the pressure was maintained for 5 s (mice) or 25 s (rats). The forceps were then removed, and a local anesthetic (proparacaine) and antibiotic ointment (gentamycin) were applied to the exposed area. Ketofen (mice and rats: 5 mg/kg) and Buprenorphine (rats only: 0.05 mg/kg) were subcutaneously injected immediately following the procedure and again at 24 and 48 h post-procedure to provide pain relief. The animals were monitored daily for signs of infection, bleeding, or loss of eye motor control.

4.4. Intravitreal Injection

The animals were first sedated and dilated with Cyclomydril (cyclopentolate/phenylephrine). A pilot hole was created in the pars plana using a 30-gauge needle. A total volume of 5 μL/eye (rats) or 1 μL/eye (mice) was injected into the vitreous using a Hamilton syringe and a blunt tip 33-gauge needle and gently dispensed over 5–10 s. Care was taken to avoid damaging the lens and retina. A topical antibiotic ointment was applied to the eye following the injection. If the optic nerve crush and intravitreal injections were performed on the same day, the injections were administered immediately following the optic nerve crush.

4.5. Growth Factor Formulation

Animal-free BDNF (#AF-450-02) was purchased from PeproTech Inc. (Cranbury, NJ, USA). BDNF was prepared at 2 μg/μL in 1% BSA-PBS, sterile-filtered, and stored at 4 °C. The formulations were stored on ice during dosing. The BDNF dose was determined based on prior in-house validation studies at EyeCRO.

4.6. Experimental Design

The optic nerve was crushed just before the intravitreal injection of the vehicle or BDNF. A second injection of the vehicle or BDNF was administered seven days after the first intravitreal injection. This injection timing was chosen based upon preliminary experiments to find the frequency that limited the potential for ocular damage caused by multiple injections. Next, optokinetic tracking was performed to measure the spatial frequency threshold to assess visual function six and thirteen days after the optic nerve crush. Fourteen days after the optic nerve crush, optical coherence tomography was performed to quantify the retinal thickness of the GCC. The animals were euthanized immediately thereafter, and the eyeballs were enucleated for flatmount immunohistochemistry to label RGCs with RBPMS and TUJ1. The time points for immunohistological assessment of

RGC soma survival, spatial frequency threshold evaluation, and OCT measurements were determined in preliminary studies. The experimental design is illustrated in Figure 1.

4.7. Retinal Flatmount Immunohistochemistry and RGC Quantification

The enucleated eyes were fixed in 4% paraformaldehyde for 2 h at 20–26 °C before retinal dissection. Free-floating retinas were double-labeled in a 48-well plate with anti-RBPMS (GeneTex, Irvine, CA, USA, Cat #GTX118619, Lot #44181; 1:500 dilution) and anti-TUJ1 (BioLegend, San Diego, CA, USA, Cat #801202, Lot #B249869; 1:1000 dilution) primary antibodies for 48 h in PBS containing 0.3% Triton-X. They were subsequently incubated in the corresponding secondary antibodies (donkey anti-rabbit 555, Thermo Fisher Scientific, Waltham, MA, USA, Cat#A31572, 1:1000 dilution, and donkey anti-mouse 488, Thermo Fisher Scientific, Cat#A21202, 1:1000 dilution) for 24 h in PBS and flatmounted onto a microscope slide. Eight evenly spaced 40× magnified images were captured from the periphery of each retinal flatmount using a Nikon epifluorescence microscope and analyzed using YOLO v1.0 (EyeCRO, LLC, Oklahoma City, OK, USA) to identify and quantify RBPMS-positive cells. The eight individual counts were then summed and used to estimate the RGC density per mm^2 for each flatmount. Representative TUJ1 images near the central retina were captured at 20× magnification and used to qualitatively assess axonal integrity. All imaging analyses were performed in a blinded manner.

4.8. Optical Coherence Tomography

Animals were anesthetized, and their eyes were dilated according to standard procedures. Following sedation and dilation, the animals were secured on a platform and assessed using a Micron IV OCT module (Phoenix-Micron, Inc., Bend, OR, USA) for imaging. The thickness of the GCC, which consists of the retinal nerve fiber layer (RNFL), ganglion cell layer (GCL), and inner plexiform layer (IPL), was measured using peripapillary circle scans acquired from each eye, and the data were plotted as the average thickness across the scans.

4.9. Optokinetic Tracking

Optokinetic tracking was performed using OptoMotry (Cerebral Mechanics Inc., Lethbridge, AB, Canada), which is designed for rodent use. For this noninvasive assessment, animals were placed on a platform surrounded by four LCD screens that resided within a light-protected box. Visual stimuli were then presented to the animal via LCD screens, and a masked observer visualized and scored the optokinetic tracking reflexes using a digital camcorder mounted on top of the box. The monitors displayed continuous vertical sine-wave gratings rotating across at 12°/s, which appeared to the animal as a virtual three-dimensional rotating sphere. The rotation of the virtual cylinder was constantly centered on the viewing position of the animal to ensure a consistent viewing distance. Tracking movements were identified as slow and steady head movements in the direction of the rotating grating. The animals were tested at spatial frequencies ranging from 0.064 to 0.514 cycles/degree to measure the spatial frequency threshold. The OptoMotry device employs a proprietary algorithm to accept input from the masked observer and automatically adjust the testing stimuli based on whether the animal exhibits a tracking reflex at a given frequency.

4.10. Statistical Analysis

For comparison between the two groups, statistical analysis was performed using the F-test, followed by Student's t-test or Aspin-Welch test using EXSUS software (version 10.0.7) and SAS version 9.4 (EPS Corporation, Tokyo, Japan) according to the manufacturers' instructions. Data are presented as the mean ± standard error of the mean. $p < 0.05$ was considered statistically significant.

5. Conclusions

Our findings could facilitate the design of preclinical neuroprotective studies of various candidate compounds for optic nerve degeneration using BDNF as a reference agent, resulting in the refinement of drug development and furthering our understanding of the mechanisms underlying optic nerve degeneration. They may also be useful to improve the possibility of applying BDNF and its relevant pathways as a neuroprotective strategy for optic nerve injury. Future research directions should include additional investigation into the optimal dose and frequency of administration of BDNF as well as potential molecular differences between optic nerve-crushed mouse and rat eyes. Ongoing work could lead to the development of new treatment options for glaucoma and other optic neuropathies.

Author Contributions: T.T., T.O. and R.R. designed, performed, and interpreted the experiments. T.T., T.O. and R.R. wrote the manuscript. N.A.S. and R.A.F. interpreted the experiments, reviewed the manuscript, and conceived the study. All authors have read and agreed to the published version of the manuscript.

Funding: This research received no external funding.

Institutional Review Board Statement: The animal study protocol was approved by the Institutional Animal Care and Use Committee of EyeCRO (#2021-07-16-001) on 16 July 2021.

Informed Consent Statement: Not applicable.

Data Availability Statement: All data generated or analyzed during this study are included in this published article.

Acknowledgments: We thank Alexander Quiambao, Amanda Rielinger, and Brittney Colvard at EyeCRO, LLC, for their experimental contributions.

Conflicts of Interest: Authors Rebecca Rausch and Rafal Farjo are employed by the company EyeCRO, LLC. Authors Takazumi Taniguchi and Takashi Ota are employed by the company Santen Pharmaceutical Co., Ltd. Najam Sharif is employed by the companies Santen, Inc., and Nanoscope Therapeutics, Inc. The authors declare that the research was conducted in the absence of any commercial or financial relationships that could be construed as a potential conflict of interest.

References

1. Weinreb, R.N.; Leung, C.K.; Crowston, J.G.; Medeiros, F.A.; Friedman, D.S.; Wiggs, J.L.; Martin, K.R. Primary open-angle glaucoma. *Nat. Rev. Dis. Primers* **2016**, *2*, 16067. [CrossRef]
2. Tham, Y.C.; Li, X.; Wong, T.Y.; Quigley, H.A.; Aung, T.; Cheng, C.Y. Global prevalence of glaucoma and projections of glaucoma burden through 2040: A systematic review and meta-analysis. *Ophthalmology* **2014**, *121*, 2081–2090. [CrossRef]
3. Sharif, N.A. Elevated Intraocular Pressure and Glaucomatous Optic Neuropathy: Genes to Disease Mechanisms, Therapeutic Drugs, and Gene Therapies. *Pharmaceuticals* **2023**, *16*, 870. [CrossRef] [PubMed]
4. Tribble, J.R.; Hui, F.; Quintero, H.; El Hajji, S.; Bell, K.; Di Polo, A.; Williams, P.A. Neuroprotection in glaucoma: Mechanisms beyond intraocular pressure lowering. *Mol. Asp. Med.* **2023**, *92*, 101193. [CrossRef] [PubMed]
5. He, S.; Stankowska, D.L.; Ellis, D.Z.; Krishnamoorthy, R.R.; Yorio, T. Targets of neuroprotection in glaucoma. *J. Ocul. Pharmacol. Ther.* **2018**, *34*, 85–106. [CrossRef] [PubMed]
6. Sharif, N.A. Therapeutic Drugs and Devices for Tackling Ocular Hypertension and Glaucoma, and Need for Neuroprotection and Cytoprotective Therapies. *Front. Pharmacol.* **2021**, *12*, 729249. [CrossRef] [PubMed]
7. Shah, M.; Cabrera-Ghayouri, S.; Christie, L.A.; Held, K.S.; Viswanath, V. Translational Preclinical Pharmacologic Disease Models for Ophthalmic Drug Development. *Pharm. Res.* **2019**, *36*, 58. [CrossRef]
8. Hu, Z.L.; Li, N.; Wei, X.; Tang, L.; Wang, T.H.; Chen, X.M. Neuroprotective effects of BDNF and GDNF in intravitreally transplanted mesenchymal stem cells after optic nerve crush in mice. *Int. J. Ophthalmol.* **2017**, *10*, 35–42.
9. Takeuchi, H.; Inagaki, S.; Morozumi, W.; Nakano, Y.; Inoue, Y.; Kuse, Y.; Mizoguchi, T.; Nakamura, S.; Funato, M.; Kaneko, H.; et al. VGF nerve growth factor inducible is involved in retinal ganglion cells death induced by optic nerve crush. *Sci. Rep.* **2018**, *8*, 16443. [CrossRef]
10. Feng, L.; Puyang, Z.; Chen, H.; Liang, P.; Troy, J.B.; Liu, X. Overexpression of brain-derived neurotrophic factor protects large retinal ganglion cells after optic nerve crush in mice. *eNeuro* **2017**, *4*, 1–8. [CrossRef]
11. Mac Nair, C.E.; Fernandes, K.A.; Schlamp, C.L.; Libby, R.T.; Nickells, R.W. Tumor necrosis factor alpha has an early protective effect on retinal ganglion cells after optic nerve crush. *J. Neuroinflamm.* **2014**, *11*, 194. [CrossRef] [PubMed]

12. Cameron, E.G.; Xia, X.; Galvao, J.; Ashouri, M.; Kapiloff, M.S.; Goldberg, J.L. Optic Nerve Crush in Mice to Study Retinal Ganglion Cell Survival and Regeneration. *Bio-Protocol* **2020**, *10*, e3559. [CrossRef]
13. Pang, I.H.; Clark, A.F. Inducible rodent models of glaucoma. *Prog. Retin. Eye Res.* **2020**, *75*, 100799. [CrossRef] [PubMed]
14. Agarwal, R.; Agarwal, P. Rodent models of glaucoma and their applicability for drug discovery. *Expert Opin. Drug Discov.* **2017**, *12*, 261–270. [CrossRef]
15. Wang, Y.; Cameron, E.G.; Li, J.; Stiles, T.L.; Kritzer, M.D.; Lodhavia, R.; Hertz, J.; Nguyen, T.; Kapiloff, M.S.; Goldberg, J.L. Muscle A-Kinase Anchoring Protein-α is an Injury-Specific Signaling Scaffold Required for Neurotrophic- and Cyclic Adenosine Monophosphate-Mediated Survival. *eBioMedicine* **2015**, *2*, 1880–1887. [CrossRef]
16. Lucas-Ruiz, F.; Galindo-Romero, C.; Salinas-Navarro, M.; González-Riquelme, M.J.; Vidal-Sanz, M.; Agudo Barriuso, M. Systemic and Intravitreal Antagonism of the TNFR1 Signaling Pathway Delays Axotomy-Induced Retinal Ganglion Cell Loss. *Front. Neurosci.* **2019**, *13*, 487979. [CrossRef]
17. Sánchez-Migallón, M.C.; Valiente-Soriano, F.J.; Nadal-Nicolás, F.M.; Vidal-Sanz, M.; Agudo-Barriuso, M. Apoptotic retinal ganglion cell death after optic nerve transection or crush in mice: Delayed RGC loss with BDNF or a caspase 3 inhibitor. *Investig. Ophthalmol. Vis. Sci.* **2016**, *57*, 81–93. [CrossRef] [PubMed]
18. Lambuk, L.; Mohd Lazaldin, M.A.; Ahmad, S.; Iezhitsa, I.; Agarwal, R.; Uskoković, V.; Mohamud, R. Brain-Derived Neurotrophic Factor-Mediated Neuroprotection in Glaucoma: A Review of Current State of the Art. *Front. Pharmacol.* **2022**, *13*, 875662. [CrossRef]
19. Friedman, W.J.; Greene, L.A. Neurotrophin Signaling via Trks and P75. *Exp. Cell Res.* **1999**, *253*, 131–142. [CrossRef]
20. Chao, M.V. The p75 neurotrophin receptor. *J. Neurobiol.* **1994**, *25*, 1373–1385. [CrossRef]
21. Geevarghese, A.; Wollstein, G.; Ishikawa, H.; Schuman, J.S. Optical Coherence Tomography and Glaucoma. *Annu. Rev. Vis. Sci.* **2021**, *7*, 693–726. [CrossRef] [PubMed]
22. Pernet, V.; Di Polo, A. Synergistic action of brain-derived neurotrophic factor and lens injury promotes retinal ganglion cell survival, but leads to optic nerve dystrophy in vivo. *Brain* **2006**, *129*, 1014–1026. [CrossRef] [PubMed]
23. Chen, H.; Weber, A.J. Brain-derived neurotrophic factor reduces TrkB protein and mRNA in the normal retina and following optic nerve crush in adult rats. *Brain Res.* **2004**, *1011*, 99–106. [CrossRef]
24. Chen, H.; Weber, A.J. Expression of glial fibrillary acidic protein and glutamine synthetase by Müller cells after optic nerve damage and intravitreal application of brain-derived neurotrophic factor. *Glia* **2002**, *38*, 115–125. [CrossRef]

pharmaceuticals

Review

Genetic Basis of Non-Syndromic Childhood Glaucoma Associated with Anterior Segment Dysgenesis: A Narrative Review

Nicola Cronbach [1], Cécile Méjécase [1,2] and Mariya Moosajee [1,2,3,*]

1 UCL Institute of Ophthalmology, London EC1V 9EL, UK; nicola.cronbach.23@ucl.ac.uk (N.C.); c.mejecase@ucl.ac.uk (C.M.)
2 Ocular Genomics and Therapeutics, The Francis Crick Institute, London NW1 1AT, UK
3 Moorfields Eye Hospital NHS Foundation Trust, London EC1V 9EL, UK
* Correspondence: m.moosajee@ucl.ac.uk

Abstract: Twenty causative genes have been reported that cause non-syndromic childhood glaucoma associated with anterior segment dysgenesis. *FOXC1*, *PAX6* and *PITX2* are the most well-known, but cases linked to *SLC4A11*, *PITX3* and *SOX11* have also been reported. As genetic testing becomes increasingly widespread and rates of molecular diagnosis rise, the extent of phenotypic overlap between the different genetic causes of non-syndromic glaucoma associated with anterior segment dysgenesis is becoming more evident. Taking aniridia as an example, whilst *PAX6* mutations remain the predominant cause, variants in *CYP1B1*, *FOXC1*, *PXDN* and *SOX11* have also been reported in patients with childhood glaucoma and aniridia. Developments in molecular-based therapies for retinal and corneal disease are advancing rapidly, and pre-clinical studies of gene-based treatments for glaucoma and aniridia are showing promising results. Use of adeno-associated viral vectors for gene delivery is most common, with improvements in intraocular pressure and retinal ganglion cell survival in Tg-$MYOC^{Y437H}$ mouse models of glaucoma, and successful correction of a germline $PAX6^{G194X}$ nonsense variant in mice using CRISPR-Cas9 gene editing. This review will explore the actions and interactions of the genetic causes of non-syndromic glaucoma associated with anterior segment dysgenesis and discuss the current developments in molecular therapies for these patients.

Keywords: childhood glaucoma; anterior segment dysgenesis; gene editing; nonsense suppression; aniridia

1. Introduction

Ocular development commences at gestational day 21 with induction of the eye field in the anterior neural plate prior to lateral evagination of the optic vesicles through the cephalic mesenchyme [1,2]. At gestational day 27, the optic vesicle contacts the surface ectoderm, and signaling through bone morphogenetic protein (BMP) and retinoic acid first from the optic vesicle and then the lens placode stimulates invagination of both structures to form the optic cup and lens vesicle, which are present by day 37 (Figure 1a) [1]. By day 47, differentiation of the optic cup into primitive neural retina and retinal pigment epithelium (RPE) has commenced, and neural crest cells from the periocular mesenchyme begin to migrate into the developing anterior segment under the control of *PAX6* (Figure 1b) [3]. *PITX2* and *FOXC1*, expressed by the presumptive cornea, and *PITX3* and *FOXE3*, expressed by the developing lens, have roles in regulating differentiation of the periocular mesenchyme into corneal keratocytes and endothelium, and detachment of the lens vesicle from the cornea, respectively (Figure 1c) [2]. From gestational week 15, *PAX6*, *BMP4* and *OTX1* stimulate elongation of the peripheral tips of the optic cup, which express *CYP1B1* and *MEIS2*, along the anterior lens surface to form the epithelium of the ciliary body and iris (Figure 1d) [2,4–6]. Concurrently, a second wave of migration of the periocular mesenchyme into the anterior segment occurs, expressing *PITX2*, *FOXC1* and *FOXC2*, which will form the ciliary body and iris stroma [2,3,5]. Trabecular meshwork development occurs from gestational week 20 with elongation, flattening and separation of mesenchymal cells in the iridocorneal angle to form lamellae, accompanied by vascular development in the adjacent sclera which becomes Schlemm's canal (Figure 1e) [2,7].

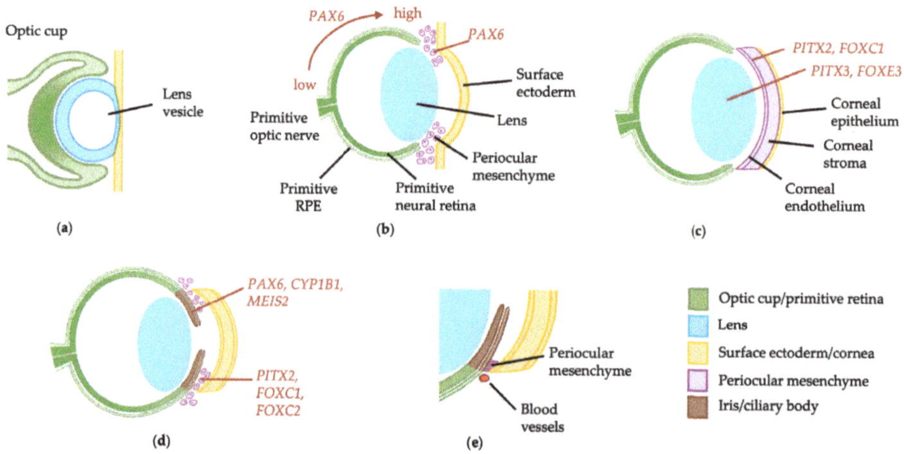

Figure 1. Anterior segment embryological development: (**a**) at day 37 of gestation, a double layered optic cup and lens vesicle can be seen; (**b**) at 47 days, the optic cup has formed primitive neural retina and retinal pigment epithelium with the optic stalk forming primitive optic nerve. Optic cup patterning is dependent on the *PAX6* expression gradient, from high expression at the peripheral optic cup tips down to low expression centrally. Cells of the central lens placode migrate to the posterior lens vesicle and elongate to form primary lens fiber cells, and the first wave of neural crest cells from the periocular mesenchyme migrates between the lens and surface ectoderm; (**c**) at 57 days, the primitive corneal epithelium is formed from the surface ectoderm, the corneal stroma and endothelium are formed from neural crest cells of the periocular mesenchyme, and there is separation of the lens stalk and formation of the anterior chamber; (**d**) by 15 weeks, elongation of optic cup tips along the anterior lens surface to form the iris and ciliary body epithelium occurs, and there is a second wave of periocular mesenchyme migration into the anterior chamber to form the iris and ciliary body stroma; (**e**) at 20 weeks, there is an accumulation of periocular mesenchyme in the iridocorneal angle which subsequently elongates and forms lamellae to become the trabecular meshwork, and vessels form in the adjacent sclera which will become Schlemm's canal. Gene expression is marked in red. RPE, retinal pigment epithelium.

Glaucoma is an overarching term for optic neuropathies resulting from retinal ganglion cell loss, which are typified by optic disc cupping and associated visual field defects. Childhood glaucoma, affecting individuals under 18 years of age, accounts for 5% of pediatric blindness worldwide [8]. In addition to optic nerve changes and visual field defects, features of childhood glaucoma include corneal enlargement and

Haab striae, increased axial length and intraocular pressure >21 mmHg; two or more features are required for diagnosis [9]. Childhood glaucoma is classified according to the underlying cause and whether this is primary or secondary to another ocular or systemic condition [9]. Primary congenital glaucoma develops before 4 years of age and may be further divided into neonatal (0–1 month), infantile (>1–24 months) and late (>24 months) onset disease. It is a primary goniodysgenesis and presents with an immature appearance of the iridocorneal angle (Figure 2a) [9,10]. Juvenile open angle glaucoma presents after 4 years of age and with normal angle appearances. Secondary causes of childhood glaucoma include onset following cataract surgery, glaucoma associated with non-acquired ocular or systemic anomalies, and glaucoma associated with acquired conditions such as trauma and uveitis.

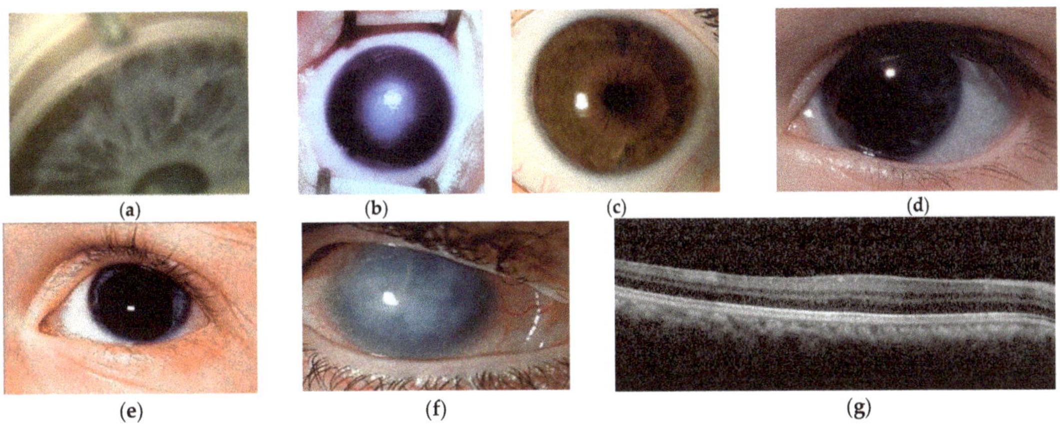

Figure 2. Clinical features of childhood glaucoma and anterior segment dysgenesis: (**a**) immature scalloped appearance of iridocorneal angle in primary congenital glaucoma; (**b**) central corneal opacity in Peters anomaly; (**c**,**d**) Axenfeld–Rieger anomaly: (**c**) iris hypoplasia and posterior embryotoxon; (**d**) corectopia; (**e**–**g**) congenital aniridia: (**e**) iris hypoplasia; (**f**) aniridia-associated keratopathy; (**g**) foveal hypoplasia.

Anterior segment dysgenesis (ASD) is a broad term that incorporates developmental anomalies involving structures in the anterior segment of the eye [11–13]. These may affect a single structure, such as microcornea, or involve multiple structures, such as the iridocorneal and keratolenticular adhesions seen in Peters anomaly (Figure 2b). Between 30 and 75% of patients with ASD develop glaucoma depending on the subtype of dysgenesis [14,15], and young patients with either glaucoma or ASD should be thoroughly examined for features of each condition [13,15–17]. Many of the anterior segment dysgeneses have variable phenotypes with considerable overlap; for example, Axenfeld–Rieger anomaly (Figure 2c,d) may include corectopia, polycoria and iris hypoplasia in association with posterior embryotoxon, and like Peters anomaly, may also feature iridocorneal adhesions [15]. Patients with congenital aniridia (Figure 2e–g) present with underdevelopment of the iris of variable severity and foveal hypoplasia; some will also have lenticular opacities and optic nerve hypoplasia from birth, and many will develop glaucoma and aniridia-associated keratopathy by adulthood [14,18].

The molecular pathways involved in anterior segment development are highly complex. There is an overlap between the known genes that cause childhood glaucoma and those that cause anterior segment dysgenesis [11–13], and there is increasing evidence of digenic and likely polygenic influences on the development process, resulting in phenotypic variation. For example, mice with either *FoxC1*$^{+/-}$ or *Cyp1b1*$^{-/-}$ mutations who also carry the tyrosinase-deficient *Tyr*$^{c-2J}$ allele develop more severe anterior segment anomalies and angle defects, respectively [19]. In humans, *CYP1B1* mutations have been shown to be modified by *PITX2*, with *CYP1B1*-related juvenile glaucoma patients carrying the *PITX2*^{P179T} allele developing earlier onset glaucoma than their *PITX2*^{wt} relatives [20]. Similarly, glaucoma onsets at an earlier age in patients with *TEK* mutations modified by *SVEP1*^{R997C} [21] and those with *MYOC* mutations modified by *COL15A1*^{R163H} [22].

However, many of the factors that contribute to anterior segment development remain unknown. Molecular diagnostic rates for childhood glaucoma and anterior segment dysgenesis remain below 25% [12,23], indicating the need for further studies to investigate additional genes and biological pathways that influence this process. Whilst certain genes, such as *CYP1B1*, *PAX6* and *FOXC1*, have been identified in multiple patients across populations, others have been reported only in single cases or small case series. Robust cellular and animal models of disease aid in establishing the phenotypic features associated with different genetic variants and allow detailed morphological and functional studies to be undertaken. For example, the development of an *Adamts18*-knockout mouse model in 2016, coupled with developments in transcriptomic analysis, has enabled extensive investigation of its role in ocular and systemic development [24]. However, similar models have not yet been developed for all genes in which pathogenic variants causing non-syndromic childhood glaucoma associated with anterior segment dysgenesis have been identified.

The aim of this review is to summarize the functions, biological relationships and associated phenotypes of the known genes that have been identified in patients with non-syndromic childhood glaucoma associated with anterior segment dysgenesis and discuss potential avenues for future treatments, to provide a platform from which to expand our knowledge of this under-investigated area.

2. Results

Twenty genes linked to childhood glaucoma associated with anterior segment dysgenesis were identified, of which twelve were also reported to cause isolated childhood glaucoma. The reported phenotypes of anterior segment dysgenesis covered the full spectrum of cornea, iris, lens and mixed anomalies (Table 1). *PAX6*, *FOXC1* and *PITX2* are expressed in multiple anterior segment tissues during ocular development; other genes, such as *CYP1B1*, have more limited expression (Figure 3).

Table 1. Genes reported to be linked to childhood glaucoma associated with anterior segment dysgenesis, and the phenotypes identified in each. AD = autosomal dominant, AR = autosomal recessive.

Gene	OMIM ID	Phenotype(s) of Anterior Segment Dysgenesis Associated with Childhood Glaucoma	Inheritance Pattern	Associated with Isolated Childhood Glaucoma
CYP1B1	601771	Peters anomaly [25–28] Axenfeld–Rieger anomaly [29] Aniridia [30] Corneal dystrophy [12,31] Unclassified ASD [32].	AR	Yes

Table 1. *Cont.*

Gene	OMIM ID	Phenotype(s) of Anterior Segment Dysgenesis Associated with Childhood Glaucoma	Inheritance Pattern	Associated with Isolated Childhood Glaucoma
FOXC1	601090	Peters anomaly [13,33] Axenfeld–Rieger anomaly [15,28,34] Aniridia [13] Unclassified ASD [12,13].	AD	Yes
PAX6	607108	Aniridia [16,34,35] Peters anomaly [33].	AD	Yes
LTBP2	602091	Weill–Marchesani syndrome [28,36] Congenital cataracts [12] Lenticular anomalies [16,37] Unclassified ASD [16].	AR	Yes
ADAMTSL4	610113	Ectopia lentis [38].	AR	Yes
CPAMD8	608841	Lenticular anomalies [39,40] Iris anomalies [40] Unclassified ASD [16,41].	AR	Yes
PXDN	605158	Peters anomaly [42] Congenital cataracts [43] Aphakia [42] Aniridia [44] Sclerocornea [33,45].	AR	Yes
TEK	600221	Sclerocornea [36].	AD	Yes
SLC4A11	610206	Congenital hereditary endothelial dystrophy [16,32,46].	AR	Yes
ADAMTS18	607512	Microcornea [47].	AR	Yes
PITX3	602669	Microcornea [48,49] Congenital cataracts [49].	AD	Yes
PITX2	601542	Peters anomaly [13,16] Axenfeld–Rieger anomaly [16,33] Sclerocornea [33].	AD	No
FOXE3	601094	Peters anomaly [50–52] Iris anomalies [50,51] Congenital cataracts/aphakia [51,53] Microphthalmia [51,54].	AD/AR	No
COL4A1	120130	Congenital cataracts [13,55,56] Axenfeld–Rieger anomaly [56] Iris anomalies [13,55].	AD	No
SOX11	600898	Peters anomaly [57] Aniridia [36,57].		No
GJA8	600897	Sclerocornea [28,33,36].	AD	No
KERA	603288	Unclassified ASD [39].	AR	No
CDH2	114020	Peters anomaly [58].	AD	No
KDM5C	314690	Peters anomaly [59].	AD	No
TFAP2A	107580	Peters anomaly [60].	AD	No

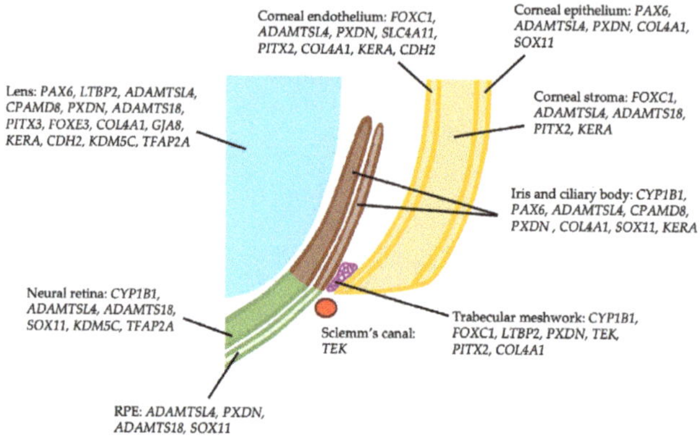

Figure 3. Ocular expression of genes causing non-syndromic childhood glaucoma and anterior segment dysgenesis.

2.1. *CYP1B1 (OMIM #601771)*

CYP1B1 encodes a 543-amino acid enzyme which is part of the cytochrome P450 superfamily and is involved in retinol [61,62], melatonin [61] and 17ß-oestradiol metabolism [62,63]. It regulates myocilin production by limiting estrogen-responsive activation of the *MYOC* promoter [63], and contributes to trabecular meshwork development through its modulation of periostin [64]. Pathogenic variants include missense, nonsense and frameshift mutations, and all have been identified throughout the gene [65]; the majority result in loss of function of metabolic activity [62,66,67], although some variants, such as c.1154C>A p.(Phe261Leu) and c.1776C>T p.(Arg469Trp) cause significantly increased retinol metabolism [62]. *CYP1B1* is the most commonly identified genetic cause of primary congenital glaucoma (PCG) [65,68], causing 15–35% of PCG in European studies [12,66], and up to 100% of PCG in some ethnic groups [69]. To date, there have been 62 *CYP1B1* variants associated with anterior segment dysgeneses, including the following: c.1064_1076del p.(Arg355Hisfs*69), c.1103G>A p.(Arg368His) and c.1169G>A p.(Arg390His) associated with Axenfeld–Rieger syndrome and Peters anomaly [25,28,29,31]; c.171G>A p.(Trp57*) and c.868dupC p.(Arg290Profs*37) associated with Peters anomaly and other corneal dystrophies [25,31,33,70]; and c.434_443del p.(Arg145Profs*4), c.756C>A p.(Asn252Lys) and c.1453T>C p.(Ser485Phe) associated with aniridia [30,71].

2.2. *FOXC1 (OMIM #601090)*

FOXC1, part of the forkhead box family of transcription factors, is expressed in the periocular mesenchyme and is necessary for separation of the cornea and lens in the developing eye [2,5,72]. It acts in conjunction with *PITX2* to promote the development of the corneal stroma and endothelium, iridocorneal angle and choroid, with co-expression of both transcription factors necessary for normal differentiation of these cellular populations [73]. In the trabecular meshwork, *FOXC1* activity is mediated by miR-204 and regulates *MEIS2* [74], a homeobox transcription factor which in turn is a direct regulator of *PAX6* [75]. *FOXC1* variants have been identified in patients with childhood glaucoma associated with Axenfeld–Rieger anomaly [15,28,34], Peters anomaly [13,33] and aniridia [13]. In the context of Axenfeld–Rieger syndrome, *FOXC1* variants cause significantly more congenital glaucoma than do *PITX2* variants, which are more associated with juvenile- and adult-onset glaucoma [15]. Whole gene deletions account for approximately one third of anterior segment dysgenesis caused by *FOXC1* [12,15]. There are 126 intragenic *FOXC1* variants associated with Axenfeld–Rieger syndrome, of which c.482T>A p. (Met161Lys) was most commonly identified [15]. In a functional study, Medina-Trillo and colleagues showed that *FOXC1* variants can result in either significantly increased or decreased transcriptional activity compared with the wild type; hypermorphic variants including c.141C>G p.(Tyr47*) and c.316C>T p.(Gln106*) were associated with isolated dominant glaucoma, whilst hypomorphic variants including c.377T>G p.(Ile126Ser) were identified in patients with Axenfeld–Rieger anomaly [76].

2.3. *PAX6 (OMIM #607108)*

Heterozygous pathogenic variants in *PAX6* are typically associated with autosomal dominant aniridia, or are part of the Wilms tumor-aniridia-genital anomalies-range of developmental delay (WAGR) syndrome if the whole gene is deleted along with *WT1* [11,18]. The PAX6 transcription factor is known to have a role in ocular development from gestational day 22 in the differentiation of the eye field and optic vesicle, through to the development of the optic cup and lens vesicle at day 37, and then in corneal, iris, ciliary body and trabecular meshwork development from day 47 onwards [2,77,78]. It has both paired and homeobox DNA-binding domains, and interacts with numerous other genes during this process, including *SOX2*, *RAX*, *MEIS2* and *ALDH1A3* [5,75,77]. Close regulation of *PAX6* expression is required throughout embryogenesis to ensure normal development; underexpression results in the aniridia spectrum of iris, foveal and optic nerve hypoplasia with associated glaucoma, cataracts and keratopathy, whilst overexpression causes microphthalmia, microcornea and ciliary body maldevelopment [79–81]. Of the 467 pathogenic and likely pathogenic *PAX6* variants reported to date [82], c.649C>T p.(Arg203*), c.760C>T p.(Arg240*), c.823C>T p.(Arg261*) and c.991C>T p.(Arg317*) are the most commonly identified [14]. Whilst there are reports implicating *PAX6* in isolated juvenile open angle glaucoma and childhood glaucoma associated with Peters anomaly, childhood glaucoma caused by *PAX6* variants is very rarely diagnosed in patients without aniridia [33,83].

2.4. *LTBP2 (OMIM #602091) and ADAMTSL4 (OMIM #610113)*

Both *LTBP2* and *ADAMTSL4* contribute to cytoskeletal development through their interactions with fibrillin-1. *LTBP2* encodes a 1821-amino acid extracellular matrix protein which interacts with fibrillin-1-containing microfibrils [84,85] and is expressed in the cornea, trabecular meshwork and lens [85]. A causative gene for Weill–Marchesani syndrome [28,36,84], which presents with short stature, brachydactyly, microspherophakia, ectopia lentis and myopia, *LTBP2* has also been linked with congenital glaucoma associated with isolated microspherophakia [16], ectopia lentis (c.4825T>A p.(Cys1609Ser) and c.529T>C p.(Trp177Arg) variants) [37] and congenital cataracts (c.895C>T p.(Arg299*) variant) [12]. The glycoprotein ADAMTSL4 binds fibrillin-1 and facilitates microfibril assembly, and is expressed in many ocular tissues including the cornea, iris and ciliary body in the anterior segment, and the retinal ganglion cells, retinal pigment epithelium, choroid and sclera in the posterior segment [86]. The missense variant c.2236C>T p.(Arg746Cys) has been linked to congenital glaucoma with ectopia lentis [38].

2.5. *CPAMD8 (OMIM #608841)*

CPAMD8 is expressed in the human eye from gestational week 9 onwards [87]. From weeks 9 through 22 of gestation, *CPAMD8* expression increases in the lens and decreases in the retina, and strong expression is also observed in the iris and cornea at week 22 [87]. The pathways by which *CPAMD8* acts to contribute to ocular development remain undefined, but recent work by Escribano and colleagues in zebrafish suggests an interaction between *cpamd8* and *adamtsl4* [88], implicating a potential contribution of *CPAMD8* to the development of fibrillar structures in the anterior segment. *CPAMD8* variants have been linked to anterior segment dysgenesis associated with both congenital and juvenile glaucoma [16,39–41], with the variants c.4298C>A p.(Thr1433Asn) and c.2352dupC p.(Arg785Glnfs*23) identified in more than one family [40].

2.6. *COL4A1 (OMIM #120130) and PXDN (OMIM #605158)*

Variants in *COL4A1* and *PXDN*, which are both basement membrane components, have been identified in patients with anterior segment dysgeneses associated with childhood glaucoma, although only the c.2159G>A p.(Gly720Asp) has been reported in more than one family [13,33,42,43,55,56]. *COL4A1* encodes the alpha-1 subunit of type IV collagen, which is the main component of endothelial and epithelial basement membranes [89]. Peroxidasin, the protein product of the *PXDN* gene, forms sulfilimine crosslinks between type IV collagen molecules, and is necessary for normal development of type IV collagen-, laminin- and fibronectin-based fibrillar networks in endothelial cells [90]. Type IV collagens are found throughout the developing eye, and by embryonic day 18.5 (E18.5), the alpha-1 isoform can be identified in all anterior segment tissues in mice [91].

2.7. *TEK (OMIM #600221)*

One of two receptor tyrosine kinases in the angiopoietin/TEK signaling pathway, autosomal dominant mutations in *TEK* have been identified in patients with primary congenital glaucoma [16] and congenital glaucoma associated with sclerocornea (specific variant not reported) [36]. In mice, *TEK* is expressed in Schlemm's canal endothelial cells, and its ligand *Angpt1* is expressed in the trabecular meshwork [92,93]. Supporting the role of *TEK* in iridocorneal angle and Schlemm's canal development, murine studies have shown that *Tek* haploinsufficiency and *Angpt1* knockout result in severe hypoplasia of both Schlemm's canal and the trabecular meshwork, whilst *Tek* knockout results in complete failure of Schlemm's canal development [93,94].

2.8. *SLC4A11 (OMIM #610206)*

Pathogenic variants in *SLC4A11* are the most commonly identified genetic cause of congenital hereditary endothelial dystrophy (CHED) [95], which presents with early onset corneal edema and, rarely, concomitant congenital glaucoma [16,32,46]. Homozygous missense variants cause approximately half of *SLC4A11*-related disease [95], and both families with congenital glaucoma whose variants have been reported had homozygous missense mutations; these were c.1343G>A p.(Gly448Asp) and c.2024A>C p.(Glu657Ala), respectively [32,46]. *SLC4A11* localizes to the corneal endothelium, where it contributes to osmotic homeostasis of the cornea through its action as a transmembrane sodium-coupled hydroxyl transporter, and thus to maintenance of corneal transparency [95,96]. *Slc4a11*-knockout mice develop progressive corneal thickening with associated endothelial cell morphological changes and loss [97], but no groups have yet reported intraocular pressure in these murine models.

2.9. *ADAMTS18 (OMIM #607512)*

Part of the ADAMTS zinc-dependent metalloproteinase family, *ADAMTS18* is expressed in adnexal, anterior segment and retinal cells during murine ocular development, with expression persisting postnatally in the lens and lacrimal gland [98,99]. It is a key factor in branching morphogenesis of the embryonic lacrimal gland [98], as well as in the development of the lung, breast and reproductive tract [24]. In relation to anterior segment development, it interacts with fibrillin-1 and is modulated by estrogen, raising the possibility of an interaction with *CYP1B1* [24]. Pathogenic variants in *ADAMTS18* cause microcornea, myopic chorioretinal atrophy, and telecanthus (MMCAT) [100] and Knobloch syndrome (high myopia, vitreoretinal degeneration, retinal detachment, cataract and occipital encephalocele) [101], but childhood glaucoma associated with microcornea has been reported in twins with compound heterozygous missense mutations of *ADAMTS18* [c.3436C>T p.(Arg1146Trp) and c.1454T>G p.(Phe485Cys)] identified through panel testing [47]. Ocular defects seen in *Adamts18* knockout mice include breaks in the lens capsule resulting in extrusion of material, but microphthalmia, microcornea and retinal anomalies were not seen [99]. However, morpholino oligonucleotide (MO) knockdown of *adamts18* in zebrafish has been shown to cause macroscopic ocular defects [102].

2.10. *PITX2 (OMIM #601542) and PITX3 (OMIM #602669)*

The bicoid-type homeodomain transcription factors *PITX2* and *PITX3* have a role in embryonic anterior segment and lens development [103,104]. As discussed above, *PITX2* interacts with *FOXC1*, and in mice, they are co-expressed in the periocular mesenchyme from E11.5 [73]. It is therefore unsurprising that the ocular phenotypes of *PITX2*-related disease are similar to those of *FOXC1* and include childhood glaucoma associated with Axenfeld–Rieger anomaly [16,33] and Peters anomaly [13,16]. As with *FOXC1*, the majority of *PITX2* variants have been identified in only one or two families, with the exception of the splicing variant c.253-11A>G and the intragenic variants c.191C>T p.(Pro64Leu), c.356del p.(Gln119Argfs*36) and c.475_476del p.(Leu159Valfs*39), which have all been reported in four or more families [15]. *PITX3* is expressed in the lens placode at murine E10 and continues to be expressed in the lens throughout development, in addition to expression in the midbrain, tongue and other embryological craniofacial tissues [104]. The c.640_656dup p.(Gly220Profs*95) variant has been identified in two families with congenital glaucoma associated with microcornea and congenital cataracts [48,49].

2.11. *FOXE3 (OMIM #601094)*

FOXE3 variants are associated with a wide spectrum of anterior segment phenotypes and have been identified in patients with childhood glaucoma associated with Peters anomaly [50–52], congenital cataracts/aphakia [51,53] and microphthalmia [51,54]. The most common variants are c.21_24del p.(Met7Ilefs*216), c.244A>G p.(Met82Val) and c.720C>A p.(Cys240*) which have all been reported in four or more families [51]. The main role of the FOXE3 transcription factor during embryogenesis is in the development of the lens, where it is expressed in mice from E12.5 onwards and directly regulates the autophagy-associated protein DNAJB1 [50]. Functional studies of *FOXE3* missense variants and small deletions isolated from patients with recessive disease showed significantly reduced transcriptional activity with homozygous or compound heterozygous variants, but no disease phenotype in their heterozygous parents, indicating that the presence of a single functional allele is sufficient for normal anterior segment development [53].

2.12. *SOX11 (OMIM #600898)*

The SOX11 SRY-Box transcription factor is a close homolog of SOX2 [105], which is known to play a key role in ocular development from induction of the eye field at gestational day 21 onwards [77]. During *Xenopus* embryogenesis, *sox11* is expressed in the optic vesicle, retina and anterior segment, and MO knockdown of *sox11* results in decreased *pax6* expression in the eye field, loss of retinal lamination and a microphthalmic phenotype [106]. In *Sox11* knockout mice, *Pax6* expression is unaffected, but *BMP7* expression is reduced, and the phenotype includes lenticular defects, Peters anomaly and anterior ocular coloboma [107]. *SOX11* variants, including c.251G>T p.(Gly84Val), have been reported in patients with congenital glaucoma associated with Peters anomaly and aniridia [36,57], although the gene is more commonly associated with intellectual developmental disorder with microcephaly with or without ocular malformations or hypogonadotropic hypogonadism (IDDMOH; Coffin Siris syndrome), of which congenital glaucoma can be a feature [28,57].

2.13. GJA8 (OMIM #600897)

GJA8 encodes connexin 50 (Cx50), a gap junction protein of lens fiber cells which is necessary for maintaining lens transparency and cellular homeostasis [28,108]. Studies of the missense mutation c.217T>C p.(Ser73Pro) in transfected human lens epithelial cells have shown aggregation of the abnormal protein in the cytosol with loss of localization to the plasma membrane, resulting in reduced function of gap junction channels and increased apoptosis [108]. Although *GJA8* variants are more commonly associated with congenital cataracts [11], the missense variants c.280G>C p.(Gly94Arg) and c.151G>A p.(Asp51Asn) have been identified in patients with congenital glaucoma associated with sclerocornea [28,33,36].

2.14. KERA (OMIM #603288)

KERA encodes keratocan, a keratan sulfate proteoglycan of the corneal stroma with a central role in maintaining collagen spacing and thus corneal transparency through its interaction with keratan sulfate glycosaminoglycans and collagen fibrils [109,110]. In adults, it is localized to stromal keratocytes [109]; during embryological development, studies of chicks by Conrad [111] and Gealy [112] have shown expression of *Kera* in the periocular mesenchyme from E4.5, with subsequent localization to the anterior corneal stroma. Keratocan-deficient mice have a thinner cornea and narrower iridocorneal angle than wild-type mice, with larger collagen fibrils in a less regular arrangement [110]. Mutations in *KERA* are predominantly linked to cornea plana, in which there is flattening of the normal convex curvature of the cornea resulting in hyperopia [113], but one patient with anterior segment dysgenesis and congenital glaucoma associated with a homozygous nonsense *KERA* mutation c.937C>T p.(Arg313*) has been reported [39].

2.15. CDH2 (OMIM #114020), KDM5C (OMIM #314690) and TFAP2A (OMIM #107580)

CDH2, *KDM5C* and *TFAP2A* are involved in neural development, with roles including axonal pathfinding [114], neuronal differentiation [115] and neural crest migration [116], respectively. All are reported to be causative of congenital glaucoma associated with Peters anomaly [58–60]. *CDH2* encodes N-cadherin, part of the cadherin family of Ca^{2+}-dependent transmembrane glycoproteins which regulate cell–cell adhesion in multiple tissues [58,114]. *CDH2* is expressed in murine lens cells and the lens stalk at E10.75-11.0, and in lens and corneal endothelial cells at E17.5 [58]. In conjunction with *CDH1* and *CDH3*, it controls separation of the lens vesicle from the surface ectoderm during anterior segment development; thus, its association with Peters anomaly is unsurprising [117]. Knockout of N-cadherin from the lens vesicle in mice results in a spectrum of anterior segment anomalies, including severe iris hyperplasia, microphthalmia and small, underdeveloped lenses with abnormal lens fiber development [117]. The missense variants c.485T>A p.(Val162Asp), c.1574A>G p.(Asp525Gly) and c.1807C>T p.(Pro603Ser), and splicing variant c.702+1G>A, have each been identified in one patient with Peters anomaly and congenital glaucoma [58].

KDM5C is a cause of X-linked syndromic intellectual developmental disorder, but the missense variant c.1204G>A p.(Asp402Asn) has been identified in one patient with Peters anomaly and congenital glaucoma [59]. Involved in transcriptional regulation through demethylation of H3K4 residues, studies of *kdm5c* in *Xenopus* have revealed that it is strongly expressed throughout neural and ocular development, particularly in the lens and retina [118]. MO knockdown studies in *Xenopus* have confirmed the essential role of *kdm5c* in ocular development, with loss of *kdm5c* resulting in microphthalmia, coloboma and abnormal retinal lamination [118].

In human eyes, expression of the transcription factor *TFAP2A* is seen in the anterior lens epithelium from gestational day 35 and in the equatorial lens epithelium, secondary lens fibers and neural retinal by day 54 [119]. Whilst the detailed mechanism by which *TFAP2A* acts has not yet been elucidated, conditional knockout of *Tcfap2a* in the lens placode of mice has been shown to result in failure of lens separation from the surface ectoderm with associated corneal anomalies, accompanied by loss of *Foxe3* and *Pitx3* expression from the lens stalk, suggesting that *TFAP2A* may be an upstream regulator of these genes as they pertain to lens separation [120]. This is consistent with the identification of a *TFAP2A* splicing variant (c.1025+2T>A) in a patient with Peters anomaly associated with congenital glaucoma [60].

3. Management of Childhood Glaucoma and Anterior Segment Dysgenesis

The current treatment options for anterior segment dysgenesis and childhood glaucoma are predominantly surgical, although refractive, medical and supportive interventions, including genetic counseling and family planning, all contribute to the holistic management of these patients and their families [17,79,121]. For congenital glaucoma, angle surgery (goniotomy and trabeculotomy) is the first line treatment; trabeculectomy with anti-fibrotic agents, glaucoma drainage devices and cyclodestructive procedures such as transscleral cyclophotocoagulation are used in refractory cases. Medical treatments (beta-blockers, carbonic anhydrase inhibitors and prostaglandin analogs) are used as adjuncts to surgery or as temporizing measures while awaiting a procedure [121,122]. In patients with aniridia, cataract surgery, and corneal and limbal grafts, alongside ocular lubricants, refraction and treatment of associated glaucoma have long been the mainstay of treatment [30,79]. Many patients with Peters anomaly require penetrating keratoplasty, although newer grafting techniques such as Descemet stripping automated endothelial keratoplasty (DSAEK) have been successfully employed in selected patients, and optical iridectomy may improve vision in patients with a smaller area of corneal opacity [17,123]. In all cases, refraction and amblyopia management should be provided as necessary, alongside educational and family support. Examination under anesthetic should be considered for young children to enable detailed examination and identification of subtle signs of anterior segment dysgenesis or associated ocular conditions. Genetic testing is paramount to enable accurate genetic counseling, family planning advice and referral to other specialties in cases that may have an associated syndromic element. Furthermore, as the potential for molecular-based therapies is beginning to translate into practice with promising research including developments in gene editing and nonsense suppression drugs, genetic testing will allow patients access to targeted research.

4. Molecular Therapies Under Development

There are currently 41 FDA-approved cellular and gene therapies [124], of which only Luxturna (voretigene neparvovec-rzyl) targets an ocular condition, *RPE65*-related retinal dystrophy. Growing research into this field as a potential source of therapy for glaucoma and anterior segment disorders has focused on animal and in vitro models, but in recent years, this has begun translating into clinical trials [79,125]. Most of these studies have investigated treatments for corneal diseases and the various methods of gene delivery. Transduction of target cells with recombinant adenoviruses (rAds) and adeno-associated viruses (rAAVs) are the most common viral vectors [125–127], although retroviruses [128] and lentiviruses [129] have been used in some studies. Non-viral vectors include metal and magnetic nanoparticles, micelles and lipid nanoparticles [125,130–132], which may be delivered topically or via injection. Alternative methods that do not require a vector include corneal electroporation and iontophoresis [133,134]. In the anterior segment, research into gene-based therapies has focused predominantly on corneal and ocular surface disease, adult-onset glaucoma and aniridia [79,125,135,136], with no published trials of gene therapies for Peters anomaly or Axenfeld–Rieger syndrome. Here, we summarize the current research into cellular and gene-based therapies for glaucoma and aniridia, and discuss the challenges faced by these approaches in the pediatric population.

4.1. Gene-Based Therapies for Glaucoma

Molecular-based therapies under development for glaucoma are predominantly targeted towards modification of the aqueous production/outflow pathways and minimizing damage to retinal ganglion cells (RGCs) [135], as these are key factors in the development of multifactorial adult-onset glaucoma. Sulak and colleagues [136] identified 153 studies into gene-based therapies for glaucoma and noted that there were twice as many therapies targeting RGC neuroprotection compared with those targeting aqueous production and outflow. Strategies to modulate the molecular pathways involved in aqueous regulation are in development, with targets including *MYOC*, *AQP1* and matrix metalloproteinases (MMPs).

Mutations in the *MYOC* gene, encoding the extracellular matrix protein myocilin, are the most commonly identified single gene cause of primary open angle glaucoma (POAG) and juvenile open angle glaucoma (JOAG), identified in 2–4% and 10–30% of cases, respectively [135]. Dominant gain-of-function mutations result in intracellular accumulation of the abnormal protein, resulting in increasing intraocular pressure (IOP) and resultant glaucomatous damage to the retinal ganglion cells and optic nerve [127]. Mice with the *MYOC* p.(Tyr437His) variant (Tg-$MYOC^{Y437H}$) can be used as a model for POAG, developing these same pathogenic features, and may be employed for in vivo confirmation of the therapeutic benefit [127,129].

Gene editing approaches targeted at specific myocilin mutations are in development. Jain [127] used a clustered regularly interspaced palindromic repeats (CRISPR)-CRISPR-associated protein 9 (Cas9) approach with an adenoviral Ad5 vector to introduce a truncating frameshift mutation into Tg-$MYOC^{Y437H}$ mice. They demonstrated significant reduction in IOP into the normal range in both younger (age $\leq$ 1 month) and older (age $\geq$ 9 months) mice, with results sustained for at least 2 months post-injection. The group have subsequently shown positive results using the same CRISPR-Cas9 system with a lentiviral vector [129], with IOP reduction into the normal range in Tg-$MYOC^{Y437H}$ mice maintained for at least 6 weeks post-injection. They additionally tested an AAV2 vector in vitro on human TM3 cells with p.(Tyr437His) or p.(Gly364Val) *MYOC* mutations, but the reduction in accumulated myocilin was only 34% lower than untreated levels with the AAV2 vector compared with 62% reduction for the lentivirus. A non-randomized interventional trial of a BD113 virus-like particle, a lentiviral-based CRISPR-Cas9 gene therapy targeted to knock down *MYOC*, has commenced recruitment of adults with myocilin-related POAG refractory to initial treatment and will primarily assess IOP reduction following a single intracameral injection [137]. Another mechanism that has been proposed for reducing the pathogenicity of *MYOC* mutations is to use small interfering RNAs (siRNAs) to disrupt the expression of the gene. The preliminary results in TM5 cells demonstrated significant suppression of both wild-type and mutant myocilin expression to 60% of control levels [138]. Whilst further work is needed, these approaches hold promise for potential gene-targeted therapies for *MYOC*-related glaucoma to become available in the future.

Another CRISPR-based approach employing a lentivirus vector carrying single guide RNA (sgRNA) to target the *TGFβ2* promoter has been developed by Mao's group [139]. Administration of the sgRNA lentivirus 2.5 weeks prior to induction of ocular hypertension prevented IOP increases, maintaining an IOP level below 20 mmHg for at least 8 weeks post-injection, compared with an increase to over 30 mmHg in controls. *AQP1*, which encodes aquaporin 1, has also been targeted with sgRNA. This transmembrane water channel is present in many tissues including the ciliary body, and deletion of *Aqp1* has been shown to significantly reduce IOP in mice [140]. Introduction of two *AQP1*-indel-inducing sgRNAs within a single AAV vector given by intravitreal injection has been shown to transduce the murine ciliary body and reduce IOP by 22% after 3 weeks, with neuroprotective effects sustained to at least 7 weeks post-injection [141]. There was also effective transfection of ex vivo human ciliary body epithelial cells. *AQP1* has therefore been proposed as a potential therapeutic target for glaucoma.

MMP activity is known to be reduced in glaucoma, and one of the key actions of the prostaglandin analog drops used to treat open angle glaucoma is to increase MMP expression in the ciliary body [142]. Using an AAV9 vector to transduce corneal endothelial cells, O'Callaghan [143] introduced tetracycline-inducible murine MMP3 into Tg-$MYOC^{Y437H}$ and dexamethasone-induced ocular hypertensive mice. In both groups, IOP 2 weeks after doxycycline induction was significantly reduced in the treated mice compared with controls. They also demonstrated significant increases in MMP3 expression and aqueous outflow in non-human primates in vivo, and increased aqueous outflow in human donor anterior segments in vitro using this approach [143]. Similar effects have been achieved using intracameral IκBα-silencing siRNA to transfect the anterior segment of nonhuman primates, with upregulation of MMP2 and MMP9 and resultant 49% reduction in IOP [144]. A lesser IOP reduction has been achieved through the use of the microRNA miR-21-5p, which is predicted to have downstream effects including the activation of MMP9. Whilst preliminary results show potential, the maximal IOP reduction following administration was 17.8% [145], lower than the 20–35% achieved by applying current topical treatments [146].

Novel gene-targeted approaches acting on established therapeutic pathways are also under development. Rho kinase inhibitors act directly on the trabecular meshwork to inhibit phosphorylation of proteins including myosin light chains, which reduces trabecular cell adhesion and resistance and promotes phagocytosis of extracellular debris by trabecular reticulum cells, thus increasing aqueous outflow and reducing IOP [147]. Current commercial formulations are administered topically, but a Phase 1 study of GVB-2001 gene therapy in adults with POAG is open to recruitment [148]. GVB-2001 consists of a human RhoA gene with a dominant negative mutation carried in an AAV vector, an approach that has been successful in transducing the trabecular meshwork and reducing nocturnal IOP rises in rats through inhibition of the Rho pathway [149].

Targeted disruption of β2-adrenoceptors has also shown promise. Loma [146] topically administered siRNAs designed to silence β2-adrenoceptors to rabbits and compared the effects to commercially available topical dorzolamide, latanoprost and timolol. The siRNAs significantly reduced IOP by $30 \pm 5\%$ compared to controls, with effects detectable 24 h after instillation and sustained for 5 days. The level of IOP reduction was comparable to that achieved by all commercial agents, but with a longer onset and offset time. A commercial compound, bamosiran, has recently completed its phase II trial. This RNA interference approach designed to prevent β2-adrenoceptor production showed non-inferiority to timolol in patients with a baseline IOP ≥ 25 mmHg, but not across all patient groups [150]. However, the positive results achieved overall by using these diverse approaches to molecular modulation of aqueous regulation lay a strong foundation for molecular therapies to become available for patients within the near future.

4.2. Gene-Based Therapies for Aniridia

A novel approach to the treatment of aniridia is nonsense suppression therapy, using drugs that impair recognition of premature termination codons (PTCs) during translation to allow the production of a full-length protein [151]. In aniridia patients in whom a genetic cause is identified, dominant pathogenic *PAX6* mutations are responsible for approximately 90% of cases [14]; of these, nonsense and frameshift mutations resulting in a PTC comprise 70% of pathogenic and likely pathogenic *PAX6* variants [82], so a treatment that targets these mutations has the potential to benefit a large proportion of these patients. Promising results using nonsense suppression therapies, which work on in-frame nonsense mutations, have been reported in vitro [152] and in the *Sey* mouse model [153,154]. *Sey* mice with a heterozygous *PAX6* p.(Gly194*) nonsense mutation treated with topical nonsense suppression drugs from postnatal day 14–60 developed partial phenotypic rescue, including reversal of corneal, lens and retinal abnormalities [153,154]. However, these benefits have not yet been translated into positive results in clinical trials. At completion, the Study of Ataluren in Participants With Nonsense Mutation Aniridia (STAR) unfortunately failed to meet its primary endpoint, with no significant difference in change in reading speed at 48 weeks between the treatment and placebo groups [155]. STAR used an oral formulation of ataluren; a topical alternative has been developed with proven physical stability and microbiological sterility over 60 days [156] which may have better ocular bioavailability and improved efficacy, and the outcomes from these assessments are awaited.

Gene editing is under investigation as another potential avenue of treatment for aniridia. Work in *Sey* mice has shown successful germline correction of the *PAX6* p.(Gly194*) mutation in vivo via microinjection of the targeted CRISPR-Cas9 complex into murine zygotes at day 0.5 post-coitus, with inheritance of the corrected gene by the offspring of these transgenic mice [157]. The progeny had normal eye development and normal *PAX6* levels in neural tissue at embryonic day 18.5, when microphthalmic changes are observed in *Sey* mice [157]. Further work has demonstrated the potential use of AAV vectors for delivery of gene therapy in aniridia, with results suggesting that the AAV9 capsid is able to safely transduce corneal and retinal cells of *Sey* mice [126].

Work to expand the therapeutic options for management of aniridia-associated keratopathy has shown that duloxetine and ritanserin, both anti-serotonin drugs which inhibit the MEK/ERK (mitogen-activated protein kinase-ERK kinase/extracellular signal related kinase) signaling pathway, can enhance endogenous *PAX6* expression in haploinsufficient corneal limbal epithelial stem cells [158–160]. A regulatory microRNA, miR-204-5p, has also been reported that upregulates *PAX6* in *PAX6*-knockdown corneal limbal stem cells and mature differentiated corneal epithelial cells in vitro; there was no significant change in *PAX6* expression in primary human limbal epithelial cells, human limbal stem cells or in the corneas of $Pax6^{(Sey/+)}$ mice transfected with miR-204-5p [161]. Further work is needed to determine the effects of these modulating factors in other ocular cell types, but they provide another potential avenue for therapeutic development. However, any strategies to increase *PAX6* expression during embryogenesis need to ensure that induced levels remain carefully modulated. Both under- and overexpression of *PAX6* can lead to ocular developmental anomalies as discussed earlier [80,81]; therefore, ensuring that corrected *PAX6* expression levels are as close as possible to the normal embryological state is essential when correcting pathogenic variants. To address this, Simpson's team [162] have designed *PAX6* MiniPromoters and reported their robust and specific expression within the *PAX6*-positive RGC, amacrine, horizontal and Müller glial cells, with significantly better regulation of *PAX6* expression than achieved by the ubiquitous smCBA promoter.

4.3. Challenges of Gene Therapy in the Pediatric Population

The anterior segment dysgeneses are congenital conditions with phenotypic features resulting from abnormal oculogenesis. Consideration therefore needs to be given to the challenges presented when administering any molecular therapies to a pediatric population and determination of trial endpoints suited to this group.

The first challenge is confirming the genetic diagnosis. Whilst molecular diagnostic rates for aniridia now reach 90–98% [163,164], rates for other ASDs remain much lower at less than 25% [12,23]. For gene-based therapies, accurate molecular diagnosis is essential, particularly given the overlapping phenotypes caused by many of the genes listed above. Secondly, these are rare diseases; therefore, identification of a sufficient number of patients for clinical trials may require collaboration across multiple centers and potentially multiple countries. Local and national rare disease registries and collaboration with charitable networks can aid in this.

Trial endpoints need to be carefully considered. Appropriate functional tests of acuity and fields vary with age, and it may be difficult to acquire quantitative results [165]. For IOP testing, Goldmann applanation tonometry remains the gold standard [166] but requires the patient to sit at the slit lamp. IOP testing with handheld devices such as the iCare tonometer can allow for monitoring in children of all ages, but their accuracy at pressures above 23 mmHg is reduced [166]; therefore, their use in trials of patients with childhood glaucoma needs to be carefully considered. For ASDs, monitoring of phenotypic changes may be necessary, and in some children, this may require examination under anesthesia, which should be incorporated into routine clinical care wherever possible.

Finally, there are the challenges inherent in the gene therapies themselves. Intracameral administration of therapies targeting the corneal endothelium and trabecular meshwork negates the need for corneal penetrance required by topical formulations and reduces the potential for systemic off-target effects, but this method of delivery to pediatric patients requires general anesthesia and introduces the potential for intraocular infection. Adenoviral vectors can provoke immune responses and inflammation whereas AAVs and lentiviruses rarely elicit these reactions [125]. Although less likely than with retroviruses, concerns have been raised regarding the potential for lentiviral vectors to cause insertional mutagenesis, or for additional mutagenic effects to occur with CRISPR-Cas9 delivery systems [125], but Jain [127] identified no alternative myocilin isoforms following treatment with a CRISPR-Cas9 approach. Studies of transduction of target cells in vitro have been promising, with up to 70% transduction efficacy of human trabecular meshwork cells by an Ad5 vector using CRISPR-Cas9 to target mutant *MYOC*, and reduction in IOP using this therapy in Tg-$MYOC^{Y437H}$ mice [127]. However, exposure to AAVs in childhood is common and therefore pre-existing circulating antibodies may reduce transduction efficacy in human subjects [165].

4.4. Future Research Directions

Future research directions in this field should focus on both targeting the underlying biological pathways affected and on improving the delivery methods in the pediatric population. To date, aniridia and adult-onset glaucoma have been the primary foci for therapeutic development, but the approaches developed in these fields have the potential to be applicable to other anterior segment dysgeneses with a similar genetic basis. For example, 13% of known pathogenic variants of *CYP1B1* are nonsense mutations, but these account for 21% of *CYP1B1*-related anterior segment dysgenesis. Studies of nonsense suppression therapies in aniridia have shown promising results, with partial phenotypic rescue in mice treated post-partum; similar trials in animal models of other nonsense-mediated anterior segment dysgeneses are an avenue that warrants further exploration. Likewise, whilst primary congenital glaucoma is thought to be predominantly driven by iridotrabeculodysgenesis and changes in Schlemm canal development, the juvenile-onset glaucoma seen in aniridia and Axenfeld–Rieger syndrome may be amenable to modulation of *TGFβ2*, *AQP1* or MMPs. If these treatments progress smoothly through Phase 1 and 2 trials in adult-onset glaucoma, trials in glaucoma secondary to anterior segment dysgeneses should be considered.

5. Methods

A combination of strategies was used to identify the genes that cause non-syndromic childhood glaucoma associated with anterior segment dysgenesis. A list was compiled in June 2024 of all green and amber genes in the NHS Genomics Medicine Service's "Structural Eye Disease" panel (v4.1) and all genes and conditions associated with the term "glaucoma" in Online Mendelian Inheritance in Man (OMIM) [167]. Each gene identified was then searched on PubMed for reports and studies detailing its association with childhood glaucoma and any anterior segment developmental anomaly. Any additional genes identified through these publications which were reported to be associated with childhood glaucoma and any anterior segment developmental anomaly were also reviewed. Inclusion criteria: all genes with a confirmed causative link to glaucoma onset under 18 years of age associated with any anterior segment developmental anomaly without syndromic features. Exclusion criteria: any genes reported only in patients with isolated childhood glaucoma, syndromic ASD or in whom glaucoma was identified subsequently to cataract surgery.

6. Conclusions

As the accessibility of genetic testing for ocular diseases has increased, so too has the number of genes identified that are linked to childhood glaucoma associated with anterior segment dysgenesis, although the molecular diagnostic rates for these conditions remain low. To date, twenty genes have been associated with this subset of ocular developmental disorders, with many more involved in the pathogenesis of primary congenital glaucoma and anterior segment dysgenesis without glaucoma [11]. The pathways governing anterior segment development are complex and mutations in different genes can cause very similar clinical features. Furthermore, there is significant variation in the phenotype resulting from mutations within individual genes. Improvements in individualized patient prognostication and management will benefit from anatomical-based reporting of clinical features and gene-based diagnosis, both for identifying the most appropriate therapies and allowing detailed genotype–phenotype correlation studies of these rare conditions in the future. Additionally, it is a step towards the change in approach to treatment required in this advent of gene therapy. As the development of these targeted treatments begins to translate into clinically available management options, knowing a patient's underlying genetic diagnosis will be key to ensuring that they receive appropriate treatment.

Author Contributions: Conceptualization, N.C. and M.M.; methodology, N.C.; writing—original draft preparation, N.C.; writing—review and editing, C.M. and M.M.; visualization, N.C.; supervision, M.M. All authors have read and agreed to the published version of the manuscript.

Funding: This work was supported by Moorfields Eye Charity, the Wellcome Trust (205174/Z/16/Z), the National Institute for Health and Care Research (NIHR), Biomedical Research Centre at Moorfields Eye Hospital NHS Foundation Trust, and UCL Institute of Ophthalmology.

Data Availability Statement: No new data were created or analyzed in this study. Data sharing is not applicable to this article.

Conflicts of Interest: The authors declare no conflicts of interest.

Abbreviations

The following abbreviations are used in this manuscript:

AD	Autosomal dominant
AR	Autosomal recessive
ASD	Anterior segment dysgenesis
BMP	Bone morphogenetic protein
CHED	Congenital hereditary endothelial dystrophy
CRISPR-Cas9	Clustered regularly interspaced palindromic repeats-CRISPR-associated protein 9
DSAEK	Descemet stripping automated endothelial keratoplasty
ECM	Extracellular matrix
FDA	US Food and Drug Administration
IOP	Intraocular pressure
JOAG	Juvenile open angle glaucoma
MEK/ERK	Mitogen-activated protein kinase-ERK kinase/extracellular signal-related kinase
MMCAT	Microcornea, myopic chorioretinal atrophy and telecanthus
MMP	Matrix metalloproteinase
MO	Morpholino oligonucleotide
OMIM	Online Mendelian Inheritance in Man
PCG	Primary congenital glaucoma
POAG	Primary open angle glaucoma
PTC	Premature termination codon
rAAV	Recombinant adeno-associated virus
rAd	Recombinant adenovirus
sgRNA	Single guide RNA
siRNA	Small interfering RNA
RGC	Retinal ganglion cell
RPE	Retinal pigment epithelium
STAR	Study of Ataluren in Participants With Nonsense Mutation Aniridia

References

1. Shaham, O.; Menuchin, Y.; Farhy, C.; Ashery-Padan, R. Pax6: A Multi-Level Regulator of Ocular Development. *Prog. Retin. Eye Res.* **2012**, *31*, 351–376. [CrossRef]
2. Cvekl, A.; Tamm, E.R. Anterior Eye Development and Ocular Mesenchyme: New Insights from Mouse Models and Human Diseases. *BioEssays News Rev. Mol. Cell. Dev. Biol.* **2004**, *26*, 374–386. [CrossRef]
3. Takamiya, M.; Stegmaier, J.; Kobitski, A.Y.; Schott, B.; Weger, B.D.; Margariti, D.; Cereceda Delgado, A.R.; Gourain, V.; Scherr, T.; Yang, L.; et al. Pax6 Organizes the Anterior Eye Segment by Guiding Two Distinct Neural Crest Waves. *PLoS Genet.* **2020**, *16*, e1008774. [CrossRef] [PubMed]
4. Stoilov, I.; Rezaie, T.; Jansson, I.; Schenkman, J.B.; Sarfarazi, M. Expression of Cytochrome P4501b1 (Cyp1b1) during Early Murine Development. *Mol. Vis.* **2004**, *10*, 629–636. [PubMed]
5. Davis-Silberman, N.; Ashery-Padan, R. Iris Development in Vertebrates; Genetic and Molecular Considerations. *Brain Res.* **2008**, *1192*, 17–28. [CrossRef] [PubMed]
6. Miesfeld, J.B.; Brown, N.L. Eye Organogenesis: A Hierarchical View of Ocular Development. *Curr. Top. Dev. Biol.* **2019**, *132*, 351–393. [CrossRef]
7. Mizokami, K.; Sugiura, T.; San Juan, R.G. The Development of Human Trabecular Meshwork. *Med. Electron. Microsc.* **1994**, *27*, 275–281. [CrossRef]
8. Marchini, G.; Toscani, M.; Chemello, F. Pediatric Glaucoma: Current Perspectives. *Pediatr. Health Med. Ther.* **2014**, *5*, 15–27. [CrossRef]
9. Thau, A.; Lloyd, M.; Freedman, S.; Beck, A.; Grajewski, A.; Levin, A.V. New Classification System for Pediatric Glaucoma: Implications for Clinical Care and a Research Registry. *Curr. Opin. Ophthalmol.* **2018**, *29*, 385–394. [CrossRef]
10. Ko, F.; Papadopoulos, M.; Khaw, P.T. Primary Congenital Glaucoma. *Prog. Brain Res.* **2015**, *221*, 177–189. [CrossRef]
11. Reis, L.M.; Seese, S.; Costakos, D.; Semina, E.V. Congenital Anterior Segment Ocular Disorders: Genotype-Phenotype Correlations and Emerging Novel Mechanisms. *Prog. Retin. Eye Res.* **2024**, *102*, 101288. [CrossRef]
12. Patel, A.; Hayward, J.D.; Tailor, V.; Nyanhete, R.; Ahlfors, H.; Gabriel, C.; Jannini, T.B.; Abbou-Rayyah, Y.; Henderson, R.; Nischal, K.K.; et al. The Oculome Panel Test: Next-Generation Sequencing to Diagnose a Diverse Range of Genetic Developmental Eye Disorders. *Ophthalmology* **2019**, *126*, 888–907. [CrossRef]
13. Reis, L.M.; Semina, E.V. Genetics of Anterior Segment Dysgenesis Disorders. *Curr. Opin. Ophthalmol.* **2011**, *22*, 314–324. [CrossRef]
14. Hingorani, M.; Hanson, I.; van Heyningen, V. Aniridia. *Eur. J. Hum. Genet.* **2012**, *20*, 1011–1017. [CrossRef]
15. Reis, L.M.; Maheshwari, M.; Capasso, J.; Atilla, H.; Dudakova, L.; Thompson, S.; Zitano, L.; Lay-Son, G.; Lowry, R.B.; Black, J.; et al. Axenfeld-Rieger Syndrome: More than Meets the Eye. *J. Med. Genet.* **2023**, *60*, 368–379. [CrossRef]
16. Knight, L.S.W.; Ruddle, J.B.; Taranath, D.A.; Goldberg, I.; Smith, J.E.H.; Gole, G.; Chiang, M.Y.; Willett, F.; D'Mellow, G.; Breen, J.; et al. Childhood and Early Onset Glaucoma Classification and Genetic Profile in a Large Australasian Disease Registry. *Ophthalmology* **2021**, *128*, 1549–1560. [CrossRef]
17. Wowra, B.; Dobrowolski, D.; Parekh, M.; Wylęgała, E. General Treatment and Ophthalmic Management of Peters' Anomaly. *J. Clin. Med.* **2024**, *13*, 532. [CrossRef] [PubMed]
18. Moosajee, M.; Hingorani, M.; Moore, A.T. PAX6-Related Aniridia. In *GeneReviews®*; Adam, M.P., Feldman, J., Mirzaa, G.M., Pagon, R.A., Wallace, S.E., Amemiya, A., Eds.; University of Washington: Seattle, WA, USA, 1993.
19. Libby, R.T.; Smith, R.S.; Savinova, O.V.; Zabaleta, A.; Martin, J.E.; Gonzalez, F.J.; John, S.W.M. Modification of Ocular Defects in Mouse Developmental Glaucoma Models by Tyrosinase. *Science* **2003**, *299*, 1578–1581. [CrossRef] [PubMed]
20. Medina-Trillo, C.; Aroca-Aguilar, J.-D.; Ferre-Fernández, J.-J.; Alexandre-Moreno, S.; Morales, L.; Méndez-Hernández, C.-D.; García-Feijoo, J.; Escribano, J. Role of FOXC2 and PITX2 Rare Variants Associated with Mild Functional Alterations as Modifier Factors in Congenital Glaucoma. *PLoS ONE* **2019**, *14*, e0211029. [CrossRef]
21. Young, T.L.; Whisenhunt, K.N.; Jin, J.; LaMartina, S.M.; Martin, S.M.; Souma, T.; Limviphuvadh, V.; Suri, F.; Souzeau, E.; Zhang, X.; et al. SVEP1 as a Genetic Modifier of TEK-Related Primary Congenital Glaucoma. *Invest. Ophthalmol. Vis. Sci.* **2020**, *61*, 6. [CrossRef]
22. Wiggs, J.L.; Howell, G.R.; Linkroum, K.; Abdrabou, W.; Hodges, E.; Braine, C.E.; Pasquale, L.R.; Hannon, G.J.; Haines, J.L.; John, S.W.M. Variations in COL15A1 and COL18A1 Influence Age of Onset of Primary Open Angle Glaucoma. *Clin. Genet.* **2013**, *84*, 167–174. [CrossRef]
23. Jackson, D.; Malka, S.; Harding, P.; Palma, J.; Dunbar, H.; Moosajee, M. Molecular Diagnostic Challenges for Non-retinal Developmental Eye Disorders in the United Kingdom. *Am. J. Med. Genet. C Semin. Med. Genet.* **2020**, *184*, 578–589. [CrossRef]
24. Nie, J.; Zhang, W. Secreted Protease ADAMTS18 in Development and Disease. *Gene* **2023**, *858*, 147169. [CrossRef] [PubMed]
25. Vincent, A.; Billingsley, G.; Priston, M.; Glaser, T.; Oliver, E.; Walter, M.; Ritch, R.; Levin, A.; Heon, E. Further Support of the Role of CYP1B1 in Patients with Peters Anomaly. *Mol. Vis.* **2006**, *12*, 506–510. [PubMed]

26. Edward, D.; Rajhi, A.A.; Lewis, R.A.; Curry, S.; Wang, Z.; Bejjani, B. Molecular Basis of Peters Anomaly in Saudi Arabia. *Ophthalmic Genet.* **2004**, *25*, 257–270. [CrossRef]
27. Churchill, A.J.; Yeung, A. A Compound Heterozygous Change Found in Peters' Anomaly. *Mol. Vis.* **2005**, *11*, 66–70.
28. Stingl, J.V.; Diederich, S.; Diel, H.; Schuster, A.K.; Wagner, F.M.; Chronopoulos, P.; Aghayeva, F.; Grehn, F.; Winter, J.; Schweiger, S.; et al. First Results from the Prospective German Registry for Childhood Glaucoma: Phenotype-Genotype Association. *J. Clin. Med.* **2021**, *11*, 16. [CrossRef]
29. Tanwar, M.; Dada, T.; Dada, R. Axenfeld-Rieger Syndrome Associated with Congenital Glaucoma and Cytochrome P4501B1 Gene Mutations. *Case Rep. Med.* **2010**, *2010*, 212656. [CrossRef]
30. Samant, M.; Chauhan, B.K.; Lathrop, K.L.; Nischal, K.K. Congenital Aniridia: Etiology, Manifestations and Management. *Expert Rev. Ophthalmol.* **2016**, *11*, 135–144. [CrossRef]
31. Franco, E.; Gagrani, M.; Scanga, H.L.; Areaux, R.G.; Chu, C.T.; Nischal, K.K. Variable Phenotype of Congenital Corneal Opacities in Biallelic CYP1B1 Pathogenic Variants. *Cornea* **2024**, *43*, 195–200. [CrossRef]
32. Al-Saei, O.; Malka, S.; Owen, N.; Aliyev, E.; Vempalli, F.R.; Ocieczek, P.; Al-Khathlan, B.; Genomics England Research Consortium; Fakhro, K.; Moosajee, M. Increasing the Diagnostic Yield of Childhood Glaucoma Cases Recruited into the 100,000 Genomes Project. *BMC Genom.* **2024**, *25*, 484. [CrossRef] [PubMed]
33. Ma, A.; Yousoof, S.; Grigg, J.R.; Flaherty, M.; Minoche, A.E.; Cowley, M.J.; Nash, B.M.; Ho, G.; Gayagay, T.; Lai, T.; et al. Revealing Hidden Genetic Diagnoses in the Ocular Anterior Segment Disorders. *Genet. Med. Off. J. Am. Coll. Med. Genet.* **2020**, *22*, 1623–1632. [CrossRef] [PubMed]
34. Thanikachalam, S.; Hodapp, E.; Chang, T.C.; Swols, D.M.; Cengiz, F.B.; Guo, S.; Zafeer, M.F.; Seyhan, S.; Bademci, G.; Scott, W.K.; et al. Spectrum of Genetic Variants Associated with Anterior Segment Dysgenesis in South Florida. *Genes* **2020**, *11*, 350. [CrossRef]
35. Collantes, E.R.A.; Delfin, M.S.; Fan, B.; Torregosa, J.M.R.; Siguan-Bell, C.; de Guzman Florcruz, N.V.; Martinez, J.M.D.; Masna-Hidalgo, B.J.; Guzman, V.P.T.; Anotado-Flores, J.F.; et al. EFEMP1 Rare Variants Cause Familial Juvenile-Onset Open Angle Glaucoma. *Hum. Mutat.* **2022**, *43*, 240–252. [CrossRef]
36. Aghayeva, F.A.; Schuster, A.K.; Diel, H.; Chronopoulos, P.; Wagner, F.M.; Grehn, F.; Pirlich, N.; Schweiger, S.; Pfeiffer, N.; Hoffmann, E.M. Childhood Glaucoma Registry in Germany: Initial Database, Clinical Care and Research (Pilot Study). *BMC Res. Notes* **2022**, *15*, 32. [CrossRef]
37. Huang, L.; Xu, T.; Gan, J.; Mao, Y.; Zhao, L.; Jiao, X.; Fan, M.; Wang, T.; Zhang, D.; Xu, M.; et al. Zonule-Associated Gene Variants in Isolated Ectopia Lentis and Glaucoma. *J. Glaucoma* **2023**, *32*, e80–e89. [CrossRef]
38. Chen, Z.-X.; Jia, W.-N.; Sun, Y.; Chen, T.-H.; Zhao, Z.-N.; Lan, L.-N.; Liu, Y.; Song, L.-H.; Jiang, Y.-X. Biallelic ADAMTSL4 Variants in a Chinese Cohort of Congenital Ectopia Lentis: Implications for Genotype-Phenotype Relationships. *Hum. Mutat.* **2022**, *43*, 2141–2152. [CrossRef]
39. Alsaif, H.S.; Khan, A.O.; Patel, N.; Alkuraya, H.; Hashem, M.; Abdulwahab, F.; Ibrahim, N.; Aldahmesh, M.A.; Alkuraya, F.S. Congenital Glaucoma and CYP1B1: An Old Story Revisited. *Hum. Genet.* **2019**, *138*, 1043–1049. [CrossRef]
40. Siggs, O.M.; Souzeau, E.; Taranath, D.A.; Dubowsky, A.; Chappell, A.; Zhou, T.; Javadiyan, S.; Nicholl, J.; Kearns, L.S.; Staffieri, S.E.; et al. Biallelic CPAMD8 Variants Are a Frequent Cause of Childhood and Juvenile Open-Angle Glaucoma. *Ophthalmology* **2020**, *127*, 758–766. [CrossRef] [PubMed]
41. Bonet-Fernández, J.-M.; Aroca-Aguilar, J.-D.; Corton, M.; Ramírez, A.-I.; Alexandre-Moreno, S.; García-Antón, M.-T.; Salazar, J.-J.; Ferre-Fernández, J.-J.; Atienzar-Aroca, R.; Villaverde, C.; et al. CPAMD8 Loss-of-Function Underlies Non-Dominant Congenital Glaucoma with Variable Anterior Segment Dysgenesis and Abnormal Extracellular Matrix. *Hum. Genet.* **2020**, *139*, 1209–1231. [CrossRef]
42. Zhu, A.Y.; Costain, G.; Cytrynbaum, C.; Weksberg, R.; Cohn, R.D.; Ali, A. Novel Heterozygous Variants in PXDN Cause Different Anterior Segment Dysgenesis Phenotypes in Monozygotic Twins. *Ophthalmic Genet.* **2021**, *42*, 624–630. [CrossRef]
43. Khan, K.; Rudkin, A.; Parry, D.A.; Burdon, K.P.; McKibbin, M.; Logan, C.V.; Abdelhamed, Z.I.A.; Muecke, J.S.; Fernandez-Fuentes, N.; Laurie, K.J.; et al. Homozygous Mutations in PXDN Cause Congenital Cataract, Corneal Opacity, and Developmental Glaucoma. *Am. J. Hum. Genet.* **2011**, *89*, 464–473. [CrossRef] [PubMed]
44. Zazo-Seco, C.; Plaisancié, J.; Bitoun, P.; Corton, M.; Arteche, A.; Ayuso, C.; Schneider, A.; Zafeiropoulou, D.; Gilissen, C.; Roche, O.; et al. Novel PXDN Biallelic Variants in Patients with Microphthalmia and Anterior Segment Dysgenesis. *J. Hum. Genet.* **2020**, *65*, 487–491. [CrossRef]
45. Choi, A.; Lao, R.; Ling-Fung Tang, P.; Wan, E.; Mayer, W.; Bardakjian, T.; Shaw, G.M.; Kwok, P.-Y.; Schneider, A.; Slavotinek, A. Novel Mutations in PXDN Cause Microphthalmia and Anterior Segment Dysgenesis. *Eur. J. Hum. Genet. EJHG* **2015**, *23*, 337–341. [CrossRef]

46. Yousaf, K.; Naz, S.; Mushtaq, A.; Wohler, E.; Sobreira, N.; Ho, B.-M.; Chen, L.-J.; Chu, W.-K.; Bashir, R. Exome Sequencing Reveals SLC4A11 Variant Underlying Congenital Hereditary Endothelial Dystrophy (CHED2) Misdiagnosed as Congenital Glaucoma. *Genes* **2023**, *14*, 310. [CrossRef]
47. Liu, X.Y.; Tao, Y.F.; Mao, Y.K.; Chen, Z.J.; Wang, Y.; Hong, Y.F.; Fan, N. A family with developmental glaucoma and microcornea due to novel ADAMTS18 gene mutations. *Zhonghua Yan Ke Za Zhi Chin. J. Ophthalmol.* **2024**, *60*, 78–83. [CrossRef]
48. Zhou, L.; Xu, Z.; Wu, Q.; Wei, X. Unilateral Buphthalmos, Corneal Staphyloma and Corneal Fistula Caused by Pathogenic Variant in the PITX3 Gene: A Case Report. *BMC Ophthalmol.* **2022**, *22*, 385. [CrossRef] [PubMed]
49. Verdin, H.; Sorokina, E.A.; Meire, F.; Casteels, I.; de Ravel, T.; Semina, E.V.; De Baere, E. Novel and Recurrent PITX3 Mutations in Belgian Families with Autosomal Dominant Congenital Cataract and Anterior Segment Dysgenesis Have Similar Phenotypic and Functional Characteristics. *Orphanet J. Rare Dis.* **2014**, *9*, 26. [CrossRef]
50. Khan, S.Y.; Vasanth, S.; Kabir, F.; Gottsch, J.D.; Khan, A.O.; Chaerkady, R.; Lee, M.-C.W.; Leitch, C.C.; Ma, Z.; Laux, J.; et al. FOXE3 Contributes to Peters Anomaly through Transcriptional Regulation of an Autophagy-Associated Protein Termed DNAJB1. *Nat. Commun.* **2016**, *7*, 10953. [CrossRef] [PubMed]
51. Plaisancié, J.; Ragge, N.K.; Dollfus, H.; Kaplan, J.; Lehalle, D.; Francannet, C.; Morin, G.; Colineaux, H.; Calvas, P.; Chassaing, N. FOXE3 Mutations: Genotype-Phenotype Correlations. *Clin. Genet.* **2018**, *93*, 837–845. [CrossRef]
52. Ormestad, M.; Blixt, A.; Churchill, A.; Martinsson, T.; Enerbäck, S.; Carlsson, P. Foxe3 Haploinsufficiency in Mice: A Model for Peters' Anomaly. *Investig. Ophthalmol. Vis. Sci.* **2002**, *43*, 1350–1357.
53. Islam, L.; Kelberman, D.; Williamson, L.; Lewis, N.; Glindzicz, M.B.; Nischal, K.K.; Sowden, J.C. Functional Analysis of FOXE3 Mutations Causing Dominant and Recessive Ocular Anterior Segment Disease. *Hum. Mutat.* **2015**, *36*, 296–300. [CrossRef] [PubMed]
54. Rashid, M.; Qasim, M.; Ishaq, R.; Bukhari, S.A.; Sajid, Z.; Ashfaq, U.A.; Haque, A.; Ahmed, Z.M. Pathogenic Variants of AIPL1, MERTK, GUCY2D, and FOXE3 in Pakistani Families with Clinically Heterogeneous Eye Diseases. *PLoS ONE* **2020**, *15*, e0239748. [CrossRef]
55. Sibon, I.; Coupry, I.; Menegon, P.; Bouchet, J.-P.; Gorry, P.; Burgelin, I.; Calvas, P.; Orignac, I.; Dousset, V.; Lacombe, D.; et al. COL4A1 Mutation in Axenfeld-Rieger Anomaly with Leukoencephalopathy and Stroke. *Ann. Neurol.* **2007**, *62*, 177–184. [CrossRef] [PubMed]
56. Coupry, I.; Sibon, I.; Mortemousque, B.; Rouanet, F.; Mine, M.; Goizet, C. Ophthalmological Features Associated with COL4A1 Mutations. *Arch. Ophthalmol.* **2010**, *128*, 483–489. [CrossRef] [PubMed]
57. Diel, H.; Ding, C.; Grehn, F.; Chronopoulos, P.; Bartsch, O.; Hoffmann, E.M. First Observation of Secondary Childhood Glaucoma in Coffin-Siris Syndrome: A Case Report and Literature Review. *BMC Ophthalmol.* **2021**, *21*, 28. [CrossRef]
58. Reis, L.M.; Houssin, N.S.; Zamora, C.; Abdul-Rahman, O.; Kalish, J.M.; Zackai, E.H.; Plageman, T.F.; Semina, E.V. Novel Variants in CDH2 Are Associated with a New Syndrome Including Peters Anomaly. *Clin. Genet.* **2020**, *97*, 502–508. [CrossRef]
59. Reis, L.M.; Atilla, H.; Kannu, P.; Schneider, A.; Thompson, S.; Bardakjian, T.; Semina, E.V. Distinct Roles of Histone Lysine Demethylases and Methyltransferases in Developmental Eye Disease. *Genes* **2023**, *14*, 216. [CrossRef]
60. Weh, E.; Reis, L.M.; Happ, H.C.; Levin, A.V.; Wheeler, P.G.; David, K.L.; Carney, E.; Angle, B.; Hauser, N.; Semina, E.V. Whole Exome Sequence Analysis of Peters Anomaly. *Hum. Genet.* **2014**, *133*, 1497–1511. [CrossRef]
61. Pan, Y.; Iwata, T. Exploring the Genetic Landscape of Childhood Glaucoma. *Children* **2024**, *11*, 454. [CrossRef]
62. Banerjee, A.; Chakraborty, S.; Chakraborty, A.; Chakrabarti, S.; Ray, K. Functional and Structural Analyses of CYP1B1 Variants Linked to Congenital and Adult-Onset Glaucoma to Investigate the Molecular Basis of These Diseases. *PLoS ONE* **2016**, *11*, e0156252. [CrossRef] [PubMed]
63. Mookherjee, S.; Acharya, M.; Banerjee, D.; Bhattacharjee, A.; Ray, K. Molecular Basis for Involvement of CYP1B1 in MYOC Upregulation and Its Potential Implication in Glaucoma Pathogenesis. *PLoS ONE* **2012**, *7*, e45077. [CrossRef]
64. Zhao, Y.; Wang, S.; Sorenson, C.M.; Teixeira, L.; Dubielzig, R.R.; Peters, D.M.; Conway, S.J.; Jefcoate, C.R.; Sheibani, N. Cyp1b1 Mediates Periostin Regulation of Trabecular Meshwork Development by Suppression of Oxidative Stress. *Mol. Cell. Biol.* **2013**, *33*, 4225–4240. [CrossRef]
65. Kumar, A.; Han, Y.; Oatts, J.T. Genetic Changes and Testing Associated with Childhood Glaucoma: A Systematic Review. *PLoS ONE* **2024**, *19*, e0298883. [CrossRef]
66. López-Garrido, M.-P.; Medina-Trillo, C.; Morales-Fernandez, L.; Garcia-Feijoo, J.; Martínez-de-la-Casa, J.-M.; García-Antón, M.; Escribano, J. Null CYP1B1 Genotypes in Primary Congenital and Nondominant Juvenile Glaucoma. *Ophthalmology* **2013**, *120*, 716–723. [CrossRef]
67. Medina-Trillo, C.; Ferre-Fernández, J.-J.; Aroca-Aguilar, J.-D.; Bonet-Fernández, J.-M.; Escribano, J. Functional Characterization of Eight Rare Missense CYP1B1 Variants Involved in Congenital Glaucoma and Their Association with Null Genotypes. *Acta Ophthalmol.* **2016**, *94*, e555–e560. [CrossRef]

68. Khan, A.O. Genetics of Primary Glaucoma. *Curr. Opin. Ophthalmol.* **2011**, *22*, 347–355. [CrossRef]
69. Plásilová, M.; Stoilov, I.; Sarfarazi, M.; Kádasi, L.; Feráková, E.; Ferák, V. Identification of a Single Ancestral CYP1B1 Mutation in Slovak Gypsies (Roms) Affected with Primary Congenital Glaucoma. *J. Med. Genet.* **1999**, *36*, 290–294. [CrossRef] [PubMed]
70. Vincent, A.; Billingsley, G.; Priston, M.; Williams-Lyn, D.; Sutherland, J.; Glaser, T.; Oliver, E.; Walter, M.A.; Heathcote, G.; Levin, A.; et al. Phenotypic Heterogeneity of CYP1B1: Mutations in a Patient with Peters' Anomaly. *J. Med. Genet.* **2001**, *38*, 324–326. [CrossRef] [PubMed]
71. Khan, A.O.; Aldahmesh, M.A.; Al-Abdi, L.; Mohamed, J.Y.; Hashem, M.; Al-Ghamdi, I.; Alkuraya, F.S. Molecular Characterization of Newborn Glaucoma Including a Distinct Aniridic Phenotype. *Ophthalmic Genet.* **2011**, *32*, 138–142. [CrossRef]
72. Kidson, S.H.; Kume, T.; Deng, K.; Winfrey, V.; Hogan, B.L. The Forkhead/Winged-Helix Gene, Mf1, Is Necessary for the Normal Development of the Cornea and Formation of the Anterior Chamber in the Mouse Eye. *Dev. Biol.* **1999**, *211*, 306–322. [CrossRef] [PubMed]
73. Berry, F.B.; Lines, M.A.; Oas, J.M.; Footz, T.; Underhill, D.A.; Gage, P.J.; Walter, M.A. Functional Interactions between FOXC1 and PITX2 Underlie the Sensitivity to FOXC1 Gene Dose in Axenfeld-Rieger Syndrome and Anterior Segment Dysgenesis. *Hum. Mol. Genet.* **2006**, *15*, 905–919. [CrossRef]
74. Paylakhi, S.H.; Moazzeni, H.; Yazdani, S.; Rassouli, P.; Arefian, E.; Jaberi, E.; Arash, E.H.; Gilani, A.S.; Fan, J.-B.; April, C.; et al. FOXC1 in Human Trabecular Meshwork Cells Is Involved in Regulatory Pathway That Includes miR-204, MEIS2, and ITGβ1. *Exp. Eye Res.* **2013**, *111*, 112–121. [CrossRef] [PubMed]
75. Zhang, X.; Friedman, A.; Heaney, S.; Purcell, P.; Maas, R.L. Meis Homeoproteins Directly Regulate Pax6 during Vertebrate Lens Morphogenesis. *Genes Dev.* **2002**, *16*, 2097–2107. [CrossRef]
76. Medina-Trillo, C.; Sánchez-Sánchez, F.; Aroca-Aguilar, J.-D.; Ferre-Fernández, J.-J.; Morales, L.; Méndez-Hernández, C.-D.; Blanco-Kelly, F.; Ayuso, C.; García-Feijoo, J.; Escribano, J. Hypo- and Hypermorphic FOXC1 Mutations in Dominant Glaucoma: Transactivation and Phenotypic Variability. *PLoS ONE* **2015**, *10*, e0119272. [CrossRef]
77. Harding, P.; Moosajee, M. The Molecular Basis of Human Anophthalmia and Microphthalmia. *J. Dev. Biol.* **2019**, *7*, 16. [CrossRef]
78. Adler, R.; Canto-Soler, M.V. Molecular Mechanisms of Optic Vesicle Development: Complexities, Ambiguities and Controversies. *Dev. Biol.* **2007**, *305*, 1–13. [CrossRef]
79. Gour, A.; Tibrewal, S.; Garg, A.; Vohra, M.; Ratna, R.; Sangwan, V.S. New Horizons in Aniridia Management: Clinical Insights and Therapeutic Advances. *Taiwan J. Ophthalmol.* **2023**, *13*, 467–478. [CrossRef]
80. Schedl, A.; Ross, A.; Lee, M.; Engelkamp, D.; Rashbass, P.; van Heyningen, V.; Hastie, N.D. Influence of PAX6 Gene Dosage on Development: Overexpression Causes Severe Eye Abnormalities. *Cell* **1996**, *86*, 71–82. [CrossRef]
81. Mort, R.L.; Bentley, A.J.; Martin, F.L.; Collinson, J.M.; Douvaras, P.; Hill, R.E.; Morley, S.D.; Fullwood, N.J.; West, J.D. Effects of Aberrant Pax6 Gene Dosage on Mouse Corneal Pathophysiology and Corneal Epithelial Homeostasis. *PLoS ONE* **2011**, *6*, e28895. [CrossRef] [PubMed]
82. "pax6"[GENE]-ClinVar-NCBI. Available online: https://www.ncbi.nlm.nih.gov/clinvar (accessed on 27 January 2025).
83. Gupta, V.; Somarajan, B.I.; Gupta, S.; Mahalingam, K.; Singh, A.; Sharma, A. A New Association of PAX6 Variation with Juvenile Onset Open Angle Glaucoma. *J. Hum. Genet.* **2023**, *68*, 355–358. [CrossRef] [PubMed]
84. Haji-Seyed-Javadi, R.; Jelodari-Mamaghani, S.; Paylakhi, S.H.; Yazdani, S.; Nilforushan, N.; Fan, J.-B.; Klotzle, B.; Mahmoudi, M.J.; Ebrahimian, M.J.; Chelich, N.; et al. LTBP2 Mutations Cause Weill-Marchesani and Weill-Marchesani-like Syndrome and Affect Disruptions in the Extracellular Matrix. *Hum. Mutat.* **2012**, *33*, 1182–1187. [CrossRef]
85. Narooie-Nejad, M.; Paylakhi, S.H.; Shojaee, S.; Fazlali, Z.; Rezaei Kanavi, M.; Nilforushan, N.; Yazdani, S.; Babrzadeh, F.; Suri, F.; Ronaghi, M.; et al. Loss of Function Mutations in the Gene Encoding Latent Transforming Growth Factor Beta Binding Protein 2, LTBP2, Cause Primary Congenital Glaucoma. *Hum. Mol. Genet.* **2009**, *18*, 3969–3977. [CrossRef] [PubMed]
86. Gabriel, L.A.R.; Wang, L.W.; Bader, H.; Ho, J.C.; Majors, A.K.; Hollyfield, J.G.; Traboulsi, E.I.; Apte, S.S. ADAMTSL4, a Secreted Glycoprotein Widely Distributed in the Eye, Binds Fibrillin-1 Microfibrils and Accelerates Microfibril Biogenesis. *Investig. Ophthalmol. Vis. Sci.* **2012**, *53*, 461–469. [CrossRef] [PubMed]
87. Cheong, S.-S.; Hentschel, L.; Davidson, A.E.; Gerrelli, D.; Davie, R.; Rizzo, R.; Pontikos, N.; Plagnol, V.; Moore, A.T.; Sowden, J.C.; et al. Mutations in CPAMD8 Cause a Unique Form of Autosomal-Recessive Anterior Segment Dysgenesis. *Am. J. Hum. Genet.* **2016**, *99*, 1338–1352. [CrossRef]
88. Escribano, J.; Tevar, A.; Bonet-Fernandez, J.-M.; Atienzar-Aroca, R.; Aroca-Aguilar, J.-D. Functional Interaction between Zebrafish Adamtsl4 and Cpamd8 Matrix Metalloproteinase-Related Genes: Implications in Ocular Anterior Segment Development and Glaucoma. *Investig. Ophthalmol. Vis. Sci.* **2024**, *65*, OD6.
89. Abreu-Velez, A.M.; Howard, M.S. Collagen IV in Normal Skin and in Pathological Processes. *N. Am. J. Med. Sci.* **2012**, *4*, 1–8. [CrossRef]

90. Lee, S.-W.; Kim, H.-K.; Naidansuren, P.; Ham, K.A.; Choi, H.S.; Ahn, H.-Y.; Kim, M.; Kang, D.H.; Kang, S.W.; Joe, Y.A. Peroxidasin Is Essential for Endothelial Cell Survival and Growth Signaling by Sulfilimine Crosslink-Dependent Matrix Assembly. *FASEB J. Off. Publ. Fed. Am. Soc. Exp. Biol.* **2020**, *34*, 10228–10241. [CrossRef]
91. Bai, X.; Dilworth, D.J.; Weng, Y.-C.; Gould, D.B. Developmental Distribution of Collagen IV Isoforms and Relevance to Ocular Diseases. *Matrix Biol. J. Int. Soc. Matrix Biol.* **2009**, *28*, 194–201. [CrossRef]
92. Kizhatil, K.; Ryan, M.; Marchant, J.K.; Henrich, S.; John, S.W.M. Schlemm's Canal Is a Unique Vessel with a Combination of Blood Vascular and Lymphatic Phenotypes That Forms by a Novel Developmental Process. *PLoS Biol.* **2014**, *12*, e1001912. [CrossRef]
93. Thomson, B.R.; Souma, T.; Tompson, S.W.; Onay, T.; Kizhatil, K.; Siggs, O.M.; Feng, L.; Whisenhunt, K.N.; Yanovitch, T.L.; Kalaydjieva, L.; et al. Angiopoietin-1 Is Required for Schlemm's Canal Development in Mice and Humans. *J. Clin. Investig.* **2017**, *127*, 4421–4436. [CrossRef] [PubMed]
94. Souma, T.; Tompson, S.W.; Thomson, B.R.; Siggs, O.M.; Kizhatil, K.; Yamaguchi, S.; Feng, L.; Limviphuvadh, V.; Whisenhunt, K.N.; Maurer-Stroh, S.; et al. Angiopoietin Receptor TEK Mutations Underlie Primary Congenital Glaucoma with Variable Expressivity. *J. Clin. Investig.* **2016**, *126*, 2575–2587. [CrossRef] [PubMed]
95. Patel, S.P.; Parker, M.D. SLC4A11 and the Pathophysiology of Congenital Hereditary Endothelial Dystrophy. *BioMed Res. Int.* **2015**, *2015*, 475392. [CrossRef]
96. Vilas, G.L.; Loganathan, S.K.; Liu, J.; Riau, A.K.; Young, J.D.; Mehta, J.S.; Vithana, E.N.; Casey, J.R. Transmembrane Water-Flux through SLC4A11: A Route Defective in Genetic Corneal Diseases. *Hum. Mol. Genet.* **2013**, *22*, 4579–4590. [CrossRef]
97. Han, S.B.; Ang, H.-P.; Poh, R.; Chaurasia, S.S.; Peh, G.; Liu, J.; Tan, D.T.H.; Vithana, E.N.; Mehta, J.S. Mice with a Targeted Disruption of Slc4a11 Model the Progressive Corneal Changes of Congenital Hereditary Endothelial Dystrophy. *Invest. Ophthalmol. Vis. Sci.* **2013**, *54*, 6179–6189. [CrossRef]
98. Wang, L.; Sun, M.; Zhang, Q.; Dang, S.; Zhang, W. ADAMTS18 Regulates Early Branching Morphogenesis of Lacrimal Gland and Has a Significant Association with the Risk of Dry Eye in Mice. *Exp. Eye Res.* **2022**, *218*, 109020. [CrossRef]
99. Ataca, D.; Caikovski, M.; Piersigilli, A.; Moulin, A.; Benarafa, C.; Earp, S.E.; Guri, Y.; Kostic, C.; Arsenijevic, Y.; Soininen, R.; et al. Adamts18 Deletion Results in Distinct Developmental Defects and Provides a Model for Congenital Disorders of Lens, Lung, and Female Reproductive Tract Development. *Biol. Open* **2016**, *5*, 1585–1594. [CrossRef]
100. Aldahmesh, M.A.; Alshammari, M.J.; Khan, A.O.; Mohamed, J.Y.; Alhabib, F.A.; Alkuraya, F.S. The Syndrome of Microcornea, Myopic Chorioretinal Atrophy, and Telecanthus (MMCAT) Is Caused by Mutations in ADAMTS18. *Hum. Mutat.* **2013**, *34*, 1195–1199. [CrossRef]
101. Aldahmesh, M.A.; Khan, A.O.; Mohamed, J.Y.; Alkuraya, H.; Ahmed, H.; Bobis, S.; Al-Mesfer, S.; Alkuraya, F.S. Identification of ADAMTS18 as a Gene Mutated in Knobloch Syndrome. *J. Med. Genet.* **2011**, *48*, 597–601. [CrossRef]
102. Lu, T.; Zhang, T.; Wang, C.; Yang, N.; Pan, Y.-H.; Dang, S.; Zhang, W. Adamts18 Deficiency in Zebrafish Embryo Causes Defective Trunk Angiogenesis and Caudal Vein Plexus Formation. *Biochem. Biophys. Res. Commun.* **2020**, *521*, 907–913. [CrossRef] [PubMed]
103. Cvekl, A.; Camerino, M.J. Generation of Lens Progenitor Cells and Lentoid Bodies from Pluripotent Stem Cells: Novel Tools for Human Lens Development and Ocular Disease Etiology. *Cells* **2022**, *11*, 3516. [CrossRef]
104. Semina, E.V.; Ferrell, R.E.; Mintz-Hittner, H.A.; Bitoun, P.; Alward, W.L.M.; Reiter, R.S.; Funkhauser, C.; Daack-Hirsch, S.; Murray, J.C. A Novel Homeobox Gene PITX3 Is Mutated in Families with Autosomal-Dominant Cataracts and ASMD. *Nat. Genet.* **1998**, *19*, 167–170. [CrossRef]
105. Dodonova, S.O.; Zhu, F.; Dienemann, C.; Taipale, J.; Cramer, P. Nucleosome-Bound SOX2 and SOX11 Structures Elucidate Pioneer Factor Function. *Nature* **2020**, *580*, 669–672. [CrossRef]
106. Cizelsky, W.; Hempel, A.; Metzig, M.; Tao, S.; Hollemann, T.; Kühl, M.; Kühl, S.J. Sox4 And Sox11 Function during Xenopus Laevis Eye Development. *PLoS ONE* **2013**, *8*, e69372. [CrossRef]
107. Tamm, E.R.; Wurm, A.; Sock, E.; Fuchshofer, R.; Wegner, M. The High–Mobility–Group Transcription Factor Sox11 Plays an Important Role During Anterior Eye Segment Development. *Investig. Ophthalmol. Vis. Sci.* **2006**, *47*, 3874.
108. Li, L.; Fan, D.-B.; Zhao, Y.-T.; Li, Y.; Yang, Z.-B.; Zheng, G.-Y. GJA8 Missense Mutation Disrupts Hemichannels and Induces Cell Apoptosis in Human Lens Epithelial Cells. *Sci. Rep.* **2019**, *9*, 19157. [CrossRef] [PubMed]
109. Kao, W.W.-Y.; Liu, C.-Y. Roles of Lumican and Keratocan on Corneal Transparency. *Glycoconj. J.* **2002**, *19*, 275–285. [CrossRef] [PubMed]
110. Liu, C.-Y.; Birk, D.E.; Hassell, J.R.; Kane, B.; Kao, W.W.-Y. Keratocan-Deficient Mice Display Alterations in Corneal Structure. *J. Biol. Chem.* **2003**, *278*, 21672–21677. [CrossRef]
111. Conrad, A.H.; Conrad, G.W. The Keratocan Gene Is Expressed in Both Ocular and Non-Ocular Tissues during Early Chick Development. *Matrix Biol.* **2003**, *22*, 323–337. [CrossRef] [PubMed]

112. Gealy, E.C.; Kerr, B.C.; Young, R.D.; Tudor, D.; Hayes, A.J.; Hughes, C.E.; Caterson, B.; Quantock, A.J.; Ralphs, J.R. Differential Expression of the Keratan Sulphate Proteoglycan, Keratocan, during Chick Corneal Embryogenesis. *Histochem. Cell Biol.* **2007**, *128*, 551–555. [CrossRef]
113. Pellegata, N.S.; Dieguez-Lucena, J.L.; Joensuu, T.; Lau, S.; Montgomery, K.T.; Krahe, R.; Kivelä, T.; Kucherlapati, R.; Forsius, H.; de la Chapelle, A. Mutations in KERA, Encoding Keratocan, Cause Cornea Plana. *Nat. Genet.* **2000**, *25*, 91–95. [CrossRef]
114. Accogli, A.; Calabretta, S.; St-Onge, J.; Boudrahem-Addour, N.; Dionne-Laporte, A.; Joset, P.; Azzarello-Burri, S.; Rauch, A.; Krier, J.; Fieg, E.; et al. De Novo Pathogenic Variants in N-Cadherin Cause a Syndromic Neurodevelopmental Disorder with Corpus Callosum, Axon, Cardiac, Ocular, and Genital Defects. *Am. J. Hum. Genet.* **2019**, *105*, 854–868. [CrossRef]
115. Karwacki-Neisius, V.; Jang, A.; Cukuroglu, E.; Tai, A.; Jiao, A.; Predes, D.; Yoon, J.; Brookes, E.; Chen, J.; Iberg, A.; et al. WNT Signalling Control by KDM5C during Development Affects Cognition. *Nature* **2024**, *627*, 594–603. [CrossRef] [PubMed]
116. Schorle, H.; Meier, P.; Buchert, M.; Jaenisch, R.; Mitchell, P.J. Transcription Factor AP-2 Essential for Cranial Closure and Craniofacial Development. *Nature* **1996**, *381*, 235–238. [CrossRef] [PubMed]
117. Pontoriero, G.F.; Smith, A.N.; Miller, L.-A.D.; Radice, G.L.; West-Mays, J.A.; Lang, R.A. Co-Operative Roles for E-Cadherin and N-Cadherin during Lens Vesicle Separation and Lens Epithelial Cell Survival. *Dev. Biol.* **2009**, *326*, 403–417. [CrossRef] [PubMed]
118. Kim, Y.; Jeong, Y.; Kwon, K.; Ismail, T.; Lee, H.-K.; Kim, C.; Park, J.-W.; Kwon, O.-S.; Kang, B.-S.; Lee, D.-S.; et al. Physiological Effects of KDM5C on Neural Crest Migration and Eye Formation during Vertebrate Development. *Epigenetics Chromatin* **2018**, *11*, 72. [CrossRef]
119. Gestri, G.; Osborne, R.J.; Wyatt, A.W.; Gerrelli, D.; Gribble, S.; Stewart, H.; Fryer, A.; Bunyan, D.J.; Prescott, K.; Collin, J.R.O.; et al. Reduced TFAP2A Function Causes Variable Optic Fissure Closure and Retinal Defects and Sensitizes Eye Development to Mutations in Other Morphogenetic Regulators. *Hum. Genet.* **2009**, *126*, 791–803. [CrossRef]
120. Pontoriero, G.F.; Deschamps, P.; Ashery-Padan, R.; Wong, R.; Yang, Y.; Zavadil, J.; Cvekl, A.; Sullivan, S.; Williams, T.; West-Mays, J.A. Cell Autonomous Roles for AP-2α in Lens Vesicle Separation and Maintenance of the Lens Epithelial Cell Phenotype. *Dev. Dyn. Off. Publ. Am. Assoc. Anat.* **2008**, *237*, 602–617. [CrossRef]
121. Mandal, A.K.; Chakrabarti, D.; Gothwal, V.K. Approach to Primary Congenital Glaucoma: A Perspective. *Taiwan J. Ophthalmol.* **2023**, *13*, 451–460. [CrossRef] [PubMed]
122. Coviltir, V.; Marinescu, M.C.; Urse, B.M.; Burcel, M.G. Primary Congenital and Childhood Glaucoma—A Complex Clinical Picture and Surgical Management. *Diagnostics* **2025**, *15*, 308. [CrossRef]
123. Hashemi, H.; Ghaffari, R.; Mohebi, M. Posterior Lamellar Keratoplasty (DSAEK) in Peters Anomaly. *Cornea* **2012**, *31*, 1201–1205. [CrossRef]
124. Approved Cellular and Gene Therapy Products—FDA. Available online: https://www.fda.gov/vaccines-blood-biologics/cellular-gene-therapy-products/approved-cellular-and-gene-therapy-products (accessed on 10 December 2024).
125. Amador, C.; Shah, R.; Ghiam, S.; Kramerov, A.A.; Ljubimov, A.V. Gene Therapy in the Anterior Eye Segment. *Curr. Gene Ther.* **2022**, *22*, 104–131. [CrossRef]
126. Mirjalili Mohanna, S.Z.; Korecki, A.J.; Simpson, E.M. rAAV-PHP.B Escapes the Mouse Eye and Causes Lethality Whereas rAAV9 Can Transduce Aniridic Corneal Limbal Stem Cells without Lethality. *Gene Ther.* **2023**, *30*, 670–684. [CrossRef]
127. Jain, A.; Zode, G.; Kasetti, R.B.; Ran, F.A.; Yan, W.; Sharma, T.P.; Bugge, K.; Searby, C.C.; Fingert, J.H.; Zhang, F.; et al. CRISPR-Cas9–Based Treatment of Myocilin-Associated Glaucoma. *Proc. Natl. Acad. Sci. USA* **2017**, *114*, 11199–11204. [CrossRef]
128. Behrens, A.; Gordon, E.M.; Li, L.; Liu, P.X.; Chen, Z.; Peng, H.; La Bree, L.; Anderson, W.F.; Hall, F.L.; McDonnell, P.J. Retroviral Gene Therapy Vectors for Prevention of Excimer Laser-Induced Corneal Haze. *Investig. Ophthalmol. Vis. Sci.* **2002**, *43*, 968–977.
129. Patil, S.V.; Kaipa, B.R.; Ranshing, S.; Sundaresan, Y.; Millar, J.C.; Nagarajan, B.; Kiehlbauch, C.; Zhang, Q.; Jain, A.; Searby, C.C.; et al. Lentiviral Mediated Delivery of CRISPR/Cas9 Reduces Intraocular Pressure in a Mouse Model of Myocilin Glaucoma. *Sci. Rep.* **2024**, *14*, 6958. [CrossRef]
130. Tandon, A.; Sharma, A.; Rodier, J.T.; Klibanov, A.M.; Rieger, F.G.; Mohan, R.R. BMP7 Gene Transfer via Gold Nanoparticles into Stroma Inhibits Corneal Fibrosis In Vivo. *PLoS ONE* **2013**, *8*, e66434. [CrossRef]
131. Tong, Y.-C.; Chang, S.-F.; Liu, C.-Y.; Kao, W.W.-Y.; Huang, C.H.; Liaw, J. Eye Drop Delivery of Nano-Polymeric Micelle Formulated Genes with Cornea-Specific Promoters. *J. Gene Med.* **2007**, *9*, 956–966. [CrossRef]
132. Yoon, K.C.; Bae, J.A.; Park, H.J.; Im, S.K.; Oh, H.J.; Lin, X.H.; Kim, M.Y.; Lee, J.H.; Lee, S.E.; Ahn, K.Y.; et al. Subconjunctival Gene Delivery of the Transcription Factor GA-Binding Protein Delays Corneal Neovascularization in a Mouse Model. *Gene Ther.* **2009**, *16*, 973–981. [CrossRef]
133. Blair-Parks, K.; Weston, B.C.; Dean, D.A. High-Level Gene Transfer to the Cornea Using Electroporation. *J. Gene Med.* **2002**, *4*, 92–100.

134. Vinciguerra, P.; Romano, V.; Rosetta, P.; Legrottaglie, E.F.; Kubrak-Kisza, M.; Azzolini, C.; Vinciguerra, R. Iontophoresis-Assisted Corneal Collagen Cross-Linking with Epithelial Debridement: Preliminary Results. *BioMed Res. Int.* **2016**, *2016*, 3720517. [CrossRef]
135. Henderson, J.; O'Callaghan, J.; Campbell, M. Gene Therapy for Glaucoma: Targeting Key Mechanisms. *Vis. Res.* **2024**, *225*, 108502. [CrossRef]
136. Sulak, R.; Liu, X.; Smedowski, A. The Concept of Gene Therapy for Glaucoma: The Dream That Has Not Come True Yet. *Neural Regen. Res.* **2023**, *19*, 92–99. [CrossRef]
137. Shanghai BDgene Co., Ltd. A Clinical Study on CRISPR/Cas9 Instantaneous Gene Editing Therapy to Primary Open-Angle Glaucoma With Elevated Intraocular Pressure and MYOC Gene Mutation. 2024. Available online: https://clinicaltrials.gov/study/NCT06465537 (accessed on 6 August 2025).
138. Huang, A.J.W. Suppression of Keratoepithelin and Myocilin by Small Interfering RNA (an American Ophthalmological Society Thesis). *Trans. Am. Ophthalmol. Soc.* **2007**, *105*, 365–378.
139. Rayana, N.P.; Sugali, C.K.; Dai, J.; Peng, M.; Liu, S.; Zhang, Y.; Wan, J.; Mao, W. Using CRISPR Interference as a Therapeutic Approach to Treat TGFβ2-Induced Ocular Hypertension and Glaucoma. *Investig. Ophthalmol. Vis. Sci.* **2021**, *62*, 7. [CrossRef]
140. Zhang, D.; Vetrivel, L.; Verkman, A.S. Aquaporin Deletion in Mice Reduces Intraocular Pressure and Aqueous Fluid Production. *J. Gen. Physiol.* **2002**, *119*, 561–569. [CrossRef]
141. Wu, J.; Bell, O.H.; Copland, D.A.; Young, A.; Pooley, J.R.; Maswood, R.; Evans, R.S.; Khaw, P.T.; Ali, R.R.; Dick, A.D.; et al. Gene Therapy for Glaucoma by Ciliary Body Aquaporin 1 Disruption Using CRISPR-Cas9. *Mol. Ther.* **2020**, *28*, 820–829. [CrossRef]
142. Yamada, H.; Yoneda, M.; Gosho, M.; Kato, T.; Zako, M. Bimatoprost, Latanoprost, and Tafluprost Induce Differential Expression of Matrix Metalloproteinases and Tissue Inhibitor of Metalloproteinases. *BMC Ophthalmol.* **2016**, *16*, 26. [CrossRef]
143. O'Callaghan, J.; Delaney, C.; O'Connor, M.; van Batenburg-Sherwood, J.; Schicht, M.; Lütjen-Drecoll, E.; Hudson, N.; Ni Dhubhghaill, S.; Humphries, P.; Stanley, C.; et al. Matrix Metalloproteinase-3 (MMP-3)–Mediated Gene Therapy for Glaucoma. *Sci. Adv.* **2023**, *9*, eadf6537. [CrossRef]
144. Sun, D.; Zhan, Z.; Zeng, R.; Liu, X.; Wang, B.; Yang, F.; Huang, S.; Li, Y.; Yang, Z.; Su, Y.; et al. Long-Term and Potent IOP-Lowering Effect of IκBα-siRNA in a Nonhuman Primate Model of Chronic Ocular Hypertension. *iScience* **2022**, *25*, 104149. [CrossRef]
145. Tan, C.; Song, M.; Stamer, W.D.; Qiao, Y.; Chen, X.; Sun, X.; Lei, Y.; Chen, J. miR-21-5p: A Viable Therapeutic Strategy for Regulating Intraocular Pressure. *Exp. Eye Res.* **2020**, *200*, 108197. [CrossRef]
146. Loma, P.; Guzman-Aranguez, A.; de Lara, M.J.P.; Pintor, J. Beta2 Adrenergic Receptor Silencing Change Intraocular Pressure in New Zealand Rabbits. *J. Optom.* **2018**, *11*, 69–74. [CrossRef]
147. Wang, J.; Wang, H.; Dang, Y. Rho-Kinase Inhibitors as Emerging Targets for Glaucoma Therapy. *Ophthalmol. Ther.* **2023**, *12*, 2943–2957. [CrossRef]
148. IVIEW Therapeutics Inc. Proof of Concept Clinical Trial of Gene Therapy GVB-2001 Delivered Via Intracameral Injection for the Treatment of Primary Open Angle Glaucoma. 2025. Available online: https://clinicaltrials.gov/study/NCT06921317 (accessed on 6 August 2025).
149. Borrás, T.; Buie, L.K.; Spiga, M.-G.; Carabana, J. Prevention of Nocturnal Elevation of Intraocular Pressure by Gene Transfer of Dominant-Negative RhoA in Rats. *JAMA Ophthalmol.* **2015**, *133*, 182–190. [CrossRef]
150. Gonzalez, V.; Moreno-Montanes, J.; Oll, M.; Sall, K.N.; Palumaa, K.; Dubiner, H.; Turman, K.; Muñoz-Negrete, F.; Ruz, V.; Jimenez, A.I. Results of Phase IIB SYLTAG Clinical Trial with Bamosiran in Patients with Glaucoma. *Investig. Ophthalmol. Vis. Sci.* **2016**, *57*, 3023.
151. Richardson, R.; Smart, M.; Tracey-White, D.; Webster, A.R.; Moosajee, M. Mechanism and Evidence of Nonsense Suppression Therapy for Genetic Eye Disorders. *Exp. Eye Res.* **2017**, *155*, 24–37. [CrossRef]
152. Lima Cunha, D.; Sarkar, H.; Eintracht, J.; Harding, P.; Zhou, J.H.; Moosajee, M. Restoration of Functional PAX6 in Aniridia Patient iPSC-Derived Ocular Tissue Models Using Repurposed Nonsense Suppression Drugs. *Mol. Ther. Nucleic Acids* **2023**, *33*, 240–253. [CrossRef]
153. Gregory-Evans, C.Y.; Wang, X.; Wasan, K.M.; Zhao, J.; Metcalfe, A.L.; Gregory-Evans, K. Postnatal Manipulation of Pax6 Dosage Reverses Congenital Tissue Malformation Defects. *J. Clin. Investig.* **2014**, *124*, 111–116. [CrossRef]
154. Wang, X.; Gregory-Evans, K.; Wasan, K.M.; Sivak, O.; Shan, X.; Gregory-Evans, C.Y. Efficacy of Postnatal In Vivo Nonsense Suppression Therapy in a Pax6 Mouse Model of Aniridia. *Mol. Ther. Nucleic Acids* **2017**, *7*, 417–428. [CrossRef]
155. PTC Therapeutics. A Phase 2, Multicenter, Randomized, Double-Masked, Placebo-Controlled Study of the Safety and Efficacy of Ataluren (PTC124) for the Treatment of Nonsense Mutation Aniridia. 2022. Available online: https://clinicaltrials.gov/study/NCT02647359 (accessed on 12 December 2024).
156. Djayet, C.; Bremond-Gignac, D.; Touchard, J.; Secretan, P.-H.; Vidal, F.; Robert, M.P.; Daruich, A.; Cisternino, S.; Schlatter, J. Formulation and Stability of Ataluren Eye Drop Oily Solution for Aniridia. *Pharmaceutics* **2020**, *13*, 7. [CrossRef]

157. Mirjalili Mohanna, S.Z.; Hickmott, J.W.; Lam, S.L.; Chiu, N.Y.; Lengyell, T.C.; Tam, B.M.; Moritz, O.L.; Simpson, E.M. Germline CRISPR/Cas9-Mediated Gene Editing Prevents Vision Loss in a Novel Mouse Model of Aniridia. *Mol. Ther. Methods Clin. Dev.* **2020**, *17*, 478–490. [CrossRef]
158. Dorot, O.; Roux, L.N.; Zennaro, L.; Oved, K.; Bremond-Gignac, D.; Pichinuk, E.; Aberdam, D. The Antipsychotropic Drug Duloxetine Rescues PAX6 Haploinsufficiency of Mutant Limbal Stem Cells through Inhibition of the MEK/ERK Signaling Pathway. *Ocul. Surf.* **2022**, *23*, 140–142. [CrossRef]
159. Oved, K.; Zennaro, L.; Dorot, O.; Zerbib, J.; Frank, E.; Roux, L.N.; Bremond-Gignac, D.; Pichinuk, E.; Aberdam, D. Ritanserin, a Potent Serotonin 2A Receptor Antagonist, Represses MEK/ERK Signalling Pathway to Restore PAX6 Production and Function in Aniridia-like Cellular Model. *Biochem. Biophys. Res. Commun.* **2021**, *582*, 100–104. [CrossRef]
160. Moustardas, P.; Abbasi, M.; Javidjam, D.; Asamoah, C.S.; Schweitzer-Chaput, A.; Cisternino, S.; Bremond-Gignac, D.; Aberdam, D.; Lagali, N. Duloxetine Enhances PAX6 Expression and Suppresses Innate Immune Responses in Murine LPS-Induced Corneal Inflammation. *Ocul. Surf.* **2024**, *34*, 225–234. [CrossRef]
161. Abbasi, M.; Amini, M.; Moustardas, P.; Gutsmiedl, Q.; Javidjam, D.; Suiwal, S.; Seitz, B.; Fries, F.N.; Dashti, A.; Rautavaara, Y.; et al. Effects of miR-204-5p Modulation on PAX6 Regulation and Corneal Inflammation. *Sci. Rep.* **2024**, *14*, 26436. [CrossRef]
162. Korecki, A.J.; Cueva-Vargas, J.L.; Fornes, O.; Agostinone, J.; Farkas, R.A.; Hickmott, J.W.; Lam, S.L.; Mathelier, A.; Zhou, M.; Wasserman, W.W.; et al. Human MiniPromoters for Ocular-rAAV Expression in ON Bipolar, Cone, Corneal, Endothelial, Müller Glial, and PAX6 Cells. *Gene Ther.* **2021**, *28*, 351–372. [CrossRef]
163. Wang, L.; Xu, Q.; Wang, W.; Sun, X.; Chen, Y. Genetic Analysis Using Next-Generation Sequencing and Multiplex Ligation Probe Amplification in Chinese Aniridia Patients. *Orphanet J. Rare Dis.* **2024**, *19*, 394. [CrossRef]
164. Hall, H.N.; Williamson, K.A.; FitzPatrick, D.R. The Genetic Architecture of Aniridia and Gillespie Syndrome. *Hum. Genet.* **2019**, *138*, 881–898. [CrossRef]
165. Daruich, A.; Robert, M.P.; Bremond-Gignac, D. Gene Therapies in Pediatric Ophthalmology. *Front. Ophthalmol.* **2023**, *3*, 1188522. [CrossRef]
166. Brusini, P.; Salvetat, M.L.; Zeppieri, M. How to Measure Intraocular Pressure: An Updated Review of Various Tonometers. *J. Clin. Med.* **2021**, *10*, 3860. [CrossRef]
167. Home-OMIM. Available online: https://omim.org/ (accessed on 1 October 2024).

pharmaceuticals

MDPI

Review

Rho Kinase (ROCK) Inhibitors in the Treatment of Glaucoma and Glaucoma Surgery: A Systematic Review of Early to Late Phase Clinical Trials

Jit Kai Tan [1,2], Peng Tee Khaw [3] and Christin Henein [3,*]

[1] Guy's Campus, King's College London, London SE1 1UL, UK
[2] Institute of Ophthalmology, University College London, London EC1V 9EL, UK
[3] National Institute for Health and Care Research Biomedical Research Centre for Ophthalmology, Moorfields Eye Hospital and University College London Institute of Ophthalmology, London EC1V 2PD, UK
* Correspondence: c.henein@ucl.ac.uk

Abstract: Background/Objectives: Primary open-angle glaucoma (POAG) is an anterior optic neuropathy that can lead to irreversible vision loss if untreated. Prostaglandin analogues are the first-line treatment, but new drug classes, such as rho kinase (ROCK) inhibitors, are being explored. This review evaluates the efficacy and safety of ROCK inhibitors in treating POAG based on completed trials, comparing results with available natural history data and identifying areas for further research. **Methods**: A systematic database search was conducted in Ovid MEDLINE and Ovid Embase on 5 April 2022 using the following keywords: 'glaucoma', 'rho kinase inhibitor', 'rho-kinase inhibitor', 'rock inhibitor', 'ripasudil', 'netarsudil', and 'fasudil'. Abstracts were screened for relevant studies and results summarized in tables. **Results**: The analysis of trials conducted for ROCK inhibitors reveals that they are a safe and efficacious drug to treat POAG, demonstrating non-inferiority to existing medical treatments. Comparison of data to natural history studies was inconclusive due to the lack of natural history studies and their limitations. The results showed ROCK inhibitors to be effective when combined with existing medical treatments. However, questions remain regarding the optimal dosage, patient selection, and cost-effectiveness. Outcome measures for future trials should be expanded to include additional methods of monitoring disease progression as well as patient quality-of-life. **Conclusions**: ROCK inhibitors have emerged with a favorable safety profile, efficaciously attenuating intraocular pressure. To elucidate their long-term therapeutic value and safety comprehensively, further independent, large-scale, prospective randomized controlled trials are warranted. Such studies are pivotal to augment our understanding of this emergent medication class.

Keywords: glaucoma; primary open-angle glaucoma; ocular hypertension; rho kinase inhibitors; glaucoma filtration surgery; fibrosis

1. Introduction

1.1. Glaucoma

In 2020, glaucoma affected over 75 million people worldwide, and projections suggest this number will escalate to 110 million by 2040. It remains the leading cause of progressive and irreversible blindness globally, with certain ethnicities displaying a predisposition to specific glaucoma types [1,2]. Glaucoma is the term used to describe a group of eye diseases that cause progressive optic neuropathy, and in which intraocular pressure (IOP) is a key modifiable factor. Glaucoma is commonly associated with raised IOP and is characterized by visual field defects and changes to the optic nerve head such as pathological cupping or, as a late sign, pallor of the optic disc. Ocular hypertension is where there is consistently or recurrently elevated IOP (greater than 21 mmHg) but with no signs of glaucoma [3].

This review examines the role of rho kinase (ROCK) inhibitors to lower the pressure in ocular hypertension (OHT) and glaucoma and in order to reduce progressive optic nerve damage and maintain visual function. We highlight their potential impact on ocular perfusion, neuroprotection, and anti-fibrotic properties after glaucoma filtration surgery. Moreover, this review critically evaluates the randomized controlled trials (RCT) conducted for this emergent class of medications.

1.1.1. Anatomy and Physiology

The aqueous humor is a transparent fluid that circulates through the anterior chamber of the eye, delivering nutrients to and eliminating metabolic waste from avascular ocular tissues such as the trabecular meshwork (TM). Synthesized by the ciliary body, this fluid is predominantly drained via the TM and the uveoscleral pathway located within the iridocorneal angle [4,5]. Structurally, the TM comprises three distinct regions: the uveal meshwork, the corneoscleral meshwork, and the juxtacanalicular tissue. These segments collectively constitute the principal resistance to aqueous outflow, working in concert with Schlemm's canal to regulate intraocular pressure [5]. The aqueous humor is secreted by the ciliary body into the posterior chamber, then flows through the pupil into the anterior chamber, and finally the majority drains via the TM at the iridocorneal angle (Figure 1).

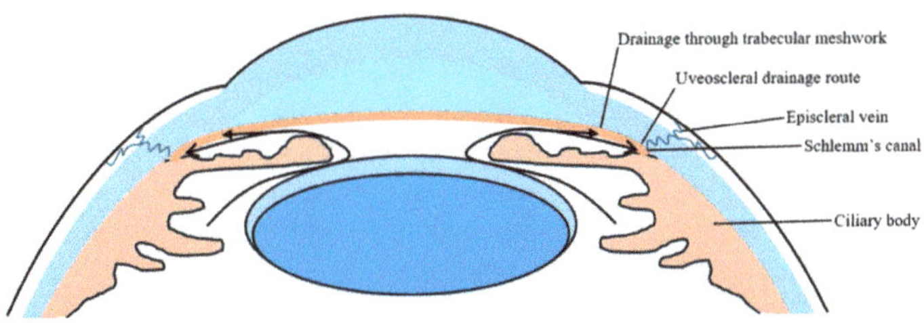

Figure 1. Image showing outflow route of aqueous humor in a healthy eye. Adapted from [6].

1.1.2. Primary Open-Angle Glaucoma

Primary open-angle glaucoma (POAG) is the most prevalent form of open-angle glaucoma characterized by optic nerve damage due to elevated intraocular pressure (IOP), despite an open iridocorneal angle facilitating aqueous humor drainage from the eye's anterior chamber. Accounting for over 70% of glaucoma cases globally, the etiology of POAG is multifaceted [7]. Significant risk factors include advanced age, heightened IOP, Black ethnicity, and a familial predisposition to the disease [8,9]. Several genetic markers, such as mutations in the myocilin and optineurin genes, have been linked to an increased susceptibility to POAG, with numerous additional associations uncovered through genome-wide association studies [10–15]. The pathogenetic mechanisms of POAG are complex, with existing genetic–environmental interactions making some individuals more at risk of developing glaucoma.

1.1.3. Pathogenesis of POAG

Glaucoma represents a spectrum of ocular conditions that, through various mechanisms, damage the optic nerve [16]. Elevated IOP is widely recognized as the major modifiable factor in the pathogenesis of POAG, believed to compromise the integrity of the lamina cribrosa, consequently impairing retinal ganglion cells (RGC), their axons, and axoplasmic transport [17]. Damage to the lamina cribrosa and the axons with their support glial cells interrupts the neurotrophic support relayed from brain target neurons to RGCs, as well as neurometabolic support from the local glial cells, potentially initiating cellular apoptosis [18,19]. These events precipitate the degeneration of RGCs, creating a deleterious milieu mediated by mechanisms such as glutamate excitotoxicity, which further exacerbates RGC loss [19,20]. The ensuing, currently irreversible damage to the optic nerve manifests as visual field deficits, significantly compromising vision.

Additionally, the pathogenesis of POAG is thought to involve a myriad of other elements including autoimmune responses, structural vulnerabilities of the lamina cribrosa, mitochondrial dysfunction, and diminished ocular blood flow stemming from possible vascular compression and reduced intracranial pressure [21–24]. Although these factors are implicated, their roles are less understood, with very limited techniques of investigation. Moreover, excessive activation of the glutaminergic system, oxidative stress with resultant free radical formation, inflammatory cytokines, and immune dysregulation are also possible contributing factors to the disease process [16].

1.1.4. Diagnosis

In POAG, pain is not a feature, and visual field defects are often detected at an advanced stage of damage, particularly where healthcare services are less well developed. Early diagnosis of POAG is crucial for preventing irreversible optic nerve damage, as symptoms become apparent only in the late stage of the disease. Diagnostic methods to detect POAG include direct ophthalmoscopy, gonioscopy, tonometry, and currently automated perimetry, with no single test being definitive in isolation [2,6]. Retinal and optic nerve imaging and in particular spectral-domain optical coherence tomography (SD-OCT), is now central to the diagnostic process, providing detailed assessments of the optic nerve and retinal nerve fiber layer (RNFL) [25,26].

1.2. The Natural History of POAG

The Early Manifest Glaucoma Trial examined progression of POAG in 46 patients with early glaucoma (including chronic simple glaucoma, normal-tension glaucoma, and exfoliative glaucoma) [27] and used the mean deviation index to determine the rate of visual field loss. The results found that patients with POAG was −1.08 dB per year, with a normal eye having 0 dB and a blind glaucoma eye having a range of −25–30 dB, giving approximately 25 years to progress from a normal eye to blindness. At the end of the study, 74% of patients showed progression in visual field loss, measured by a significant loss in light sensitivity in three consecutive tests at the same three or more test point locations [28].

A historical cohort study conducted in St. Lucia studied 287 eyes of 155 subjects and noted that 153 untreated eyes had worsened in visual field scores over one year according to the Advanced Glaucoma Intervention Study (AGIS) scoring system. The majority of these eyes were classified as having mild-moderate glaucoma. There was a significant association with increasing age, higher IOP, and progression of visual field defects [29].

1.2.1. Medical Treatments for POAG

There are currently six classes of conventional medical treatments for POAG, which are prostaglandin analogues; beta blockers; carbonic anhydrase inhibitors; alpha agonists; parasympathomimetics; and more recently, rho kinase (ROCK) inhibitors. These drug classes act in different ways, ranging from increasing aqueous outflow to decreasing aqueous humor production. Over the past decade, there have been changing trends in prescribing practices with an increasing use of preservative-free preparations and combination therapies, with the aim to improve patient compliance through a simpler treatment regimen.

Prostaglandin analogues (PGA) such as latanoprost are the first-line medical treatment and act through increasing uveoscleral outflow and therefore lower IOP. Side effects of this medication include eyelash elongation, periocular pigmentation and conjunctival hyperemia, iris cysts, cystoid macular oedema, and peri-orbital fat atrophy [2,30]. Beta blockers (BB) such as timolol are usually added on to medical therapy if IOP is uncontrolled with a prostaglandin analogue alone, and they work through decreasing aqueous humor production [2]. Whilst the ocular side effect profile is low, there can be serious systemic side effects such as bronchospasm, hypotension, impotence, and insomnia [31].

Carbonic anhydrase inhibitors such as acetazolamide rapidly lower aqueous humor production through decreasing bicarbonate secretion into ciliary epithelial cells [32]. Side effects can range from mild, such as stinging and metallic taste in the mouth, to serious electrolyte imbalances and Stevens–Johnson syndrome [30]. Alpha agonists include brimonidine and work to lower IOP by both lowering aqueous production and raising uveoscleral outflow by influencing the sympathetic tone of the ciliary process [30]. It is important to note that this class of medication crosses the blood–brain barrier and can lead to central nervous system and respiratory depression [32]. The final class of drugs used in conventional medical treatment are parasympathomimetics such as pilocarpine that increase aqueous outflow through the TM [32]. Side effects of parasympathomimetics include miosis, brow aches, accommodative spasm, and induced myopia [33].

1.2.2. Surgical Treatments for POAG

There are a variety of surgical treatments that are currently used to treat POAG, ranging from external aqueous filtering operations such as trabeculectomy to minimally invasive glaucoma surgery (MIGS). Trabeculectomies are the gold-standard surgical treatment for POAG, usually reserved in cases where IOP is insufficiently controlled by medication alone. Formation of a filtering bleb allows aqueous humor to flow from the anterior chamber to subconjunctival space, lowering IOP. However, subconjunctival fibrosis remains the leading cause of bleb failure and is the biggest challenge to successful filtration surgery [34]. Medical management tends to be used until it is definitively managed surgically. There is a growing use of preservative-free preparations of eyedrops in patients that are likely to require filtration surgery and improve bleb survival rate [35]. MIGS procedures tend to be indicated for mild to moderate glaucoma, with the aim of reducing medication burden, as well as the need for filtration surgery. New innovations are implantable devices that have sustained release of ocular antihypertensive agents such as travoprost [36].

1.3. Approved ROCK Inhibitors

ROCK inhibitors represent an innovative category of ocular hypotensive agents and are the first new class of anti-glaucoma medications to receive approval from the U.S. Food and Drug Administration (FDA) since latanoprost in the 1990s, indicated for the treatment of POAG and OHT [37]. Ripasudil (K-115) emerged as the inaugural ROCK inhibitor sanctioned for POAG treatment in Japan in 2014 (Figure 2). Subsequently, Rhopressa, which contains netarsudil (AR-13324), was granted FDA approval in the United States in 2017 for POAG management [38]. Additionally, Rhokiinsa, another formulation with netarsudil, received authorization in the European Union in 2019 for treating POAG and OHT [39]. Roclanda, a combination drop with netarsudil and latanoprost, has recently received approval in the European Union in 2021 for treatment of POAG and OHT [40]. Numerous other ROCK inhibitors are currently undergoing trials in both human and animal studies [41–43]. Chemical formulations of netarsudil, ripasudil, and sovesudil are shown in Figure 3.

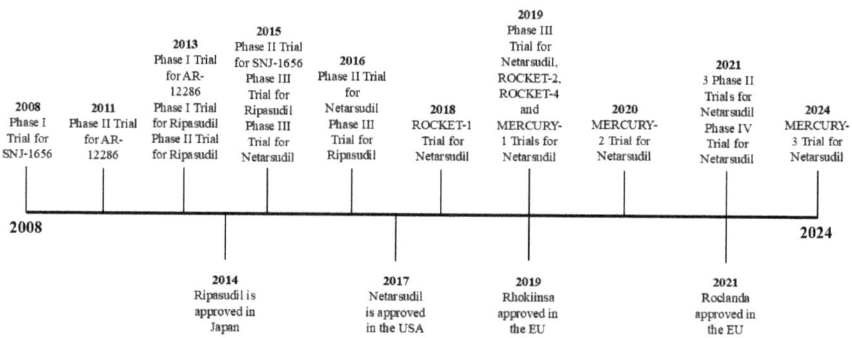

Figure 2. Timeline of ROCK inhibitor trials and years they were approved.

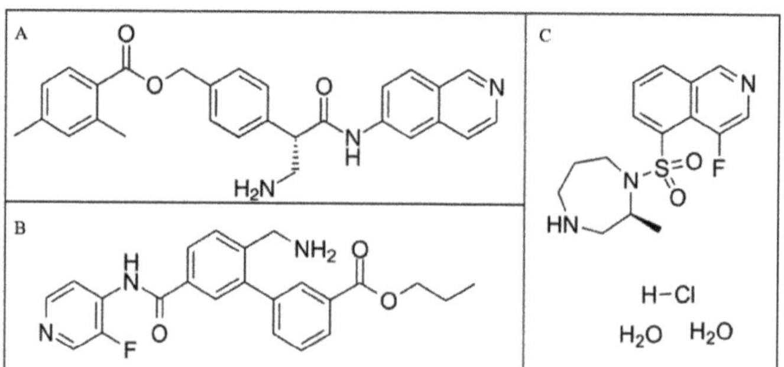

Figure 3. Chemical structures of (**A**) netarsudil, (**B**) sovesudil, (**C**) ripasudil. Netarsudil has the molecular formula $C_{28}H_{27}N_3O_3$ and weighs 453.5 g/mol; ripasudil has the molecular formula $C_{15}H_{18}FN_3O_2S$ and weighs 323.4 g/mol; sovesudil has the molecular formula $C_{23}H_{22}FN_3O_3$ and weighs 407.4 g/mol.

1.3.1. Mechanism of Action of ROCK Inhibitors

ROCK inhibitors function by impeding the activity of ROCK1 and ROCK2, which are the key effectors within the rho family of G-proteins, comprising RhoA, RhoB, and RhoC. These proteins are active when bound to guanosine triphosphate (GTP) and rendered inactive when bound to guanosine diphosphate [44,45]. The activation of these GTP-bound proteins leads to the stimulation of ROCK, which in turn influences a range of downstream effector tissues. Through this pathway, ROCK inhibitors can modulate various cellular functions, including cell adhesion, migration, smooth muscle contraction, proliferation, and the regulation of cell polarity [46–49].

1.3.2. Increased Ocular Perfusion

Reduced blood flow to the optic nerve is linked to glaucomatous damage, with high IOP exacerbating optic nerve atrophy by disturbing perfusion and energy metabolism [50]. Additionally, low perfusion pressure and nocturnal blood pressure dips have been implicated in further optic nerve damage [51]. The ROCK pathway, integral to vascular tone regulation, is a therapeutic target to potentially improve ocular blood flow and mitigate vascular damage, as evidenced by animal studies showing increased blood flow with ROCK inhibitors [43,52–54]. Despite promising animal studies, the vasoprotective effects of ROCK inhibitors in humans remain to be confirmed, necessitating more comprehensive research to validate these findings and their clinical application.

1.3.3. Neuroprotective Effects of ROCK Inhibitors

Raised IOP is believed to compress neuronal tissue at the lamina cribrosa, inhibiting growth of axons through inhibitory proteins such as myelin-associated glycoprotein. Neurometabolically supportive glial cells in the optic nerve head and the axons and cytoskeletal contraction may further this [18,19]. POAG upregulates the ROCK pathway, contributing to this inhibition, impacting cytoskeletal dynamics and cellular motility. ROCK inhibitors can counteract these effects, promoting axon regeneration and neuronal survival by influencing critical molecular pathways [51,55–57].

Animal studies have suggested that ROCK inhibitors reduce factors contributing to neuronal loss, such as reactive gliosis and leucocyte infiltration. Through these factors, axonal regeneration possibly following central nervous system injuries may be promoted by inhibition of the ROCK pathway [58–63]. As further research into the neuroprotective capabilities of ROCK inhibitors is conducted, caution must be exercised regarding their potential. Factors such as the limited regenerative capacity of adult neurons and the complexity of the neuronal growth environment must not be overlooked [64].

There is a paucity of trials combining ROCK inhibitors and brimonidine. Neuroprotective benefits of this combined agent are still unknown. Future studies should include optic nerve head thinning as an endpoint to assess the neuroprotective potential of this combination therapy. It would also be crucial to be stringent with the population recruited in terms of diet and any nutritional deficiencies which could reduce any neuroprotective effects.

1.3.4. Cellular Effects of ROCK Inhibitors

Through inhibition of the rho family of G-proteins, ROCK inhibitors have a variety of effects on different cell types in the eye [61]. They have been reported to regulate cell morphology, adhesion, and apoptosis and alter IOP through modulation of the actin cytoskeleton [65–68]. Modification of ocular blood flow to the optic nerve head [53,69], reduction of scar formation, and inflammation following glaucoma surgery are also theorized effects of ROCK inhibitors [70,71].

Calcium-independent and -dependent pathways affect the trabecular meshwork. ROCK inhibitors only affect the calcium-independent pathway [54]. Activation of the ROCK pathway indirectly results in raised levels of actin filament density, thereby increasing stability and rigidity of the actin cytoskeleton [72]. Through these effector pathways, ROCK has demonstrated the ability to regulate cellular, migration, life cycle, and growth directly or indirectly through muscle cell contractility and the actin cytoskeleton [73].

Increased outflow resistance in SC is a downstream effect of this, reducing aqueous humor drainage in the juxtacanalicular connective tissue of SC, an effect called 'funneling' [37,74,75]. This effect is shown in Figure 4.

ROCK inhibitors have been shown to alter SC's actin cytoskeleton through opposition of the ROCK pathway, reducing synthesis of the extracellular matrix, density of actin stress fibers, and cell tension and stiffness, easing and aiding aqueous humor outflow [37,76,77]. This is theorized to reduce the 'funneling' effect and outflow resistance, through greater pore formation in the juxtacanalicular region of SC from actin cytoskeleton modification [37].

ROCK inhibitors have a similar effect in the TM, due to the smooth-muscle-like properties of the TM and expression of α-smooth muscle actin [76]. ROCK inhibitors influence TM cells through the calcium-independent pathway [78], playing a critical role in myosin light chain (MLC) kinase and MLC phosphatase regulation [79]. ROCK inhibitor administration results in reduced levels of MLC phosphorylation in TM cells and tissue, lowering smooth muscle response to contractile agonists [76,80].

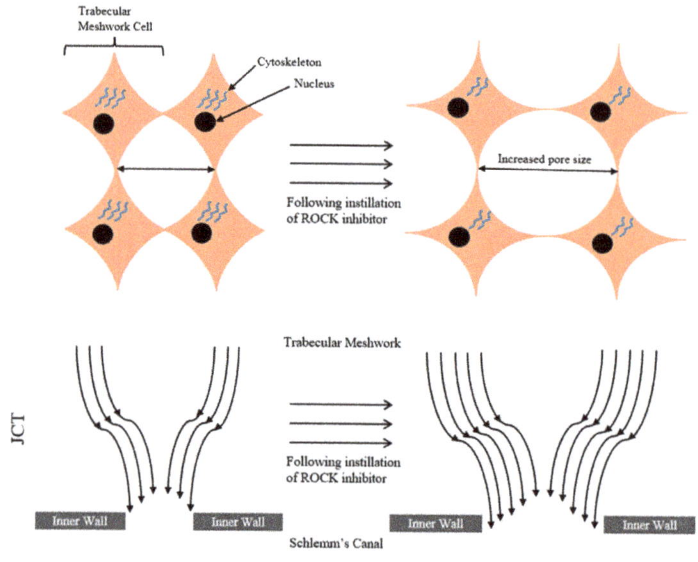

Figure 4. The 'funneling' effect seen in the TM.

MLC is a regulator of myosin II, which can influence cell adhesions, shaping and initiating actin cross-bridging and the formation of cell-to-cell junctions when phosphorylated [79], raising resistance to aqueous humor outflow. ROCK inhibitors have been demonstrated to be able to reversibly oppose these changes and suppress prolonged contraction of actin stress fiber focal adhesions and TM cells facilitating aqueous humor outflow [80,81].

Inhibition of the ROCK pathway influences endothelin-1-induced ROCK signaling, in turn resulting in reduced vasoconstriction of retinal blood vessels and vasomotor tone through lowering smooth muscle contraction [82,83].

This increased perfusion to the optic nerve head, theorized to aid in glaucomatous disease [84]. ROCK inhibitors can also lower episcleral venous pressure [38], a contributing factor to increased IOP, as demonstrated by the modified Goldmann equation [85]. However, its significance in influencing IOP through this mechanism is not yet understood, as SC is the primary point of outflow resistance for aqueous humor.

1.4. Review Aims

This review aims to evaluate the safety and efficacy of ROCK inhibitors in the treatment of POAG, with a particular focus on their IOP-lowering effects based on clinical trial data. Additionally, it will compare the safety and efficacy profiles of ROCK inhibitors with existing medical treatments for POAG and explore their potential as adjunctive therapies. The review will also discuss current challenges in the field, including patient selection, optimal dosing strategies, and broader clinical applications, while identifying key areas for future research.

2. Results

Following searching three databases, 751 articles were identified for screening, and 232 full test articles were assessed (Figure 5). Of these, 161 met the eligibility criteria. One relevant article was manually added after reviewing a reference list. A total of 71 articles were excluded. Reasons for exclusion were 34 animal studies, 24 conference abstracts, 3 expert commentaries, 2 review articles, and 8 non-English language studies. No articles were excluded based on publication date. A total of 27 RCTs were identified, including 3 phase I trials, 8 phase II trials, 15 phase III trials, and 1 phase IV trial. Of these, 11 were conducted in Japan, 14 in the United States, and 1 in India, and 1 was a multicenter trial in the European Union. Two trials investigated SNJ-1656 (Senju Pharmaceutical Co. Ltd., Osaka, Japan), two examined AR-12286 (Aerie Pharmaceuticals, Inc., Durham, NC, USA), eight focused on ripasudil (Kowa Company, Ltd., Nagoya, Japan), and eleven evaluated netarsudil (Aerie Pharmaceuticals, Inc., Durham, NC, USA). Change in IOP from baseline was the primary endpoint in the majority of studies.

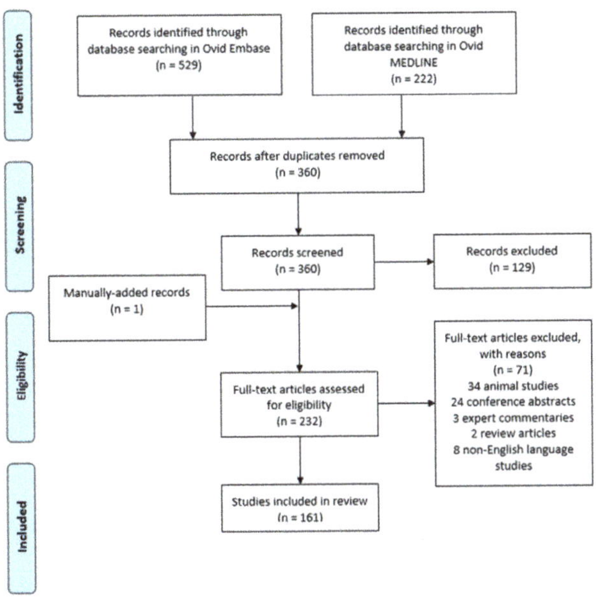

Figure 5. PRISMA diagram summarizing the steps for selection of studies [86].

2.1. ROCK Inhibitor Trials

There have been numerous trials that look at ROCK inhibitors in the context of glaucoma, the first being SNJ-1656 in 2008 (Figure 2). In 2008, the first human trial that examined the safety and effectiveness of ROCK inhibitors in lowering IOP was the phase I trial for SNJ-1656, an ophthalmic solution of Y-39983, which is a selective ROCK inhibitor [87]. The trial was a randomized, double-masked, group-comparison study with a total of 45 patients divided into 5 groups that were instilled with SNJ-1656 in increasing doses (0.003%, 0.01%, 0.03%, 0.05%, 0.1%). After this trial was conducted, trials for netarsudil and ripasudil were conducted to evaluate the efficacy and safety of those drugs [88].

The first human trial for netarsudil was conducted in 2015 to compare efficacy with latanoprost and was a randomized, double-masked study conducted over 28 days with a total of 213 patients that completed the study [89]. Patients were randomized to receive netarsudil 0.01%, 0.02%, or 0.005% latanoprost. Table 1 summarizes the clinical trials that have been conducted to date. It should be noted that AR-12286 was abandoned by Aerie Pharmaceuticals in 2017 as netarsudil was judged to have a longer duration of action [44].

Table 1. Summary of trials conducted examining the efficacy of ROCK inhibitors on POAG.

Study	Phase	Indication	Study Design	Drug	Study Size	Duration of Treatment
SNJ-1656						
Tanihara et al. [87]	I	Healthy volunteers	Randomized, double-masked, group-comparison, placebo–control	Placebo (OD) SNJ-1656 0.003% (OD) SNJ-1656 0.01% (OD) SNJ-1656 0.03% (OD) SNJ-1656 0.05% (OD) SNJ-1656 0.1% (OD)	45 patients	7 days followed by 7 days of repeated instillation
Inoue et al. [90]	II	POAG OHT	Randomized, double-masked, multicenter, placebo–control	Placebo (BD) SNJ-1656 0.03% (BD) SNJ-1656 0.05% (BD) SNJ-1656 0.1% (BD)	63 (initially 66 patients)	7 days
AR-12286						
Kopczynski et al. [91]	I	Healthy volunteers	Randomized, double-masked, crossover study	AR-12286 0.5% with 0.0065% BC (OD) AR-12286 0.5% with 0.015% BC (OD)	18 patients	2 periods of 8 days with a 7-day washout period in between
Williams et al. (NCT00902200) [92]	II	POAG OHT	Randomized, double-masked, multicenter, vehicle–control	Vehicle (OD for 14 days, then BD for 7 days) AR-12286 0.05% (OD for 14 days, then BD for 7 days) AR-12286 0.1% (OD for 14 days, then BD for 7 days) AR-12286 0.25% (OD for 14 days, then BD for 7 days)	89 patients	3 consecutive 7-day periods
Ripasudil						
Tanihara et al. [88]	I	Healthy volunteers	Randomized, double-masked, placebo–control, group comparison	Placebo (OD) Ripasudil 0.05% (OD) Ripasudil 0.1% (OD) Ripasudil 0.2% (OD) Ripasudil 0.4% (OD) Ripasudil 0.8% (OD)	50 patients	7 days followed by 7 days of repeated instillation
Tanihara et al. [93]	II	POAG OHT	Randomized, double-masked, multicenter, placebo–control, group comparison	Placebo (BD) Ripasudil 0.1% (BD) Ripasudil 0.2% (BD) Ripasudil 0.4% (BD)	210 patients	8 weeks
Tanihara et al. [94]	III	POAG OHT	Randomized, open label, Latin-square crossover	Placebo (BD) Ripasudil 0.2% (BD) Ripasudil 0.4% (BD)	28 (initially 43 patients)	3 periods of 2 days each
Tanihara et al. [95]	III	POAG OHT Exfoliation glaucoma Pigmentary glaucoma	Non-randomized, multicenter, open-label	Ripasudil 0.4% (BD) Ripasudil 0.4% with PGA (BD) Ripasudil 0.4% with BB (BD) Ripasudil 0.4% with FC PGA and BB (BD)	388 patients	52 weeks
Tanihara et al. (JAPIC111700) [96]	III	POAG OHT	Randomized, double-masked, multicenter, placebo–control	Placebo (BD) Ripasudil 0.4% (BD)	208 patients	8 weeks
Tanihara et al. (JAPIC111700) [96]	III	POAG OHT	Randomized, double-masked, multicenter, placebo–control	Placebo (BD) Ripasudil 0.4% (BD)	205 patients	8 weeks
Muhlisah et al. (UMIN000019017) [97]	III	POAG	Randomized, multicenter, open-label	Ripasudil (BD)	90 (initially 122 patients)	3 months
Tanihara et al. (Crossover Study) (jRCT2080225220) [98]	III	Healthy volunteers	Randomized, single-center, open-label	RBFC (BD) Ripasudil (BD) Brimonidine	18 patients	8 days

Table 1. *Cont.*

Study	Phase	Indication	Study Design	Drug	Study Size	Duration of Treatment
Netarsudil						
Bacharach et al. (NCT01731002) [89]	III	OAG OHT	Randomized, double-masked, multicenter	Latanoprost 0.005% (OD) Netarsudil 0.01% (OD) Netarsudil 0.02% (OD)	213 (initially 224 patients)	28 days
Aerie (NCT03310580) [99]	II	OAG OHT	Randomized, double-masked, multicenter	Netarsudil 0.02% (OD) Netarsudil 0.04% (OD)	42 patients	28 days
Serle et al. (NCT02207491) [100]	III	OAG OHT	Randomized, double-masked, multicenter	Timolol 0.5% (BD) Netarsudil 0.02% (OD)	411 patients	3 months
Kahook et al. (NCT02207621) [101]	III	OAG OHT	Randomized, double-masked, multicenter	Timolol 0.5% (BD) Netarsudil 0.02% (OD) Netarsudil 0.02% (BD)	756 patients	3-month data reported for 12-month trial
Lewis et al. (NCT02057575) [102]	II	OAG OHT	Randomized, double-masked, multicenter	Latanoprost 0.005% and Netarsudil 0.01% (OD) Latanoprost 0.005% and Netarsudil 0.02% (OD) Latanoprost 0.005% (OD) Netarsudil 0.02% (OD)	292 (initially 298 patients)	28 days
Aerie (NCT02246764) [103]	III	OAG OHT	Randomized, double-masked, multicenter	Netarsudil 0.02% and Placebo (OD) Netarsudil 0.02% (BD) Timolol 0.5% (BD)	83 patients	12 months
Peace et al. (NCT02874846) [104]	II	OAG OHT	Randomized, double-masked, vehicle-control	Vehicle (OD) Netarsudil 0.02% (OD)	12 patients	9 days
Sit et al. (NCT03233308) [105]	II	POAG OHT	Randomized, double-masked, vehicle–control	Vehicle (OD) Netarsudil 0.02% (OD)	40 patients	7 days
Araie et al. (NCT03844945) [99]	II	POAG OHT	Randomized, double-masked, parallel-group, placebo–control	Placebo (OD) Netarsudil 0.01% (OD) Netarsudil 0.02% (OD) Netarsudil 0.04% (OD)	207 (initially 215 patients)	4 weeks
Khouri et al. (NCT02558374) [106]	III	OAG OHT	Randomized, double-masked	Timolol 0.5% (BD) Netarsudil 0.02% (OD)	708 patients	6 months
Asrani et al. (NCT02558400) [107]	III	OAG OHT	Randomized, double-masked	Latanoprost 0.005%and Netarsudil 0.02% FC (OD) Latanoprost 0.005% (OD) Netarsudil 0.02% (OD)	718 patients	3-month data reported for 12-month trial
Walters et al. (NCT02674854) [108]	III	OAG OHT	Randomized, double-masked	Latanoprost 0.005% FC and Netarsudil 0.02% (OD) Latanoprost 0.005% (OD) Netarsudil 0.02% (OD)	750 patients	3 months
Stalmans et al. (NCT03284853) [109]	III	OAG OHT	Randomized, double-masked, parallel-group, multicenter, active-control	Latanoprost 0.005% and netarsudil 0.02% FC (OD) Bimatoprost 0.03% and timolol 0.5% FC (OD)	430 patients	6 months
Zaman et al. (NCT03808688) [110]	IV	OAG OHT	Prospective, interventional, open-label, multicenter	Netarsudil 0.02% (OD)	242 (initially 262 patients)	12 weeks
Shahid et al. [111]	III	OAG OHT	Randomized, parallel-group, open-label	Netarsudil 0.02% (OD) Netarsudil 0.02% and bimatoprost 0.01% (OD) Bimatoprost 0.01% (OD)	109 (initially 133 patients)	12 weeks

Abbreviations: OD = once-daily, POAG = primary open-angle glaucoma, OHT = ocular hypertension, BD = twice daily, BC = benzalkonium chloride, PGA = prostaglandin analogue, BB = beta blocker, FC = fixed combination, RBFC = ripasudil–brimonidine fixed-dose combination, OAG = open-angle glaucoma.

2.2. *Patient Characteristics*

The trials for ripasudil were focused on adults with a diagnosis of OHT or POAG [88]. The patients were excluded if they had any ocular comorbidities; undergone any ocular surgeries (other than cataract surgery more than 1 year ago, retinal laser treatment and yttrium–aluminum–garnet laser posterior capsulotomy more than 90 days ago, and eyelid surgery more than 120 days ago) [93,95]; severe visual field defects; impaired visual acuity; narrow iridocorneal angles; and any impairment to the liver, kidneys, heart, or endocrine system diseases. Baseline IOPs in the trials examined ranged from 13 to 35 mmHg. Baseline visual fields were not mentioned.

The trials for netarsudil similarly included adults with a diagnosis of OHT or POAG, with an IOP of >17 mmHg. Best corrected visual acuity was required to be at least 20/200 Snellen, following guidelines from the Early Treatment Diabetic Retinopathy Study. Patients were excluded if they had pseudo-exfoliation or pigment dispersion glaucoma, a history of angle closure or narrow iridocorneal angles, or previous glaucoma surgeries. Individuals with evidence of any ocular inflammation, clinically significant blepharitis, or conjunctivitis as well as women of childbearing potential who were pregnant, nursing, planning a pregnancy, or not using a medically acceptable form of birth control were excluded from these studies. Baseline IOPs in the trials examined ranged from 14 to 36 mmHg. Baseline visual fields were not mentioned.

2.3. *Efficacy and Safety of ROCK Inhibitors*

The most common adverse event reported was conjunctival hyperemia, which is a vasodilatory response of microvasculature in the conjunctiva due to increasing vascular smooth muscle relaxation from ROCK inhibitors [80]. The most serious adverse events in the trials examined included corneal erosion [95], visual field defects [96], conjunctival hemorrhage [89], reduced visual acuity [100], and non-infectious conjunctivitis [110]. However, these serious adverse events had a low incidence (<11.0%). A recent Cochrane review found that once-daily netarsudil 0.02% had an increased adverse event rate of 66 more events per 100 person-months when compared to placebo [112]. It also found that twice-daily ripasudil 0.4% + timolol 0.5% had a raised adverse event rate of 35 more events per 100 person-months compared to twice-daily 0.5% timolol monotherapy [112]. This is summarized in Table 2. It is important to note that most studies only reported conjunctival hyperemia; where other side effects have been mentioned, these are also included in the table. There has also been anecdotal evidence of bullous keratopathy following netarsudil use, which was reversed with steroid use and discontinuation of netarsudil [113].

Table 2. Summary of efficacy and safety of ROCK inhibitor trials.

Study	Duration of Follow-Up	Baseline IOP Inclusion Range	Drug (Number of Participants)	Baseline IOP mmHg (SD)	Efficacy mmHg (SD)	Frequent Adverse Events
SNJ-1656						
Tanihara et al. [87]	7 days	-			Change in IOP from baseline after 2 h	Conjunctival hyperemia
			Placebo (15)	14.1 (2.5)	−0.9	0.0%
			SNJ-1656 0.003% (6)	14.1 (1.4)	−1.2	16.7%
			SNJ-1656 0.01% (6)	13.7 (1.5)	−1.5	0.0%
			SNJ-1656 0.03% (6)	13.7 (2.2)	−2.2	33.3%
			SNJ-1656 0.05% (6)	13.2 (1.4)	−1.5	83.3%
			SNJ-1656 0.1% (6)	13.4 (2.7)	−2.0	100.0%
Inoue et al. [90]	7 days	22 ≤ IOP ≤ 31 mmHg			Change in IOP from baseline at peak	Conjunctival hyperemia
			Placebo (16)	~22.5	−1.5 (2.2)	Not reported
			SNJ-1656 0.03% (15)		−5.0 (2.4)	60.0%
			SNJ-1656 0.05% (14)		−4.4 (2.7)	100.0%
			SNJ-1656 0.1% (18)		−4.5 (1.9)	83.0%

Table 2. *Cont.*

Study	Duration of Follow-Up	Baseline IOP Inclusion Range	Drug (Number of Participants)	Baseline IOP mmHg (SD)	Efficacy mmHg (SD)	Frequent Adverse Events
AR-12286						
Kopczynski et al. [91]	8 days	14 ≤ IOP ≤ 20 mmHg			Change in IOP from baseline at peak	Conjunctival hyperemia
			AR-12286 0.5% with 0.015% BC (9)	17.0 (2.3)	−7.2 (2.0)	6.0%
			AR-12286 0.5% with 0.0065% BC (9)		−6.9 (1.8)	0.0%
Williams et al. (NCT00902200) [92]	22 days	21 ≤ IOP ≤ 24 mmHg			Change in IOP from baseline at peak	Conjunctival hyperemia
			Vehicle (22)	26.3 (2.5)	−2.4	9.1%
			AR-12286 0.05% (22)	26.0 (2.2)	−4.1	27.3%
			AR-12286 0.1% (23)	27.3 (3.2)	−5.0	39.1%
			AR-12286 0.25% (22)	26.9 (2.0)	−6.0	59.1%
Ripasudil						
Tanihara et al. [88]	7 days	≥13 mmHg			Change in IOP from baseline after 2 h	Conjunctival hyperemia
			Placebo (9)	15.1 (2.3)	−1.6	0.0%
			0.05% Ripasudil (8)	14.3 (2.5)	−3.4	62.5%
			Ripasudil 0.1% (7)	13.1 (1.8)	−2.2	57.1%
			Ripasudil 0.2% (8)	14.5 (2.5)	−2.6	62.5%
			Ripasudil 0.4% (7)	13.9 (1.5)	−4.0	28.6%
			Ripasudil 0.8% (8)	13.1 (1.4)	−4.3	87.5%
Tanihara et al. [93]	8 weeks	21 < IOP < 35 mmHg			Change in IOP from baseline at peak	Conjunctival hyperemia
			Placebo (54)	23.0 (2.1)	−2.5	13%
			Ripasudil 0.1% (53)	23.3 (2.4)	−3.7	43%
			Ripasudil 0.2% (54)	23.2 (2.0)	−4.2	57%
			Ripasudil 0.4% (49)	23.2 (1.9)	−4.5	65%
Tanihara et al. [94]	2 days	21 < IOP < 30 mmHg			Change in IOP from baseline after 2 h	Conjunctival hyperemia
			Placebo (28)	22.0 (2.1)	−4.1	10.7%
			Ripasudil 0.2% (28)		−6.8	78.6%
			Ripasudil 0.4% (28)		−7.3	96.4%
Tanihara et al. [95]	52 weeks	15 < IOP < 35 mmHg			Change in IOP from baseline at peak	Conjunctival hyperemia (74.6%) Blepharitis (20.6%) Allergic conjunctivitis (17.2%)
			Ripasudil 0.4% (173)	19.3 (2.7)	−3.7	
			Ripasudil 0.4% + with PGA (62)	17.6 (2.0)	−2.4	
			Ripasudil 0.4% + BB (60)	18.2 (2.3)	−2.0	
			Ripasudil 0.4% + FC PGA + BB (59)	17.6 (2.0)	−1.7	
Tanihara et al. (JAPIC111700) [96]	8 weeks	IOP ≥ 18 mmHg on Timolol			Change in IOP from baseline at peak	Conjunctival hyperemia
			Placebo (104)	19.7 (1.7)	−1.3	5.8%
			Ripasudil 0.4% (104)	19.9 (1.9)	−2.9	65.4%
Tanihara et al. (JAPIC111701) [96]	8 weeks	IOP ≥ 18 mmHg on Latanoprost			Change in IOP from baseline at peak	Conjunctival hyperemia
			Placebo (103)	19.6 (1.9)	−1.8	8.7%
			Ripasudil 0.4% (102)	20.1 (1.9)	−3.2	55.9%
Muhlisah et al. (UMIN000019017) [97]	3 months	-			Change in IOP from baseline at end of study	Hypotony, choroidal detachment
			Control (52)	16.2 (4.4)	−5.3	19.2%, 3.8%
			Ripasudil (38)	16.8 (5.0)	−6.2	18.4%, 10.5%
Tanihara et al. (Crossover Trial) (jRCT2080225220) [98]	8 days	IOP > 15 mmHg			Change in IOP from baseline at end of study	Conjunctival hyperemia
			RBFC (6)	12.7 (2.4)	−3.7	100.0%
			Ripasudil (6)	Not recorded	Not recorded	100.0%
			Brimonidine (6)	Not recorded	Not recorded	58.8%

Table 2. *Cont.*

Study	Duration of Follow-Up	Baseline IOP Inclusion Range	Drug (Number of Participants)	Baseline IOP mmHg (SD)	Efficacy mmHg (SD)	Frequent Adverse Events
Netarsudil						
Bacharach et al. (NCT01731002) [89]	28 days	24 ≤ IOP ≤ 36 mmHg			Change in IOP from baseline after 28 days	Conjunctival hyperemia
			Netarsudil 0.01% (74)	25.8	−5.4	52.0%
			Netarsudil 0.02% (72)	25.6	−5.9	57.0%
			Latanoprost 0.005% (77)	25.5	−6.8	16.0%
Aerie (NCT03310580) [99]	28 days	15 ≤ IOP < 30 mmHg			Mean diurnal iop at 28 days	Conjunctival hyperemia
			Placebo (15)		17.3	0.0%
			Netarsudil 0.02% (15)		14.4	66.7%
			Netarsudil 0.04% (12)		14.3	71.4%
Serle et al. (NCT02207491) [100]	3 months	17 < IOP < 27 mmHg			Change in IOP from baseline after 3 months (diurnal range)	Conjunctival hyperemia, conjunctival hemorrhage
			Netarsudil 0.02% (202)	22.5	−3.3 to −5.0	53.0%, 13.0%
			Timolol 0.5% (209)	22.3	−3.7 to −5.1	7.0%, 0.5%
Kahook et al. (NCT02207621) [101]	3 months	17 < IOP < 27 mmHg			Change in IOP from baseline after 3 months (diurnal range)	Conjunctival hyperemia, conjunctival hemorrhage, corneal verticillata
			Netarsudil 0.02% (251)	21.4	−3.3 to −4.6	50.0%, 15.0%, 9.0%
			Netarsudil 0.02% (254)	21.5	−4.1 to −5.4	59.0%, 17.0%, 15.0%
			Timolol 0.5% (251)	21.5	−3.7 to −5.1	10.0%, 0.0%, 0.4%
Lewis et al. (NCT02057575) [102]	28 days	24 ≤ IOP < 36 mmHg			Change in IOP from baseline after 28 days	Conjunctival hyperemia
			Latanoprost and netarsudil 0.01% (74)	25.1 (2.3)	−7.8	41.0%
			Latanoprost and netarsudil 0.02% (73)	25.1 (2.4)	−8.6	40.0%
			Latanoprost (73)	26.0 (2.8)	−7.6	14.0%
			Netarsudil 0.02% (78)	25.4 (2.7)	−6.3	40.0%
Peace et al. (NCT02874846) [104]	9 days	17 < IOP < 30 mmHg			Change in IOP from nocturnal baseline	None
			Vehicle (4)	22.9 (1.3)	−3.5	
			Netarsudil 0.2% (8)	22.4 (2.1)	−0.4	
Sit et al. (NCT03233308) [105]	7 days	17 < IOP < 30 mmHg			Change in IOP from baseline after 7 days	Conjunctival hyperemia
			Vehicle (18)	23.0 (1.4)	−4.5	0.0%
			Netarsudil 0.2% (18)	22.9 (1.6)	−1.0	72.2%
Araie et al. (NCT03844945) [99]	4 weeks	14 < IOP < 30 mmHg			Change in IOP from baseline after 4 weeks	Conjunctival hyperemia
			Placebo (55)	21.1 (3.7)	−1.7 (1.8)	1.8%
			Netarsudil 0.01% (54)	20.5 (2.8)	−4.1 (2.1)	23.6%
			Netarsudil 0.02% (51)	20.3 (2.8)	−4.8 (1.8)	37.0%
			Netarsudil 0.04% (55)	20.8 (3.2)	−4.8 (2.2)	56.9%
Khouri et al. (NCT02558374) [106]	6 months	20 < IOP < 30 mmHg			Change in IOP from baseline after 6 months	Conjunctival hyperemia
			Netarsudil 0.02% OD (351)	22.4	−3.9 to −4.7	47.9%
			Timolol 0.5% BD (357)	22.4	−3.8 to −5.2	9.2%

Table 2. *Cont.*

Study	Duration of Follow-Up	Baseline IOP Inclusion Range	Drug (Number of Participants)	Baseline IOP mmHg (SD)	Efficacy mmHg (SD)	Frequent Adverse Events
Asrani et al. (NCT02558400) [107]	3 months	20 < IOP < 36 mmHg			Change in IOP from baseline after 3 months (mean diurnal)	Conjunctival hyperemia
			Latanoprost 0.005% and netarsudil 0.02% FC OD (238)	23.7	−8.1	53.4%
			Latanoprost 0.005% OD (243)	23.6	−5.5	41.0%
			Netarsudil 0.02% OD (237)	23.5	−6.4	14.0%
Walters et al. (NCT02674854) [108]	3 months	20 < IOP < 36 mmHg			Change in IOP from baseline after 3 months (mean diurnal)	Conjunctival hyperemia
			Latanoprost 0.005% and netarsudil 0.02% FC OD (245)	23.5	−7.6	54.5%
			Latanoprost 0.005% OD (255)	23.6	−5.0	42.7%
			Netarsudil 0.02% OD (250)	23.5	−6.0	22.3%
Stalmans et al. (NCT03284853) [109]	6 months	IOP $\geq$ 17 mmHg in at least one eye IOP < 28 mmHg in both eyes			Change in IOP from baseline after 3 months	Conjunctival hyperemia, cornea verticillata
			Latanoprost 0.005% and netarsudil 0.02% FC OD (218)	25.1 (3.4)	−9.9	30.7%, 11.0%
			Bimatoprost 0.03% and timolol 0.5% FC OD (212)	24.8 (3.3)	−10.4	9.0%, 0.0%
Zaman et al. (NCT03808688) [110]	12 weeks	-			Change in IOP from Baseline after 12 weeks	Conjunctival hyperemia
			Netarsudil 0.02% (24)	19.4 (4.5)	−3.9 (3.6)	22.2%
			Netarsudil 0.02% replacing PGA (57)		−0.6 (3.1)	
			Netarsudil 0.02% replacing 2 concomitant therapies (6)		0.0 (1.6)	
			Netarsudil 0.02% replacing non-PGA monotherapy (4)		0.5 (1.7)	
			Netarsudil 0.02% and PGA (55)	20.3 (5.0)	−5.5 (5.0)	19.9%
			Netarsudil 0.02% and BB (2)		−4.3 (2.8)	
			Netarsudil 0.02% replacing $\geq$ 1 classes of concomitant therapy (64)		−0.4 (2.5)	
			Netarsudil 0.02% and $\geq$ 2 classes of concomitant therapy (64)		−4.5 (4.1)	
Shahid et al. [111]	12 weeks	21 < IOP < 32 mmHg			Change in IOP from baseline after 12 weeks	Conjunctival hyperemia
			Netarsudil 0.02% (43)	26.6 (4.8)	-	71.4%
			Netarsudil 0.02% and bimatoprost 0.01% (32)	24.6 (3.0)	-	83.3%
			Bimatoprost 0.01% (34)	24.6 (4.6)	-	15.3%

Abbreviations: IOP = intraocular pressure, SD = standard deviation, BC = benzalkonium chloride, PGA = prostaglandin analogue, BB = beta blocker, FC = fixed combination, RBFC = ripasudil–brimonidine fixed-dose combination, OD = once daily, BD = twice daily.

2.4. Efficacy of ROCK Inhibitors

The primary efficacy endpoint across all but one study examined was IOP reduction from baseline measured either with a Goldmann or Perkins tonometer. A Cochrane review found that once-daily netarsudil 0.02% was more effective at reducing IOP by about 3.11 mmHg at <6 months when compared to placebo, and twice-daily ripasudil 0.4% + timolol 0.5% was more effective than twice-daily timolol 0.5% monotherapy by about 0.75 mmHg at <6 months [112].

2.4.1. ROCK Inhibitors vs. Placebo

The first trial conducted for SNJ-1656 found significant changes in IOP changes between placebo and treatment eyes in the 0.03% and 0.1% solutions. The maximal change in IOP 4 h post-instillation of 0.1% SNJ-1656 was greater than all lower concentrations of SNJ-1656 [87]. Confidence intervals, baseline demographic characteristics, and individual patient results were not reported in this study. The first study for AR-12286 found that all three concentrations of the medication were statistically significant for reduction in IOP, with the peak effect resulting between 2 to 4 h after instillation. The largest IOP reduction was from the 0.25% AR-12286 formulation, given twice daily, up to −6.8 mmHg [92]. Again, confidence intervals and individual patient results were not reported.

A phase II study for sovesudil utilized three arms to investigate the efficacy of low-dose (0.25%) and high-dose (0.5%) sovesudil against a placebo over 4 weeks. The high-dose group demonstrated a statistically significant reduction in diurnal IOP at all timepoints compared to the placebo ($p = 0.0146$), although the difference between the low-dose and placebo groups was not significant ($p = 0.1350$) [114]. The phase I trial of ripasudil studied five groups of ten participants each, with eight participants receiving the treatment and two participants receiving a placebo once daily. The concentrations of ripasudil instilled were 0.05%, 0.1%, 0.2%, 0.4%, and 0.8%. The main findings of the trial were that repeated instillation did not increase IOP reduction for at least 7 days, and that twice-daily dosing at higher concentrations of ripasudil were clinically appropriate for use. Conjunctival hyperemia occurred in what appeared to be a dose-dependent manner, although severity and duration did not increase with repeated instillation [88]. In a phase II trial, 210 participants were divided randomly into four groups, to receive one of placebo, 0.1%, 0.2%, or 0.4% concentration of ripasudil to be instilled in both eyes twice daily. Significant differences in the adjusted mean of IOP change were found in all time points for the 0.1% and 0.4% concentration groups and at 2 h following instillation for the 0.2% group ($p < 0.05$) [93].

Another phase II trial that administered placebo, 0.2%, and 0.4% ripasudil to patients after a washout period between 5 and 30 days found that there were statistically significant differences in IOP reductions in both ripasudil dosages and placebo at all investigated timepoints except for the placebo at 24 h after first instillation ($p < 0.05$) [94]. One open-label phase III trial assessed long-term IOP reductions following trabeculectomy or trabeculectomy combined with cataract surgery with 3-month postoperative treatment with ripasudil. Peak reduction in IOP was found between baseline and month 12, which were −5.4 and −5.0 mmHg in the treatment and non-treatment groups, respectively. No statistically significant reductions in IOP were found throughout the follow-up period [97].

In the first phase II trial for netarsudil, it was found that netarsudil met the primary efficacy endpoint, reducing mean IOP by −3.5 mmHg, and the secondary efficacy endpoint of the study, which was a significant reduction in mean diurnal IOP change of −3.5 mmHg ($p < 0.001$) at the end of the study [104]. Another phase II trial found netarsudil 0.2% treatment resulted in an absolute change of −4.5 mmHg from baseline compared to an IOP change of −0.98 mmHg in the vehicle group, a statistically significant

IOP reduction from baseline ($p < 0.0001$) [105]. A phase II trial found that mean reductions of IOP from baseline at the end of the study were −1.7, −4.1, −4.8, and −4.8 for the placebo, 0.01%, 0.02%, and 0.04%, respectively—a statistically significant difference in IOP reduction ($p < 0.0001$) at all concentrations. Time-matched analysis of change in IOP also found that each post-treatment time point in all netarsudil concentrations was lower than the placebo [99].

2.4.2. ROCK Inhibitor Monotherapy vs. Other

A one-year phase III trial (Table 2) [95] enrolled 354 participants and divided them into four groups to receive either monotherapy of ripasudil 0.4%, additive therapy to PGA, additive therapy to BB, or fixed combination (FC) drugs. The peak mean IOP reductions at week 52 were −3.7 mmHg in the monotherapy group, −2.4 mmHg in the additive to PGA group, −3.0 mmHg in the additive to BB group, and −1.7 mmHg in the FC drug group. At week 52, monotherapy still yielded the highest mean IOP reduction when compared to troughs of all other treatment groups. Statistically significant differences in IOP reduction were found in all treatment groups at all timepoints ($p < 0.01$) [95].

In the first phase III trial for netarsudil (Table 2) [89], latanoprost 0.005% was used as a comparator ocular hypotensive drug against netarsudil 0.01% and 0.02%. At the endpoint of the study, mean diurnal IOP was reduced by 5.5, 5.7, and 6.8 mmHg in the netarsudil 0.01%, 0.02%, and latanoprost groups, respectively. Netarsudil failed to meet the criteria for non-inferiority to latanoprost, which was an upper 95% confidence limit of <1.5 mmHg for the difference between netarsudil and latanoprost. The main limitation of this study was that due to the relatively short study period, the long-term efficacy and safety profile of netarsudil could not be determined [89].

The ROCKET-1 study (Table 2) [100] was a phase III trial that compared efficacy of netarsudil 0.02% to timolol 0.5% in 411 participants. The mean changes in IOP from baseline ranged from −3.3 to −5.0 mmHg and −3.7 to −5.1 mmHg for netarsudil and timolol, respectively, which was significant at all treatment time points ($p < 0.0001$). However, netarsudil did not meet the criteria for non-inferiority to timolol [100]. The main limitation of the ROCKET-1 study is that half of the patients that were screened were on prostaglandin therapy [100].

The ROCKET-2 study (Table 2) [100] examined the safety and efficacy of netarsudil 0.02% and timolol 0.5% in 756 participants. Both treatment groups for netarsudil met the criteria for non-inferiority to timolol in this study, although both concentrations of netarsudil produced similar IOP reductions to timolol at all points of IOP measurement [100].

The ROCKET-4 study (Table 2) [106] compared the efficacy of once-daily netarsudil against twice-daily timolol in 708 participants for 6 months. Netarsudil met the criteria for non-inferiority to timolol, as the mean differences in IOP were within 1.5 mmHg at all timepoints of the study [106]. Only 186 patients from each treatment arm were included in the primary efficacy analysis, introducing the possibility of selection bias.

2.4.3. ROCK Inhibitor Combination Therapy vs. Other

A phase III trial that compared the effects of combined ripasudil with either latanoprost or timolol to compare difference in efficacy of lowering IOP enrolled 413 participants that were divided into groups of 205 and 208 into ripasudil–latanoprost and ripasudil–timolol groups, respectively. The mean IOP reductions in the ripasudil–timolol study 2 h after instillation were −2.9 and −1.3 mmHg for the treatment and placebo groups, respectively. This represented a statistically significant difference in IOP reduction of −1.6 mmHg (95% CI, 1.1–2.1 mmHg) ($p < 0.001$). In the ripasudil–latanoprost study, the mean IOP reductions at baseline IOP were −3.2 and −1.8 mmHg for the treatment and placebo groups, respectively. This was a statistically significant difference of −1.4 mmHg (95% CI, 0.9–1.9 mmHg) ($p < 0.001$) [96]. Despite a significant difference during statistical analysis, the authors stated that they were not confident in the differences found in the ripasudil-latanoprost groups.

Another clinical trial observed that netarsudil, netarsudil/bimatoprost, and bimatoprost were all efficacious in IOP reduction. The mean IOP ranges for the netarsudil, netarsudil/bimatoprost, and bimatoprost groups were 17.5–18.6 mmHg (baseline 26.6 ± 4.8), 15.8–16.5 mmHg (from 24.6 ± 3.0), and 14.0–14.9 mmHg (baseline 24.6 ± 4.6), respectively, which were all statistically significant reductions in IOP ($p < 0.001$). Mean reductions in IOP from baseline were not provided [111].

A real-world trial examined netarsudil 0.02% therapy either as a replacement for existing medical therapies or as an addition to IOP-lowering medication. This study found that of the subgroups receiving netarsudil 0.02% monotherapy as a replacement for other IOP-lowering medication, the treatment-naïve subgroup (newly diagnosed previously untreated) had the greatest reduction in IOP by the end of study, which was −3.9 mmHg. The largest IOP reduction in the subgroups receiving netarsudil 0.02% as a complement to PGA therapy, lowering IOP by −5.5 mmHg [110].

In a 28-day trial of fixed-dose netarsudil–latanoprost (PG324) 0.01% and 0.02% against monotherapy of latanoprost 0.005% and netarsudil 0.02% for 292 participants, it was found that there were changes in IOP of −7.8, −8.6, −7.6, and −6.3 mmHg for PG324 0.01%, 0.02%, latanoprost, and netarsudil, respectively. At the end of the study, the group that received PG324 0.02% had 69% of patients with a mean diurnal IOP $\leq$ 18 mmHg, the highest percentage of participants out of all groups [102].

The MERCURY-1 trial examined the difference in efficacy between FC netarsudil/latanoprost netarsudil 0.02% monotherapy and latanoprost 0.005% monotherapy. Netarsudil/latanoprost also met the criteria for superiority to each individual medication at all nine points of measurement by lowering IOP by an additional 1.8–3.0 mmHg versus netarsudil and an additional 1.3–2.5 mmHg versus latanoprost [107]. The MERCURY-2 trial, which examined the same aims as the MERCURY-1 trial, found that netarsudil/latanoprost met the criteria for superiority to each active component at all timepoints measured by lowering IOP by an additional 2.2–3.3 mmHg versus netarsudil and an additional 1.5–2.4 mmHg versus latanoprost [108]. The limitations of the MERCURY-1 and -2 trials were that both trials were conducted for 3 months, making it challenging to fully characterize the safety and efficacy profile of netarsudil/latanoprost and provide insight as to how netarsudil/latanoprost compares with other FC products long-term or when used in conjunction with other ocular hypotensive drugs. Two-thirds of the patients enrolled across MERCURY-1 and -2 were undergoing prior ocular anti-hypotensive therapy, which could affect the results of the trials.

MERCURY-3 investigated the difference between netarsudil/latanoprost against bimatoprost/timolol for a total of 6 months. Netarsudil/latanoprost achieved non-inferiority to bimatoprost/timolol at all timepoints with a between-treatment IOP reduction of $\leq$1.5 mmHg through to month 3 of treatment ($p < 0.05$) [109]. MERCURY-3 had strict inclusion and exclusion criteria, which may have limited the variety of patients recruited, affecting the generalizability of the results to other populations.

2.4.4. Adverse Events

Early trials showed researchers that the most common adverse event was conjunctival hyperemia, across all chemical formulations of ROCK inhibitors, which appeared to be largely self-limiting and remains the most common adverse event in the trials that follow. It is important to note that preservatives used in eye drops can also be a cause of conjunctival hyperemia. In summary, the conjunctival hyperemia rates for netarsudil and ripasudil monotherapy were 14.0–72.2% and 28.6–96.4%, respectively. The conjunctival hyperemia rates for netarsudil combination therapies were 30.7–83.3%. Conjunctival hyperemia rates for ripasudil monotherapy were not reported [95]. Comparatively, the rates of conjunctival hyperemia of PGA were 14.0–42.7%. It should be noted that some formulations of netarsudil and ripasudil contain benzalkonium chloride as a preservative [95,115], which is well-documented to cause ocular surface irritation [116].

In a trial examining the efficacy of sovesudil in normal tension glaucoma, a total of 43 patients out of 119 (36.1%) reported adverse events. Incidences of conjunctival hyperemia were 2.6%, 17.5%, and 24.4% in the placebo, low-dose, and high-dose sovesudil groups, respectively, with four subjects (9.8%) in the high-dose group and one subject (2.5%) in the low-dose group withdrawing from the study because of ocular adverse events that were unspecified [114].

Likewise, the most common adverse event in ripasudil trials was conjunctival hyperemia. Two phase III trials reported conjunctival hyperemia as the most common adverse event [94,95]. Three participants discontinued one of the trials solely due to conjunctival hyperemia. A total of 51 participants discontinued the trial due to blepharitis and/or allergic conjunctivitis, which resolved after discontinuation of ripasudil administration. Further analysis revealed that pollinosis-induced complications and monotherapy of ripasudil 0.4% were two significant factors related to patients discontinuing that study [95].

In a trial comparing combinations of ripasudil–latanoprost against ripasudil–timolol, the most common adverse event was conjunctival hyperemia (125/208). It should be noted that the addition of ripasudil to first-line drugs did not cause an increase in conjunctival hyperemia [96]. Another clinical study found that the most common adverse event following trabeculectomy or trabeculectomy combined with cataract surgery was hypotony in both treatment (7/38) and non-treatment groups (10/52). No cases of conjunctival hyperemia were reported in either group [97].

Conjunctival hyperemia was the most common adverse event in the trials for netarsudil. In the first phase III study conducted for netarsudil, 92 out of 224 participants reported conjunctival hyperemia. There were no changes of note to visual acuity or any other ocular or systemic measure [89]. Another trial found conjunctival hyperemia to be significantly increased in the netarsudil (31/43) and netarsudil/bimatoprost (26/32) groups ($p < 0.001$), with a higher severity than the bimatoprost monotherapy group ($p < 0.001$) [111].

In the ROCKET-1 and ROCKET-2 studies, the most common adverse event in the netarsudil treatment groups was conjunctival hyperemia (383/707). In the ROCKET-2 trial, it was noted that twice-daily instillation of netarsudil resulted in more frequent adverse events with moderate severity. Other common adverse events without visual sequalae in the ROCKET-1 and ROCKET-2 trials were corneal verticillata (70/707) and conjunctival hemorrhage (107/707); however, no netarsudil-related systemic issues were reported [101]. The ROCKET-4 trial saw a high incidence of conjunctival hyperemia in the treatment group compared to the control (47.9% vs. 9.2%), although it should be noted that some of these patients had conjunctival hyperemia at baseline [106].

The safety results of the MERCURY-1 and 2 studies were similar to those of the ROCKET trials, with conjunctival hyperemia being the most commonly reported adverse event, although it led to discontinuation of the study for 29 patients in the MERCURY-1 study [107,108]. MERCURY-3 also reported conjunctival hyperemia as the most common adverse event, followed by corneal verticillata. A total of 67 participants (30.7%) in the netarsudil/latanoprost and 19 participants (9.0%) in the bimatoprost/timolol group suffered from conjunctival hyperemia ($p < 0.0001$). An overall discontinuation rate in the study of 11.2% due to adverse events was observed, with 14 patients in the netarsudil/latanoprost group being discontinued due to conjunctival hyperemia [109]. Adverse events that occurred in the rest of trials include blurry vision, corneal deposits, increased lacrimation, reduced visual acuity, and conjunctival oedema [100,105,110].

2.5. Anti-Fibrotic Effects of ROCK Inhibitors After Glaucoma Filtration Surgery

Tissue growth factor beta (TGF-ß) signals trans-differentiation of conjunctival fibroblasts into myofibroblasts, focal cell adhesions, and cell contraction, and it is characterized by uncontrolled production, degradation of the extracellular matrix, and increased contractility [117].

Some studies have shown that administration of ROCK inhibitors following trabeculectomy can inhibit TGF-ß. This reduces expression of α-smooth muscle actin and lowers post-surgical scarring [76,117]. It was previously found that latanoprost, a first-line medication, induces collagen contraction, inhibiting post-surgical wound healing [118], but was suppressed by ROCK inhibitor administration [70].

Currently, there have not been any RCTs conducted examining the anti-fibrotic effect of ROCK inhibitors on trabeculectomy success, despite a previous study revealing no significant difference [119]. Muhlisa and colleagues observed the effects of ripasudil on trabeculectomy success in terms of reduction in post-operative procedures such as bleb needling. They reported that bleb needling was performed in 9/38 and 12/52 patients in the ripasudil-treated and non-ripasudil groups, respectively [97]. Should the anti-fibrotic effect of ROCK inhibitors prove significant, they could potentially serve a dual purpose in both reducing fibrosis and IOP.

2.6. The Economy of ROCK Inhibitors

A recent cost analysis based on the MERCURY-1 trial found that the costs per patient were USD 1673.61, USD 1654.73, and USD 1570.83 for netarsudil/latanoprost, netarsudil monotherapy, and latanoprost monotherapy, respectively, after 3 months of treatment in the United States. A cost effectiveness ratio of USD 46.17 per mmHg reduction in IOP was found, with netarsudil/latanoprost being the most cost-effective option for lowering IOP [120]. To our knowledge, this is the only cost analysis conducted for ROCK inhibitors. Although this study supports the implementation of ROCK inhibitors as a cost-effective treatment, further considerations such as the benefit in the best and worst seeing eye are required. For the UK, cost effectiveness should be expressed as incremental cost per quality-adjusted life year. Further consideration should be given to low-income countries with a high disease burden of POAG.

3. Discussion

General Limitations of the Trials

The trials reviewed consistently measured IOP reduction but often neglected functional outcomes like OCT and visual field tests, which are pertinent to patients with POAG. Notably, all studies focused solely on patients with POAG or OHT. There was a lack of standardized reporting for adverse events, leading to inconsistent data across trials. None of the studies had a standardized method of adverse events reporting, such as the Consolidated Standards of Reporting Trials [121], which led to inconsistent reporting of adverse events. Throughout the studies reviewed, none of the studies conducted a direct comparison of different ROCK inhibitors, which is an area that warrants research.

Evidence suggests that ROCK inhibitors might offer neuroprotection and facilitate axonal repair, yet RCTs have not assessed their effects on structural or functional optic nerve head damage. The impact of ROCK inhibitors on post-surgical fibrosis and corneal endothelium remains unexplored in RCTs.

Ethnic diversity in trial participants was limited, with 66.7% (14/21) of RCTs reporting predominantly White cohorts. Previous studies have shown that ethnic minorities are heavily underrepresented in POAG trials [122]. As POAG disproportionately affects certain ethnic groups, it is crucial for future research to include a broader ethnic spectrum to assess differential drug responses.

Patience adherence is a vital point of consideration when assessing the clinical utility of this new class of medication. Side effects such as conjunctival hyperemia, while usually transient and self-resolving, can still play a role in negatively impacting adherence to treatment regimen [81]. However, a standardized tool to evaluate the impact of such factors on adherence remains to be seen. A lack of consensus on POAG-specific patient-reported outcome measures is still unclear, with 41 separate quality-of-life questionnaires identified, underscoring the need for a universally accepted method of measuring patient-reported outcome measures [123].

Conflicts of interest were prevalent, as most trial investigators had pharmaceutical affiliations, underscoring the importance of independent studies for unbiased results. These limitations are summarized in Table 3, using the Risk of Bias 2 Tool [124].

Table 3. Summary of critical appraisal of trials included in this study following the Risk of Bias 2 Tool and conflict of interest declarations.

Study	Randomization Process	Deviation from Intended Interventions	Missing Outcome Data	Measurement of the Outcome	Selection of the Reported Results	Conflict of Interest	Overall
Tanihara et al. [87]	Low risk	Low risk	Low risk	Low risk	Low risk	High risk	Some concerns
Inoue et al. [90]	Low risk	Low risk	Low risk	Low risk	Some concerns	High risk	Some concerns
Williams et al. [92]	Low risk	Low risk	Low risk	Low risk	Low risk	High risk	Some concerns
Kopczynski et al. [91]	Low risk	Low risk	Low risk	Low risk	Low risk	High risk	Some concerns
Tanihara et al. [88]	Low risk	Low risk	Low risk	Low risk	Low risk	High risk	Some concerns
Tanihara et al. [93]	Low risk	Some concerns	Low risk	Low risk	Low risk	High risk	Some concerns
Tanihara et al. [94]	Low risk	Low risk	Low risk	Low risk	Low risk	High risk	Some concerns
Tanihara et al. [95]	High risk	Low risk	Low risk	Low risk	Low risk	High risk	Some concerns
Tanihara et al. [96]	Low risk	Low risk	Low risk	Low risk	Low risk	High risk	Some concerns
Tanihara et al. [96]	Low risk	Low risk	Low risk	Low risk	Low risk	High risk	Some concerns
Muhlisah et al. [97]	Low risk	Some concerns	Low risk	Low risk	Low risk	Low risk	Some concerns
Tanihara et al. (Crossover Study) [98]	Low risk	Low risk	Low risk	Low risk	Low risk	High risk	Some concerns
Bacharach et al. [89]	Some concerns	Low risk	Low risk	Low risk	Low risk	High risk	Some concerns
Serle et al. [100]	Low risk	High risk	High risk	Low risk	Low risk	High risk	High risk
Kahook et al. [101]	Low risk	High risk	High risk	Low risk	Low risk	High risk	High risk
Lewis et al. [102]	Low risk	Low risk	Low risk	Low risk	Low risk	High risk	Some concerns
Peace et al. [104]	Low risk	Low risk	Low risk	Low risk	Low risk	High risk	Some concerns
Sit et al. [105]	Low risk	Some concerns	Low risk	Low risk	Low risk	High risk	Some concerns
Araie et al. [99]	Low risk	Low risk	Low risk	Low risk	Low risk	High risk	Some concerns
Khouri et al. [106]	Low risk	High risk	High risk	Low risk	Low risk	High risk	High risk
Asrani et al. [107]	Low risk	Low risk	Low risk	Low risk	Low risk	High risk	Some concerns
Walters et al. [108]	Low risk	Low risk	Low risk	Low risk	Low risk	High risk	Some concerns
Stalmans et al. [109]	Some concerns	Low risk	Low risk	Low risk	Low risk	High risk	Some concerns
Shahid et al. [111]	Low risk	Low risk	Low risk	Low risk	Low risk	Low risk	Low risk
Zaman et al. [110]	High risk	Low risk	High risk	Low risk	Low risk	High risk	High risk

Low risk: The study is judged to be at low risk of bias for all domains for this result. Some concerns: The study is judged to raise some concerns in at least one domain for this result, but not to be at high risk of bias for any domain. High risk: The study is judged to be at high risk of bias in at least one domain for this result, or the study is judged to have some concerns for multiple domains in a way that substantially lowers confidence in the result.

Although the trials demonstrated non-inferiority of IOP-lowering efficacy of ROCK inhibitors to conventional first-line medical treatments for POAG, it is still unclear whether ROCK inhibitors can exist as an option for monotherapy, with the quality of evidence rated low or very low by a recent Cochrane review [112]. It is clear from the studies that ROCK inhibitors can enhance IOP-lowering potential of conventional medications when combined, with a generally mild side effect profile that is usually transient. However, due to the chronic nature of POAG and the fact that many patients will remain on medical therapy in the long run, it is paramount to thoroughly investigate any long-term drug interactions between ROCK inhibitors and existing IOP-lowering therapy. Given the paucity of literature in this aspect due to the recent approval of ROCK inhibitors, the MERCURY trials are the only large RCTs that examined the efficacy and safety profile of ROCK inhibitors combined with conventional POAG medications. The MERCURY-2 trial was only conducted for 1 year, which is an inadequate amount of time in the context of a side-effect profile for a medication that is likely to be taken for decades. There are some studies that have designed a series of urea-based ROCK inhibitors to increase efficacy and lower adverse events, although more research on its toxicology is required [125].

There are a multitude of prospective and retrospective studies that examine the safety and efficacy profile of using ROCK inhibitors as an adjunctive therapy to first-line medications [126–129] and have found favorable results in efficacy and adverse events. However, because of their study design and smaller sample sizes, research of a higher level of evidence is required to fully assess the efficacy and safety profile of combining ROCK inhibitors and conventional treatment in the long term.

4. Conclusions

This review primarily focuses on adult patients with OHT and mild to moderate POAG with limited inclusion of other glaucoma subtypes. Despite extensive research on ROCK inhibitors, further insights could be gained through functional assessments (e.g., visual field testing, intraocular pressure measurements) and structural evaluations (e.g., spectral-domain optical coherence tomography for retinal nerve fiber layer thickness, angiography for ocular perfusion). Additionally, incorporating patient-reported outcomes on treatment adherence and acceptability would provide a more comprehensive understanding of the therapeutic impact.

ROCK inhibitors have been validated as effective and well-tolerated, with an overall favorable safety profile despite some reported serious adverse events. Selection and reporting biases, along with inconsistent adverse event reporting, are areas needing refinement in future trials. While the long-term role of ROCK inhibitors in clinical practice appears promising, further research is required to fully establish their therapeutic position, particularly in combination with existing treatment options.

Future Directions

Significant progress has been made in understanding the therapeutic effects of ROCK inhibitors in lowering IOP. However, several challenges remain, and further research is needed to explore their full potential and determine additional clinical benefits. ROCK inhibitors have demonstrated IOP-lowering efficacy, but their impact on patients with varying severities of POAG remains unclear. It is hypothesized that these inhibitors may be more effective in the early stages of POAG due to their mechanism of action, which modifies the TM structure and function. In advanced POAG, irreversible TM damage and Schlemm's canal collapse may limit their therapeutic utility [112,130]. However, the specific stage beyond which ROCK inhibitors lose effectiveness has yet to be determined.

Optimizing dosage and treatment regimens remains a key area for further research. While higher doses of ROCK inhibitors have been linked to greater IOP reduction, they are also associated with an increased risk of adverse events. Determining an optimal dosing strategy could help minimize cosmetically undesirable side effects while maintaining therapeutic efficacy. Additionally, the development of an intraocular drug-releasing implant could offer a viable alternative for patients who face challenges with self-administering eye drops, ensuring more consistent and sustained drug delivery.

5. Materials and Methods

The key terms 'rho kinase inhibitor', 'ripasudil', 'netarsudil', 'fasudil', and 'glaucoma' were used, along with MeSH terms and free-text searches, to identify relevant concepts. A comprehensive literature search was conducted on 20 May 2024, using the MEDLINE (1946–present) and EMBASE (1974–present) databases. One reviewer (J.K.T.) independently screened the titles and abstracts of the retrieved articles using EndNote Software (Version 20.6, Bld 17174). Full-text articles of relevant studies were then analyzed, along with citations from their reference lists. Only randomized controlled trials were included. Two reviewers (J.K.T., C.H.) independently extracted key data, including study title, year of publication, author(s), number of participants, study design, type of intervention, primary outcomes measured, and main results. Non-English-language articles were excluded. Clinical trials involving human participants investigating ROCK inhibitors were included. Additionally, animal studies were considered if they provided novel insights into the management of key conditions. Studies involving patients under the age of 18 were excluded. Search strategies are provided in Sections A and B.

Inclusion and Exclusion Criteria

Studies were included if they evaluated ROCK inhibitors as a treatment option and exclusively involved patients diagnosed with OHT or POAG. Trials involving patients with other forms of glaucoma or those under the age of 18 were excluded.

The main themes of this review are as follows:

- Safety and efficacy outcomes of rho kinase inhibitor trials.
- Comparison of rho kinase inhibitors with conventional treatments for primary open-angle glaucoma.
- Additional benefits of rho kinase inhibitors in glaucoma management, including neuroprotective and anti-fibrotic effects.

Author Contributions: J.K.T.: data curation; formal analysis; investigation; methodology; project administration; resources; software; validation; visualization; roles/writing—original draft; and writing—review & editing. P.T.K.: Formal analysis; investigation; methodology; supervision; and writing—review & editing. C.H.: conceptualization; data curation; formal analysis; funding acquisition; investigation; methodology; project administration; resources; software; supervision; and writing—review & editing. All authors have read and agreed to the published version of the manuscript.

Funding: This research received no external funding.

Data Availability Statement: No new data were created or analyzed in this study. Data sharing is not applicable to this article.

Acknowledgments: The authors thank Desta Bokre for assistance in database search strategy and obtaining full text papers. Peng Khaw is supported by the Helen Hamlyn Trust, the Nolan Family, the

Katz Foundation, Moorfields Eye Charity and the NIHR Biomedical Research Centre at Moorfields Eye Hospital and the UCL Institute of Ophthalmology.

Conflicts of Interest: P.T.K.: consultant and lecturer at Aerie Pharmaceuticals; consultant at Novartis, Thea, Santen, CMER Hospital Group Hong Kong Alcon, DrugTech. C.H.: grants from Glaucoma UK Award Number 183772; funding from Health Education England/National Institute for Health Research (NIHR) for this research project (Clinical Lectureship CL-2020-18-009). The views expressed in this publication are those of the author(s) and not necessarily those of the NIHR, (Partner Name), NHS, or the UK Department of Health and Social Care.

Appendix A

Embase 1974–5 April 2022		
Search Lines	**Search Terms**	**Search Results**
1	glaucoma *	100,842
2	exp glaucoma/	91,390
3	1 or 2	105,022
4	rho kinase inhibitor	4177
5	rho-kinase inhibitor	4177
6	rock inhibitor	2146
7	ripasudil *	219
8	netarsudil *	229
9	fasudil *	3030
10	5 or 6 or 7 or 8 or 9	7594
11	3 or 10	112,087
12	3 and 10	529

* Was the symbol used for truncated search terms

Appendix B

Medline 1946–5 April 2022		
Search Lines	**Search Terms**	**Search Results**
1	glaucoma *	77,322
2	exp glaucoma/	56,949
3	1 or 2	77,541
4	rho kinase inhibitor	1589
5	rho-kinase inhibitor	1589
6	rock inhibitor	1366
7	ripasudil *	131
8	netarsudil *	113
9	fasudil *	1374
10	5 or 6 or 7 or 8 or 9	3699
11	3 or 10	81,018
12	3 and 10	222

* Was the symbol used for truncated search terms

References

1. Bourne, R.R.A.; Stevens, G.A.; White, R.A.; Smith, J.L.; Flaxman, S.R.; Price, H.; Jonas, J.B.; Keeffe, J.; Leasher, J.; Naidoo, K.; et al. Causes of vision loss worldwide, 1990–2010: A systematic analysis. *Lancet Glob. Health* **2013**, *1*, e339–e349. [CrossRef]
2. Jonas, J.B.; Aung, T.; Bourne, R.R.; Bron, A.M.; Ritch, R.; Panda-Jonas, S. Glaucoma. *Lancet* **2017**, *390*, 2183–2193. [CrossRef]
3. National Institute for Health and Care Excellence. Glaucoma. Available online: https://cks.nice.org.uk/topics/glaucoma/background-information/definition/ (accessed on 12 January 2025).
4. Goel, M.; Picciani, R.G.; Lee, R.K.; Bhattacharya, S.K. Aqueous humor dynamics: A review. *Open Ophthalmol. J.* **2010**, *4*, 52–59. [CrossRef]
5. Keller, K.E.; Acott, T.S. The Juxtacanalicular Region of Ocular Trabecular Meshwork: A Tissue with a Unique Extracellular Matrix and Specialized Function. *J. Ocul. Biol.* **2013**, *1*, 3.
6. Kang, J.M.; Tanna, A.P. Glaucoma. *Med. Clin. N. Am.* **2021**, *105*, 493–510. [CrossRef]
7. Quigley, H.A.; Broman, A.T. The number of people with glaucoma worldwide in 2010 and 2020. *Br. J. Ophthalmol.* **2006**, *90*, 262–267. [CrossRef] [PubMed]
8. Kass, M.A.; Heuer, D.K.; Higginbotham, E.J.; Johnson, C.A.; Keltner, J.L.; Miller, J.P.; Parrish, R.K., 2nd; Wilson, M.R.; Gordon, M.O. The Ocular Hypertension Treatment Study: A randomized trial determines that topical ocular hypotensive medication delays or prevents the onset of primary open-angle glaucoma. *Arch. Ophthalmol.* **2002**, *120*, 701–713, discussion 829–730. [CrossRef]
9. Rudnicka, A.R.; Mt-Isa, S.; Owen, C.G.; Cook, D.G.; Ashby, D. Variations in primary open-angle glaucoma prevalence by age, gender, and race: A Bayesian meta-analysis. *Investig. Ophthalmol. Vis. Sci.* **2006**, *47*, 4254–4261. [CrossRef]
10. Abu-Amero, K.; Kondkar, A.A.; Chalam, K.V. An Updated Review on the Genetics of Primary Open Angle Glaucoma. *Int. J. Mol. Sci.* **2015**, *16*, 28886–28911. [CrossRef]
11. Craig, J.E.; Han, X.; Qassim, A.; Hassall, M.; Cooke Bailey, J.N.; Kinzy, T.G.; Khawaja, A.P.; An, J.; Marshall, H.; Gharahkhani, P.; et al. Multitrait analysis of glaucoma identifies new risk loci and enables polygenic prediction of disease susceptibility and progression. *Nat. Genet.* **2020**, *52*, 160–166. [CrossRef]
12. Hardy, K.M.; Hoffman, E.A.; Gonzalez, P.; McKay, B.S.; Stamer, W.D. Extracellular trafficking of myocilin in human trabecular meshwork cells. *J. Biol. Chem.* **2005**, *280*, 28917–28926. [CrossRef]
13. Khawaja, A.P.; Cooke Bailey, J.N.; Wareham, N.J.; Scott, R.A.; Simcoe, M.; Igo, R.P., Jr.; Song, Y.E.; Wojciechowski, R.; Cheng, C.Y.; Khaw, P.T.; et al. Genome-wide analyses identify 68 new loci associated with intraocular pressure and improve risk prediction for primary open-angle glaucoma. *Nat. Genet.* **2018**, *50*, 778–782. [CrossRef]
14. Rezaie, T.; Child, A.; Hitchings, R.; Brice, G.; Miller, L.; Coca-Prados, M.; Heon, E.; Krupin, T.; Ritch, R.; Kreutzer, D.; et al. Adult-onset primary open-angle glaucoma caused by mutations in optineurin. *Science* **2002**, *295*, 1077–1079. [CrossRef]
15. Wiggs, J.L.; Pasquale, L.R. Genetics of glaucoma. *Hum. Mol. Genet.* **2017**, *26*, R21–R27. [CrossRef]
16. Weinreb, R.N.; Khaw, P.T. Primary open-angle glaucoma. *Lancet* **2004**, *363*, 1711–1720. [CrossRef]
17. Burgoyne, C.F.; Downs, J.C.; Bellezza, A.J.; Suh, J.K.; Hart, R.T. The optic nerve head as a biomechanical structure: A new paradigm for understanding the role of IOP-related stress and strain in the pathophysiology of glaucomatous optic nerve head damage. *Prog. Retin. Eye Res.* **2005**, *24*, 39–73. [CrossRef]
18. Li, Y.; Li, D.; Ying, X.; Khaw, P.T.; Raisman, G. An energy theory of glaucoma. *Glia* **2015**, *63*, 1537–1552. [CrossRef]
19. Quigley, H.A. Glaucoma. *Lancet* **2011**, *377*, 1367–1377. [CrossRef]
20. Hare, W.; WoldeMussie, E.; Lai, R.; Ton, H.; Ruiz, G.; Feldmann, B.; Wijono, M.; Chun, T.; Wheeler, L. Efficacy and safety of memantine, an NMDA-type open-channel blocker, for reduction of retinal injury associated with experimental glaucoma in rat and monkey. *Surv. Ophthalmol.* **2001**, *45* (Suppl. S3), S284–S289, discussion S295-286. [CrossRef]
21. Findl, O.; Strenn, K.; Wolzt, M.; Menapace, R.; Vass, C.; Eichler, H.G.; Schmetterer, L. Effects of changes in intraocular pressure on human ocular haemodynamics. *Curr. Eye Res.* **1997**, *16*, 1024–1029. [CrossRef]
22. Grus, F.H.; Joachim, S.C.; Wuenschig, D.; Rieck, J.; Pfeiffer, N. Autoimmunity and glaucoma. *J. Glaucoma* **2008**, *17*, 79–84. [CrossRef]
23. Mursch-Edlmayr, A.S.; Bolz, M.; Strohmaier, C. Vascular Aspects in Glaucoma: From Pathogenesis to Therapeutic Approaches. *Int. J. Mol. Sci.* **2021**, *22*, 4662. [CrossRef]
24. Wang, L.; Wei, X. T Cell-Mediated Autoimmunity in Glaucoma Neurodegeneration. *Front. Immunol.* **2021**, *12*, 803485. [CrossRef]
25. Chauhan, B.C.; O'Leary, N.; AlMobarak, F.A.; Reis, A.S.C.; Yang, H.; Sharpe, G.P.; Hutchison, D.M.; Nicolela, M.T.; Burgoyne, C.F. Enhanced detection of open-angle glaucoma with an anatomically accurate optical coherence tomography-derived neuroretinal rim parameter. *Ophthalmology* **2013**, *120*, 535–543. [CrossRef]
26. Yu, M.; Lin, C.; Weinreb, R.N.; Lai, G.; Chiu, V.; Leung, C.K. Risk of Visual Field Progression in Glaucoma Patients with Progressive Retinal Nerve Fiber Layer Thinning: A 5-Year Prospective Study. *Ophthalmology* **2016**, *123*, 1201–1210. [CrossRef]

27. Leske, M.C.; Heijl, A.; Hyman, L.; Bengtsson, B. Early Manifest Glaucoma Trial: Design and baseline data. *Ophthalmology* **1999**, *106*, 2144–2153. [CrossRef]
28. Heijl, A.; Bengtsson, B.; Hyman, L.; Leske, M.C.; Early Manifest Glaucoma Trial, G. Natural history of open-angle glaucoma. *Ophthalmology* **2009**, *116*, 2271–2276. [CrossRef]
29. Wilson, M.R.; Kosoko, O.; Cowan, C.L., Jr.; Sample, P.A.; Johnson, C.A.; Haynatzki, G.; Enger, C.; Crandall, D. Progression of visual field loss in untreated glaucoma patients and glaucoma suspects in St. Lucia, West Indies. *Am. J. Ophthalmol.* **2002**, *134*, 399–405. [CrossRef]
30. Lee, A.J.; Goldberg, I. Emerging drugs for ocular hypertension. *Expert. Opin. Emerg. Drugs* **2011**, *16*, 137–161. [CrossRef]
31. Fogagnolo, P.; Rossetti, L. Medical treatment of glaucoma: Present and future. *Expert. Opin. Investig. Drugs* **2011**, *20*, 947–959. [CrossRef]
32. Cvenkel, B.; Kolko, M. Current Medical Therapy and Future Trends in the Management of Glaucoma Treatment. *J. Ophthalmol.* **2020**, *2020*, 6138132. [CrossRef]
33. Lusthaus, J.; Goldberg, I. Current management of glaucoma. *Med. J. Aust.* **2019**, *210*, 180–187. [CrossRef]
34. Skuta, G.L.; Parrish, R.K., 2nd. Wound healing in glaucoma filtering surgery. *Surv. Ophthalmol.* **1987**, *32*, 149–170. [CrossRef]
35. Lee, S.; Park, D.Y.; Huh, M.G.; Cha, S.C. Influence of preoperative glaucoma medication on long-term outcomes of trabeculectomy. *Sci. Rep.* **2024**, *14*, 28341. [CrossRef]
36. Glaukos. iDose® TR Prescribing Information. Available online: https://www.glaukos.com/prescribing-information/idosetr/ (accessed on 15 January 2025).
37. Tanna, A.P.; Johnson, M. Rho Kinase Inhibitors as a Novel Treatment for Glaucoma and Ocular Hypertension. *Ophthalmology* **2018**, *125*, 1741–1756. [CrossRef]
38. Moshirfar, M.; Parker, L.; Birdsong, O.C.; Ronquillo, Y.C.; Hofstedt, D.; Shah, T.J.; Gomez, A.T.; Hoopes, P.C., Sr. Use of Rho kinase Inhibitors in Ophthalmology: A Review of the Literature. *Med. Hypothesis Discov. Innov. Ophthalmol.* **2018**, *7*, 101–111.
39. Agency, E.M. Rhokiinsa. Available online: https://www.ema.europa.eu/en/medicines/human/EPAR/rhokiinsa (accessed on 25 March 2024).
40. Agency, E.M. Roclanda. Available online: https://www.ema.europa.eu/en/medicines/human/EPAR/roclanda (accessed on 11 December 2024).
41. Gieser, D.K.; Tracy Williams, R.; O'Connell, W.; Pasquale, L.R.; Rosenthal, B.P.; Walt, J.G.; Katz, L.M.; Siegartel, L.R.; Wang, L.; Rosenblatt, L.C.; et al. Costs and utilization of end-stage glaucoma patients receiving visual rehabilitation care: A US multisite retrospective study. *J. Glaucoma* **2006**, *15*, 419–425. [CrossRef]
42. Rosenthal, R.; Choritz, L.; Schlott, S.; Bechrakis, N.E.; Jaroszewski, J.; Wiederholt, M.; Thieme, H. Effects of ML-7 and Y-27632 on carbachol- and endothelin-1-induced contraction of bovine trabecular meshwork. *Exp. Eye Res.* **2005**, *80*, 837–845. [CrossRef]
43. Waki, M.; Yoshida, Y.; Oka, T.; Azuma, M. Reduction of intraocular pressure by topical administration of an inhibitor of the Rho-associated protein kinase. *Curr. Eye Res.* **2001**, *22*, 470–474. [CrossRef]
44. Al-Humimat, G.; Marashdeh, I.; Daradkeh, D.; Kooner, K. Investigational Rho Kinase Inhibitors for the Treatment of Glaucoma. *J. Exp. Pharmacol.* **2021**, *13*, 197–212. [CrossRef]
45. Nourinia, R.; Nakao, S.; Zandi, S.; Safi, S.; Hafezi-Moghadam, A.; Ahmadieh, H. ROCK inhibitors for the treatment of ocular diseases. *Br. J. Ophthalmol.* **2018**, *102*. [CrossRef] [PubMed]
46. Carta, F.; Supuran, C.T.; Scozzafava, A. Novel therapies for glaucoma: A patent review 2007–2011. *Expert Opin. Ther. Pat.* **2012**, *22*, 79–88. [CrossRef]
47. Etienne-Manneville, S.; Hall, A. Rho GTPases in cell biology. *Nature* **2002**, *420*, 629–635. [CrossRef]
48. Loge, C.; Wallez, V.; Scalbert, E.; Cario-Tourmaniantz, C.; Loirand, G.; Pacaud, P.; Lesieur, D. Rho-kinase inhibitors: Pharmacomodulations on the lead compound Y-32885. *J. Enzym. Inhib. Med. Chem.* **2002**, *17*, 381–390. [CrossRef]
49. Moura-Coelho, N.; Tavares Ferreira, J.; Bruxelas, C.P.; Dutra-Medeiros, M.; Cunha, J.P.; Pinto Proenca, R. Rho kinase inhibitors-a review on the physiology and clinical use in Ophthalmology. *Graefe's Arch. Clin. Exp. Ophthalmol.* **2019**, *257*, 1101–1117. [CrossRef]
50. Stefansson, E.; Pedersen, D.B.; Jensen, P.K.; la Cour, M.; Kiilgaard, J.F.; Bang, K.; Eysteinsson, T. Optic nerve oxygenation. *Prog. Retin. Eye Res.* **2005**, *24*, 307–332. [CrossRef] [PubMed]
51. Fechtner, R.D.; Weinreb, R.N. Mechanisms of optic nerve damage in primary open angle glaucoma. *Surv. Ophthalmol.* **1994**, *39*, 23–42. [CrossRef] [PubMed]
52. Budzyn, K.; Marley, P.D.; Sobey, C.G. Targeting Rho and Rho-kinase in the treatment of cardiovascular disease. *Trends Pharmacol. Sci.* **2006**, *27*, 97–104. [CrossRef]
53. Tokushige, H.; Waki, M.; Takayama, Y.; Tanihara, H. Effects of Y-39983, a selective Rho-associated protein kinase inhibitor, on blood flow in optic nerve head in rabbits and axonal regeneration of retinal ganglion cells in rats. *Curr. Eye Res.* **2011**, *36*, 964–970. [CrossRef]

54. Uehata, M.; Ishizaki, T.; Satoh, H.; Ono, T.; Kawahara, T.; Morishita, T.; Tamakawa, H.; Yamagami, K.; Inui, J.; Maekawa, M.; et al. Calcium sensitization of smooth muscle mediated by a Rho-associated protein kinase in hypertension. *Nature* **1997**, *389*, 990–994. [CrossRef]
55. Goldhagen, B.; Proia, A.D.; Epstein, D.L.; Rao, P.V. Elevated levels of RhoA in the optic nerve head of human eyes with glaucoma. *J. Glaucoma* **2012**, *21*, 530–538. [CrossRef]
56. Mimura, F.; Yamagishi, S.; Arimura, N.; Fujitani, M.; Kubo, T.; Kaibuchi, K.; Yamashita, T. Myelin-associated glycoprotein inhibits microtubule assembly by a Rho-kinase-dependent mechanism. *J. Biol. Chem.* **2006**, *281*, 15970–15979. [CrossRef]
57. Van de Velde, S.; De Groef, L.; Stalmans, I.; Moons, L.; Van Hove, I. Towards axonal regeneration and neuroprotection in glaucoma: Rho kinase inhibitors as promising therapeutics. *Prog. Neurobiol.* **2015**, *131*, 105–119. [CrossRef] [PubMed]
58. Bertrand, J.; Di Polo, A.; McKerracher, L. Enhanced survival and regeneration of axotomized retinal neurons by repeated delivery of cell-permeable C3-like Rho antagonists. *Neurobiol. Dis.* **2007**, *25*, 65–72. [CrossRef]
59. Dergham, P.; Ellezam, B.; Essagian, C.; Avedissian, H.; Lubell, W.D.; McKerracher, L. Rho signaling pathway targeted to promote spinal cord repair. *J. Neurosci.* **2002**, *22*, 6570–6577. [CrossRef] [PubMed]
60. Hirata, A.; Inatani, M.; Inomata, Y.; Yonemura, N.; Kawaji, T.; Honjo, M.; Tanihara, H. Y-27632, a Rho-associated protein kinase inhibitor, attenuates neuronal cell death after transient retinal ischemia. *Graefes Arch. Clin. Exp. Ophthalmol.* **2008**, *246*, 51–59. [CrossRef] [PubMed]
61. Shaw, P.X.; Sang, A.; Wang, Y.; Ho, D.; Douglas, C.; Dia, L.; Goldberg, J.L. Topical administration of a Rock/Net inhibitor promotes retinal ganglion cell survival and axon regeneration after optic nerve injury. *Exp. Eye Res.* **2017**, *158*, 33–42. [CrossRef]
62. Tura, A.; Schuettauf, F.; Monnier, P.P.; Bartz-Schmidt, K.U.; Henke-Fahle, S. Efficacy of Rho-kinase inhibition in promoting cell survival and reducing reactive gliosis in the rodent retina. *Investig. Ophthalmol. Vis. Sci.* **2009**, *50*, 452–461. [CrossRef]
63. Yamamoto, K.; Maruyama, K.; Himori, N.; Omodaka, K.; Yokoyama, Y.; Shiga, Y.; Morin, R.; Nakazawa, T. The novel Rho kinase (ROCK) inhibitor K-115: A new candidate drug for neuroprotective treatment in glaucoma. *Investig. Ophthalmol. Vis. Sci.* **2014**, *55*, 7126–7136. [CrossRef]
64. Thomas, N.M.; Nagrale, P. Rho Kinase Inhibitors as a Neuroprotective Pharmacological Intervention for the Treatment of Glaucoma. *Cureus* **2022**, *14*, e28445. [CrossRef]
65. Chircop, M. Rho GTPases as regulators of mitosis and cytokinesis in mammalian cells. *Small GTPases* **2014**, *5*, e29770. [CrossRef] [PubMed]
66. Keller, K.E.; Kopczynski, C. Effects of netarsudil on actin-driven cellular functions in normal and glaucomatous trabecular meshwork cells: A live imaging study. *J. Clin. Med.* **2020**, *9*, 3524. [CrossRef]
67. Tian, B.; Kaufman, P.L. Comparisons of actin filament disruptors and Rho kinase inhibitors as potential antiglaucoma medications. *Expert. Rev. Ophthalmol.* **2012**, *7*, 177–187. [CrossRef]
68. Wang, W.; Halasz, E.; Townes-Anderson, E. Actin Dynamics, Regulated by RhoA-LIMK-Cofilin Signaling, Mediates Rod Photoreceptor Axonal Retraction After Retinal Injury. *Investig. Ophthalmol. Vis. Sci.* **2019**, *60*, 2274–2285. [CrossRef] [PubMed]
69. Sugiyama, T.; Shibata, M.; Kajiura, S.; Okuno, T.; Tonari, M.; Oku, H.; Ikeda, T. Effects of fasudil, a Rho-associated protein kinase inhibitor, on optic nerve head blood flow in rabbits. *Investig. Ophthalmol. Vis. Sci.* **2011**, *52*, 64–69. [CrossRef]
70. Ibrahim, D.G.; Ko, J.A.; Iwata, W.; Okumichi, H.; Kiuchi, Y. An in vitro study of scarring formation mediated by human Tenon fibroblasts: Effect of Y-27632, a Rho kinase inhibitor. *Cell Biochem. Funct.* **2019**, *37*, 113–124. [CrossRef]
71. Ko, J.; Ibrahim, D.G.; Kiuchi, Y. Effects of rho-associated protein kinase inhibitor Y-27632 on scarring formation after glaucoma filtration surgery. Molecular Biology of the Cell. *Conference* **2017**, *28*. [CrossRef]
72. Huang, T.Y.; DerMardirossian, C.; Bokoch, G.M. Cofilin phosphatases and regulation of actin dynamics. *Curr. Opin. Cell Biol.* **2006**, *18*, 26–31. [CrossRef] [PubMed]
73. Guan, G.; Cannon, R.D.; Coates, D.E.; Mei, L. Effect of the Rho-Kinase/ROCK Signaling Pathway on Cytoskeleton Components. *Genes* **2023**, *14*, 272. [CrossRef]
74. Johnson, M.; Shapiro, A.; Ethier, C.R.; Kamm, R.D. Modulation of outflow resistance by the pores of the inner wall endothelium. *Investig. Ophthalmol. Vis. Sci.* **1992**, *33*, 1670–1675.
75. Overby, D.R.; Zhou, E.H.; Vargas-Pinto, R.; Pedrigi, R.M.; Fuchshofer, R.; Braakman, S.T.; Gupta, R.; Perkumas, K.M.; Sherwood, J.M.; Vahabikashi, A.; et al. Altered mechanobiology of Schlemm's canal endothelial cells in glaucoma. *Proc. Natl. Acad. Sci. USA* **2014**, *111*, 13876–13881. [CrossRef] [PubMed]
76. Honjo, M.; Tanihara, H. Impact of the clinical use of ROCK inhibitor on the pathogenesis and treatment of glaucoma. *Jpn. J. Ophthalmol.* **2018**, *62*, 109–126. [CrossRef] [PubMed]
77. Rao, P.V.; Pattabiraman, P.P.; Kopczynski, C. Role of the Rho GTPase/Rho kinase signaling pathway in pathogenesis and treatment of glaucoma: Bench to bedside research. *Exp. Eye Res.* **2017**, *158*, 23–32. [CrossRef] [PubMed]

78. Wiederholt, M.; Thieme, H.; Stumpff, F. The regulation of trabecular meshwork and ciliary muscle contractility. *Prog. Retin. Eye Res.* **2000**, *19*, 271–295. [CrossRef]
79. Somlyo, A.P.; Somlyo, A.V. Ca2+ sensitivity of smooth muscle and nonmuscle myosin II: Modulated by G proteins, kinases, and myosin phosphatase. *Physiol. Rev.* **2003**, *83*, 1325–1358. [CrossRef]
80. Rao, V.P.; Epstein, D.L. Rho GTPase/Rho kinase inhibition as a novel target for the treatment of glaucoma. *BioDrugs* **2007**, *21*, 167–177. [CrossRef]
81. Challa, P.; Arnold, J.J. Rho-kinase inhibitors offer a new approach in the treatment of glaucoma. *Expert. Opin. Investig. Drugs* **2014**, *23*, 81–95. [CrossRef]
82. Hein, T.W.; Rosa, R.H., Jr.; Yuan, Z.; Roberts, E.; Kuo, L. Divergent roles of nitric oxide and rho kinase in vasomotor regulation of human retinal arterioles. *Investig. Ophthalmol. Vis. Sci.* **2010**, *51*, 1583–1590. [CrossRef]
83. Ohta, Y.; Takaseki, S.; Yoshitomi, T. Effects of ripasudil hydrochloride hydrate (K-115), a Rho-kinase inhibitor, on ocular blood flow and ciliary artery smooth muscle contraction in rabbits. *Jpn. J. Ophthalmol.* **2017**, *61*, 423–432. [CrossRef]
84. Grieshaber, M.C.; Flammer, J. Blood flow in glaucoma. *Curr. Opin. Ophthalmol.* **2005**, *16*, 79–83. [CrossRef]
85. Lee, S.S.; Robinson, M.R.; Weinreb, R.N. Episcleral Venous Pressure and the Ocular Hypotensive Effects of Topical and Intracameral Prostaglandin Analogs. *J. Glaucoma* **2019**, *28*, 846–857. [CrossRef]
86. Moher, D.; Liberati, A.; Tetzlaff, J.; Altman, D.G.; Group, P. Preferred reporting items for systematic reviews and meta-analyses: The PRISMA statement. *BMJ* **2009**, *339*, b2535. [CrossRef]
87. Tanihara, H.; Inatani, M.; Honjo, M.; Tokushige, H.; Azuma, J.; Araie, M. Intraocular pressure-lowering effects and safety of topical administration of a selective ROCK inhibitor, SNJ-1656, in healthy volunteers. *Arch. Ophthalmol.* **2008**, *126*, 309–315. [CrossRef] [PubMed]
88. Tanihara, H.; Inoue, T.; Yamamoto, T.; Kuwayama, Y.; Abe, H.; Araie, M. Phase 1 clinical trials of a selective rho kinase inhibitor, k-115. *JAMA Ophthalmol.* **2013**, *131*, 1288–1295. [CrossRef] [PubMed]
89. Bacharach, J.; Dubiner, H.B.; Levy, B.; Kopczynski, C.C.; Novack, G.D. Double-masked, randomized, dose-response study of AR-13324 versus latanoprost in patients with elevated intraocular pressure. *Ophthalmology* **2015**, *122*, 302–307. [CrossRef]
90. Inoue, T.; Tanihara, H.; Tokushige, H.; Araie, M. Efficacy and safety of SNJ-1656 in primary open-angle glaucoma or ocular hypertension. *Acta Ophthalmol.* **2015**, *93*, e393–e395. [CrossRef]
91. Kopczynski, C.; Novack, G.D.; Swearingen, D.; van Haarlem, T. Ocular hypotensive efficacy, safety and systemic absorption of AR-12286 ophthalmic solution in normal volunteers. *Br. J. Ophthalmol.* **2013**, *97*, 567–572. [CrossRef] [PubMed]
92. Williams, R.D.; Novack, G.D.; Van Haarlem, T.; Kopczynski, C. Ocular hypotensive effect of the Rho kinase inhibitor AR-12286 in patients with glaucoma and ocular hypertension. *Am. J. Ophthalmol.* **2011**, *152*, 834–841.e831. [CrossRef]
93. Tanihara, H.; Inoue, T.; Yamamoto, T.; Kuwayama, Y.; Abe, H.; Araie, M. Phase 2 randomized clinical study of a Rho kinase inhibitor, k-115, in primary open-angle glaucoma and ocular hypertension. *Am. J. Ophthalmol.* **2013**, *156*, 731–736.e732. [CrossRef]
94. Tanihara, H.; Inoue, T.; Yamamoto, T.; Kuwayama, Y.; Abe, H.; Suganami, H.; Araie, M. Intra-ocular pressure-lowering effects of a Rho kinase inhibitor, ripasudil (K-115), over 24 h in primary open-angle glaucoma and ocular hypertension: A randomized, open-label, crossover study. *Acta Ophthalmol.* **2015**, *93*, e254–e260. [CrossRef]
95. Tanihara, H.; Inoue, T.; Yamamoto, T.; Kuwayama, Y.; Abe, H.; Fukushima, A.; Suganami, H.; Araie, M. One-year clinical evaluation of 0.4% ripasudil (K-115) in patients with open-angle glaucoma and ocular hypertension. *Acta Ophthalmol.* **2016**, *94*, e26–e34. [CrossRef] [PubMed]
96. Tanihara, H.; Inoue, T.; Yamamoto, T.; Kuwayama, Y.; Abe, H.; Suganami, H.; Araie, M. Additive intraocular pressure-lowering effects of the rho kinase inhibitor ripasudil (K-115) combined with timolol or latanoprost: A report of 2 randomized clinical trials. *JAMA Ophthalmol.* **2015**, *133*, 755–761. [CrossRef]
97. Muhlisah, A.; Hirooka, K.; Nurtania, A.; Onoe, H.; Okumichi, H.; Nitta, E.; Baba, T.; Tanito, M.; Matsuoka, Y.; Nakakura, S.; et al. Effect of ripasudil after trabeculectomy with mitomycin C: A multicentre, randomised, prospective clinical study. *BMJ Open Ophthalmol.* **2024**, *9*. [CrossRef] [PubMed]
98. Tanihara, H.; Yamamoto, T.; Aihara, M.; Koizumi, N.; Minami, H.; Kojima, S.; Isobe, T.; Kanazawa, M.; Suganami, H.; Group, K.C.S. Crossover Randomized Study of Pharmacologic Effects of Ripasudil-Brimonidine Fixed-Dose Combination Versus Ripasudil or Brimonidine. *Adv. Ther.* **2023**, *40*, 3559–3573. [CrossRef] [PubMed]
99. Araie, M.; Sugiyama, K.; Aso, K.; Kanemoto, K.; Kothapalli, K.; Kopczynski, C.; Senchyna, M.; Hollander, D.A. Phase 2 Randomized Clinical Study of Netarsudil Ophthalmic Solution in Japanese Patients with Primary Open-Angle Glaucoma or Ocular Hypertension. *Adv. Ther.* **2021**, *38*, 1757–1775. [CrossRef]
100. Serle, J.B.; Katz, L.J.; McLaurin, E.; Heah, T.; Ramirez-Davis, N.; Usner, D.W.; Novack, G.D.; Kopczynski, C.C. Two Phase 3 Clinical Trials Comparing the Safety and Efficacy of Netarsudil to Timolol in Patients with Elevated Intraocular Pressure: Rho Kinase Elevated IOP Treatment Trial 1 and 2 (ROCKET-1 and ROCKET-2). *Am. J. Ophthalmol.* **2018**, *186*, 116–127. [CrossRef]

101. Kahook, M.Y.; Serle, J.B.; Mah, F.S.; Kim, T.; Raizman, M.B.; Heah, T.; Ramirez-Davis, N.; Kopczynski, C.C.; Usner, D.W.; Novack, G.D. Long-term Safety and Ocular Hypotensive Efficacy Evaluation of Netarsudil Ophthalmic Solution: Rho Kinase Elevated IOP Treatment Trial (ROCKET-2). *Am. J. Ophthalmol.* **2019**, *200*, 130–137. [CrossRef]
102. Lewis, R.A.; Levy, B.; Ramirez, N.; Kopczynski, C.C.; Usner, D.W.; Novack, G.D. Fixed-dose combination of AR-13324 and latanoprost: A double-masked, 28-day, randomised, controlled study in patients with open-angle glaucoma or ocular hypertension. *Br. J. Ophthalmol.* **2016**, *100*, 339–344. [CrossRef]
103. Pharmaceuticals, A. Study of Netarsudil (AR-13324) Ophthalmic Solution in Patients with Glaucoma or Ocular Hypertension (ROCKET-3). Available online: https://classic.clinicaltrials.gov/ct2/show/NCT02246764 (accessed on 24 December 2024).
104. Peace, J.H.; McKee, H.J.; Kopczynski, C.C. A Randomized, Phase 2 Study of 24-h Efficacy and Tolerability of Netarsudil in Ocular Hypertension and Open-Angle Glaucoma. *Ophthalmol. Ther.* **2021**, *10*, 89–100. [CrossRef]
105. Sit, A.J.; Gupta, D.; Kazemi, A.; McKee, H.; Challa, P.; Liu, K.C.; Lopez, J.; Kopczynski, C.; Heah, T. Netarsudil Improves Trabecular Outflow Facility in Patients with Primary Open Angle Glaucoma or Ocular Hypertension: A Phase 2 Study. *Am. J. Ophthalmol.* **2021**, *226*, 262–269. [CrossRef]
106. Khouri, A.S.; Serle, J.B.; Bacharach, J.; Usner, D.W.; Lewis, R.A.; Braswell, P.; Kopczynski, C.C.; Heah, T.; Benza, R.; Boyle, J.W.; et al. Once-Daily Netarsudil Versus Twice-Daily Timolol in Patients With Elevated Intraocular Pressure: The Randomized Phase 3 ROCKET-4 Study. *Am. J. Ophthalmol.* **2019**, *204*, 97–104. [CrossRef]
107. Asrani, S.; Robin, A.L.; Serle, J.B.; Lewis, R.A.; Usner, D.W.; Kopczynski, C.C.; Heah, T.; Ackerman, S.L.; Alpern, L.M.; Bashford, K.; et al. Netarsudil/Latanoprost Fixed-Dose Combination for Elevated Intraocular Pressure: Three-Month Data from a Randomized Phase 3 Trial. *Am. J. Ophthalmol.* **2019**, *207*, 248–257. [CrossRef] [PubMed]
108. Walters, T.R.; Ahmed, I.I.K.; Lewis, R.A.; Usner, D.W.; Lopez, J.; Kopczynski, C.C.; Heah, T.; Group, M.-S. Once-Daily Netarsudil/Latanoprost Fixed-Dose Combination for Elevated Intraocular Pressure in the Randomized Phase 3 MERCURY-2 Study. *Ophthalmol. Glaucoma* **2019**, *2*, 280–289. [CrossRef] [PubMed]
109. Stalmans, I.; Lim, K.S.; Oddone, F.; Fichtl, M.; Belda, J.I.; Hommer, A.; Laganovska, G.; Schweitzer, C.; Voykov, B.; Zarnowski, T.; et al. MERCURY-3: A randomized comparison of netarsudil/latanoprost and bimatoprost/timolol in open-angle glaucoma and ocular hypertension. *Graefes Arch. Clin. Exp. Ophthalmol.* **2024**, *262*, 179–190. [CrossRef] [PubMed]
110. Zaman, F.; Gieser, S.C.; Schwartz, G.F.; Swan, C.; Williams, J.M. A multicenter, open-label study of netarsudil for the reduction of elevated intraocular pressure in patients with open-angle glaucoma or ocular hypertension in a real-world setting. *Curr. Med. Res. Opin.* **2021**, *37*, 1011–1020. [CrossRef]
111. Shahid, S.; Rizvi, S.W.A.; Khan, A.A.; Ashraf, H.; Akhter, A. Comparison of safety and efficacy of Netarsudil 0.02% and Bimatoprost 0.01% monotherapy and combination therapy in primary open-angle glaucoma and ocular hypertension. *Indian. J. Ophthalmol.* **2024**, *72*, 427–431. [CrossRef]
112. Clement Freiberg, J.; von Spreckelsen, A.; Kolko, M.; Azuara-Blanco, A.; Virgili, G. Rho kinase inhibitor for primary open-angle glaucoma and ocular hypertension. *Cochrane Database Syst. Rev.* **2022**, *6*, CD013817. [CrossRef]
113. Ramakrishnan, M.S.; Addis, V.M.; Lehman, A.Y.; Sankar, P.S. Netarsudil-associated epithelial keratopathy. *Am. J. Ophthalmol. Case Rep.* **2020**, *19*, 100800. [CrossRef]
114. Ha, A.; Kim, Y.K.; Jeoung, J.W.; Satyal, S.; Kim, J.; Kim, S.; Park, K.H. Sovesudil (locally acting rho kinase inhibitor) for the treatment of normal-tension glaucoma: The randomized phase II study. *Acta Ophthalmol.* **2022**, *100*, e470–e477. [CrossRef]
115. Agency, E.M. Roclanda: EPAR—Product information. Available online: https://www.ema.europa.eu/en/documents/product-information/roclanda-epar-product-information_en.pdf (accessed on 12 February 2025).
116. Goldstein, M.H.; Silva, F.Q.; Blender, N.; Tran, T.; Vantipalli, S. Ocular benzalkonium chloride exposure: Problems and solutions. *Eye* **2022**, *36*, 361–368. [CrossRef]
117. Futakuchi, A.; Inoue, T.; Fujimoto, T.; Inoue-Mochita, M.; Kawai, M.; Tanihara, H. The effects of ripasudil (K-115), a Rho kinase inhibitor, on activation of human conjunctival fibroblasts. *Exp. Eye Res.* **2016**, *149*, 107–115. [CrossRef] [PubMed]
118. Liu, Y.; Ko, J.A.; Yanai, R.; Kimura, K.; Chikama, T.; Sagara, T.; Nishida, T. Induction by latanoprost of collagen gel contraction mediated by human tenon fibroblasts: Role of intracellular signaling molecules. *Investig. Ophthalmol. Vis. Sci.* **2008**, *49*, 1429–1436. [CrossRef]
119. Mizuno, Y.; Okada, N.; Onoe, H.; Tokumo, K.; Okumichi, H.; Hirooka, K.; Kiuchi, Y. Effect of the rho-kinase inhibitor ripasudil in needling with mitomycin C for the failure of filtering bleb after trabeculectomy: A cross-sectional study. *BMC Ophthalmol.* **2022**, *22*, 433. [CrossRef]
120. Adjei, K.; Ali, A.A. PSS4 A Cost Effectiveness Analysis of Latanoprost Monotherapy, Netarsudil Monotherapy and Fixed Dose Combination Latanoprost/Netarsudil in Management of Primary Open Angle Glaucoma in USA. *Value Health* **2021**, *24* (Suppl. S1), S220–S221. [CrossRef]

121. Schulz, K.F.; Altman, D.G.; Moher, D.; Group, C. *CONSORT* 2010 statement: Updated guidelines for reporting parallel group randomised trials. *BMJ* **2010**, *340*, c332. [CrossRef]
122. Allison, K.; Patel, D.G.; Greene, L. Racial and Ethnic Disparities in Primary Open-Angle Glaucoma Clinical Trials: A Systematic Review and Meta-analysis. *JAMA Netw. Open* **2021**, *4*, e218348. [CrossRef] [PubMed]
123. Gazzard, G.; Kolko, M.; Iester, M.; Crabb, D.P.; Educational Club of Ocular, S.; Glaucoma, M. A Scoping Review of Quality of Life Questionnaires in Glaucoma Patients. *J. Glaucoma* **2021**, *30*, 732–743. [CrossRef]
124. Sterne, J.A.C.; Savovic, J.; Page, M.J.; Elbers, R.G.; Blencowe, N.S.; Boutron, I.; Cates, C.J.; Cheng, H.Y.; Corbett, M.S.; Eldridge, S.M.; et al. RoB 2: A revised tool for assessing risk of bias in randomised trials. *BMJ* **2019**, *366*, l4898. [CrossRef]
125. Yin, Y.; Cameron, M.D.; Lin, L.; Khan, S.; Schroter, T.; Grant, W.; Pocas, J.; Chen, Y.T.; Schurer, S.; Pachori, A.; et al. Discovery of Potent and Selective Urea-Based ROCK Inhibitors and Their Effects on Intraocular Pressure in Rats. *ACS Med. Chem. Lett.* **2010**, *1*, 175–179. [CrossRef]
126. Bahr, T.; Woolf, S.; Favre, H.; Waldman, C. Comparison of netarsudil and latanoprostene bunod as adjuncts to maximum medical therapy in primary open-angle glaucoma. *Can. J. Ophthalmol.* **2022**. [CrossRef]
127. Inoue, K.; Okayama, R.; Shiokawa, M.; Ishida, K.; Tomita, G. Efficacy and safety of adding ripasudil to existing treatment regimens for reducing intraocular pressure. *Int. Ophthalmol.* **2018**, *38*, 93–98. [CrossRef] [PubMed]
128. Tsukahara, S.; Enomoto, N.; Ishida, K.; Anraku, A.; Tomita, G. One-Year Efficacy and Safety Assessment of Ripasudil, a Rho Kinase Inhibitor, in an Addition to or Replacing Existing Treatment Regimens: A Retrospective Study. *J. Ocul. Pharmacol. Ther.* **2020**, *36*, 512–521. [CrossRef] [PubMed]
129. Villegas, N.C.; Lee, W.S. Effectiveness of netarsudil as an additional therapy for glaucoma in patients already on maximally tolerated medical therapy. *Clin. Ophthalmol.* **2021**, *15*, 4367–4372. [CrossRef]
130. Andres-Guerrero, V.; Garcia-Feijoo, J.; Konstas, A.G. Targeting Schlemm's Canal in the Medical Therapy of Glaucoma: Current and Future Considerations. *Adv. Ther.* **2017**, 1–21. [CrossRef] [PubMed]

pharmaceuticals

MDPI

Review

Ischemic Optic Neuropathy: A Review of Current and Potential Future Pharmacotherapies

Omar Badla [1,2], Beshr Abdulaziz Badla [3], Amr Almobayed [1,2], Carlos Mendoza [1], Krishna Kishor [1,2] and Sanjoy K. Bhattacharya [1,2,4,*]

1 Department of Ophthalmology, Bascom Palmer Eye Institute, University of Miami Miller School Medicine, Miami, FL 33136, USA; axa4141@miami.edu (A.A.)
2 Miami Integrative Metabolomics Research Center, Miami, FL 33136, USA
3 College of Medicine, Mohammed Bin Rashid University of Medicine and Health Sciences, Dubai P.O. Box 505055, United Arab Emirates; beshr.badla@students.mbru.ac.ae
4 Graduate Program in Molecular and Cellular Pharmacology, University of Miami Miller School of Medicine, Miami, FL 33136, USA
* Correspondence: sbhattacharya@med.miami.edu

Abstract: The treatment of arteritic anterior ischemic optic neuropathy (AAION), non-arteritic ischemic optic neuropathy (NAAION), and posterior ischemic optic neuropathy (PION) is a topic of ongoing research with mixed evidence on some pharmacotherapies and a need for more consensus. This manuscript provides an overview of these conditions' current, potential future, and attempted pharmacotherapies. AAION's current treatment regimen consists of high-dose steroids, with methotrexate, tocilizumab, and abatacept, being the most viable steroid-sparing therapy candidates. As for NAAION, the treatments being tried are vast, with mixed evidence supporting each modality. Similarly, despite the various treatment options explored, there still needs to be a universally effective therapy for PION. More research is needed to formulate an agreed-upon treatment regimen for these conditions.

Keywords: ischemic optic neuropathy; ION; AION; AAION; NAAION; NAION; PION; management; pharmacotherapy

1. Introduction

Ischemic optic neuropathy (ION) encompasses a spectrum of diseases, each with its own causes, symptoms, and treatments [1]. Broadly, the condition can be divided into anterior (AION) and posterior (PION) optic neuropathies [1]. The subtypes of AION and PION differ and are detailed in Figure 1. Arteritic anterior ischemic optic neuropathy (AAION) accounts for 5–10% [2] of AION cases. AAION is essentially a manifestation of giant cell arteritis (GCA). However, other types of vasculitis, like systemic lupus erythematosus, polyarteritis nodosa, and herpes zoster, can also be a cause [2]. Non-arteritic anterior ischemic optic neuropathy (NAAION) is an important cause of vision loss in patients aged 50 and older [3]. While the exact pathogenesis remains elusive, it is hypothesized to be caused by decreased perfusion by the short posterior ciliary arteries (SPCAs) to the optic nerve head (ONH) leading to ischemia, edema, and compression of the optic nerve (ON) with subsequent retinal ganglion cell (RGC) death [3–5]. Vision loss in PION is due to ischemia of the posterior part of the ON. The optic disc looks normal during the acute phase because the site of the ischemia is farther back in the ON; with the advancement of the ischemic degeneration, descending optic atrophy will occur [6]. The following sections will delve into the current, potential future, and attempted pharmacotherapies for these conditions.

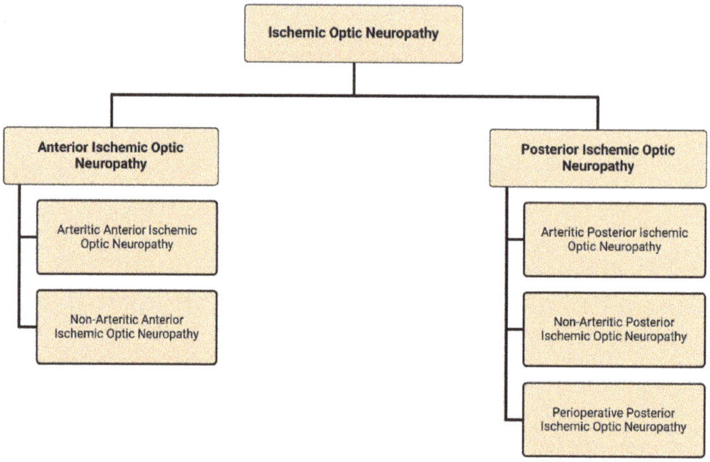

Figure 1. The subtypes of AION and PION. Created with Biorender.com.

2. Anterior Ischemic Optic Neuropathy

2.1. Arteritic Anterior Ischemic Optic Neuropathy

The pathogenesis of AAION involves inflammation of the SPCAs leading to their thrombosis and infarction of the ONH [2]. Therefore, anti-inflammatory strategies are commonly applied to treat patients with AAION using pharmacotherapies such as corticosteroids, tocilizumab, methotrexate, abatacept, and tumor necrosis factor-alpha (TNF-α) inhibitors. Table 1 acts as a summary of the treatment strategies of AAION.

2.1.1. Steroid Treatment Strategies

There are multiple treatment regimens utilizing corticosteroids, which all aim to reduce the inflammation in the vessels leading to AAION. They induce an anti-inflammatory effect by crossing into the cell nucleus and increasing the transcription of anti-inflammatory genes, thereby preventing further thrombosis in the inflamed arteries of the retina caused by GCA and reducing the resultant edema [7]. An acute attack of AAION is an ophthalmological emergency as blindness in the affected eye can rapidly develop and is an even bigger emergency in the case of bilateral AAION as this can lead to total blindness. This is why despite different corticosteroid regimens, it is essential in the treatment of AAION to administer a very large dose quickly upon presentation in patients with visual manifestations. Hayreh et al. described the following regimen: intravenous (IV) methylprednisolone (MP) (1 g/kg/day) administered for 3 days followed by at least 4–6 weeks of oral prednisolone at 1 mg/kg/day [8]. The dose is then only tapered guided by levels of erythrocyte sedimentation rate (ESR) and C-reactive protein (CRP) [2,8]. Tapering needs to be a slow delicate process, as flare-ups can occur. Recommendations for tapering are as follows: reducing the dose by 10 mg per month till 20 mg/day is reached, to then by 5 mg every month, and then by 1 mg per month until a dose of 10 mg/day is reached [9]. To highlight the approach of rheumatologists to GCA, the recommended regimen by the British Society for Rheumatology is as follows: in the absence of ischemic visual manifestations, an initial dose of 40–60 mg oral prednisolone daily, in the presence of ischemic visual manifestations, 500 mg^{-1}g IV MP daily for up to 3 days before commencing oral prednisolone therapy [10]. The IV dosage is lower than that described before; however, the guidelines demonstrate that even rheumatological guidelines consider high-dose glucocorticoids necessary for the treatment of GCA with visual manifestations. The tapering process can take at least 1–2 years [9] or, as reported by Hayreh et al., a median of 48.7 months (95% CI: 34.6–71.4 months) [2]. Once the lowest dose for each individual patient is reached, it is considered from an ophthalmologist's view that treatment to prevent lifelong risk of blindness has to also be lifelong [2,8,9]; while according to the British Society for Rheumatology, glucocorticoid dose should be tapered to zero over 12–18 months [10]. This highlights the difference between a rheumatologist's approach and an ophthalmologist's approach to GCA treatment, a gap that should be bridged as GCA patients can be primarily under a rheumatologist's care, and preventing the significant morbidity for patients should be paramount. IV steroid induction in place of oral steroids leads to no reduction of total steroid dose by 1 year of treatment [11]; however, it leads to an increased likelihood of improvement in patients receiving IV steroid induction (40% vs. 13%) [12].

2.1.2. Steroid Sparing Treatment Strategies

The previously mentioned long-term treatment of AAION as a manifestation of GCA involves a corticosteroid regimen over months and years. Well-known side effects of steroids include osteoporosis, steroid-induced myopathy, insulin resistance, hypertension, gastritis, and even psychiatric side effects such as anxiety and depression [13]. Strategies to decrease the exposure to corticosteroids, and therefore, their side effects, are described below.

- Tocilizumab

Tocilizumab is an anti-interleukin (IL)-6 receptor monoclonal antibody, which is an FDA-approved treatment for GCA. It is important, however, to recognize that treating GCA and preventing blindness due to GCA from AAION are two different entities. Randomized control trials looking at the effectiveness of both subcutaneous doses every week or every other week of tocilizumab as a steroid-sparing therapy for GCA found higher rates of sustained remission when combined with a 26-week course of prednisone (cumulative median dose = 1862 mg) at 52 weeks (56% and 53%, respectively) duration compared to a larger course of prednisone (cumulative median dose = 3818 mg) taken over 52 weeks at the same 52-week mark (18%) [14]. Other studies [15–20] have demonstrated similar steroid-sparing properties and effectiveness in the treatment of GCA; however, this effectiveness is studied over periods that do not demonstrate the life-long safety necessity of treating GCA to prevent blindness from AAION. This is further complicated by the side effect profile of long-term tocilizumab therapy such as cytopenias, upper respiratory tract infections, bronchitis, and liver toxicity [21]. This, at the same time, does not preclude the use of tocilizumab in the context of AAION. It could be used as a steroid-sparing therapy up until proven safety periods, as Unizony et al. demonstrated a decreased incidence of visual manifestations for GCA patients on tocilizumab [22] and as reported by Vionnet et al., as a salvage therapy after treatment failure in patients with corticosteroid-resistant progressive AAION [23].

- Methotrexate

Methotrexate has also been considered as a steroid-sparing strategy candidate. Its proposed mechanism in treating AAION is decreasing the inflammation through multiple mechanisms including repressing T-cell activation, down-regulating B cells, and its antifolate antimetabolite action that all act to suppress the immune system, and therefore, the inflammation in AAION [24]. The adverse effects of methotrexate include gastrointestinal manifestations, hepatotoxicity, bone marrow suppression, and teratogenicity [24]. For GCA, Mahr et al. found that methotrexate did decrease the exposure to corticosteroids and also reduced the risk of relapses when used as an adjunctive treatment [25,26]. Again, the treatment of GCA and preventing visual loss from it should be considered distinct; however, one can infer the effectiveness of decreasing relapses to directly decrease the disease activity leading to visual loss.

- Abatacept

Abatacept, a cytotoxic T-lymphocyte-associated protein 4 (CTLA-4) immunoglobulin, is another candidate for steroid-sparing strategies and has been in use for the treatment of other diseases with a known side effect profile. Abatacept has not been studied specifically in the prevention of long-term visual losses in AAION due to GCA; however, Langford et al. conducted a randomized, double-blind trial of abatacept for the treatment of GCA. In that trial, abatacept and glucocorticoids were started until the 12th week of treatment, when patients free of remission were randomized to keep receiving abatacept monthly until 12 months with a tapered glucocorticoid dose until the 28th week or a tapered glucocorticoid dose only until the 28th week. Relapse-free (defined in this study as the need to increase prednisone dose or restart prednisone after discontinuation) rates at 12 months in the abatacept-treated group were significantly higher at 48% compared to the prednisone-only group at 31% as well as a significantly longer median duration of remission (9.9 months vs. 3.9 months) [27]. They found no significant difference in adverse effects between the two groups, making abatacept a viable candidate for future investigation looking at long-term ocular manifestations of GCA and if an abatacept–corticosteroid regimen with lower steroid load is safe.

- TNF-α inhibitors

Another one of the biological treatments is TNF-α inhibitors, which are used in the treatment of multiple inflammatory and autoimmune disorders such as inflammatory bowel disease, psoriasis, rheumatoid arthritis, ankylosing spondylitis, and more [28]. In the case of GCA, however, they have not shown significant benefits or efficacy in its treatment [10,29].

- Prostaglandin E1

Prostaglandin E1 (PGE1) is a vasodilator whose effects can be theoretically used to help restore circulation and blood flow through the PCAs and prevent ischemia of the ONH. Adding PGE1 to the initial high-dose corticosteroid regimen has been attempted in a case report by Steigerwalt et al. who found improved and stable visual acuity (VA) in these two patients who remained stable on follow-up examinations [30]. This, however, is a small sample size, and larger studies are necessary to evaluate its true effects in AAION patients.

Table 1. Summary table of the discussed treatment strategies of AAION.

Pharmacotherapy	Design of Studies Analyzed	Year of Publication	Conclusions
Corticosteroids	Review [2].	2009	Treatment with a rapidly administered large dose of corticosteroids followed by a long tapering steroid regimen remains the mainstay of AAION treatment. There remains a gap between the length of corticosteroid regimens recommended by ophthalmologists and rheumatologists.
	Review [8].	2011	
	Review [9].	2021	
	Guideline Study/Review [10].	2020	
	Clinical Trial [11].	2000	
	Retrospective Study [12].	2001	
Tocilizumab	Clinical Trial [14].	2017	An FDA-approved for the treatment of GCA, tocilizumab has demonstrated effectiveness as a steroid-sparing therapy. Further studies must be conducted looking at tocilizumab in the context of AAION due to GCA past the currently studied timelines.
	Randomized, Double-Blind, Placebo-Controlled Trial [15].	2016	
	Retrospective Study [16].	2016	
	Case Series [17].	2011	
	Retrospective Study [18].	2015	
	Retrospective Study [19].	2012	
	Case Report [20].	2014	
	Retrospective Study [22].	2021	
	Review [23].	2017	

Table 1. *Cont.*

Pharmacotherapy	Design of Studies Analyzed	Year of Publication	Conclusions
Methotrexate	Meta-Analysis [25].	2007	Methotrexate was reported to effectively reduce the risk of relapse and corticosteroid exposure in GCA patients. More should be studied about its role in the context of AAION and vision protection.
Abatacept	Randomized, Double-Blind Trial [27].	2018	Demonstrated lower remission rates and corticosteroid exposure in the treatment of GCA with a good adverse effect profile. Viable candidate for a future investigation looking at long-term ocular manifestations of GCA.
TNF-α inhibitors	Guideline Study/Review [10].	2020	Has not demonstrated significant benefits or efficacy in the treatment of GCA and, therefore, its ocular manifestations.
	Review [29].	2022	
Prostaglandin E1	Case Report [30].	2010	In the referenced case report, PGE1 was found to improve and stabilize visual acuity (VA) in two patients on follow-up. Larger studies are necessary to determine its efficacy.

2.2. Non-Arteritic Anterior Ischemic Optic Neuropathy

Currently, there is no widely accepted treatment for this NAAION, and management is limited to risk factor mitigation [3,4]. In the following sections, we will provide an overview of this condition's current and potential future pharmacotherapies, ranging from pharmacotherapies tested in vitro to pharmacotherapies tested in humans (Table 2).

2.2.1. Corticosteroids and Associated Pharmacotherapies

- Corticosteroids:

Corticosteroids are hypothesized to treat NAAION by reducing disc swelling and preventing free-radical-induced damage [31]. Experiments conducted on rat models of non-arteritic anterior ischemic optic neuropathy (rAION) have tested MP's anti-inflammatory and neuroprotective efficacy on ONs and RGCs [5,32]. In two experiments, intraperitoneal MP was administered to experimental rats, and control rats received intraperitoneal phosphate-buffered saline (PBS) [5,32]. Both studies reported the neuroprotective effects of MP treatment. Specifically, the early administration of MP was associated with anti-inflammatory effects [5] and increased RGC survival and ON function [32]. However, in studies on human patients, a single-masked, randomized clinical trial assessing the effect of 500 mg of IV MP, given twice a day for three days and followed by 2 weeks of oral prednisolone, found that this regimen does not improve the final visual and structural outcome in NAAION patients compared to controls [33]. Likewise, a prospective interventional comparative case series tested the same dose of IV MP followed by oral dosing. This study concluded that this regimen did not produce a beneficial effect compared to controls [34]. A randomized, double-blind, controlled trial conducted on nondiabetic NAAION patients evaluated the role of oral corticosteroids in treating this condition [31]. Nineteen patients were given oral prednisolone (80 mg) for 2 weeks, tapered to 70 mg, 60 mg, then by 5 mg at five-day intervals until completion. Another 19 patients were on placebo. The primary outcome measures included best-corrected visual acuity (BCVA), changes in the retinal nerve fiber layer (RNFL) on optical coherence tomography (OCT), and visual evoked response (VER) [31]. In the steroid group, there were statistically significant improvements in VER parameters and BCVA coupled with a quicker resolution of disc edema on OCT on a 1-month follow-up. Despite that, these improvements were deemed clinically insignificant, as there was no significant difference in VA improvement in the steroid group compared to controls at 6 months [31]. Similarly, another randomized controlled trial evaluated the use of 75 mg of oral prednisolone on NAAION patients within five days of diagnosis and recommended against its use [35]. A retrospective analysis compared the visual outcomes of four patient groups: a control group, a group on oral prednisone, a group on 250 units of MP followed by oral prednisone, and a group on 500 units of MP followed by oral prednisone [3]. This study noted a statistically significant improvement in BCVA in all steroid groups compared to baseline. However, it reported no significant differences in the final visual outcomes when comparing the four groups [3]. One notable finding highlighted in this study is the short-term improvement in BCVA in patients receiving 500 units of MP during the 7- to 14-day follow-up period, compared to baseline [3]. Lastly, a systematic review and meta-analysis conducted on studies dating until the 10th of June 2019 assessed different therapies for NAAION and concluded that steroids do not significantly improve VA and visual field (VF) [4]. While experimental studies on the neuroprotective effects of MP on rAION models proved effective, studies on human patients have failed to showcase a clinically significant improvement in final VA compared to controls.

- Triamcinolone:

Intravitreal triamcinolone (IVTA) has been tried for its anti-inflammatory effects and to avoid the systemic side effects associated with IV or oral administration of corticosteroids [36]. Experiments conducted on rAION models tested the efficacy of this treatment modality. One experiment utilized three groups of female Wistar rats, with group one receiving no treatment, group two receiving 56 μg of IVTA, and group three receiving intravitreal saline [37]. The treatment was administered within 10 min of rAION induction, and the rats were euthanized 30 days later, after which the RGC nuclei were counted [37]. This study found no significant difference in RGC count across the three groups and concluded that this treatment was not beneficial [37]. Another study utilized the rAION model but had three experimental groups that received 0.32 mg/2 μL of IVTA after 1, 7, and 14 days of rAION induction, respectively [38]. The control group received PBS. The results revealed that using ITVA within a week post-induction leads to increased RGC survival, visual evoked potential (VEP) electrophysiological amplitude, and decreased immune cell infiltration of the ON [38]. The difference in conclusions between the two studies could be due to the different dosing scales and outcome measurement techniques. As for human studies, a systematic review and meta-analysis evaluated the use of IVTA, and while the treatment led to an improvement in VA and VF, neither of the two studies were trials and had a relatively small number of cases [4].

2.2.2. Neuroprotective Agents

- EPO:

Erythropoietin (EPO) is expressed by various tissues, including neurons and glial cells, and has been tested for NAAION for its neuroprotective in vitro effects in neuronal injury animal models [34]. In a prospective interventional comparative case series, IV recombinant human EPO was administered with systemic IV corticosteroids to patients with NAAION within 14 days of diagnosis [34]. This combination had no beneficial effect compared to the control group or the group that received steroids alone [34]. The subsequent study performed by the same group of researchers evaluated IV EPO independently and did so for cases within five days of diagnosis to study the drug's effect at a narrower therapeutic window [35]. In this randomized control trial, the EPO group received 100,00 units twice daily for three days. At the 6-month follow-up, the EPO group had a statistically significant difference in improvement in BCVA when compared to the steroid group and controls [35]. Moreover, the patients in the EPO group experienced a lesser decrease in the thickness of the peripapillary RNFL compared to the two other groups [35]. These findings suggest that IV EPO exerts a functional and structural neuroprotective effect on the ONs of NAAION patients if used within five days of diagnosis [35].

- G-CSF:

Granulocyte colony-stimulating factor (G-CSF) is a hematopoietic growth factor used to treat neutropenia [39]. Recently, studies have delved into its neuroprotective in Alzheimer's and Parkinson's diseases [39]. The G-CSF receptor is present in RGCs and other neural cells [39], which has led to its application in rAION models to test its efficacy in treating NAAION. Experiments involving this rat model have demonstrated G-CSF's anti-inflammatory and antiapoptotic effects by preserving RGC density, decreasing ED-1-positive cell infiltration [40], and reducing pro-inflammatory cytokine expression [41]. Moreover, combining G-CSF with meloxicam, an anti-inflammatory cyclooxygenase two inhibitor, resulted in similar anti-inflammatory and antiapoptotic effects in the rAION as mentioned above, with the additional effect of decreasing GCS-F-associated leukocytosis [42]. More recently, it was discovered that G-CSF exerts one of its neuroprotective antiapoptotic effects by modulating the TAF9-P53-TRIAP1-CASP3 pathway [39]. In a prospective interventional case series pilot study, an intravitreal G-CSF injection was administered at 60 μg per 0.1 mL within 2 weeks of NAAION onset [43]. The results of this study revealed an improvement in BCVA at the 1-month follow-up, but the final BCVA showed no significant improvement compared to the baseline [43]. The study concluded that the injection is safe, but the effect lasts a month and then declines [43]. The next step would be to perform a controlled trial with more cases and potentially test the use of multiple injections to maintain the effect.

- Citicoline:

Citicoline has been considered for NAAION due to its neuroprotective effects [44]. The mechanism by which the drug exerts its neuroprotective effect is by targeting the undamaged axons [44]. A randomized pilot study enrolled 36 NAAION patients and 20 age-matched controls to test the effect of a 500 mg/day oral solution of citicoline on VA, pattern electroretinogram (PERG), VEP, Humphrey 24-2 visual field mean deviation (HFA MD), and RNFL thickness [45]. The results of this study revealed a significant difference in all the criteria mentioned above between the groups at 6 months; therefore, the conclusion is that citicoline exerts a neuro-enhancing and neuroprotective effect in NAAION patients [45]. These results are promising but need to be supported by a study with a larger cohort and long-term follow-ups to assess the longevity of the drug's effect.

- Trabodenoson:

Trabodenoson has been theorized to treat NAAION by activating the A1 receptor (A1R), an adenosine receptor that exerts a neuroprotective effect [46]. One study tested the efficacy of this drug in a topical daily 3% formulation on a rodent NAION model (rNAION). The drug was started 3 days before and 21 days after ischemia induction [46]. This formulation successfully reduced ON edema and preserved RGC count and ON axons in rNAION vs. control rodents [46]. These results are promising and showcase the neuroprotective effects of topical trabodenoson. However, this formulation must be evaluated in human NAAION patients before it can be considered for this condition.

- Vincamine:

Vincamine is an extract from the Apocynaceae Vinca plant, and it has been tested on a rAION model for its potential neuroprotective qualities [47]. This study found that intragastric Vincamine has an antiapoptotic effect on RGCs of rAION rodents. The proposed mechanism involved in its neuroprotective role is the PI3K/Akt/eNOS pathway. This study was limited by the small number of rodents per group, the short duration of the follow-up, and the lack of pAKT/eNOS testing in the 4 weeks leading up to the first measurement [47]. In a study conducted on 27 human NAAION patients and 15 age-matched controls, vincamine was given to the patients in the treatment group, and both groups received steroid pulse therapy and neurotrophic treatment [48]. The treatment group had a statistically significant larger improvement in mean deviation of the visual field and RNFL and ganglion cell complex thickness [48]. These results are promising but need to be verified with a trial containing more patients.

- CNTF:

Ciliary-derived neurotrophic factor (CNTF) is comparable to IL-6 in its antiapoptotic effect on RGCs in response to injury [49]. Hence, an experimental study tested its potential neuroprotective effect in an rAION model [50]. This study evaluated the effect of an intravitreal injection of 0.75 μg one day after rAION induction, with control mice receiving a sham injection. Fifteen days after ischemic induction, the rodents were euthanized, and their RGCs were counted by identifying cyan fluorescent protein (CFP) cells using stereology in a flat-mounted retina [50]. Results revealed a higher number of CFP-positive cells in CNTF-treated rodents than in sham-treated rodents and a lower decrease in CFP-positive cell density in experimental vs. control RGC layers [50]. These results are promising and showcase the neuroprotective effect of this drug. However, for this effect to apply to NAAION patients, further studies need to be performed to showcase the sustainability of this effect over a more extended follow-up period using tests that can observe visual function over that period, such as PERG. Moreover, human studies need to be performed too down the line.

- BDNF/LM22A-4:

Brain-derived neurotrophic factor (BDNF) exerts a neurotrophic effect that enhances RGC survival by activating tropomyosin-related kinase B (TrkB) receptors [51]. However, BDNF poses some possible limitations, including infection, inflammation, and poor blood–brain barrier penetration [51]. Therefore, safer activation of the TrkB receptors can be achieved by using (N,N′,N′-tris [2-hydroxyethyl])-1,3,5-benzene tricarboxamide (LM22A-4), a partial agonist specific to the TrkB receptors [51]. The effects of LM22A-4 were tested in this study in vitro by assessing its effects on the survival of immunopanned RGC cultures and in vitro by assessing the effect of its intravitreal injection and 3-week systemic administration on RGC density and serial OCTs of a murine AION model [51]. The results revealed a significant difference in all measured parameters, which indicates a neuroprotective and beneficial effect for NAAION [51]. Another tested application for BDNF in the treatment of NAAION is its incorporation with bone marrow-derived stem cells and their combined effect in the affected eyes of the rAION model [52]. BDNF improved the quantity and quality of engrafted stem cells and the neuroglial differentiation of the stem cells within the ischemic retina [52]. Overall, BDNF and LM22A-4 have promising beneficial direct and indirect effects on animal models of NAAION. However, more extensive animal studies need to be conducted with extended follow-up periods to monitor this treatment's long-term effects before advancing to human studies.

- Memantine:

Memantine is an NMDA receptor antagonist that blocks glutamate-mediated neurotoxicity in RGCs [4]. A systematic review and meta-analysis found that memantine improved VA when analyzed as a continuous variable but not as a categorical variable. Additionally, the drug was found not to improve VF [4].

- Minocycline:

Minocycline, a tetracycline derivative, has been shown to be neuroprotective in hypoxic and ischemic CNS models [53]. This effect can be explained by its inflammation-modulatory effect [53]. Additionally, it has been reported that it has neuroprotective effects in animal glaucomatous and traumatic optic neuropathies [53]. These findings prompted the trial of daily intraperitoneal minocycline (33 mg/kg) in an rNAION model [53]. While this study's results highlighted the immunomodulatory effect of minocycline, the RGC survival did not improve with the administration of minocycline compared to controls [53]. Therefore, compared to previously mentioned neuroprotective drugs, minocycline does not show promise in its ability to preserve RGCs.

- Butylidenephthalide:

Butylidenephthalide (BP) is a component of Angelica, a Chinese medicine with many applications [54]. Research on BP has found it to be effective for its antitumor, anti-inflammatory, and neuroprotective qualities [54]. These qualities made BP a potentially valuable agent for an ischemic injury model; therefore, its effects on RGC survival, visual function, and inflammatory response inhibition were tested in the rAION model [54]. For this purpose, three rAION groups were used: sham, PBS-treated, and BP-treated. The BP group received 10 mg/kg of the drug intraperitoneally for seven consecutive days, after which different parameters were assessed [54]. BP was found to increase RGC survival rates, prevent visual loss on flash visual evoked potential (FVEP) stimulation, improve optic disc edema, preserve RNFL thickness on OCT, and decrease macrophage infiltration into the ON compared to controls [54]. One of the mechanisms by which BP is thought to diminish the inflammatory response is through the NF-κB pathway [54]. While the early experimental results are promising, it is essential to follow the experimental evidence with human studies to validate the effect of this drug in NAAION patients.

- Bardoxolone Methyl and Omaveloxolone:

Similarly, bardoxolone methyl (RTA 402) and omaveloxolone (RTA 408), two synthetic oleanane triterpenoids, also inhibit the NF-κB pathway, which accounts for their anti-inflammatory and antioxidative properties [55]. These properties prompted the trial of RTA 402 and RTA 408 for NAAION in an rAION model [55]. The study included six groups: an rAION + RTA 402 (20 mg/kg) group, an rAION + RTA 402 (40 mg/kg) group, an rAION + RTA 408 (10 mg/kg) group, an rAION + RTA 408 (20 mg/kg), an rAION + PBS group, and a sham group [55]. Each group includes 18 rodents, six of which will be tested using FG retrograde labeling, six will undergo OCT, VEP, terminal deoxynucleotidyl transferase- (TdT-) dUTP nick end labeling (TUNEL), and IHC, and six will undergo immunoblotting analysis [55]. The results for RTA 402 revealed anti-apoptotic, anti-inflammatory, antioxidative stress, and myelin-preserving effects on RGCs coupled with a visual function-preserving effect [55]. These results were not found with RTA 408 [55]. The evidence points towards RTA 402's ability to modulate the Nrf2 and nuclear transcription factor nuclear factor kappa beta (NFκB) pathways as the reason for its RGC protective effect [55]. Overall, the results show that RTA 402 is a potential treatment for NAAION, but more comprehensive studies involving human patients are needed.

- Prostaglandin J_2 and MAGL/COX inhibitors:

Prostaglandin J2 (PGJ2) has been shown in rNAION models to preserve RGCs, RGC function, and ON function compared to controls injected with PBS [56]. PGJ2 antagonizes NFκB and activates nuclear factor peroxisomal proliferator-activated receptor-gamma (PPARγ), thus decreasing inflammation and neuronal apoptosis, respectively [56]. Miller et al. assessed the toxicity and efficacy of IV and intravitreal (IVT) PGJ2 in the treatment of a nonhuman primate model of non-arteritic anterior ischemic optic neuropathy (pNAAION). They found a significant reduction in histological damage as measured by axon counts ($p = 0.05$) and preserved electrophysiological function as measured by VEP and PERG ($p = 0.03$) in pNAAION treated with a single dose of IVT PGJ2 [56]. As for toxicity, no evidence of persistent ocular toxicity was found at the dose of IVT PGJ2 chosen for efficacy experiments (50 μg); however, at higher doses, transient ON function loss was observed, which resolved within a week. This toxicity must be considered for any future human studies evaluating the safety and efficacy of IVT PGJ2 in human NAAION. Mehrabian et al. investigated the addition of a monoacylglycerol lipase (MAGL) inhibitor KML29 to decrease the production of arachidonic acid (AA) and, therefore, downstream inflammation. In addition to KML29, they also investigated the addition of downstream inhibition of inflammation using cyclooxygenase 1/2(COX1/2) inhibitor meloxicam. They however found that neither KML29 nor meloxicam improved ONH edema or RGC survival in combination with PGJ2 better than reported PGJ2 alone [57]. They also reported that meloxicam and KML29 alone improve RGC survival compared to the vehicle alone [57].

- QPI-1007

QPI-1007 is a caspase-2 inhibitor and was considered a potential treatment for NAAION as it would inhibit apoptosis, increase RGC survivability, and therefore act as a neuroprotector. A phase 2/3 RCT of the efficacy of intravitreal QPI-1007, measuring the primary outcome of VA loss as compared to the sham intravitreal injection group in the treatment of human NAAION, was based on results of a phase 1 trial where a small sample of patients had no serious adverse effects and better preservation of VA compared to ION historical controls [58]. The phase I trial compared its small cohorts to bigger cohorts in historical controls, making the comparison weaker. It also had multiple investigators at 28 sites making it more difficult to ensure the validity of the results. Nevertheless, the phase 2/3 RCT was conducted and then terminated as the results were not positive. From the data released in the EU clinical trials register, no significant difference in the primary outcome between any of the four treatment regimens with QPI-1007 as compared to the control group was found. This result highlights the difficulty of translating in vitro or rat model results into real-world treatments but also the importance of trials taking place to test potential treatments [59].

- RPh201

RPh201 is an extract from gum mastic that has been studied in both phase 1 and phase 2 studies in the treatment of NAAION. Rath et al. claimed that RPh201 induces neuronal differentiation, synaptogenesis, immunomodulation, and neuroprotective effects in vitro and in vivo in unpublished data [60]. Based on this, a phase 1 trial was conducted, which established the safety of subcutaneous RPh201 in healthy volunteers [61]. A phase 2a single-site, prospective, randomized, placebo-controlled, double-masked trial in 20 patients was then conducted to assess the safety and changes in visual function and structure in patients diagnosed with NAAION for at least 6 months and a maximum of 3 years. They found RPh201 to be safe and a statistically insignificant improvement in BCVA from baseline as compared to the placebo group (14.8 $\pm$ 15.8 letters for RPh201 and 6.6 $\pm$ 15.3 for placebo, $p = 0.27$) [60]. The sample size is small, no statistically significant improvement was found, and the rationale for treatment after 6 months of diagnosis is questionable due to the irreversible nerve damage that would have occurred by then [62]. Nevertheless, a phase 3 double-masked clinical study evaluating the efficacy and safety of RPh201 in patients with previous NAAION was started in 2018 and completed in 2020, with results unpublished on ClinicalTrials.gov (NCT03547206). With unpublished results, no conclusion can definitely be drawn about the efficacy of RPh201, and it remains to be seen if the results will be published or not.

- Vitamin B3

Vitamin B3, or niacin, acts as a precursor to the coenzymes nicotinamide adenine dinucleotide (NAD+) and nicotinamide adenine dinucleotide phosphate (NADP+), which are needed in the synthesis of ATP. Vitamin B3 has been reported to prevent RGC and axonal loss in a glaucoma rat model [63]. Chen et al. looked at the neuroprotective properties of vitamin B3 in a rat model of NAAION and found that compared to the saline-treated rats, B3-supplemented rats had significantly higher P1-N2 flash visual evoked potential (FVEP) (62.9 $\pm$ 10.37 μV vs. 32.84 $\pm$ 7.62 μV), significantly higher RGC density (1086.7 $\pm$ 133.4/mm^2 vs. 647.2 $\pm$ 98.5/mm^2), and a significant reduction in oxidative damage [63]. Niacin is a commonly available vitamin supplement with proven safety, so trials in humans supplementing existing treatment regimens are an option to further understand its role in neuroprotection in NAAION patients.

- M01, a HECT domain-E3 ubiquitin ligase inhibitor:

Based on the studies showing the role of E3 ligase activity suppression in prolonging the survival of axons following ON injury in rNAION models [64], Chien et al. investigated the role of M01, a homologous E6-associated protein carboxyl terminus (HECT) domain-E3 ubiquitin ligase inhibitor in neuroprotection of RGCs in the ON and retina of rNAION models. They elucidated that M01 upregulated nuclear factor erythroid 2-related factor 2 (NRF2), a transcription factor involved in the regulation of antioxidant proteins, which subsequently decreased levels of Thioredoxin interacting protein (TXNIP), an oxidative stress mediator, and NLR family pyrin domain containing 3 (NLRP3), an inflammasome. These actions lead to a decrease in TUNEL-positive cells, which indicate apoptotic RGCs, post NAAION induction in the M01 treated group (7.1 ± 3.9/HPF) compared to the saline group (15.3 ± 4.5/HPF) [64]. They also found decreased edema, inflammatory factors, demyelination, and an increase in M2-subtype microglial polarization. These findings present the modulation of HECT domain-E3 ubiquitin ligase pathways as a new approach toward the treatment of NAAION that needs to be further investigated to fully identify its efficacy and safety.

- Brimonidine

Brimonidine tartrate is an alpha-adrenergic agonist, which has been studied in humans as a treatment for NAAION. Brimonidine had promising reports of neuroprotective ability in animal studies [65] and is safe for human use, so it was a good candidate for human studies. Despite its promising results in animal studies, topical brimonidine has been demonstrated to not improve VA or VF significantly in human trials [4,49,66–68]. This result highlights the need for human trials to fully evaluate a candidate drug that was found to be beneficial in animal trials.

- Progesterone:

The hormone progesterone has been demonstrated in modes of central nervous system injury and cerebral ischemia models to provide neuroprotection and reduce infarct volume leading to functional recovery, respectively [69]. On this basis, Allen et al. aimed to investigate the role of progesterone treatment in rat models of NAAION and of middle cerebral artery occlusion (MCAO). These represent two different models of ocular ischemia, and the results were different for either model. In the rAION model, progesterone showed no effect on VEP reduction or RGC loss; however, in the MCAO model, progesterone reduced electroretinogram deficits and RGC loss [69]. While the results are not promising in the context of NAAION, progesterone remains a target for future research surrounding neuroprotection in other contexts.

- PLGA-Icariin

Sourced from the Epimedium spp., icariin is a flavonoid glucoside which is a natural product that has been found to have anti-inflammatory properties [70]. Desai and Wen et al. studied the long-term therapeutic effects of icariin in a rat NAAION model. Icariin's short half-life was prolonged with the use of poly(lactide-co-glycolide) (PLGA) microspheres to have a longer-lasting release of intravitreally administrated icariin. The anti-inflammatory effect was determined by them to be through induction of endogenous G-CSF expression, which activated the non-canonical NF-kB signaling pathway by promoting phosphorylation of inhibitor of nuclear factor kappa-B kinase subunit beta (IKK-β) [70]. This anti-inflammatory effect of icariin was found to significantly reduce ON edema, macrophage infiltration, and increased RGC density as well as P1-N2 amplitude evidenced by FVEP measurements at day 28 after rAION induction [70]. The elucidated pathway and the neuroprotective results of this study provide possible future targets and a basis for intravitreal icariin to be further studied in the setting of human studies.

- Puerarin:

Puerarin is a bioactive ingredient extracted from the root of Pueraria lobata and has been found to have vasodilatory, neuroprotective, anti-ischemic, and anti-inflammatory activity through multiple mechanisms in animal models [71]. MANN Le et al. studied these effects of puerarin in a rat model of NAAION and measured the FVEP, RGCs, and inflammatory response. They report a significant 2.3-fold higher P1-N2 amplitude in FVEP analysis, a 1.6-fold higher RGC density, and 2.8-fold lower apoptotic RGCs in the puerarin-treated group than the saline group [72]. These promising results offer a base for targeting the specific mechanism that acts as neuroprotection or the usage of puerarin itself in future studies for the treatment of NAAION.

- miR-124:

The microRNA miR-124 was found in experimental lab models by Chen et al. to have neuroprotective effects on RGCs due to increased survivability and an improved visual function measured by FVEP by altering gene expression towards upregulating anti-inflammatory reactions through their capacity to activate anti-inflammatory M2 macrophages [73]. This novel approach requires human testing to be evaluated as a treatment in patients suffering from NAAION.

2.2.3. Stem Cell Pharmacotherapies

- Mesenchymal Stem Cells:

Mesenchymal stem cells (MSCs) have applications in various research and therapeutic settings due to their immunomodulatory, tissue-regenerative, and ischemic tissue repair effects [74,75]. An experimental study assessed the safety of MSCs by injecting human Wharton's jelly intravitreally into a standard set of Wistar rats [74]. The injections led to retinal venous congestion, damage, inflammation, and impaired visual function [74]. That being said, a recent prospective, non-randomized phase II study conducted on five NAAION patients ascertained that the treatment was safe and well tolerated except for the development of an epiretinal membrane in one patient [76]. Four out of five patients experienced visual improvement. Additionally, the P100 VEP amplitudes of three patients improved [76]. These findings need to be supported by larger studies.

- Mesenchymal Stem Cell Exosome:

Due to the potential limitations associated with MSC use, such as the ones discussed above, other MSC-derived treatments are being considered. One of these treatments is MSC-exosome, which has emerged as a promising therapy in managing various ischemic and neurodegenerative diseases and is more stable and less immunogenic than MSCs [75]. MSC-exosome has been theorized to treat NAAION by promoting neural plasticity and angiogenesis and reducing inflammation, immunity, oxidative stress, and apoptosis [75]. This modality has yet to be tested on any NAAION animal models, which is a primary step that needs to be taken to assess its safety and efficacy.

- Mesenchymal Stem Cell-Derived Medium:

Another MSC-derived treatment that has been considered for treating NAAION is MSC-derived conditioned medium (MDCM) [74]. This treatment incorporates soluble molecules derived from MSCs that retain MSC's advantageous effects while limiting their unwanted side effects [74]. The efficacy of MDCM was assessed in the rAION model and was found to preserve visual function and RGC density and reduce inflammation in the ON [74]. These findings are promising but need to be supported by human studies.

2.2.4. Anti-VEGF Pharmacotherapies

- Anti-VEGF:

This class of drugs can theoretically treat NAAION by decreasing the vasogenic edema associated with the condition by decreasing microvascular permeability [49,77]. This section will discuss the current evidence on three anti-vascular endothelial growth factor (VEGF) drugs.

- Bevacizumab:

A prospective trial tested intravitreal bevacizumab for NAAION and found no advantage for the drug compared to controls [78]. Additionally, a study reported that NAAION manifested in several patients who were administered intravitreal injections of an anti-VEGF treatment for age-related macular degeneration [79]. Only two case reports have noted the beneficial effect of intravitreal bevacizumab on NAAION. The first case report indicated that the treatment led to a rapid and significant resolution of the optic disc edema in NAAION [80]. The second case report evaluated the use of intravitreal bevacizumab in a case of NAAION with macular edema and subretinal fluid and reported a positive visual outcome [81]. As discussed, the evidence for this drug has been mixed, but the larger studies indicate that bevacizumab is ineffective for NAAION.

- Ranibizumab:

Intravitreal ranibizumab is another anti-VEGF agent that has been tried for NAAION, primarily in animal models. Studies in rAION and primate non-arteritic anterior ischemic optic neuropathy (pNAION) models have reported that intravitreal ranibizumab is ineffective for NAAION in these models [82,83].

- Aflibercept:

Conversely, a case report assessed the effect of a single intravitreal aflibercept, another anti-VEGF agent, and found that a 2 mg injection a day after the onset of NAAION significantly increased VA within the first week after the injection, reduced mean RNFL thickness by 22.7 μm, and improved VF at the 3-month follow-up [84]. Similarly, a retrospective trial assessed the effect of an intravitreal aflibercept injection of 2 mg/0.05 mL in 15 cases compared to 10 conservatively treated cases [85]. The results point towards a potential role for this treatment modality in acute NAAION in improving disc edema and VA, which need to be verified by conducting large-scale studies [85].

2.2.5. Anti-Parkinson Pharmacotherapies

- Levodopa/Carbidopa:

Levodopa and Carbidopa act synergistically to increase dopamine levels in the central nervous system, which is said to have neuromodulatory and neuroprotective effects [4]. The effect of this combination has been tested for NAAION patients in multiple studies [4]. There have been conflicting results regarding the efficacy of treatment modality for NAAION. A retrospective study found that central VA improves if levodopa is used within 15 days of disease onset, which correlates with a maculopapular RGC-sparing effect [86]. Another retrospective study found that levodopa and carbidopa together led to an improvement in VA in NAAION patients with a VA of 20/40 or worse [87]. That being said, the 18 patients in the control group showed no improvement, which calls selection bias into question [4,87]. A randomized controlled trial of the same combination was performed on NAAION patients with long-standing vision impairment, and improvement was found in VA. However, the visual improvement could also be accounted for by the spontaneous improvement as NAAION resolved [4,88]. Additionally, another randomized placebo-controlled trial that used levodopa and carbidopa found the combination ineffective in correcting long-standing visual loss [89]. Overall, there seems to be contradicting evidence; therefore, more studies must be performed to support one conclusion.

2.2.6. Blood-Associated Pharmacotherapies

- Aspirin:

Aspirin was attempted for the treatment and prevention of NAAION due to the condition's association with vascular risk factors [49]. As a treatment option, aspirin was ineffective at improving VA and field outcomes in a retrospective study [49]. Aspirin was also assessed as a tool for preventing NAAION in the second eye with mixed results [49,79]. That being said, the largest retrospective studies indicate no long-term benefit to using aspirin in preventing NAAION in the second eye [49,79,90]. Despite the lack of evidence supporting its use in the prevention of NAAION in the second eye, experts still recommend aspirin after an episode of NAAION for its preventative effect of lowering stroke and myocardial infarction risk in this vasculopathic patient population [79].

- Platelet Rich Plasma:

Platelet-rich plasma (PRP) has also been attempted to treat NAAION [91]. In a prospective nonrandomized controlled trial, 12 NAAION patients received two tenon capsule injections of PRP and were compared to 13 control patients. Both groups received basic treatment with IV iodine hydrobromide and butylphthalide-sodium for 10 consecutive days. The two PRP injections were administered on days 1 and 10 of the basic treatment [91]. This trial showed that the PRP patient group's BCVA improved between days 1 and 30. However, there was no significant difference between the PRP and control groups for the same parameters [91].

- Heparin-Induced Extracorporeal LDL/Fibrinogen Precipitation:

Heparin-induced extracorporeal LDL/fibrinogen precipitation (HELP) decreases plasma viscosity and improves microcirculation [4]. When analyzed in a systematic review and meta-analysis, HELP did not improve VA as a categorical variable [4].

- Anticoagulants:

Heparin and warfarin have shown no benefit in patients with NAAION [4].

2.2.7. Miscellaneous Pharmacotherapies

- A Multivitamin, Mineral, Carotenoid, and Antioxidant Supplement Regimen:

Vega et al. conducted a retrospective case series study looking at the effect of a multivitamin, mineral, carotenoid, and antioxidant supplement regimen (including Vit A, B1, B2, B3, B6, B12, C + Zinc, selenium, manganese + lutein, zeaxanthin + glutathione, flavonoids, coenzyme Q10, with the stroke patients receiving aspirin) on VF measurements in 48 patients diagnosed with retinal vascular disease including NAAION, RAO, and homonymous hemaniopa or quadrantopia following stroke. [92]. In the NAAION patients, their visual field index (VFI) progressed at a rate of +11.5 $\pm$ 15% per year for the NAAION patients (n = 18, $p < 0.0001$), with most improvement occurring after 2 months of follow-up, meaning improvement is more likely due to supplementation than spontaneous resolution as part of natural disease history [92]. However, this case series has a small sample size with no comparison group, and the time of onset of visual loss was not known for each patient. There is also no way of knowing which of these supplements is contributing to the improvement. The improvements, however, are significant, and further studies evaluating such treatment regimens must be evaluated in randomized, double-blind, controlled, prospective clinical studies to fully support the use of these regimens in NAAION patients.

- 4-PBA:
- 4-phenylbutyric acid (4-PBA) is a chemical chaperone that has been studied in the treatment of cystic fibrosis, liver injury, and animal models of vision loss including glaucoma [93]. The unfolded protein response pathway is used by cells to control the endoplasmic reticulum (ER) and, in the case of cellular insult, to initially act as a defense line activating pro-survival pathways. However, after prolonged endoplasmic reticulum stress, pro-apoptotic pathways are upregulated, and this leads to cell death. Kumar et al. aimed to use intraperitoneally administered 4-BPA to reduce this ER stress and therefore preserve cell survivability in a mouse model of NAAION [93]. 4-PBA-treated NAAION eyes had a significant 22% higher number of RGC, and a significantly higher (5-μm) ganglion cell complex thickness on OCT imaging after induction of NAAION compared to the saline-treated group [93]. These results highlight the unfolded protein response pathway and decreasing ER stress using 4-PBA as valid therapeutic target candidates to be evaluated in human studies and serve in the treatment of NAAION.
- Endothelin Receptor Antagonists: Bosentan:
- Endothelin, a vasoconstrictive peptide released by both endothelial cells and vascular smooth muscle, is strongly implicated in cardiovascular disorders and in obstructive sleep apnea, a condition present in up to 70–85% of patients with NAAION [94]. Studies looking at blocking the effects of endothelin with endothelin receptor antagonist

bosentan have found that it increased retinal blood flow at the ONH in both healthy and glaucoma patients [94]. For these reasons, Chiquet et al. considered bosentan a good candidate to target the vasoconstriction caused by endothelin in the acute phase of NAAION and are conducting a multicenter randomized controlled trial looking at change in VF, VA, quality of life, and macular ganglion cell layer thickness. As of 1 July 2024, the results of the clinical trial have not been published, but it remains a promising treatment option for NAAION, which should be followed up.

- Omega-3 Polyunsaturated Fatty Acids:
- Omega-3 polyunsaturated fatty acids (ω-3 PUFAs) are a common food supplement demonstrated to modulate different signaling pathways and inhibit both inflammation and cell apoptosis in multiple models [95,96]. This was studied by Georgiou et al. in rat models of AION. The rats were given ω-3 PUFAs by gavage for 10 days and compared to those given saline. Rats receiving the ω-3 PUFAs were then found to have higher RGC densities, higher amplitudes of FVEP, lower numbers of apoptotic cells in the RGC layer, reduced macrophage recruitment at the ON, and increased M2 macrophage anti-inflammatory markers than in the saline group [97]. This is yet to be evaluated in humans, and further experiments need to be performed to elucidate its uses in NAAION.
- Bioengineered Algae Oil:
- ω-3 PUFAs can contain different ratios of docosahexaenoic acid (DHA) to eicosapentaenoic acid (EPA), and with the background that one study showed that pure DHA or a combination containing more DHA than EPA promoted more expression of neurotrophins and their receptors in neuron cell lines [98], Huang et al. investigated the effects of algae oil from bioengineered marine microalgae Schizochytrium sp., which is a DHA-rich ω-3 PUFA in a model of rAION [99]. Huang et al. found significantly higher FVEP and density of RGCs in the algae oil-treated group [99]. In future human studies, the importance of DHA/EPA ratios in ω-3 PUFAs should be investigated as different ratios could play a role in treatment effectiveness.
- P-Selectin:
- P-selectin plays a role in the recruitment of leukocytes to platelet aggregates and in inflammatory leukocyte extravasations [100]; therefore, targeting this pathway could decrease the number of ischemic injuries in NAAION patients. Kapupara et al. investigated soluble recombinant P-selectin immunoglobulin G chimeric fusion protein in rat AION models and demonstrated an increased RGC survival rate through stabilization of the blood–ON barrier and increased Nrf2 transcription factors levels and activating its signaling pathway [100]. This target needs to be further evaluated in human experiments to assess its efficacy in NAAION patients.
- Anti-Nogo Antibody:
- Nogo-A is an inhibitory protein in the central nervous system that prevents the continued expansion of neurons at the end of development [101]. Johnson et al. investigated the role of anti-NOGO receptor monoclonal antibody 11C7mAb in a rat model of non-arteritic anterior ischemic optic neuropathy and found a higher rate of FVEP preservation and a reduction in microglia, extrinsic macrophages with axon sparing, decreased extracellular debris, and less myelin damage in those receiving the antibody versus the group receiving the vehicle only [101]. Further human studies are required to evaluate 11C7mAb as a treatment for NAAION.

2.2.8. Future Pharmacotherapy Study Targets

- As an effective treatment for NAAION has not been conclusively found, studies looking at potential targets for the treatment of NAAION must be highlighted as only further research can achieve a breakthrough.
- A proteomics study of systemic inflammatory markers in acute and chronic NAAION patients conducted by Mesentier-Luoro et al. identified with immunoprofiling a multitude of markers in both acute and chronic NAAION patients which were significantly

unique to each group when compared to the controls, with some overlap between the acute and chronic patients. Since multiple aforementioned studies on corticosteroids aiming to decrease the inflammation component seen in NAAION have not been successful as treatments, a closer, more targeted approach to blunt the inflammatory response could be a possible treatment. Candidate novel specific targets found by Mesentier-Luoro et al. most notably included Eotaxin-3, MCP-2, TPO, and TRAIL in acute NAAION patients and in chronic NAAION, IL-1α, and CXCL10 [102]. These biomarkers reveal more specifics about the systemic inflammation profile of NAAION patients and could be targeted for treatment and help treat the inflammatory component of NAAION. It is important to note the small sample size of the study and the need for a natural history study to have a longitudinal follow-up of patients and try to decrease the effect of inter-patient variability.

- In the previously mentioned study by Kumar et al. (Section 2.2.7, 4-PBA), the unfolded protein response pathway could also serve as a promising area for future research. Kumar et al. identified within those pathways increased expressions of pro-apoptotic transcriptional regulator C/EBP homologous protein (CHOP) and decreased pro-survival chaperon glucose-regulated protein 78 (GRP78) levels in both the ON and RGCs after NAAION induction in mouse models [93]. These elucidated pathways can be further studied or targeted to help further our understanding of treating NAAION.
- When investigating the effects of M01 as a neuroprotector in the aforementioned study, Chien et al. (Section 2.2.2, M01, a HECT domain-E3 ubiquitin ligase inhibitor) found that the protective effect of M01 on RGCs following ON ischemia through upregulating Nr2 was independent of the pathway they hypothesized would be involved, specifically the NEDD4 protein, as it is a known down-regulator of Nrf2. As mentioned by Chien et al., the surprising result calls for more investigation of the E3 ubiquitin ligase inhibitor pathway, or possibly another treatment that could achieve a more potent neuroprotective effect through action on NEDD4.
- Polyamidoamine Dendrimer Nanoparticles:

To target the ischemic lesion, Guo et al. investigated the use of polyamidoamine dendrimer nanoparticles, which are non-biodegradable biocompatible molecules, to target the injured cells in both rNAION and pNAION models and, therefore, acting as a drug carrier when linked to biologically active compounds [103]. They reported that following NAAION induction, Cy-5 dendrimers selectively accumulated in astrocytes and circulating macrophages and that systemic administration offered better penetration into the eye and ON than intravitreal administration [103]. The use of dendrimers to target the ischemic lesions in NAAION may provide a novel treatment route that should be investigated further.

Table 2. Summary table elucidating which of the discussed treatments for NAAION have been tried in human patients and their efficacy.

Pharmacotherapy	Has This Pharmacotherapy Been Used in Human Patients?	Summary
Corticosteroids	Yes [3,4,31,33–35]	Potential benefit in improving BCVA in the acute phase with MP. Otherwise, no clinically significant benefit improvement in outcome measures.
Triamcinolone	Yes [4]	A systematic review and meta-analysis found this drug to improve VA and VF in two studies, which had a relatively small number of cases. Larger, more comprehensive studies are needed to support this data.
EPO	Yes [34,35]	EPO administration within five days of NAAION diagnosis led to a functional and structural neuroprotective effect on the ONs at the 6-month follow-up.

Table 2. *Cont.*

Pharmacotherapy	Has This Pharmacotherapy Been Used in Human Patients?	Summary
G-CSF	Yes [43]	In this study, intravitreal injection of G-CSF within 2 weeks of NAAION onset resulted in a BCVA improvement at the 1-month follow-up, but this effect was not seen in the final BCVA measurement indicating the short-term effect of this drug.
Citicoline	Yes [45]	A 500 mg/day oral solution of citicoline exerted a neuro-enhancing and neuroprotective effect in a randomized pilot study that enrolled 36 NAAION patients and 20 age-matched controls. These results are promising and need to be verified with larger studies.
Trabodenoson	No	Topical trabodenoson has shown promising results in a rodent NAION model but has yet to be tried on human NAAION patients [46].
Vincamine	Yes [48]	Vincamine led to statistically significant improvement in mean deviation of the visual field and RNFL and ganglion cell complex thickness in a study with 27 NAAION patients and 15 age-matched controls. These promising results must be verified with a larger trial.
CNTF	No	The positive neuroprotective effects of this drug have been shown in an rAION model but are yet to be verified in human studies [50].
BDNF/LM22A-4	No	BDNF and LM22A-4 have promising beneficial direct and indirect effects on animal models of NAAION, but no human studies have been performed yet [51,52].
Memantine	Yes [4]	A systematic review and meta-analysis found this drug to only improve VA when analyzed as a continuous variable but not as a categorical variable. Additionally, no improvement in VF was found.
Minocycline	No	Compared to previously mentioned neuroprotective drugs, minocycline does not show promise in its ability to preserve RGC in an rNAION model [53].
Butylidenephthalide	No	While the early experimental results in an rAION model are promising, it is essential to follow the experimental evidence with human studies to validate the effect of this drug in NAAION patients [54].
Bardoxolone methyl and omaveloxolone	No	A study of these two treatments has been conducted in an rAION and revealed that out of the two drugs, bardoxolone methyl could be a potential treatment for NAAION, but this needs to be verified with human studies [55].
Prostaglandin J_2 and MAGL/COX inhibitors	No	PGJ2 has been shown to be neuroprotective in rNAION and pNAAION models only [56]. MAGL/COX inhibitors are neuroprotective in rNAION models only when used independently [57].
QPI-1007	Yes [58,59]	Phase 1 studies showed some promise for QPI-1007 in improving VA [58]; however, phase 2/3 RCTs were terminated and data showed no significant difference [59].
RPh201	Yes [60]	Phase 1 studies established the safety of RPh201 [61], and Phase 2a demonstrated non-statistically significant improvement of BCVA [60]. Completed Phase 3 study results are unpublished and cannot be assessed.

Table 2. *Cont.*

Pharmacotherapy	Has This Pharmacotherapy Been Used in Human Patients?	Summary
Vitamin B3	Yes, as part of a multivitamin regimen, not alone [92]	Vitamin B3 showed neuroprotective effects in rat models of NAAION [63]. Evaluated as part of a multivitamin, mineral, and carotenoid regimen in a case series where VFI improved in NAAION patients. However, bigger studies with a comparison group must be held, and vitamin B3 alone has not been assessed [92].
M01, a HECT domain-E3 ubiquitin ligase inhibitor	No	Findings in rNAION models present the modulation of HECT domain-E3 ubiquitin ligase pathways as a new approach toward the treatment of NAAION that needs to be investigated in humans [64].
Brimonidine	Yes [4,49,66–68]	Neuroprotective ability in animal studies [65]; however, human trials did not find significant improvement of VA or VF [4,49,66–68].
Progesterone	No	Progesterone showed no neuroprotective effects in models of NAAION [69].
PLGA-Icariin	No	Demonstrated a neuroprotective effect in rat NAAION models only [70].
Puerarin	No	Demonstrated a neuroprotective effect in rat NAAION models only [72].
miR-124	No	Demonstrated a neuroprotective effect in rat NAAION models only [73].
Mesenchymal stem cells	Yes [76]	A prospective, non-randomized phase II study conducted on five NAAION patients ascertained that the treatment was safe, generally well tolerated, and showed positive results, albeit in a limited number of patients. More extensive studies are needed to verify these findings.
Mesenchymal stem cell exosome	No	This treatment has yet to be tested on animal NAAION models or human NAAION patients.
Mesenchymal stem cell-derived medium	No	The efficacy of MDCM was assessed in the rAION model and was found to preserve visual function and RGC density and reduce inflammation in the ON [74]. These findings are promising but need to be supported by human studies.
Bevacizumab	Yes [78,80,81]	The evidence for this drug in human patients is mixed, but the larger studies indicate that bevacizumab is ineffective for NAAION.
Ranibizumab	No	Studies in rAION and pNAION models have reported the drug to be ineffective for NAAION [82,83].
Aflibercept	Yes [84–86]	Human studies have shown promising results, which need to be supported by larger studies.
Levodopa/Carbidopa	Yes [4,90–93]	Overall, there seems to be contradicting evidence within human studies; therefore, more studies must be carried out to support one conclusion.
Aspirin	Yes [49,79,90]	Aspirin was found to be ineffective as a treatment for NAAION. As a preventative measure for the development of NAAION in the second eye, aspirin was found to have mixed results but is still recommended after an episode of NAAION in patients with vasculopathic risk factors.
Platelet-rich plasma	Yes. [91]	A prospective nonrandomized controlled trial revealed that this modality is ineffective in NAAION patients when compared to controls.

Table 2. *Cont.*

Pharmacotherapy	Has This Pharmacotherapy Been Used in Human Patients?	Summary
Heparin-induced extracorporeal LDL/fibrinogen precipitation	Yes [4]	This modality did not improve VA as a categorical variable when assessed in a systematic review and meta-analysis.
Anticoagulants	Yes [4]	This modality was ineffective in patients with NAAION.
A multivitamin, mineral, carotenoid, and antioxidant supplement regimen	Yes [92]	Evaluated in a case series where VFI improved in NAAION patients. However, bigger studies with a comparison group must be held [92].
4-PBA	No	Demonstrated a neuroprotective effect in rat NAAION models only [93].
Bosentan	In progress [94]	Ongoing multicenter randomized controlled trial on bosentan in NAAION [94].
Omega-3 polyunsaturated fatty acid	No	Demonstrated a neuroprotective effect in rat NAAION models only [97].
Bioengineered algae oil	No	Demonstrated a neuroprotective effect in rat NAAION models only [99].
P-Selectin	No	Demonstrated a neuroprotective effect in rat NAAION models only [100].
Anti-Nogo antibody	No	Demonstrated a neuroprotective effect in rat NAAION models only [101].

3. Posterior Ischemic Optic Neuropathy

PION is classified based on etiology into arteritic, non-arteritic, and perioperative [8,104,105]. Table 3 provides an overview of the treatments attempted for PION subtypes.

3.1. Arteritic Posterior Ischemic Optic Neuropathy

Arteritic PION is a complication of GCA, and its management follows a similar approach to A-AION, as discussed above. Treatment involves prompt and aggressive systemic steroid therapy aimed at preventing further visual loss, rather than restoring lost vision [8,104].

3.2. Non-Arteritic Posterior Ischemic Optic Neuropathy

Disorders other than GCA can cause non-arteritic PION. Multiple systemic diseases are associated with non-arteritic PION [104], including arteriosclerosis, atherosclerosis, arterial hypertension, diabetes mellitus [106,107], hypotension [106,108], hemodialysis [109,110], and epidural hematoma [111]. There is currently no standard treatment for non-arteritic PION. However, high-dose steroid therapy showed significant VA and VF improvement compared to untreated eyes [8,104]. This treatment is only sometimes successful [112]. Some case reports discuss additional agents to steroids, which could enhance the improvement in vision. For example, in one case, high-dose steroids were used on the first day of vision loss, and PGE1 was started the very next day. A marked improvement occurred within one day [112]. Another case of vision loss after hemodialysis and hypotension reported usage of EPO with prednisone for three days with considerable improvement in vision. They concluded that treatment with EPO within the first five days of injury is recommended, specifically in similar situations involving vision loss related to procedural-based hypotension [110].

3.3. Perioperative Posterior Ischemic Optic Neuropathy

Perioperative PION can happen with any surgery, but it most commonly occurs with spine surgeries due to the prone position for an extended period [113]. There is no effective treatment for improving vision after perioperative PION [114–116]. The primary approach is to focus on all the expected risk factors and solve them before surgery. There are multiple risk factors for perioperative ION, which may be influenced by patient-specific susceptibilities, including long duration in the prone position, excessive blood loss, hypotension, anemia [105,117], male sex, obesity, use of the Wilson frame, and lower percent colloid administration [118,119]. However, the risk factors for any given patient or procedure may vary and are likely multifactorial [117]. A case report considered using prednisolone tablets daily and showed mild improvement in vision [120]. In another report, the author used IV MP after cosmetic blepharoplasty with clinical improvement but presented with a pale optic disc and an ON-related VF defect [121] One more case report combined hyperbaric oxygen therapy and steroids successfully restoring vision after postoperative PION [122].

Table 3. Summary of drug treatments for posterior ischemic optic neuropathy (PION) by type, mechanism of action, and effectiveness.

Drug	Mechanism of Action	Effectiveness	Arteritic PION	Non-Arteritic PION	Perioperative PION	References
Systemic steroids (e.g., prednisone)	Reduces inflammation	Prevents further visual deterioration	✓			[8,104]
High-dose steroids	Reduces inflammation	Significant improvement in visual acuity and fields, though not always effective		✓		[8,104,112]
PGE1 and high-dose steroids	Vasodilation and neuroprotection	Marked vision improvement within one day when started early		✓		[112]
EPO with prednisone	Enhances oxygen delivery to damaged tissue	Improvement in vision when administered within five days		✓		[110]
IV methylprednisolone (MP)	Reduces inflammation	Mild visual improvement, though inconsistent			✓	[120,121]
Hyperbaric oxygen therapy with steroids	Enhances oxygenation	Successfully restored vision postoperatively			✓	[122]

3.4. Potential Future Treatments

Currently, there are no active trials or experiments specific to PION. However, a rat model has been developed to simulate PION using a photochemical procedure [123]. Therefore, interventions can be assessed using this model in the future.

4. Conclusions

As discussed, the treatment of AION and PION is a topic of active research. Our manuscript summarizes the current landscape in treating AAION, NAAION, and PION. The treatment of AAION currently consists mainly of high-dose steroid regimens, with methotrexate, tocilizumab, and abatacept being the most viable candidates for steroid-sparing therapy. These steroid-sparing strategies must be studied further in the context of preventing ocular vision loss instead of just in the treatment of GCA. As for NAAION, some pharmacotherapeutic approaches have shown promise in preliminary studies, such as animal studies and early human studies; there is still a long way to go before a specific unanimous treatment regimen is established. Similarly, despite the various treatment options explored, there is still no universally effective therapy for PION. Future research, mainly using animal models such as the photochemical rat model, offers the potential for developing new standard treatments to better manage and potentially prevent this condition.

Author Contributions: Conceptualization, O.B., B.A.B., A.A., C.M., K.K. and S.K.B.; methodology, O.B., B.A.B. and A.A; writing—first draft, O.B., B.A.B. and A.A.; writing—review and editing, O.B., B.A.B. and A.A.; supervision, C.M., K.K. and S.K.B. All authors have read and agreed to the published version of the manuscript.

Funding: This research received no external funding.

Conflicts of Interest: The authors declare no conflicts of interest.

Abbreviations

AAION (arteritic anterior ischemic optic neuropathy); NAAION (non-arteritic ischemic optic neuropathy); PION (posterior ischemic optic neuropathy); ION (ischemic optic neuropathy); AION (anterior ischemic optic neuropathy); GCA (giant cell arteritis); SPCA (short posterior ciliary arteries); ONH (optic nerve head); ON (optic nerve); RGC (retinal ganglion cell); TNF-α (tumor necrosis factor-alpha); MP (methylprednisolone); ESR (erythrocyte sedimentation rate); CRP (C-reactive protein); IV (intravenous); IL (interleukin); CTLA-4 (cytotoxic T-lymphocyte-associated protein 4); PGE1 (prostaglandin E1); rAION (rat model of anterior ischemic optic neuropathy); PBS (phosphate-buffered saline); BCVA (best corrected visual acuity); RNFL (retinal nerve fiber layer); OCT (optical coherence tomography); VER (visual evoked response); VA (visual acuity); VF (visual field); IVTA (intravitreal triamcinolone); VEP (visual evoked potential); EPO (erythropoietin); G-CSF (granulocyte colony-stimulating factor); PERG (pattern electroretinogram); HFA MD (Humphrey 24-2 visual field mean deviation); A1R (A1 receptor); rNAION (rodent NAION model); CNTF (ciliary-derived neurotrophic factor); CFP (cyan fluorescent protein); BDNF (brain-derived neurotrophic factor); TrkB (tropomyosin-related kinase B); LM22A-4 ((N,N′,N′-tris [2-hydroxyethyl])-1,3,5-benzene tricarboxamide); BP (butylidenephthalide); RTA 402 (bardoxolone methyl); RTA 408 (omaveloxolone); NFκB 9 nuclear transcription factor nuclear factor kappa beta); PGJ2 (prostaglandin J2); PPARγ (nuclear factor peroxisomal proliferator-activated receptor-gamma); IVT (intravitreal); pNAAION (primate model of non-arteritic anterior ischemic optic neuropathy); MAGL (monoacylglycerol lipase); AA (arachidonic acid); COX1/2 (cyclooxygenase 1/2); NAD+ (nicotinamide adenine dinucleotide); NADP+ (nicotinamide adenine dinucleotide phosphate); FVEP (flash visual evoked potential); HECT (homologous E6-associated protein carboxyl terminus); NRF2 (nuclear factor erythroid 2-related factor 2); TXNIP (thioredoxin interacting protein); NLRP3 (NLR family pyrin domain containing 3); MCAO (middle cerebral artery occlusion); PLGA (poly(lactide-co-glycolide)); IKK-β (inhibitor of nuclear factor kappa-B kinase subunit beta); MSCs (mesenchymal stem cells); MDCM (MSC-derived conditioned medium); VEGF (vascular endothelial growth factor); pNAION (primate non-arteritic anterior ischemic optic neuropathy model); PRP (platelet-rich plasma); HELP (heparin-induced extra-corporeal LDL/fibrinogen precipitation); 4-PBA (4-phenylbutyric acid); ER (endoplasmic reticulum); ω-3 PUFA (omega-3 polyunsaturated fatty acids); DHA (docosahexaenoic acid); EPA (eicosapentaenoic acid); CHOP (C/EBP homologous protein); GRP78 (glucose-regulated protein 78).

References

1. Salvetat, M.L.; Pellegrini, F.; Spadea, L.; Salati, C.; Zeppieri, M. Non-Arteritic Anterior Ischemic Optic Neuropathy (NA-AION): A Comprehensive Overview. *Vision* **2023**, *7*, 72. [CrossRef] [PubMed]
2. Hayreh, S.S. Ischemic optic neuropathy. *Prog. Retin. Eye Res.* **2009**, *28*, 34–62. [CrossRef] [PubMed]
3. Zhao, F.F.; Chen, Y.; Li, T.P.; Wang, Y.; Lin, H.J.; Yang, J.F.; Chen, L.; Tan, S.Y.; Liang, J.J.; Cen, L.P. Visual outcome of various dose of glucocorticoids treatment in nonarteritic anterior ischemic optic neuropathy—A retrospective analysis. *BMC Ophthalmol.* **2024**, *24*, 100. [CrossRef]
4. Lantos, K.; Dömötör, Z.R.; Farkas, N.; Kiss, S.; Szakács, Z.; Garami, A.; Varga, G.; Lujber, L.; Kanaan, R.; Hegyi, P.; et al. Efficacy of Treatments in Nonarteritic Ischemic Optic Neuropathy: A Systematic Review and Meta-Analysis. *Int. J. Environ. Res. Public Health* **2022**, *19*, 2718. [CrossRef] [PubMed]
5. Huang, T.L.; Wen, Y.T.; Chang, C.H.; Chang, S.W.; Lin, K.H.; Tsai, R.K. Early Methylprednisolone Treatment Can Stabilize the Blood-Optic Nerve Barrier in a Rat Model of Anterior Ischemic Optic Neuropathy (rAION). *Investig. Ophthalmol. Vis. Sci.* **2017**, *58*, 1628–1636. [CrossRef]

6. Teja, S.; Patel, V.R. Ischemic Optic Neuropathies: Diagnosis and Management. *Int. Ophthalmol. Clin.* **2019**, *59*, 39–57. [CrossRef] [PubMed]
7. Chourpiliadis, C.; Aeddula, N.R. Physiology, Glucocorticoids. In *StatPearls [Internet]*; StatPearls Publishing: Treasure Island, FL, USA, 2024.
8. Hayreh, S. Management of ischemic optic neuropathies. *Indian J. Ophthalmol.* **2011**, *59*, 123. [CrossRef]
9. Bajpai, V.; Madan, S.; Beri, S. Arteritic anterior ischaemic optic neuropathy: An update. *Eur. J. Ophthalmol.* **2021**, *31*, 2818–2827. [CrossRef]
10. Mackie, S.L.; Dejaco, C.; Appenzeller, S.; Camellino, D.; Duftner, C.; Gonzalez-Chiappe, S.; Mahr, A.; Mukhtyar, C.; Reynolds, G.; de Souza, A.W.S.; et al. British Society for Rheumatology guideline on diagnosis and treatment of giant cell arteritis: Executive summary. *Rheumatology* **2020**, *59*, 487–494. [CrossRef]
11. Chevalet, P.A.; Barrier, J.H.; Pottier, P.I.; Magadur-Joly, G.E.; Pottier, M.A.; Hamidou, M.O.; Planchon, B.E.; El Kouri, D.O.; Connan, L.O.; Dupond, J.L.; et al. A randomized, multicenter, controlled trial using intravenous pulses of methylprednisolone in the initial treatment of simple forms of giant cell arteritis: A one year followup study of 164 patients. *J. Rheumatol.* **2000**, *27*, 1484–1491.
12. Chan, C.C.K. Steroid management in giant cell arteritis. *Br. J. Ophthalmol.* **2001**, *85*, 1061–1064. [CrossRef] [PubMed]
13. Yasir, M.; Goyal, A.; Sonthalia, S. Corticosteroid Adverse Effects. In *StatPearls [Internet]*; StatPearls Publishing: Treasure Island, FL, USA, 2024.
14. Stone, J.H.; Tuckwell, K.; Dimonaco, S.; Klearman, M.; Aringer, M.; Blockmans, D.; Brouwer, E.; Cid, M.C.; Dasgupta, B.; Rech, J.; et al. Trial of Tocilizumab in Giant-Cell Arteritis. *N. Engl. J. Med.* **2017**, *377*, 317–328. [CrossRef] [PubMed]
15. Villiger, P.M.; Adler, S.; Kuchen, S.; Wermelinger, F.; Dan, D.; Fiege, V.; Bütikofer, L.; Seitz, M.; Reichenbach, S. Tocilizumab for induction and maintenance of remission in giant cell arteritis: A phase 2, andomized, double-blind, placebo-controlled trial. *Lancet* **2016**, *387*, 1921–1927. [CrossRef] [PubMed]
16. Régent, A.; Redeker, S.; Deroux, A.; Kieffer, P.; Ly, K.H.; Dougados, M.; Liozon, E.; Larroche, C.; Guillevin, L.; Bouillet, L.; et al. Tocilizumab in Giant Cell Arteritis: A Multicenter Retrospective Study of 34 Patients. *J. Rheumatol.* **2016**, *43*, 1547–1552. [CrossRef] [PubMed]
17. Seitz, M.; Reichenbach, S.; Bonel, H.; Adler, S.; Wermelinger, F.; Villiger, P. Rapid induction of remission in large vessel vasculitis by IL-6 blockade. *Swiss Med. Wkly.* **2011**, *141*, w13156.
18. Loricera, J.; Blanco, R.; Hernández, J.L.; Castañeda, S.; Mera, A.; Pérez-Pampín, E.; Peiró, E.; Humbría, A.; Calvo-Alén, J.; Aurrecoechea, E.; et al. Tocilizumab in giant cell arteritis: Multicenter open-label study of 22 patients. *Semin. Arthritis Rheum.* **2015**, *44*, 717–723. [CrossRef] [PubMed]
19. Unizony, S.; Arias-Urdaneta, L.; Miloslavsky, E.; Arvikar, S.; Khosroshahi, A.; Keroack, B.; Stone, J.R.; Stone, J.H. Tocilizumab for the treatment of large-vessel vasculitis (giant cell arteritis, Takayasu arteritis) and polymyalgia rheumatica. *Arthritis Care Res.* **2012**, *64*, 1720–1729. [CrossRef]
20. Oliveira, F.; Butendieck, R.R.; Ginsburg, W.W.; Parikh, K.; Abril, A. Tocilizumab, an effective treatment for relapsing giant cell arteritis. *Clin. Exp. Rheumatol.* **2014**, *32* (Suppl. 82), 76–78.
21. Preuss, C.V.; Anjum, F. Tocilizumab. StatPearls [Internet]. 12 February 2024. Available online: https://www.ncbi.nlm.nih.gov/books/NBK570644/ (accessed on 1 August 2024).
22. Unizony, S.; McCulley, T.J.; Spiera, R.; Pei, J.; Sidiropoulos, P.N.; Best, J.H.; Birchwood, C.; Pavlov, A.; Stone, J.H. Clinical outcomes of patients with giant cell arteritis treated with tocilizumab in real-world clinical practice: Decreased incidence of new visual manifestations. *Arthritis Res. Ther.* **2021**, *23*, 8. [CrossRef]
23. Vionnet, J.; Buss, G.; Mayer, C.; Sokolov, A.A.; Borruat, F.X.; Spertini, F. Tocilizumab for giant cell arteritis with corticosteroid-resistant progressive anterior ischemic optic neuropathy. *Jt. Bone Spine* **2017**, *84*, 615–619. [CrossRef]
24. Hanoodi, M.; Mittal, M. Methotrexate. Neuroimaging Pharmacopoeia, Second Edition [Internet]. 16 August 2023; pp. 149–161. Available online: https://www.ncbi.nlm.nih.gov/books/NBK556114/ (accessed on 1 August 2024).
25. Mahr, A.D.; Jover, J.A.; Spiera, R.F.; Hernández-García, C.; Fernández-Gutiérrez, B.; LaValley, M.P.; Merkel, P.A. Adjunctive methotrexate for treatment of giant cell arteritis: An individual patient data meta-analysis. *Arthritis Rheum.* **2007**, *56*, 2789–2797. [CrossRef]
26. Leon, L.; Rodriguez-Rodriguez, L.; Morado, I.; Rosales, Z.; Vadillo, C.; Freites, D.; Macarron, P.; Fernandez-Gutierrez, B.; Blanco, M.; Jover, J.A.; et al. Treatment with methotrexate and risk of relapses in patients with giant cell arteritis in clinical practice. *Clin. Exp. Rheumatol.* **2018**, *36* (Suppl. 111), 121–128. [PubMed]
27. Langford, C.A.; Cuthbertson, D.; Ytterberg, S.R.; Khalidi, N.; Monach, P.A.; Carette, S.; Seo, P.; Moreland, L.W.; Weisman, M.; Koening, C.L.; et al. A Randomized, Double-Blind Trial of Abatacept (CTLA-4Ig) for the Treatment of Giant Cell Arteritis. *Arthritis Rheumatol.* **2017**, *69*, 837–845. [CrossRef]
28. Gerriets, V.; Goyal, A.; Khaddour, K. Tumor Necrosis Factor Inhibitors. In *StatPearls [Internet]*; StatPearls Publishing: Treasure Island, FL, USA, 2024.
29. Vilares-Morgado, R.; Nunes, H.M.M.; dos Reis, R.S.; Barbosa-Breda, J. Management of ocular arterial ischemic diseases: A review. *Graefe's Arch. Clin. Exp. Ophthalmol.* **2022**, *261*, 1–22. [CrossRef] [PubMed]
30. Steigerwalt, R.D.; Cesarone, M.R.; Belcaro, G.; Pascarella, A.; De Angelis, M.; Gattegna, R.; Nebbioso, M. Arteritic anterior ischemic optic neuropathy treated with intravenous prostaglandin E_1 and steroids. *Int. J. Angiol.* **2010**, *19*, e113–e115. [CrossRef]

31. Saxena, R.; Singh, D.; Sharma, M.; James, M.; Sharma, P.; Menon, V. Steroids versus No Steroids in Nonarteritic Anterior Ischemic Optic Neuropathy. *Ophthalmology* **2018**, *125*, 1623–1627. [CrossRef]
32. Huang, T.L.; Huang, S.P.; Chang, C.H.; Lin, K.H.; Chang, S.W.; Tsai, R.K. Protective effects of systemic treatment with methylprednisolone in a rodent model of non-arteritic anterior ischemic optic neuropathy (rAION). *Exp. Eye Res.* **2015**, *131*, 69–76. [CrossRef] [PubMed]
33. Pakravan, M.; Sanjari, N.; Esfandiari, H.; Pakravan, P.; Yaseri, M. The effect of high-dose steroids, and normobaric oxygen therapy, on recent onset non-arteritic anterior ischemic optic neuropathy: A randomized clinical trial. *Graefe's Arch. Clin. Exp. Ophthalmol.* **2016**, *254*, 2043–2048. [CrossRef]
34. Pakravan, M.; Esfandiari, H.; Hassanpour, K.; Razavi, S.; Pakravan, P. The Effect of Combined Systemic Erythropoietin and Steroid on Non-arteritic Anterior Ischemic Optic Neuropathy: A Prospective Study. *Curr. Eye Res.* **2017**, *42*, 1079–1084. [CrossRef]
35. Nikkhah, H.; Golalipour, M.; Doozandeh, A.; Pakravan, M.; Yaseri, M.; Esfandiari, H. The effect of systemic erythropoietin and oral prednisolone on recent-onset non-arteritic anterior ischemic optic neuropathy: A randomized clinical trial. *Graefe's Arch. Clin. Exp. Ophthalmol.* **2020**, *258*, 2291–2297. [CrossRef]
36. Radoi, C.; Garcia, T.; Brugniart, C.; Ducasse, A.; Arndt, C. Intravitreal triamcinolone injections in non-arteritic anterior ischemic optic neuropathy. *Graefe's Arch. Clin. Exp. Ophthalmol.* **2014**, *252*, 339–345. [CrossRef]
37. Pereira, L.S.; Ávila, M.P.; Salustiano, L.X.; Paula, A.C.; Arnhold, E.; McCulley, T.J. Intravitreal Triamcinolone Acetonide Injection in a Rodent Model of Anterior Ischemic Optic Neuropathy. *J. Neuro-Ophthalmol. Off. J. N. Am. Neuro-Ophthalmol. Soc.* **2018**, *38*, 561–565. [CrossRef] [PubMed]
38. Huang, T.L.; Wen, Y.T.; Chang, C.H.; Chang, S.W.; Lin, K.H.; Tsai, R.K. Efficacy of Intravitreal Injections of Triamcinolone Acetonide in a Rodent Model of Nonarteritic Anterior Ischemic Optic Neuropathy. *Investig. Opthalmol. Vis. Sci.* **2016**, *57*, 1878. [CrossRef] [PubMed]
39. Tsai, R.K.; Lin, K.L.; Huang, C.T.; Wen, Y.T. Transcriptomic Analysis Reveals That Granulocyte Colony-Stimulating Factor Trigger a Novel Signaling Pathway (TAF9-P53-TRIAP1-CASP3) to Protect Retinal Ganglion Cells after Ischemic Optic Neuropathy. *Int. J. Mol. Sci.* **2022**, *23*, 8359. [CrossRef] [PubMed]
40. Chang, C.H.; Huang, T.L.; Huang, S.P.; Tsai, R.K. Neuroprotective effects of recombinant human granulocyte colony-stimulating factor (G-CSF) in a rat model of anterior ischemic optic neuropathy (rAION). *Exp. Eye Res.* **2014**, *118*, 109–116. [CrossRef] [PubMed]
41. Wen, Y.T.; Huang, T.L.; Huang, S.P.; Chang, C.H.; Tsai, R.K. Early applications of granulocyte colony-stimulating factor (G-CSF) can stabilize the blood-optic nerve barrier and further ameliorate optic nerve inflammation in a rat model of anterior ischemic optic neuropathy (rAION). *Dis. Models Mech.* **2016**, *9*, 1193–1202. [CrossRef]
42. Liu, P.K.; Wen, Y.T.; Lin, W.; Kapupara, K.; Tai, M.; Tsai, R.K. Neuroprotective effects of low-dose G-CSF plus meloxicam in a rat model of anterior ischemic optic neuropathy. *Sci. Rep.* **2020**, *10*, 10351. [CrossRef]
43. Abri Aghdam, K.; Aghajani, A.; Ashraf Khorasani, M.; Soltan Sanjari, M.; Chaibakhsh, S.; Habibi, A.; Falavarjani, K.G. Intravitreal Injection Of The Granulocyte-Colony Stimulating Factor For The Treatment Of Non-Arteritic Anterior Ischemic Optic Neuropathy: A Pilot Study. *Semin. Ophthalmol.* **2021**, *36*, 649–657. [CrossRef]
44. Oddone, F.; Rossetti, L.; Parravano, M.; Sbardella, D.; Coletta, M.; Ziccardi, L.; Roberti, G.; Carnevale, C.; Romano, D.; Manni, G.; et al. Citicoline in Ophthalmological Neurodegenerative Disease: A Comprehensive Review. *Pharmaceuticals* **2021**, *14*, 281. [CrossRef]
45. Parisi, V.; Barbano, L.; di Renzo, A.; Coppola, G.; Ziccardi, L. Neuroenhancement and neuroprotection by oral solution citicoline in non-arteritic ischemic optic neuropathy as a model of neurodegeneration: A randomized pilot study. *PLoS ONE* **2019**, *14*, e0220435. [CrossRef]
46. Guo, Y.; Mehrabian, Z.; Johnson, M.A.; Albers, D.S.; Rich, C.C.; Baumgartner, R.A.; Bernstein, S.L. Topical Trabodenoson Is Neuroprotective in a Rodent Model of Anterior Ischemic Optic Neuropathy (rNAION). *Transl. Vis. Sci. Technol.* **2019**, *8*, 47. [CrossRef]
47. Li, L.; Su, Y.; Liu, J.; Chen, C. Efficacy of Vincamine treatment in a rat model of anterior ischemic optic neuropathy. *Eur. J. Ophthalmol.* **2021**, *31*, 3442–3449. [CrossRef] [PubMed]
48. Liang, C.Q.; Chen, C.Z.; Su, Y.; Yi, Z.H. Therapeutic effect and safety of vincamine in anterior non-arteritic ischemic optic neuropathy. *Guoji Yanke Zazhi (Int. Eye Sci.)* **2017**, *17*, 1845–1848.
49. Peeler, C.; Cestari, D.M. Non-Arteritic Anterior Ischemic Optic Neuropathy (NAION): A Review and Update on Animal Models. *Semin. Ophthalmol.* **2016**, *31*, 99–106. [CrossRef] [PubMed]
50. Mathews, M.K.; Guo, Y.; Langenberg, P.; Bernstein, S.L. Ciliary neurotrophic factor (CNTF)-mediated ganglion cell survival in a rodent model of non-arteritic anterior ischaemic optic neuropathy (NAION). *Br. J. Ophthalmol.* **2015**, *99*, 133–137. [CrossRef] [PubMed]
51. Ali Shariati, M.; Kumar, V.; Yang, T.; Chakraborty, C.; Barres, B.A.; Longo, F.M.; Liao, Y.J. A Small Molecule TrkB Neurotrophin Receptor Partial Agonist as Possible Treatment for Experimental Nonarteritic Anterior Ischemic Optic Neuropathy. *Curr. Eye Res.* **2018**, *43*, 1489–1499. [CrossRef]
52. Goldenberg-Cohen, N.; Avraham-Lubin, B.C.R.; Sadikov, T.; Askenasy, N. Effect of Coadministration of Neuronal Growth Factors on Neuroglial Differentiation of Bone Marrow–Derived Stem Cells in the Ischemic Retina. *Investig. Opthalmol. Vis. Sci.* **2014**, *55*, 502. [CrossRef]

53. Mehrabian, Z.; Guo, Y.; Weinreich, D.; Bernstein, S.L. Oligodendrocyte death, neuroinflammation, and the effects of minocycline in a rodent model of nonarteritic anterior ischemic optic neuropathy (rNAION). *Mol. Vis.* **2017**, *23*, 963–976.
54. Chou, Y.Y.; Chien, J.Y.; Ciou, J.W.; Huang, S.P. The Protective Effects of n-Butylidenephthalide on Retinal Ganglion Cells during Ischemic Injury. *Int. J. Mol. Sci.* **2022**, *23*, 2095. [CrossRef]
55. Chien, J.Y.; Chou, Y.Y.; Ciou, J.W.; Liu, F.Y.; Huang, S.P. The Effects of Two Nrf2 Activators, Bardoxolone Methyl and Omaveloxolone, on Retinal Ganglion Cell Survival during Ischemic Optic Neuropathy. *Antioxidants* **2021**, *10*, 1466. [CrossRef]
56. Miller, N.R.; Johnson, M.A.; Nolan, T.; Guo, Y.; Bernstein, A.M.; Bernstein, S.L. Sustained Neuroprotection From a Single Intravitreal Injection of PGJ $_2$ in a Nonhuman Primate Model of Nonarteritic Anterior Ischemic Optic Neuropathy. *Investig. Opthalmol. Vis. Sci.* **2014**, *55*, 7047–7056. [CrossRef]
57. Mehrabian, Z.; Guo, Y.; Miller, N.R.; Henderson, A.D.; Roth, S.; Bernstein, S.L. Approaches to Potentiated Neuroprotective Treatment in the Rodent Model of Ischemic Optic Neuropathy. *Cells* **2021**, *10*, 1440. [CrossRef] [PubMed]
58. Antoszyk, A.; Katz, B.; Singh, R.; Gurses-Ozden, R.; Erlich, S.; Rothenstein, D.; Sharon, N.; Hodge, J.; Levin, L.; Miller, N. A Phase I Open Label, Dose Escalation Trial Of QPI-1007 Delivered By A Single Intravitreal (IVT) Injection To Subjects with Low Visual Acuity And Acute Non-Arteritic Anterior Ischemic Optic Neuropathy (NAION). *Investig. Ophthalmol. Vis. Sci.* **2013**, *54*, 4575.
59. EU Clinical Trials Register [Internet]. Available online: https://www.clinicaltrialsregister.eu/ctr-search/trial/2015-003079-31/DE (accessed on 1 August 2024).
60. Rath, E.Z.; Hazan, Z.; Adamsky, K.; Solomon, A.; Segal, Z.I.; Levin, L.A. Randomized Controlled Phase 2a Study of RPh201 in Previous Nonarteritic Anterior Ischemic Optic Neuropathy. *J. Neuro-Ophthalmol.* **2019**, *39*, 291–298. [CrossRef] [PubMed]
61. Hazan, Z.; Adamsky, K.; Lucassen, A.; Levin, L.A. A First-in-Human Phase 1 Randomized Single and Multiple Ascending Dose Study of RPh201 in Healthy Volunteers. *Clin. Pharmacol. Drug Dev.* **2020**, *9*, 366–374. [CrossRef]
62. Hayreh, S.S.; Zimmerman, M.B. Nonarteritic Anterior Ischemic Optic Neuropathy. *Ophthalmology* **2008**, *115*, 298–305.e2. [CrossRef]
63. Chen, T.W.; Wu, P.Y.; Wen, Y.T.; Desai, T.D.; Huang, C.T.; Liu, P.K.; Tsai, R.K. Vitamin B3 Provides Neuroprotection via Antioxidative Stress in a Rat Model of Anterior Ischemic Optic Neuropathy. *Antioxidants* **2022**, *11*, 2422. [CrossRef]
64. Chien, J.Y.; Ciou, J.W.; Yen, Y.; Huang, S.P. Protective effects of compound M01 on retinal ganglion cells in experimental anterior ischemic optic neuropathy by inhibiting TXNIP/NLRP3 inflammasome pathway. *Biomed. Pharmacother.* **2023**, *169*, 115861. [CrossRef]
65. Maciulaitiene, R.; Kalesnykas, G.; Pauza, D.H.; Januleviciene, I. A combination of topical and systemic administration of brimonidine is neuroprotective in the murine optic nerve crush model. *PLoS ONE* **2024**, *19*, e0308671. [CrossRef]
66. Berry, S.; Lin, W.; Sadaka, A.; Lee, A. Nonarteritic anterior ischemic optic neuropathy: Cause, effect, and management. *Eye Brain* **2017**, *9*, 23–28. [CrossRef]
67. Hayreh, S.S. Controversies on neuroprotection therapy in non-arteritic anterior ischaemic optic neuropathy. *Br. J. Ophthalmol.* **2020**, *104*, 153–156. [CrossRef]
68. Wilhelm, B.; Lüdtke, H.; Wilhelm, H. Efficacy and tolerability of 0.2% brimonidine tartrate for the treatment of acute non-arteritic anterior ischemic optic neuropathy (NAION): A 3-month, double-masked, andomized, placebo-controlled trial. *Graefe's Arch. Clin. Exp. Ophthalmol.* **2006**, *244*, 551–558.
69. Allen, R.S.; Olsen, T.W.; Sayeed, I.; Cale, H.A.; Morrison, K.C.; Oumarbaeva, Y.; Lucaciu, I.; Boatright, J.H.; Pardue, M.T.; Stein, D.G. Progesterone Treatment in Two Rat Models of Ocular Ischemia. *Investig. Opthalmol. Vis. Sci.* **2015**, *56*, 2880–2891. [CrossRef] [PubMed]
70. Desai, T.D.; Wen, Y.T.; Daddam, J.R.; Cheng, F.; Chen, C.C.; Pan, C.L.; Lin, K.L.; Tsai, R.K. Long term therapeutic effects of icariin-loaded PLGA microspheres in an experimental model of optic nerve ischemia via modulation of CEBP-β/G-CSF /noncanonical NF κB axis. *Bioeng. Transl. Med.* **2022**, *7*, e10289. [CrossRef] [PubMed]
71. Zhou, Y.; Zhang, H.; Peng, C. Puerarin: A Review of Pharmacological Effects. *Phytother. Res.* **2014**, *28*, 961–975. [CrossRef]
72. Nguyen Ngo Le, M.A.; Wen, Y.T.; Ho, Y.C.; Kapupara, K.; Tsai, R.K. Therapeutic Effects of Puerarin Against Anterior Ischemic Optic Neuropathy Through Antiapoptotic and Anti-Inflammatory Actions. *Investig. Opthalmol. Vis. Sci.* **2019**, *60*, 3481. [CrossRef]
73. Chen, Y.H.; Huang, Y.C.; Chen, C.H.; Wen, Y.T.; Tsai, R.K.; Chen, C. Investigation of the Protective Effect of Extracellular Vesicle miR-124 on Retinal Ganglion Cells Using a Photolabile Paper-Based Chip. *Investig. Opthalmol. Vis. Sci.* **2023**, *64*, 17. [CrossRef]
74. Wen, Y.T.; Ho, Y.C.; Lee, Y.C.; Ding, D.C.; Liu, P.K.; Tsai, R.K. The Benefits and Hazards of Intravitreal Mesenchymal Stem Cell (MSC) Based-Therapies in the Experimental Ischemic Optic Neuropathy. *Int. J. Mol. Sci.* **2021**, *22*, 2117. [CrossRef]
75. Li, D.; Gong, Y. A Promising Strategy for Non-Arteritic Anterior Ischemic Optic Neuropathy: Intravitreal Mesenchymal Stem Cell Exosome. *Curr. Stem Cell Res. Ther.* **2021**, *16*, 109–114. [CrossRef]
76. Pastor, J.C.; Pastor-Idoate, S.; López-Paniagua, M.; Para, M.; Blazquez, F.; Murgui, E.; García, V.; Coco-Martín, R.M. Intravitreal allogeneic mesenchymal stem cells: A non-randomized phase II clinical trial for acute non-arteritic optic neuropathy. *Stem Cell Res. Ther.* **2023**, *14*, 261. [CrossRef]
77. Gaier, E.D.; Torun, N. The enigma of nonarteritic anterior ischemic optic neuropathy. *Curr. Opin. Ophthalmol.* **2016**, *27*, 498–504. [CrossRef]
78. Rootman, D.B.; Gill, H.S.; Margolin, E.A. Intravitreal bevacizumab for the treatment of nonarteritic anterior ischemic optic neuropathy: A prospective trial. *Eye* **2013**, *27*, 538–544. [CrossRef] [PubMed]
79. Miller, N.R.; Arnold, A.C. Current concepts in the diagnosis, pathogenesis and management of nonarteritic anterior ischaemic optic neuropathy. *Eye* **2015**, *29*, 65–79. [CrossRef] [PubMed]

80. Bennett, J.L.; Thomas, S.; Olson, J.L.; Mandava, N. Treatment of Nonarteritic Anterior Ischemic Optic Neuropathy with Intravitreal Bevacizumab. *J. Neuro-Ophthalmol.* **2007**, *27*, 238–240. [CrossRef] [PubMed]
81. Dave, V.; Pappuru, R. An unusual presentation of nonarteritic ischemic optic neuropathy with subretinal fluid treated with intravitreal bevacizumab. *Indian J. Ophthalmol.* **2016**, *64*, 87–88. [CrossRef] [PubMed]
82. Huang, T.L.; Chang, C.H.; Chang, S.W.; Lin, K.H.; Tsai, R.K. Efficacy of Intravitreal Injections of Antivascular Endothelial Growth Factor Agents in a Rat Model of Anterior Ischemic Optic Neuropathy. *Investig. Ophthalmol. Vis. Sci.* **2015**, *56*, 2290–2296. [CrossRef]
83. Miller, N.R.; Johnson, M.A.; Nolan, T.; Guo, Y.; Bernstein, S.L. A Single Intravitreal Injection of Ranibizumab Provides No Neuroprotection in a Nonhuman Primate Model of Moderate-to-Severe Nonarteritic Anterior Ischemic Optic Neuropathy. *Investig. Opthalmol. Vis. Sci.* **2015**, *56*, 7679–7686. [CrossRef]
84. Ayhan, Z.; Kocaoğlu, G.; Yaman, A.; Bajin, M.S.; Saatci, A.O. Single intravitreal aflibercept injection for unilateral acute nonarteritic ischemic optic neuropathy. *Case Rep. Ophthalmol. Med.* **2015**, *2015*, 783241. [CrossRef]
85. Cheng, K.C.; Chiu, C.C.; Chen, K.J.; Chang, Y.C. Intravitreal Aflibercept for Patients with Acute Nonarteritic Anterior Ischemic Optic Neuropathy: A Retrospective Trial. *J. Clin. Med.* **2023**, *12*, 4868. [CrossRef]
86. Lyttle, D.P.; Johnson, L.N.; Margolin, E.A.; Madsen, R.W. Levodopa as a possible treatment of visual loss in nonarteritic anterior ischemic optic neuropathy. *Graefe's Arch. Clin. Exp. Ophthalmol.* **2016**, *254*, 757–764. [CrossRef]
87. Johnson, L.N.; Guy, M.E.; Krohel, G.B.; Madsen, R.W. Levodopa may improve vision loss in recent-onset, nonarteritic anterior ischemic optic neuropathy. *Ophthalmology* **2000**, *107*, 521–526. [CrossRef]
88. Johnson, L.N.; Gould, T.J.; Krohel, G.B. Effect of Levodopa and Carbidopa on Recovery of Visual Function in Patients with Nonarteritic Anterior Ischemic Optic Neuropathy of Longer Than Six Months' Duration. *Am. J. Ophthalmol.* **1996**, *121*, 77–83. [CrossRef] [PubMed]
89. Simsek, T.; Eryilmaz, T.; Acaroglu, G. Efficacy of levodopa and carbidopa on visual function in patients with non-arteritic anterior ischaemic optic neuropathy. *Int. J. Clin. Pract.* **2005**, *59*, 287–290. [CrossRef] [PubMed]
90. Bialer, O.Y.; Stiebel-Kalish, H. Evaluation and management of nonarteritic anterior ischemic optic neuropathy: A national survey. *Graefe's Arch. Clin. Exp. Ophthalmol.* **2024**, *Online ahead of print*. [CrossRef]
91. Jin, X.; Fu, J.; Lv, R.; Hao, X.; Wang, S.; Sun, M.; Xu, G.; Zhang, Q.; Zhang, L.; Li, Y.; et al. Efficacy and safety of platelet-rich plasma for acute nonarteritic anterior ischemic optic neuropathy: A prospective cohort study. *Front. Med.* **2024**, *11*, 1344107. [CrossRef] [PubMed]
92. Fernández-Vega, B.; Nicieza, J.; Álvarez-Barrios, A.; Álvarez, L.; García, M.; Fernández-Vega, C.; Vega, J.A.; González-Iglesias, H. The Use of Vitamins and Coenzyme Q10 for the Treatment of Vascular Occlusion Diseases Affecting the Retina. *Nutrients* **2020**, *12*, 723. [CrossRef] [PubMed]
93. Kumar, V.; Mesentier-Louro, L.A.; Oh, A.J.; Heng, K.; Shariati, M.A.; Huang, H.; Hu, Y.; Liao, Y.J. Increased ER Stress after Experimental Ischemic Optic Neuropathy and Improved RGC and Oligodendrocyte Survival after Treatment with Chemical Chaperon. *Investig. Opthalmol. Vis. Sci.* **2019**, *60*, 1953–1966. [CrossRef]
94. Chiquet, C.; Vignal, C.; Gohier, P.; Heron, E.; Thuret, G.; Rougier, M.B.; Lehmann, A.; Flet, L.; Quesada, J.L.; Roustit, M.; et al. Treatment of nonarteritic anterior ischemic optic neuropathy with an endothelin antagonist: ENDOTHELION (ENDOTHELin antagonist receptor in Ischemic Optic Neuropathy)—A multicentre andomized controlled trial protocol. *Trials* **2022**, *23*, 916. [CrossRef]
95. Ghasemi Fard, S.; Wang, F.; Sinclair, A.J.; Elliott, G.; Turchini, G.M. How does high DHA fish oil affect health? A systematic review of evidence. *Crit. Rev. Food Sci. Nutr.* **2019**, *59*, 1684–1727. [CrossRef]
96. Chen, X.; Chen, C.; Fan, S.; Wu, S.; Yang, F.; Fang, Z.; Fu, H.; Li, Y. Omega-3 polyunsaturated fatty acid attenuates the inflammatory response by modulating microglia polarization through SIRT1-mediated deacetylation of the HMGB1/NF-κB pathway following experimental traumatic brain injury. *J. Neuroinflamm.* **2018**, *15*, 116. [CrossRef]
97. Georgiou, T.; Wen, Y.T.; Chang, C.H.; Kolovos, P.; Kalogerou, M.; Prokopiou, E.; Neokleous, A.; Huang, C.T.; Tsai, R.K. Neuroprotective Effects of Omega-3 Polyunsaturated Fatty Acids in a Rat Model of Anterior Ischemic Optic Neuropathy. *Investig. Opthalmol. Vis. Sci.* **2017**, *58*, 1603. [CrossRef]
98. Zhang, Y.P.; Brown, R.E.; Zhang, P.C.; Zhao, Y.T.; Ju, X.H.; Song, C. DHA, EPA and their combination at various ratios differently modulated Aβ25-35-induced neurotoxicity in SH-SY5Y cells. *Prostaglandins Leukot. Essent. Fat. Acids* **2018**, *136*, 85–94. [CrossRef] [PubMed]
99. Huang, T.L.; Wen, Y.T.; Ho, Y.C.; Wang, J.K.; Lin, K.H.; Tsai, R.K. Algae Oil Treatment Protects Retinal Ganglion Cells (RGCs) via ERK Signaling Pathway in Experimental Optic Nerve Ischemia. *Mar. Drugs* **2020**, *18*, 83. [CrossRef] [PubMed]
100. Kapupara, K.; Wen, Y.T.; Tsai, R.K.; Huang, S.P. Soluble P-selectin promotes retinal ganglion cell survival through activation of Nrf2 signaling after ischemia injury. *Cell Death Dis.* **2017**, *8*, e3172. [CrossRef]
101. Johnson, M.A.; Mehrabian, Z.; Guo, Y.; Ghosh, J.; Brigell, M.G.; Bernstein, S.L. Anti-NOGO Antibody Neuroprotection in a Rat Model of NAION. *Transl. Vis. Sci. Technol.* **2021**, *10*, 12. [CrossRef] [PubMed]
102. Mesentier-Louro, L.A.; Stell, L.; Yan, Y.; Montague, A.A.; de Jesus Perez, V.; Liao, Y.J. Immunoprofiling of Nonarteritic Anterior Ischemic Optic Neuropathy. *Transl. Vis. Sci. Technol.* **2021**, *10*, 17. [CrossRef] [PubMed]

103. Guo, Y.; Johnson, M.A.; Mehrabian, Z.; Mishra, M.K.; Kannan, R.; Miller, N.R.; Bernstein, S.L. Dendrimers Target the Ischemic Lesion in Rodent and Primate Models of Nonarteritic Anterior Ischemic Optic Neuropathy. *PLoS ONE* **2016**, *11*, e0154437. [CrossRef]
104. Hayreh, S.S. Posterior ischaemic optic neuropathy: Clinical features, pathogenesis, and management. *Eye* **2004**, *18*, 1188–1206. [CrossRef]
105. Brewer, R.; Sadun, A.A. Posterior ischemic optic neuropathy: Perioperative risk factors. *Taiwan J. Ophthalmol.* **2020**, *10*, 167–173. [CrossRef] [PubMed] [PubMed Central]
106. Isayama, Y.; Takahashi, T.; Inoue, M.; Jimura, T. Posterior Ischemic Optic Neuropathy. *Ophthalmologica* **1983**, *187*, 141–147. [CrossRef]
107. Inoue, M. Vascular Optic Neuropathy in Diabetes Mellitus. *Jpn. J. Ophthalmol.* **1997**, *41*, 328–331. [CrossRef]
108. Yonezawa, Y.; Koga, K.; Higashi, Y.; Hasebe, M.; Fukushima, C.; Omiya, C.; Nishioka, K.; Yahata, K. A Successfully Treated Case of Posterior Ischemic Optic Neuropathy That Developed during Antihypertensive Therapy for Hypertensive Emergency. *Intern. Med.* **2024**, *63*, 527–532. [CrossRef] [PubMed]
109. Buono, L.M.; Foroozan, R.; Savino, P.J.; Danesh-Meyer, H.v.; Stanescu, D. Posterior ischemic optic neuropathy after hemodialysis. *Ophthalmology* **2003**, *110*, 1216–1218. [CrossRef] [PubMed]
110. Paez-Escamilla, M.; Abo-Zed, A.; Abramovitz, B.; Stefko, S.T.; Waxman, E. Recovery of vision after treatment of hemodialysis related bilateral optic nerve ischemia. *Am. J. Ophthalmol. Case Rep.* **2022**, *25*, 101373. [CrossRef] [PubMed]
111. Gerber, C.J.; Neil-Dwyer, G.; Kennedy, P. Posterior ischaemic optic neuropathy after a spontaneous extradural haematoma. *J. Neurol. Neurosurg. Psychiatry* **1992**, *55*, 630. [CrossRef]
112. Steigerwalt, R.D.; Cesarone, M.R.; Belcaro, G.; de Angelis, M.; Pascarella, A.; Nebbioso, M. Non-arteritic Posterior Ischaemic Optic Neuropathy Treated with Intravenous Prostaglandin E1 and Oral Corticosteroids. *Neuro-Ophthalmol.* **2011**, *35*, 81–84. [CrossRef]
113. Sperber, J.; Owolo, E.; Zachem, T.J.; Bishop, B.; Johnson, E.; Lad, E.M.; Goodwin, C.R. Perioperative Blindness in Spine Surgery: A Scoping Literature Review. *J. Clin. Med.* **2024**, *13*, 1051. [CrossRef]
114. Hayreh, S.S. Ischemic optic neuropathies—Where are we now? *Graefe's Arch. Clin. Exp. Ophthalmol.* **2013**, *251*, 1873–1884. [CrossRef]
115. Srinivasan, S.; Moorthy, S.; Sreekumar, K.; Kulkarni, C. Diffusion-weighted MRI in acute posterior ischemic optic neuropathy. *Indian J. Radiol. Imaging* **2012**, *22*, 106–107. [CrossRef]
116. Yang, T.H.; Lin, M.C. Using diffuse weighted image and apparent diffusion coefficient in MRI for diagnosis of posterior ischemic optic neuropathy in a young male: A case report and literature review. *BMC Ophthalmol.* **2022**, *22*, 168. [CrossRef]
117. Newman, N.J. Perioperative Visual Loss After Nonocular Surgeries. *Am. J. Ophthalmol.* **2008**, *145*, 604–610.e1. [CrossRef]
118. Postoperative Visual Loss Study Group. Risk Factors Associated with Ischemic Optic Neuropathy after Spinal Fusion Surgery. *Anesthesiology* **2012**, *116*, 15–24. [CrossRef] [PubMed]
119. Ibrahim, T.F.; Sweis, R.T.; Nockels, R.P. Reversible postoperative blindness caused by bilateral status epilepticus amauroticus following thoracolumbar deformity correction: Case report. *J. Neurosurg. Spine* **2017**, *27*, 63–67. [CrossRef] [PubMed]
120. Shifa, J.; Abebe, W.; Bekele, N.; Habte, D. A case of bilateral visual loss after spinal cord surgery. *Pan Afr. Med. J.* **2016**, *23*, 119. [CrossRef] [PubMed]
121. der Kelen, L.; Mommaerts, M. Visual loss after cosmetic blepharoplasty using local anaesthesia containing epinephrine—A case series. *Ann. Maxillofac. Surg.* **2021**, *11*, 340. [CrossRef]
122. Allashem, H.M.; Sward, D.G.; Sethuraman, K.; Matthews, M.K. Hyperbaric oxygen therapy for perioperative posterior ischemic optic neuropathy: A case report. *Undersea Hyperb. Med.* **2019**, *46*, 701–707. [CrossRef]
123. Wang, Y.; Brown, D.P.; Watson, B.D.; Goldberg, J.L. Rat Model of Photochemically-Induced Posterior Ischemic Optic Neuropathy. *J. Vis. Exp.* **2015**, *105*, 52402.

pharmaceuticals

MDPI

Review

Advances in Neuroprotection in Glaucoma: Pharmacological Strategies and Emerging Technologies

Li-Hsin Wang [1], Chun-Hao Huang [2] and I-Chan Lin [2,3,*]

1 School of Medicine, College of Medicine, Taipei Medical University, Taipei 110301, Taiwan; b101107053@tmu.edu.tw
2 Department of Ophthalmology, Wan Fang Hospital, Taipei Medical University, Taipei 110301, Taiwan; 110072@w.tmu.edu.tw
3 Department of Ophthalmology, School of Medicine, College of Medicine, Taipei Medical University, Taipei 11031, Taiwan
* Correspondence: 111122@w.tmu.edu.tw

Abstract: Glaucoma is a major global health concern and the leading cause of irreversible blindness worldwide, characterized by the progressive degeneration of retinal ganglion cells (RGCs) and their axons. This review focuses on the need for neuroprotective strategies in glaucoma management, addressing the limitations of current treatments that primarily target intraocular pressure (IOP) reduction. Despite effective IOP management, many patients continue to experience RGC degeneration, leading to irreversible blindness. This review provides an overview of both pharmacological interventions and emerging technologies aimed at directly protecting RGCs and the optic nerve, independent of IOP reduction. Pharmacological agents such as brimonidine, neurotrophic factors, memantine, Ginkgo biloba extract, citicoline, nicotinamide, insulin, and resveratrol show promise in preclinical and early clinical studies for their neuroprotective properties. Emerging technologies, including stem cell therapy, gene therapy, mitochondrial-targeted therapies, and nanotechnologies, offer innovative approaches for neuroprotection and regeneration of damaged RGCs. While these interventions hold significant potential, further research and clinical trials are necessary to confirm their efficacy and establish their role in clinical practice. This review highlights the multifaceted nature of neuroprotection in glaucoma, aiming to guide future research and clinical practice toward more effective management of glaucoma-induced neurodegeneration.

Keywords: neuroprotection; glaucoma; retinal neurodegeneration

1. Introduction

Glaucoma is a major global health concern and is recognized as one of the primary causes of irreversible blindness worldwide. The condition is typified by the gradual deterioration of retinal ganglion cells (RGCs) and their axons, resulting in the loss of the visual field and ultimately irreversible blindness. With over 70 million individuals affected globally, the impact of glaucoma is substantial and is predicted to increase to 111.8 million affected individuals by 2040 [1].

While current treatments primarily focus on lowering intraocular pressure (IOP), RGC loss can occur even with normal IOP levels, and a significant proportion of patients continue to experience vision loss despite effective IOP management. There is increasing evidence that other mechanisms, such as oxidative stress, excitotoxicity, vascular dysregulation, and neuroinflammation, also play a significant role in the pathogenesis of the disease [2–4]. This has led to a growing interest in neuroprotective strategies that aim to directly preserve and protect RGCs and the optic nerve from neurodegenerative processes, independent of reducing IOP, therefore slowing down the decline in visual function. By focusing on neuroprotection, researchers hope to develop therapies that can halt or even reverse the damage to the optic nerve, providing a significant advancement in the management of glaucoma.

This review aims to provide a comprehensive overview of the current state of neuroprotective strategies for glaucoma, focusing on both pharmacological treatments and emerging technologies. By examining a wide range of approaches, this review seeks to highlight the multifaceted nature of neuroprotection. This holistic perspective is essential for understanding the potential benefits and limitations of each strategy, guiding future research and clinical practice towards a more effective management of glaucoma-induced neurodegeneration.

2. Mechanisms of Loss of Retinal Functions in Glaucoma

The key mechanisms underlying RGC death in glaucoma involve a complex interplay of various pathological processes, including neurotrophin deprivation, excitotoxicity, oxidative stress, mitochondrial dysfunction, inflammation, and apoptosis (Figure 1).

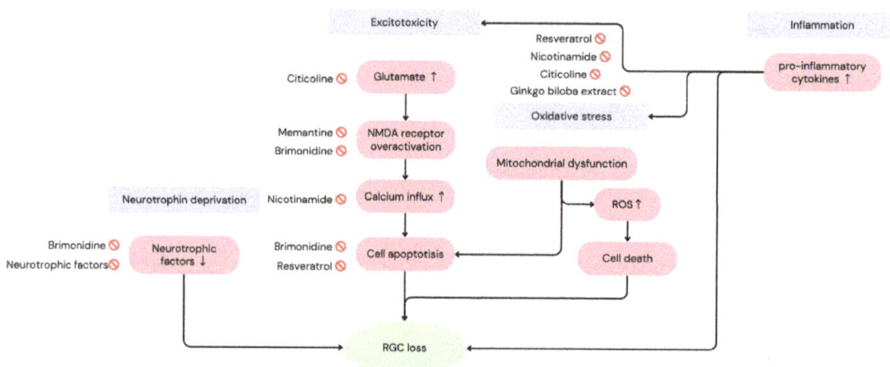

Figure 1. The mechanisms of glaucoma inducing the loss of retinal ganglion cells and the targets of each neuroprotective compound [5].

Neurotrophins, such as ciliary neurotrophic factor (CNTF) and brain-derived neurotrophic factor (BDNF), are crucial in the survival and maintenance of RGCs. In glaucoma, the deprivation of these neurotrophins due to impaired axonal transport and receptor expression leads to RGC degeneration [6].

Excitotoxicity, primarily driven by elevated levels of the neurotransmitter glutamate, leads to the overactivation of N-methyl-D-aspartate (NMDA) receptors on RGCs, causing a harmful influx of calcium ions that triggers apoptotic pathways. While the role of glutamate in glaucoma is complicated and not entirely understood, it is clear that its dysregulation contributes significantly to RGC death [2].

Oxidative stress is another critical factor in glaucomatous neurodegeneration. An imbalance between reactive oxygen species (ROS) production and the body's antioxidant defenses leads to oxidative damage in retinal ganglion cells (RGCs), affecting proteins, lipids, and DNA. This oxidative stress can be exacerbated by mitochondrial dysfunction, as RGCs have high energy demands, and any impairment in mitochondrial function can lead to energy deficits, further compromising cell viability. The release of cytochrome c from damaged mitochondria into the cytoplasm is a key event that activates the apoptotic machinery in RGCs [3].

Inflammation also plays a significant role in glaucoma, with pro-inflammatory cytokines like tumor necrosis factor-alpha (TNF-α) being upregulated in the retina and optic nerve head. These cytokines can induce RGC apoptosis either directly or by exacerbating excitotoxicity and oxidative stress. Additionally, the glial cells in the retina, including astrocytes, Müller cells, and microglia, contribute to both protective and detrimental responses in glaucoma. While they can help clear debris and produce neurotrophic factors, their overactivation can lead to chronic inflammation and further RGC damage [4].

Apoptosis, or programmed cell death, is the final common pathway through which many of these mechanisms converge to cause RGC loss in glaucoma. Both intrinsic and extrinsic apoptotic pathways are activated in response to the various stressors affecting RGCs. Key molecules in these pathways include the Bcl-2 family of proteins, which control mitochondrial membrane permeability, and caspases, which are the executioners of apoptosis [7].

3. Pharmacological Interventions (Summary in Table 1)

3.1. Brimonidine

Brimonidine, an alpha-2 adrenergic agonist, widely used to reduce IOP, has demonstrated notable neuroprotective properties in the treatment of glaucoma. While its IOP-lowering effects are well documented, emerging evidence highlights its potential to protect RGCs through various mechanisms independent of IOP reduction.

Table 1. Summary of neuroprotective compounds for glaucoma with clinical trials.

Active Compound	Mechanism of Action	Route of Administration	Status in Clinical Studies	Clinical Trials
Brimonidine	Binds alpha-2 adrenergic receptors, regulates apoptotic proteins, upregulates neurotrophic factors, reduces NMDA [1] receptor excitotoxicity	Topical	FDA-approved for IOP [2] reduction, neuroprotective effects under investigation	Krupin et al., 2011 [8]; De Moraes et al., 2012 [9]
Neurotrophic Factors (BDNF [3], CNTF [4], NGF [5])	Binds to specific receptors (TrkB and TrkA), promotes survival of neurons, enhances axon regeneration, modulates apoptosis	Implants (CTNF), topical (rhNGF [6])	Ongoing Phase II trials (CTNF), Phase Ib trial showed safety (rhNGF)	Goldberg et al., 2023 [10]; Beykin et al., 2022 [11]
Memantine	Non-competitive NMDA receptor antagonist, reduces calcium influx, protects neurons from glutamate-induced excitotoxicity	Oral	Failed large-scale clinical trials	Weinreb et al., 2018 [12]
GBE [7]	Scavenges ROS, reduces oxidative stress, stabilizes mitochondria, antagonizes PAF	Oral	Mixed results in clinical trials	Quaranta et al., 2003 [13]; Lee et al., 2013 [14]; Guo et al., 2014 [15]
Citicoline	Serves as precursor for phospholipids, enhances neurotransmitter synthesis	Oral, topical	Ongoing large-scale clinical trials	NCT05315206 NCT05710198
Nicotinamide	Replenishes NAD levels, supports mitochondrial function, regulates calcium homeostasis, reduces oxidative stress	Oral	Ongoing large-scale long-term trials	NCT05275738 NCT05405868

[1] N-methyl-D-aspartate, [2] intraocular pressure, [3] brain-derived neurotrophic factor, [4] ciliary neurotrophic factor, [5] nerve growth factor, [6] recombinant human nerve growth factor, [7] ginkgo biloba extract.

Brimonidine has a quinoxaline chemical structure with an imidazoline ring that allows it to bind to alpha-2 adrenergic receptors in the eye. Several preclinical studies have explored the neuroprotective mechanisms of brimonidine. It has been shown to enhance the survival of RGCs and protect RGCs from various types of optic nerve injuries, including NMDA-induced neurotoxicity, ischemia, optic neuritis, and ocular hypertension [16–23]. It achieves these protective effects by regulating anti- or pro-apoptotic proteins, such as Bcl-2/BxL or Bax, and upregulating neurotrophic factors like brain-derived neurotrophic factor (BDNF), fibroblast growth factor (FGF), and their receptors [24–26]. Brimonidine also alleviates neuronal death by altering NMDA receptors and reducing glutamate accumulation post-injury, thereby decreasing excitotoxicity [27–29]. Additionally, brimonidine interferes with the amyloid-β pathway, reducing amyloid-β levels, which are implicated in RGC death [30, 31]. Improvement in impaired vascular autoregulation, protection of the retina from ischemic injury, and support of neuronal regeneration were also observed [16,19,20,22–24,32–34]. These effects suggest that brimonidine's neuroprotection is not solely reliant on its ability to lower IOP. Recent studies have also shown additive protective effects when combining brimonidine and ripasudil. This combined treatment modulates multiple signaling pathways, including the suppression of proinflammatory mediators and the increase in trophic factors, leading to enhanced RGC survival compared to single-agent treatments [35,36].

Clinical studies have provided mixed but promising results regarding the neuroprotective effects of brimonidine. The Low-pressure Glaucoma Treatment Study (LoGTS) compared the effects of brimonidine and timolol in patients with low-pressure glaucoma. Although both medications were similarly effective in lowering IOP, visual field deterioration was notably less frequent in the brimonidine group compared to the timolol group. However, caution is advised when interpreting the results due to the high dropout rate, especially among those taking brimonidine [8,9]. Topical brimonidine also enhanced contrast sensitivity regardless of its effects on IOP, whereas timolol showed no such improvement [37]. Another study suggests that brimonidine may preserve retinal nerve fiber layer (RNFL) independent of its IOP-lowering effects [38].

Despite these findings, the evidence for brimonidine's neuroprotective effects in clinical settings remains inconclusive. The high attrition rates and adverse effects reported in some studies raise concerns about the consistency and applicability of these results. Large double-blinded, randomized controlled trials are needed to definitively determine the role of brimonidine in the treatment of glaucoma [39].

3.2. Neurotrophic Factors

Neurotrophic factors (NTFs) are essential molecules involved in neuron development, survival, and repair. They bind to specific receptors, activating tyrosine kinase signaling pathways, which lead to various neuroprotective actions, such as the promotion of axon regeneration and enhancement of neuronal cell function [40–42].

Neurotrophic factors such as nerve growth factor (NGF), BDNF, CNTF, FGF-2, glial cell-line-derived neurotrophic factor (GDNF), neurturin (NRTN), and neuritin are particularly relevant to glaucoma. These factors have been shown to prevent RGC death in several rodent models of glaucoma [43–50].

BDNF plays a crucial role in the survival of retinal ganglion cells (RGCs) by preventing apoptosis through the activation of extracellular signal-regulated kinases (Erk1/2) and c-jun, while inhibiting caspase 2 via its receptor, TrkB [51,52]. Research has shown significantly reduced levels of BDNF in the serum and aqueous humor of patients with normal tension glaucoma (NTG) and primary open-angle glaucoma (POAG) [44,53,54]. BDNF has also been shown to improve RGC survival and prevent RGC death induced by amyloid-β-triggered apoptosis in rat models [40,55,56]. Nonetheless, further studies are needed to clarify the causal link between BDNF and glaucoma and to assess the effectiveness of BDNF supplementation as a neuroprotective treatment.

CNTF is produced locally by RGCs, and its levels are also found to be decreased in the aqueous humor and lacrimal fluid of patients with POAG [57]. CNTF exerts its effects through a heterotrimeric receptor complex that includes CNTF receptor α, gp130, and leukemia inhibitory factor receptor β [58]. It can protect damaged RGCs directly by binding to receptors on their surface, or indirectly by influencing glial cell activity via the JAK/STAT, PI3K/AKT, and MAPK signaling pathways [59–62]. Preclinical studies have shown a promising neuroprotective ability of CNTF with single intravitreal injection [63]. The combination of CNTF with other drugs such as cyclic adenosine monophosphate (cAMP) and Rho kinase inhibitor was also investigated, with benefits shown in axon regeneration [64–66]. The NT-501 device, a polymer capsule housing a genetically modified human cell line that releases CNTF, has been designed for long-term delivery. Ongoing clinical trials are evaluating the therapeutic effectiveness of NT-501-encapsulated cell therapy in the treatment of glaucoma. Phase I trials showed that NT-501 implants were safe, and ongoing Phase II trials aim to evaluate the efficacy of improving visual fields and retinal structure [10,67–69].

NGF promotes the survival and regeneration of retinal and brain cells [70]. It works by interacting with TrkA and p75 neurotrophin receptors (p75NTR). When NGF binds to TrkA, it activates the PI3K/Akt pathway, which boosts Bcl-2 levels and lowers Bax proteins, ultimately inhibiting apoptosis [71]. It also activates the RAS and PLC pathways, which are involved in regulating cell survival and differentiation [72]. Topical application of NGF preserved RGCs and their axons in a model of ocular hypertension [43,73]. It has been shown that in advanced glaucoma patients, topical treatment with NGF improved optic nerve functions, including visual acuity, visual field, and contrast sensitivity [43]. In preclinical studies, recombinant human NGF (rhNGF) enhanced neuroprotection, neuroregeneration, and axon growth were associated with reduced p75NTR and increased TrkA phosphorylation [71,74–76]. RhNGF has been evaluated in clinical trials, showing safety in healthy volunteers and some neuroprotective effects in glaucoma patients [11,77].

Despite promising preclinical results, translating these findings into clinical practice poses several challenges. The effective and long-lasting delivery of NTFs to the retina is a significant hurdle. Intravitreal injections, though effective, may not be practical for chronic conditions like glaucoma that require long-term treatment. Future research should focus on long-term studies and innovative delivery systems to fully harness the potential of neurotrophic factors in glaucoma therapy.

3.3. Memantine

Memantine is a voltage-dependent, non-competitive NMDA receptor antagonist with low affinity that selectively binds to activated glutamatergic receptors [78,79], reducing excessive glutamatergic activity, calcium influx, and the activation of pro-apoptotic cascades without affecting normal neurotransmission [80,81], thereby protecting RGCs from excitotoxic damage. Its rigid tricyclic amine structure (1-amino-3,5-dimethyladamantane) allows it to effectively block the NMDA receptor channel by interacting with critical asparagine residues (N-sites) like N616 in the NR1 and NR2 subunits. This binding inhibits excessive calcium influx while preserving normal receptor activity. Key amino acids involved in memantine's channel blockade include N616, A645, and Y647 in NR1, and N615 in NR2B.

In animal models, memantine has demonstrated promising neuroprotective effects. Intraperitoneal, subcutaneous, and oral administrations of memantine in monkeys and rats improve RGC survival, prevent optic nerve fiber loss, and maintain retinal and visual pathway function without adverse effects on normal eyes [80,82,83]. However, the role of elevated glutamate in glaucoma remains debated and inconsistent [84–86].

Although preclinical studies have shown promising results, large-scale clinical trials have not been able to confirm significant benefits of memantine in the treatment of glaucoma. Two double-masked, placebo-controlled, multicenter Phase III clinical trials (NCT00141882 and NCT00168350) involving 2296 patients with POAG over four years did not show that memantine delayed visual field progression compared to the placebo [12].

To better evaluate memantine's potential in glaucoma treatment, future studies should consider including patients with early-stage glaucoma and a less heterogeneous population regarding progression risk factors. Longer study durations, higher doses, and alternative methods of medication delivery should be explored. Additionally, using more advanced imaging techniques could provide more objective structural parameters for assessing early glaucoma progression [12,87].

3.4. Ginkgo Biloba Extract

Ginkgo biloba has been used in traditional Chinese medicine for hundreds of years. Modern research has explored the neuroprotective properties of Ginkgo biloba extract (GBE), particularly in the context of neurodegenerative diseases like glaucoma [88,89].

GBE contains active compounds such as flavonoid glycosides and terpene lactones, including ginkgolides A, B, C, and bilobalide, which contribute to its neuroprotective effects. The flavonoid glycosides, with their polyphenolic structure with two aromatic rings and glycosidic attachments, exhibit antioxidant properties by scavenging ROS and reducing oxidative stress. Terpene lactones, particularly ginkgolide B with its cage-like structure, antagonize the platelet activator factor (PAF), reducing trauma-induced apoptosis and enhancing recovery from ischemic injury. Bilobalide, another key terpene with its trilactone rings, stabilizes mitochondria, further mitigating ROS production. GBE's ability to reduce lipid peroxidation helps protect cellular membranes and enhance erythrocyte deformability and blood perfusion. Additionally, both flavonoids and terpenoids in GBE contribute to vasodilation, supporting improved blood flow in tissues, and further reinforcing its neuroprotective properties [90].

In preclinical studies, GBE has demonstrated significant neuroprotective effects, highlighting GBE's potential to enhance RGC survival and protect against glaucomatous damage [91–93]. Clinical trials investigating the effects of GBE on glaucoma patients have yielded mixed results. One study noted improvements in preexisting visual field defects in patients with NTG after one month of taking GBE oral capsules [13]. Another 4-year long-term study found that taking GBE significantly slowed the progression of visual field defects in NTG patients without affecting IOP [14]. However, a randomized controlled trial (RCT) showed no effect on visual field performance or contrast sensitivity in NTG patients of the same GBE dosage and duration as the first study [15], highlighting the variability in clinical outcomes. Numerous studies have shown increased blood flow velocity and decreased vascular resistance in the retrobulbar and peripapillary vascular systems [94,95].

Despite promising preclinical findings, the clinical evidence for GBE's neuroprotective effects in glaucoma remains inconclusive [90]. The discrepancies in clinical trial results underscore the need for further research to identify the specific conditions and patient populations that may benefit from GBE treatment. Future studies should consider enrolling patients with early-stage glaucoma, controlling for hypotensive treatments, and utilizing advanced imaging techniques. In summary, GBE holds promise as a neuroprotective treatment for glaucoma, but larger, well-designed clinical trials are needed to validate its effectiveness and determine its place in clinical practice.

3.5. Citicoline

Citicoline, also known as cytidine 5′-diphosphocholine, a naturally occurring compound, has garnered attention for its potential neuroprotective effects in the context of glaucoma.

The chemical structure of citicoline consists of cytidine linked to a diphosphate bridge and choline, allowing it to serve as a precursor for the synthesis of important phospholipids such as phosphatidylcholine, phosphatidylethanolamine, sphingomyelin, and cardiolipin, which are essential components of neuronal cell membranes [96]. These phospholipids are integral for the formation of synaptic membranes, enhancing the release and recycling of neurotransmitters like acetylcholine, dopamine, serotonin, and norepinephrine, which are crucial for normal neuronal function [97]. The choline component also facilitates the synthesis of acetylcholine. While stabilizing neuronal membranes and enhancing important neurotransmitter synthesis, citicoline reduces glutamate excitotoxicity, improves axonal transport deficits, and lowers oxidative stress by producing glutathione [98].

In animal models, citicoline has demonstrated significant neuroprotective effects. Studies have shown that the administration of citicoline can attenuate the decrease in RGC density and improve visual acuity and the integrity of the pre-chiasmatic white matter of the optic nerve without impacting IOP, indicating a neuroprotective effect [99].

Early studies demonstrated that intramuscular (IM) citicoline improved visual field performance in patients with POAG, with long-term benefits in visual field preservation [100]. Other studies have also shown improvements in visual evoked potentials (VEP) and pattern electroretinograms (PERG) with IM citicoline [101]. Studies evaluating the effects of oral citicoline reported similar benefits on retinocortical function [102,103].

Longitudinal studies have also shown that oral citicoline can slow the progression of glaucomatous damage, with significant improvements in RNFL and ganglion cell complex thickness [104]. The topical administration of citicoline eyedrops has also shown similar effects on retinocortical function, as measured by PERG and VEP [105–108], but with more adverse effects compared to oral formulation [109]. Other studies investigating the combination therapy of citicoline with drugs including docosahexaenoic acid, homotaurine, and vitamin E also showed significant visual improvement [110,111]. Citicoline-loaded liposomes were also investigated, showing promising results for improved bioavailability and sustained drug release in ocular applications [112].

The neuroprotective potential of citicoline in glaucoma has sparked several ongoing clinical trials to explore its therapeutic benefits. One such trial is a randomized controlled trial evaluating the effects of oral citicoline over a year in patients with OAG (NCT05315206). Furthermore, a large Phase III randomized controlled trial is being planned to determine the efficacy of citicoline eyedrops in slowing visual field deterioration and structural changes in glaucoma patients (NCT05710198).

Citicoline has become a potential neuroprotective treatment for glaucoma. Although initial preclinical and early clinical studies have shown promising outcomes, larger and more thorough trials are necessary to validate its effectiveness and define its place in clinical practice. The current research will offer important insights into the viability of citicoline as a standard therapeutic option for glaucoma patients.

3.6. Nicotinamide

Nicotinamide, also known as vitamin B3 or NAM, has a simple chemical structure consisting of a pyridine ring with an amide group at the 3-position. This structure allows it to serve as a precursor to nicotinamide adenine dinucleotide (NAD), a vital coenzyme in cellular metabolic processes. NAD is essential for mitochondrial function, as it supports oxidative phosphorylation and ATP production, which are critical for maintaining neuronal health and energy homeostasis. As cellular concentrations of NAD decline with age, mitochondrial dysfunction, inflammation, and oxidative stress worsen, leading to neurodegeneration [113–115]. Nicotinamide replenishes NAD levels, supporting mitochondrial function and cellular energy metabolism, while also influencing calcium homeostasis, vascular regulation, and neuronal health, making it a promising candidate for neuroprotection in glaucoma [116–118].

A study revealed that the reduced retinal NAD levels caused by aging in mice could be mitigated by high-dose oral nicotinamide [119,120]. Another study showed that oral nicotinamide inhibits RGC loss and nuclear shrinkage [121]. These studies showed that nicotinamide protected RGCs from glaucomatous damage by improving mitochondrial health and reducing oxidative stress. A recent study demonstrated that a combination of NADPH and NAC synergistically reduced apoptosis, axonal damage, and peroxidation in retinal ganglion cells while inhibiting gliosis and p38/MAPK activation in Müller cells, suggesting their potential as a neuroprotective and anti-inflammatory treatment for glaucoma [122]. Novel delivery methods were also investigated and demonstrated a positive result [123]. Similar findings were observed with genetic approaches to overexpress NAD-producing enzymes, such as nicotinamide mononucleotide adenylyl transferase 1 (NMNAT1), NMNAT2, and NMNAT3 [120,124,125]. On the other hand, epigallocatechin gallate (EGCG), a polyphenol found in green tea, promotes NAD production through a mechanism dependent on NMN and NMNAT2, making it a useful compound for developing new drugs [125].

Human studies have also supported the potential of nicotinamide in glaucoma treatment. A study reported significantly reduced plasma NAD levels in patients with primary POAG compared to controls, suggesting a systemic NAD deficiency in glaucoma [126]. This finding aligns with preclinical data and underscores the need for therapeutic strategies to restore NAD levels. A randomized crossover clinical trial (ACTRN12617000809336) showed that oral nicotinamide supplementation enhanced RGC function, as assessed by

ERG, regardless of IOP levels [127]. Additionally, a Phase II randomized clinical trial (NCT03797469) found short-term improvements in visual function in patients with moderate OAG who received a combination of nicotinamide and pyruvate [128].

Ongoing clinical trials are exploring the long-term effects of nicotinamide supplementation in glaucoma patients. The Glaucoma Nicotinamide Trial (TGNT) is investigating the neuroprotective benefits of nicotinamide in 660 OAG patients, with an anticipated completion date by the end of 2026 (NCT05275738). Additionally, a Phase III RCT is evaluating the effect of oral nicotinamide on visual field progression over 27 months, with results expected by 2026 (NCT05405868). Another study is examining the neuroprotective potential of nicotinamide riboside, a more bioavailable precursor, over 24 months in 125 POAG patients [129,130]. As research progresses, nicotinamide could become an integral part of glaucoma management, offering hope for improved patient outcomes.

3.7. Insulin

The mTOR pathway is crucial for RGC energy metabolism, with insulin serving as a primary activator of both mTORC1 and mTORC2 [131]. Insulin crosses the blood-brain/retinal barrier and impacts neuronal survival, neurotransmission, and glucose uptake. Impaired insulin signaling is associated with neurodegenerative diseases, including glaucoma [132]. The research indicates that insulin enhances glucose transport, promotes dendritic regeneration, and supports neuronal survival. The activation of mTORC1/2 is vital for insulin-mediated dendrite repair and synapse restoration in RGCs, positioning insulin as a promising therapeutic target [133].

Intranasal insulin has emerged as a promising method for bypassing the blood–brain barrier without causing systemic side effects like hypoglycemia [134]. Research on intranasal insulin in Alzheimer's and mild cognitive impairment shows that it is a safe and effective way to target insulin signaling in neurodegeneration [135]. Although clinical trials in glaucoma are yet to be conducted, preclinical evidence suggests that exogenous insulin may help preserve RGCs [132]. A Phase 1 trial investigating topical insulin eye drops for glaucoma was conducted and showed safety in patients with glaucoma [136]. Overall, intranasal and topical insulin presents a promising treatment strategy for neurodegenerative diseases, though further research is required to validate its efficacy.

3.8. Resveratrol

Resveratrol (RES), a polyphenolic compound with antioxidant properties, has shown potential in slowing glaucoma progression by supporting RGC health [137]. RES stimulates cell growth, reduces apoptosis, and alleviates oxidative stress, especially in RGCs exposed to hydrogen peroxide [138].

Additionally, RES protects RGC axons by inhibiting JNK protein phosphorylation through Sirt1 activation and prevents oxidative stress-related damage by suppressing the MAPK pathway [139]. RES preserves retinal function by modulating hypoxia-inducible factor-1 alpha (HIF-1α), vascular endothelial growth factor (VEGF), and the p38/p53 pathways, while concurrently activating the PI3K/Akt pathway to promote cell survival [140]. These mechanisms demonstrate RES's therapeutic potential in protecting RGCs from degeneration and managing glaucoma progression [141]. Intravitreal administration of RES successfully protected RGCs from high IOP-induced cell death, though further research is needed to fully evaluate its efficacy [142].

3.9. Other Novel Drugs

Recent advancements in neuroprotective therapies for glaucoma have highlighted several promising compounds and mechanisms.

Dopamine (DA), produced by dopaminergic amacrine cells (DACs), plays key roles in light adaptation and circadian rhythms [143]. Increasing DA release from DACs, along with overexpressing the Drd1 receptor in all RGCs, significantly improves RGC survival, encourages axon regeneration after optic nerve damage, and protects vision in glaucoma models [144]. Future research using high-throughput single-cell sequencing will be important for understanding the cellular pathways activated by DRD1 [145].

Rho kinase (ROCK) inhibitors, such as ripasudil and fasudil, have shown promising neuroprotective effects in the treatment of glaucoma. These inhibitors not only reduce IOP by enhancing aqueous humor outflow through the trabecular meshwork but also exhibit neuroprotective properties by improving blood flow [146] to the optic nerve and promoting the survival of RGCs [147].

Omidenepag (OMD), an E prostanoid receptor 2 (EP2) agonist, demonstrated neuroprotective effects by suppressing excitotoxic RGC death and modulating glia–neuron interactions via cAMP signaling pathways [148]. The bioprecursor prodrug 10β,17β-dihydroxyestra-1,4-dien-3-one (DHED) selectively generates 17β-estradiol (E2) in the retina following topical application, effectively protecting RGCs and their axons in a male rat glaucoma model without raising systemic E2 levels [149]. The P2X7 receptor antagonist A740003 provided neuroprotection by inhibiting microglial activation, reducing retinal inflammation, and enhancing RGC survival in a chronic intraocular hypertension model [150].

Irisin, an exercise-induced myokine, showed neuroprotective properties by attenuating neuroinflammation and promoting autophagy through integrin αVβ5/AMPK signaling in an acute ocular hypertension model [151]. The novel small molecule H7E, an HDAC8 inhibitor, protected against glaucoma damage by inhibiting Müller glial activation and preventing retinal cell death from oxidative stress [152]. Finally, the Gramine derivative ITH12657 demonstrated significant neuroprotection against excitotoxicity-induced RGC death in a rat model, with specific protection observed in Brn3a+ RGCs and α-ONs-RGCs [153].

Collectively, these studies underscore the potential of diverse neuroprotective strategies in mitigating glaucomatous damage and preserving vision.

4. Emerging Technologies

4.1. Stem Cell Therapy

Stem cell therapy is a promising new approach for treating neurodegenerative diseases like glaucoma. It provides a dual therapeutic benefit by regenerating RGCs and differentiating them into new cell types, while also creating a neurotrophic environment that supports damaged RGCs [154,155].

Mesenchymal stem cells (MSCs), which are multipotent and can differentiate into various cell types such as neurons and glial cells, offer neuroprotective and regenerative effects. They promote neuronal growth, regulate inflammation and immune responses, stimulate angiogenesis, and help reduce demyelination and apoptosis [156–158]. MSCs also secrete platelet-derived growth factor (PDGF) and neurotrophic factors (NTFs) like CNTF, FGF-2, GDNF, neuritin, and BDNF, which enhance cell survival and foster the development of other cells [155,159–162].

Animal models and preclinical studies have shown that MSCs are effective in promoting RGC survival, reducing RGC loss, boosting growth factor expression, enhancing anti-inflammatory properties, and protecting trabecular meshwork tissue when administered intravitreally or intracamerally [160,163–169]. However, despite these encouraging results in animal models, clinical trials have encountered difficulties. For example, one trial found no significant improvement in visual performance after the intravitreal injection of autologous bone marrow-derived MSCs in a patient with advanced glaucoma. Additionally, another participant in the study developed retinal detachment with proliferative vitreoretinopathy, raising safety concerns [170]. Currently, the Intravitreal Mesenchymal Stem Cell Transplantation in Advanced Glaucoma Study (NCT02330978), a Phase I trial, is assessing the safety of intravitreal MSC injections in advanced glaucoma patients. Other trials, such as the Stem Cell Ophthalmology Treatment Study and its follow-up (NCT01920867, NCT03011541), are examining the variability in patient outcomes based on different MSC delivery techniques.

Human embryonic stem cells (hESCs) are pluripotent and can differentiate into any cell type, making them a potential source for RGCs [171]. Experimental studies have developed protocols for differentiating hESCs into RGCs, which have shown successful integration and the mediation of light responses in preclinical studies involving rat and monkey eyes [172–174]. However, the use of hESCs is ethically controversial and scientifically challenging. Other types of stem cells, such as oligodendrocyte precursor cells (OPCs), human neuronal progenitor cells, and retinal stem cells, have also demonstrated potential in protecting RGCs in animal models [175,176]. Mouse-induced pluripotent stem cell (miPSC)/mouse embryonic stem cell (mESC)-derived RGC and spermatogonial stem cell-derived RGC were also investigated [177,178]. Adipose tissue-derived regenerative cells, with their multipotent properties, are being evaluated in clinical trials (NCT02144103) for their safety and efficacy in glaucoma treatment [179].

Despite the potential neuroprotective abilities of stem cells, safety issues remain a significant concern. It is important to carefully evaluate the balance between graft survival and the risk of tumor development, as prolonged stem cell survival raises the likelihood of tumor formation. Moreover, the implanted cells might secrete harmful agents, affecting the microenvironment of RGCs [180]. Adverse effects, including reactive gliosis, vitreous clumping, extensive inflammation, and epiretinal membrane thickening are also concerns [181].

Despite the challenges, stem cell therapy holds significant potential for glaucoma treatment. Ongoing research aims to improve the safety and effectiveness of these treatments. Future investigations should prioritize ensuring the survival and function of transplanted cells, supporting their integration into retinal and brain networks, and maintaining long-term safety. Continued research and larger clinical trials are essential to fully realize the potential of stem cell therapy in glaucoma treatment, paving the way for its integration into clinical practice.

4.2. Gene Therapy

Gene therapy has become a promising strategy for the neuroprotection and treatment of glaucoma. Although currently restricted in clinical use, experimental studies and preclinical research have shown significant potential for gene therapy in managing and potentially reversing glaucomatous damage.

Gene therapy strategies for glaucoma focus on modifying specific genes associated with the disease's pathogenesis and delivering neuroprotective factors to preserve RGCs. Studies have demonstrated that clustered regularly interspaced short-palindromic repeat (CRISPR)-mediated genome editing of the myocilin (MYOC) gene effectively lowers IOP and inhibits glaucomatous damage in mouse models [182]. Another genetic target under investigation is the tunica interna endothelial cell kinase (TEK), which plays a role in Schlemm's canal development [183,184]. Additionally, genetic constructs designed to overexpress neuroprotective factors such as BDNF and its receptor TrkB have shown significant neuroprotective effects, enhancing RGC survival and preserving visual function in rodent models. This effect may prove to be a feasible therapeutic strategy in the planned Phase I/II trials [185–189]. In addition to BDNF, other neurotrophic factors such as CNTF have been encoded into genetic constructs to support RGC survival [45,190–192].

Other gene targets of different functions have also been investigated [193]. A study reactivating calcium/calmodulin-stimulated protein kinase II (CaMKII) activity in diseased mice resulted in the protection of RGCs and the preservation of visual function [156]. The overexpression of the complement C3 inhibitor shows a neuroprotective effect of RGC axons and somata [194]. The transduction of VEGF variants by VEGFR2 and PI3K signaling promotes synaptogenesis and increases the length of neurites and axons [195]. Combining AAV-γ-synuclein (mScng) promotor with CRISPR/Cas9 gene editing knocks down pro-degenerative genes, preserving the acutely injured RGC somata and axons [196]. The overexpression of Bcl-XL attenuates both RGC soma pathology and axonal degeneration in the optic nerve [197]. The overexpression of NMNAT1 as well as NMNAT2 restores the decreased NAD levels, showing a significant neuroprotective effect of RGC soma, axons, and the maintenance of visual function [120,198]. The overexpression of myc-associated protein X (MAX) prevents RGC death and protects optic nerve axons [199]. The overexpression of the X-linked inhibitor of apoptosis (XIAP) blocks the activation of apoptosis, providing both functional and structural protection of RGCs [200]. The overexpression of SOD2 reduced malonaldehyde, protecting RGCs from oxidative stress [201]. The overexpression of the adaptor molecule Protrudin significantly enhances central nervous system regeneration by promoting the accumulation of essential growth molecules and organelles in distal axons, both in vitro and in vivo. This regeneration is facilitated by Protrudin's ability to link axonal organelles, motors, and membranes, highlighting its potential as a therapeutic target for axonal injury [202]. NFATc4 knockout in mice enhances RGC survival and delays axonal degeneration following optic nerve injury by suppressing pro-apoptotic signaling, highlighting NFATc4 as a potential therapeutic target for optic neuropathies [203]. Tau overexpression exacerbates retinal degeneration, while Tau silencing offers significant protection, highlighting the critical role of Tau in retinal integrity and its potential as a therapeutic target in glaucoma [204].

Despite promising results in preclinical models, gene therapy for glaucoma faces several challenges. The multifactorial and polygenic nature of glaucoma complicates the identification of effective genetic targets. Additionally, issues related to gene transfer efficiency, site-specific binding, and the potential for mutagenesis need to be addressed. Whole-genome sequencing and advancements in genome editing technology hold the potential to identify new therapeutic targets and refine existing gene therapy approaches. The development of more efficient viral vectors and delivery methods is crucial for translating preclinical successes into clinical practice. With continued advancements in genetic research and therapy, gene therapy may lead to breakthroughs in preserving vision and improving the quality of life for glaucoma patients.

4.3. Mitochondrial-Targeted Therapies and Transplantation

Mitochondria are essential for bioenergetics and play key roles in calcium regulation, cell signaling, apoptosis, and synaptic support. RGCs, with their large dendritic trees and unmyelinated axons in the retina, rely heavily on mitochondria to meet their high metabolic demands. As a result, these cells are highly susceptible to mitochondrial dysfunction, as seen in disorders like Leber hereditary optic neuropathy (LHON) and autosomal dominant optic atrophy [205] In glaucoma models, mitochondrial abnormalities appear before optic nerve degeneration, suggesting that metabolic stress plays a key role in RGC injury and degeneration [206].

Various neuroprotective, mitochondria-targeted therapies for glaucoma have been proposed, including dietary modifications, antioxidant supplementation, stem cell therapy, gene therapy, and mitochondrial transplantation [207]. Citicoline supports mitochondrial function, potentially preventing cell damage, promoting the growth of axons and dendrites, and improving dopaminergic and cholinergic activity [208]. Nicotinamide, through its role in regulating metabolism, has shown significant protection against optic nerve degeneration in animal models of glaucoma [209]. Though clinical evidence for metabolic therapies in glaucoma is limited due to the disease's slow progression, small studies suggest neuroprotective benefits [128,210].

Mitochondrial transplantation, or mitotherapy, involves transferring functional mitochondria to cells with mitochondrial dysfunction, offering potential treatment for diseases linked to mitochondrial impairment [211]. Studies have shown promising results in animal models, including enhanced oxidative metabolism, neuroprotection, and axon regrowth in RGCs [212]. Human in vitro studies have also demonstrated mitotherapy's ability to restore cellular function in diseases like LHON [213]. The success of mitochondrial transplantation is influenced by factors like the source and delivery method of the mitochondria, both of which require further investigation.

4.4. Nanotechnologies

Exosomes and nanoparticles are emerging as significant tools in neuroprotective and regenerative strategies for glaucoma treatment. Exosomes, extracellular vesicles released by different cell types, range from 30 to 150 nm in diameter and carry a variety of biologically active substances such as proteins, mRNA, miRNAs, and lipids. They act as intracellular signaling organelles and are involved in numerous pathological processes, including nerve injury and repair, vascular regeneration, and immune response. Their potential therapeutic applications in glaucoma are being increasingly recognized [214,215].

Exosomes derived from MSCs have demonstrated significant neuroprotective effects in glaucoma models. Studies using rat models revealed that bone marrow MSC (BMSC)-derived exosomes significantly increased the survival and axonal regeneration of RGCs through miRNA-dependent mechanisms [166,216]. These exosomes integrated into the inner retinal layers, highlighting the crucial role of miRNAs in their neuroprotective benefits. Additionally, human umbilical cord MSC-derived exosomes have been reported to promote RGC survival and glial activation in rat models [217]. These exosomes also preserved RGCs in human retinal explants post-axotomy by releasing growth factors, especially PDGF [160]. Another recent study demonstrates that small extracellular vesicles (sEVs) derived from MSCs overexpressing microRNA-22-3p (miR22) protect retinal RGCs from apoptosis and preserve retinal function in an NMDA-induced RGC injury model by inhibiting the MAPK signaling pathway. The findings suggest that miR22-sEVs could be a promising therapeutic approach for glaucoma and other diseases involving RGC damage [218]. Despite these promising findings, treatments involving BMSC-derived exosomes and human MSC-derived exosomes have been associated with extensive gliosis and inflammation, necessitating further research to optimize their therapeutic potential and minimize adverse effects [169].

Nanoparticles offer another promising approach for neuroprotection in glaucoma. These tiny particles can encapsulate drugs, siRNA, mRNA, or other nucleic acids, providing a hydrophobic environment that improves the solubility of drugs with low water solubility. Nanoparticles protect encapsulated drugs from hydrolytic degradation and improve their transport across biological barriers, making them effective drug carriers. The use of pharmaceutical nanoparticles also reduces the side effects associated with traditional topical treatments [219].

However, most nano eye drops designed for neuroprotection are still in the animal experimental stage. A study developed magnetic nanoparticles conjugated with neurotrophins. These nanoparticles protect neurotrophins from rapid degradation and facilitate their localization in the retina [220]. Other studies on rodent models suggested that GDNF delivered by nanoparticles could serve as a neuroprotective tool for treating glaucomatous optic neuropathy [221,222]. Another study demonstrates that circular RNA (circRNA)-based therapy using lipid nanoparticle (LNP)-formulated circNGF provides prolonged NGF protein expression and superior protection for RGCs compared to traditional NGF protein therapy, without retinal toxicity [223]. The other study created a topical formulation using tocopheryl polyethylene glycol succinate (TPGS) and Pluronic F127 curcumin nanocarriers, which demonstrated neuroprotective effects in rodent models of ocular hypertension, optic nerve disease, and partial optic nerve transection [224]. Additionally, a study explores the development of polymeric microparticles to sustain the release of hydrogen sulfide (H_2S) for treating glaucoma. The research identifies critical process parameters that affect particle size, entrapment efficiency, and release profile, successfully optimizing a formulation that provides sustained H_2S release for up to 30 days, with potential for both ocular hypotensive and neuroprotective effects [225]. Another study demonstrated that NAM-loaded extracellular vesicles (NAM-EVs) can effectively deliver NAM to RGCs, providing a neuroprotective effect and maintaining the health and function of RGCs [123]. The other study investigated the use of liposomal carriers to enhance the delivery and therapeutic efficacy of citicoline in treating glaucoma, showing promising results for improved bioavailability and sustained drug release in ocular applications [112].

Exosomes and nanoparticles represent innovative and promising strategies for neuroprotection and regeneration in glaucoma treatment. Further human studies are essential to fully elucidate the therapeutic potential of these technologies and optimize their applications in clinical settings, ultimately aiming to improve outcomes for glaucoma patients.

5. Conclusions

Glaucoma continues to pose a major global health challenge, requiring the development of innovative neuroprotective strategies to complement existing IOP-lowering treatments. Pharmacological interventions such as brimonidine, neurotrophic factors, memantine, Ginkgo biloba extract, citicoline, nicotinamide, insulin, and resveratrol have demonstrated promising neuroprotective effects in preclinical studies, with some showing positive results in early clinical trials. Emerging technologies, including stem cell therapy, gene therapy, mitochondrial-targeted therapies, and nanotechnologies, offer innovative approaches to neuroprotection and RGC regeneration. Despite the promising potential of these strategies, further research is required to address the challenges associated with their clinical application. Larger, long-term clinical trials are essential to confirm the efficacy and safety of these treatments, ensuring their successful integration into clinical practice.

The outlook for neuroprotective therapies in glaucoma is promising, but significant hurdles remain in translating preclinical success into routine clinical use. As these novel therapies continue to be explored, it is likely that targeted treatments combining pharmacological agents and cutting-edge technologies could become part of a comprehensive glaucoma management plan. In the future, personalized medicine approaches based on genetic and molecular profiling may further refine these treatments, offering more tailored and effective solutions for preventing vision loss in glaucoma patients.

Funding: This work was supported in part by a grant from Taipei Medical University (TMU112-AE1-B04) and Wan Fang Hospital (113-wf-eva-16).

Institutional Review Board Statement: Not applicable.

Informed Consent Statement: Not applicable.

Conflicts of Interest: The authors declare no conflicts of interest.

Abbreviations

RGCs	Retinal Ganglion Cells
IOP	Intraocular Pressure
BDNF	Brain-Derived Neurotrophic Factor
CNTF	Ciliary Neurotrophic Factor
NMDA	N-Methyl-D-Aspartate
ROS	Reactive Oxygen Species
TNF-α	Tumor Necrosis Factor-Alpha
FGF	Fibroblast Growth Factor
RNFL	Retinal Nerve Fiber Layer
NTG	Normal Tension Glaucoma
POAG	Primary Open-Angle Glaucoma
AH	Aqueous Humor
MAPK	Mitogen-Activated Protein Kinase
NGF	Nerve Growth Factor
p75NTR	p75 Neurotrophin Receptor
rhNGF	Recombinant Human Nerve Growth Factor
VEGF	Vascular Endothelial Growth Factor
NAD	Nicotinamide Adenine Dinucleotide
NADPH	Nicotinamide Adenine Dinucleotide Phosphate Hydrogen
DA	Dopamine
DACs	Dopaminergic Amacrine Cells
DRD1	Dopamine Receptor D1

ROCK	Rho-Associated Protein Kinase
EP2	E Prostanoid Receptor 2
CAMKII	Calcium/Calmodulin-Dependent Protein Kinase II
CRISPR	Clustered Regularly Interspaced Short Palindromic Repeats
MYOC	Myocilin
TEK	Tunica Interna Endothelial Cell Kinase
AAV	Adeno-Associated Virus
XIAP	X-Linked Inhibitor of Apoptosis
SOD2	Superoxide Dismutase 2
LHON	Leber Hereditary Optic Neuropathy
MSCs	Mesenchymal Stem Cells
PDGF	Platelet-Derived Growth Factor
hESCs	Human Embryonic Stem Cells
OPCs	Oligodendrocyte Precursor Cells
miPSC	Mouse-Induced Pluripotent Stem Cell
mESC	Mouse Embryonic Stem Cell
BMSC	Bone Marrow Mesenchymal Stem Cells
sEVs	Small Extracellular Vesicles
LNP	Lipid Nanoparticle
TPGS	Tocopheryl Polyethylene Glycol Succinate
EVs	Extracellular Vesicles

References

1. Tham, Y.C.; Li, X.; Wong, T.Y.; Quigley, H.A.; Aung, T.; Cheng, C.Y. Global prevalence of glaucoma and projections of glaucoma burden through 2040: A systematic review and meta-analysis. *Ophthalmology* **2014**, *121*, 2081–2090. [CrossRef] [PubMed]
2. Casson, R.J. Possible role of excitotoxicity in the pathogenesis of glaucoma. *Clin. Exp. Ophthalmol.* **2006**, *34*, 54–63. [CrossRef] [PubMed]
3. Fan Gaskin, J.C.; Shah, M.H.; Chan, E.C. Oxidative Stress and the Role of NADPH Oxidase in Glaucoma. *Antioxidants* **2021**, *10*, 238. [CrossRef] [PubMed]
4. Vohra, R.; Tsai, J.C.; Kolko, M. The role of inflammation in the pathogenesis of glaucoma. *Surv. Ophthalmol.* **2013**, *58*, 311–320. [CrossRef]
5. Chen, S.D.; Wang, L.; Zhang, X.L. Neuroprotection in glaucoma: Present and future. *Chin. Med. J.* **2013**, *126*, 1567–1577. [CrossRef]
6. Pasutto, F.; Matsumoto, T.; Mardin, C.Y.; Sticht, H.; Brandstatter, J.H.; Michels-Rautenstrauss, K.; Weisschuh, N.; Gramer, E.; Ramdas, W.D.; van Koolwijk, L.M.; et al. Heterozygous NTF4 mutations impairing neurotrophin-4 signaling in patients with primary open-angle glaucoma. *Am. J. Hum. Genet.* **2009**, *85*, 447–456. [CrossRef]
7. Malik, J.M.; Shevtsova, Z.; Bahr, M.; Kugler, S. Long-term in vivo inhibition of CNS neurodegeneration by Bcl-XL gene transfer. *Mol. Ther.* **2005**, *11*, 373–381. [CrossRef] [PubMed]
8. Krupin, T.; Liebmann, J.M.; Greenfield, D.S.; Ritch, R.; Gardiner, S.; Low-Pressure Glaucoma Study Group. A randomized trial of brimonidine versus timolol in preserving visual function: Results from the Low-Pressure Glaucoma Treatment Study. *Am. J. Ophthalmol.* **2011**, *151*, 671–681. [CrossRef]
9. De Moraes, C.G.; Liebmann, J.M.; Greenfield, D.S.; Gardiner, S.K.; Ritch, R.; Krupin, T.; Low-pressure Glaucoma Treatment Study Group. Risk factors for visual field progression in the low-pressure glaucoma treatment study. *Am. J. Ophthalmol.* **2012**, *154*, 702–711. [CrossRef]
10. Goldberg, J.L.; Beykin, G.; Satterfield, K.R.; Nunez, M.; Lam, B.L.; Albini, T.A. Phase I NT-501 Ciliary Neurotrophic Factor Implant Trial for Primary Open-Angle Glaucoma: Safety, Neuroprotection, and Neuroenhancement. *Ophthalmol. Sci.* **2023**, *3*, 100298. [CrossRef]
11. Beykin, G.; Stell, L.; Halim, M.S.; Nunez, M.; Popova, L.; Nguyen, B.T.; Groth, S.L.; Dennis, A.; Li, Z.; Atkins, M.; et al. Phase 1b Randomized Controlled Study of Short Course Topical Recombinant Human Nerve Growth Factor (rhNGF) for Neuroenhancement in Glaucoma: Safety, Tolerability, and Efficacy Measure Outcomes. *Am. J. Ophthalmol.* **2022**, *234*, 223–234. [CrossRef] [PubMed]
12. Weinreb, R.N.; Liebmann, J.M.; Cioffi, G.A.; Goldberg, I.; Brandt, J.D.; Johnson, C.A.; Zangwill, L.M.; Schneider, S.; Badger, H.; Bejanian, M. Oral Memantine for the Treatment of Glaucoma: Design and Results of 2 Randomized, Placebo-Controlled, Phase 3 Studies. *Ophthalmology* **2018**, *125*, 1874–1885. [CrossRef] [PubMed]
13. Quaranta, L.; Bettelli, S.; Uva, M.G.; Semeraro, F.; Turano, R.; Gandolfo, E. Effect of Ginkgo biloba extract on preexisting visual field damage in normal tension glaucoma. *Ophthalmology* **2003**, *110*, 359–362; discussion 362–354. [CrossRef]
14. Lee, J.; Sohn, S.W.; Kee, C. Effect of Ginkgo biloba extract on visual field progression in normal tension glaucoma. *J. Glaucoma* **2013**, *22*, 780–784. [CrossRef]

15. Guo, X.; Kong, X.; Huang, R.; Jin, L.; Ding, X.; He, M.; Liu, X.; Patel, M.C.; Congdon, N.G. Effect of Ginkgo biloba on visual field and contrast sensitivity in Chinese patients with normal tension glaucoma: A randomized, crossover clinical trial. *Investig. Ophthalmol. Vis. Sci.* **2014**, *55*, 110–116. [CrossRef]
16. Guo, X.; Namekata, K.; Kimura, A.; Noro, T.; Azuchi, Y.; Semba, K.; Harada, C.; Yoshida, H.; Mitamura, Y.; Harada, T. Brimonidine suppresses loss of retinal neurons and visual function in a murine model of optic neuritis. *Neurosci. Lett.* **2015**, *592*, 27–31. [CrossRef]
17. Lambert, W.S.; Ruiz, L.; Crish, S.D.; Wheeler, L.A.; Calkins, D.J. Brimonidine prevents axonal and somatic degeneration of retinal ganglion cell neurons. *Mol. Neurodegener.* **2011**, *6*, 4. [CrossRef]
18. Lindsey, J.D.; Duong-Polk, K.X.; Hammond, D.; Chindasub, P.; Leung, C.K.; Weinreb, R.N. Differential protection of injured retinal ganglion cell dendrites by brimonidine. *Investig. Ophthalmol. Vis. Sci.* **2015**, *56*, 1789–1804. [CrossRef] [PubMed]
19. Lafuente, M.P.; Villegas-Perez, M.P.; Mayor, S.; Aguilera, M.E.; Miralles de Imperial, J.; Vidal-Sanz, M. Neuroprotective effects of brimonidine against transient ischemia-induced retinal ganglion cell death: A dose response in vivo study. *Exp. Eye Res.* **2002**, *74*, 181–189. [CrossRef]
20. Vidal-Sanz, M.; Lafuente, M.P.; Mayor-Torroglosa, S.; Aguilera, M.E.; Miralles de Imperial, J.; Villegas-Perez, M.P. Brimonidine's neuroprotective effects against transient ischaemia-induced retinal ganglion cell death. *Eur. J. Ophthalmol.* **2001**, *11* (Suppl. S2), S36–S40. [CrossRef]
21. Chrysostomou, V.; Rezania, F.; Trounce, I.A.; Crowston, J.G. Oxidative stress and mitochondrial dysfunction in glaucoma. *Curr. Opin. Pharmacol.* **2013**, *13*, 12–15. [CrossRef] [PubMed]
22. Goldenberg-Cohen, N.; Dadon-Bar-El, S.; Hasanreisoglu, M.; Avraham-Lubin, B.C.; Dratviman-Storobinsky, O.; Cohen, Y.; Weinberger, D. Possible neuroprotective effect of brimonidine in a mouse model of ischaemic optic neuropathy. *Clin. Exp. Ophthalmol.* **2009**, *37*, 718–729. [CrossRef]
23. Aktas, Z.; Gurelik, G.; Akyurek, N.; Onol, M.; Hasanreisoglu, B. Neuroprotective effect of topically applied brimonidine tartrate 0.2% in endothelin-1-induced optic nerve ischaemia model. *Clin. Exp. Ophthalmol.* **2007**, *35*, 527–534. [CrossRef] [PubMed]
24. Gao, H.; Qiao, X.; Cantor, L.B.; WuDunn, D. Up-regulation of brain-derived neurotrophic factor expression by brimonidine in rat retinal ganglion cells. *Arch. Ophthalmol.* **2002**, *120*, 797–803. [CrossRef]
25. Jung, K.I.; Kim, J.H.; Han, J.S.; Park, C.K. Exploring Neuroprotective Effects of Topical Brimonidine in Experimental Diabetic Retinopathy. *Vivo* **2024**, *38*, 1609–1620. [CrossRef]
26. Semba, K.; Namekata, K.; Kimura, A.; Harada, C.; Mitamura, Y.; Harada, T. Brimonidine prevents neurodegeneration in a mouse model of normal tension glaucoma. *Cell Death Dis.* **2014**, *5*, e1341. [CrossRef] [PubMed]
27. Jung, K.I.; Kim, J.H.; Park, C.K. alpha2-Adrenergic modulation of the glutamate receptor and transporter function in a chronic ocular hypertension model. *Eur. J. Pharmacol.* **2015**, *765*, 274–283. [CrossRef]
28. Lee, D.; Kim, K.Y.; Noh, Y.H.; Chai, S.; Lindsey, J.D.; Ellisman, M.H.; Weinreb, R.N.; Ju, W.K. Brimonidine blocks glutamate excitotoxicity-induced oxidative stress and preserves mitochondrial transcription factor a in ischemic retinal injury. *PLoS ONE* **2012**, *7*, e47098. [CrossRef]
29. Lee, K.Y.; Nakayama, M.; Aihara, M.; Chen, Y.N.; Araie, M. Brimonidine is neuroprotective against glutamate-induced neurotoxicity, oxidative stress, and hypoxia in purified rat retinal ganglion cells. *Mol. Vis.* **2010**, *16*, 246–251.
30. Goldblum, D.; Kipfer-Kauer, A.; Sarra, G.M.; Wolf, S.; Frueh, B.E. Distribution of amyloid precursor protein and amyloid-beta immunoreactivity in DBA/2J glaucomatous mouse retinas. *Investig. Ophthalmol. Vis. Sci.* **2007**, *48*, 5085–5090. [CrossRef]
31. Nizari, S.; Guo, L.; Davis, B.M.; Normando, E.M.; Galvao, J.; Turner, L.A.; Bizrah, M.; Dehabadi, M.; Tian, K.; Cordeiro, M.F. Non-amyloidogenic effects of alpha2 adrenergic agonists: Implications for brimonidine-mediated neuroprotection. *Cell. Death Dis.* **2016**, *7*, e2514. [CrossRef] [PubMed]
32. Feke, G.T.; Bex, P.J.; Taylor, C.P.; Rhee, D.J.; Turalba, A.V.; Chen, T.C.; Wand, M.; Pasquale, L.R. Effect of brimonidine on retinal vascular autoregulation and short-term visual function in normal tension glaucoma. *Am. J. Ophthalmol.* **2014**, *158*, 105–112.e101. [CrossRef]
33. Lonngren, U.; Napankangas, U.; Lafuente, M.; Mayor, S.; Lindqvist, N.; Vidal-Sanz, M.; Hallbook, F. The growth factor response in ischemic rat retina and superior colliculus after brimonidine pre-treatment. *Brain Res. Bull.* **2006**, *71*, 208–218. [CrossRef] [PubMed]
34. Ben Simon, G.J.; Bakalash, S.; Aloni, E.; Rosner, M. A rat model for acute rise in intraocular pressure: Immune modulation as a therapeutic strategy. *Am. J. Ophthalmol.* **2006**, *141*, 1105–1111. [CrossRef] [PubMed]
35. Namekata, K.; Noro, T.; Nishijima, E.; Sotozono, A.; Guo, X.; Harada, C.; Shinozaki, Y.; Mitamura, Y.; Nakano, T.; Harada, T. Drug combination of topical ripasudil and brimonidine enhances neuroprotection in a mouse model of optic nerve injury. *J. Pharmacol. Sci.* **2024**, *154*, 326–333. [CrossRef]
36. Otsubo, M.; Sase, K.; Tsukahara, C.; Fujita, N.; Arizono, I.; Tokuda, N.; Kitaoka, Y. Axonal protection by combination of ripasudil and brimonidine with upregulation of p-AMPK in TNF-induced optic nerve degeneration. *Int. Ophthalmol.* **2024**, *44*, 173. [CrossRef]
37. Evans, D.W.; Hosking, S.L.; Gherghel, D.; Bartlett, J.D. Contrast sensitivity improves after brimonidine therapy in primary open angle glaucoma: A case for neuroprotection. *Br. J. Ophthalmol.* **2003**, *87*, 1463–1465. [CrossRef]
38. Tsai, J.C.; Chang, H.W. Comparison of the effects of brimonidine 0.2% and timolol 0.5% on retinal nerve fiber layer thickness in ocular hypertensive patients: A prospective, unmasked study. *J. Ocul. Pharmacol. Ther.* **2005**, *21*, 475–482. [CrossRef]

39. Scuteri, D.; Bagetta, G.; Nucci, C.; Aiello, F.; Cesareo, M.; Tonin, P.; Corasaniti, M.T. Evidence on the neuroprotective properties of brimonidine in glaucoma. *Prog. Brain Res.* **2020**, *257*, 155–166. [CrossRef] [PubMed]
40. Kimura, A.; Namekata, K.; Guo, X.; Harada, C.; Harada, T. Neuroprotection, Growth Factors and BDNF-TrkB Signalling in Retinal Degeneration. *Int. J. Mol. Sci.* **2016**, *17*, 1584. [CrossRef]
41. Chang, E.E.; Goldberg, J.L. Glaucoma 2.0: Neuroprotection, neuroregeneration, neuroenhancement. *Ophthalmology* **2012**, *119*, 979–986. [CrossRef] [PubMed]
42. Saragovi, H.U.; Hamel, E.; Di Polo, A. A neurotrophic rationale for the therapy of neurodegenerative disorders. *Curr. Alzheimer Res.* **2009**, *6*, 419–423. [CrossRef] [PubMed]
43. Lambiase, A.; Aloe, L.; Centofanti, M.; Parisi, V.; Bao, S.N.; Mantelli, F.; Colafrancesco, V.; Manni, G.L.; Bucci, M.G.; Bonini, S.; et al. Experimental and clinical evidence of neuroprotection by nerve growth factor eye drops: Implications for glaucoma. *Proc. Natl. Acad. Sci. USA* **2009**, *106*, 13469–13474. [CrossRef] [PubMed]
44. Oddone, F.; Roberti, G.; Micera, A.; Busanello, A.; Bonini, S.; Quaranta, L.; Agnifili, L.; Manni, G. Exploring Serum Levels of Brain Derived Neurotrophic Factor and Nerve Growth Factor Across Glaucoma Stages. *PLoS ONE* **2017**, *12*, e0168565. [CrossRef]
45. Pease, M.E.; Zack, D.J.; Berlinicke, C.; Bloom, K.; Cone, F.; Wang, Y.; Klein, R.L.; Hauswirth, W.W.; Quigley, H.A. Effect of CNTF on retinal ganglion cell survival in experimental glaucoma. *Investig. Ophthalmol. Vis. Sci.* **2009**, *50*, 2194–2200. [CrossRef]
46. Grozdanic, S.D.; Lazic, T.; Kuehn, M.H.; Harper, M.M.; Kardon, R.H.; Kwon, Y.H.; Lavik, E.B.; Sakaguchi, D.S. Exogenous modulation of intrinsic optic nerve neuroprotective activity. *Graefes Arch. Clin. Exp. Ophthalmol.* **2010**, *248*, 1105–1116. [CrossRef]
47. Parrilla-Reverter, G.; Agudo, M.; Sobrado-Calvo, P.; Salinas-Navarro, M.; Villegas-Perez, M.P.; Vidal-Sanz, M. Effects of different neurotrophic factors on the survival of retinal ganglion cells after a complete intraorbital nerve crush injury: A quantitative in vivo study. *Exp. Eye Res.* **2009**, *89*, 32–41. [CrossRef] [PubMed]
48. Weber, A.J.; Viswanathan, S.; Ramanathan, C.; Harman, C.D. Combined application of BDNF to the eye and brain enhances ganglion cell survival and function in the cat after optic nerve injury. *Investig. Ophthalmol. Vis. Sci.* **2010**, *51*, 327–334. [CrossRef]
49. Flachsbarth, K.; Jankowiak, W.; Kruszewski, K.; Helbing, S.; Bartsch, S.; Bartsch, U. Pronounced synergistic neuroprotective effect of GDNF and CNTF on axotomized retinal ganglion cells in the adult mouse. *Exp. Eye Res.* **2018**, *176*, 258–265. [CrossRef]
50. Hameed, S.S.; Bodi, N.E.; Miller, R.C.; Sharma, T.P. Neuritin 1 Drives Therapeutic Preservation of Retinal Ganglion Cells in an Ex Vivo Human Glaucoma Model. *J. Ocul. Pharmacol. Ther.* **2024**. *ahead of print*. [CrossRef]
51. Johnson, T.V.; Bull, N.D.; Martin, K.R. Neurotrophic factor delivery as a protective treatment for glaucoma. *Exp. Eye Res.* **2011**, *93*, 196–203. [CrossRef]
52. Vecino, E.; Garcia-Crespo, D.; Garcia, M.; Martinez-Millan, L.; Sharma, S.C.; Carrascal, E. Rat retinal ganglion cells co-express brain derived neurotrophic factor (BDNF) and its receptor TrkB. *Vis. Res.* **2002**, *42*, 151–157. [CrossRef] [PubMed]
53. Cha, Y.W.; Kim, S.T. Serum and aqueous humor levels of brain-derived neurotrophic factor in patients with primary open-angle glaucoma and normal-tension glaucoma. *Int. Ophthalmol.* **2021**, *41*, 3869–3875. [CrossRef] [PubMed]
54. Uzel, M.M.; Elgin, U.; Boral, B.; Cicek, M.; Sen, E.; Sener, B.; Yilmazbas, P. The effect of trabeculectomy on serum brain-derived neurotrophic factor levels in primary open-angle glaucoma. *Graefes Arch. Clin. Exp. Ophthalmol.* **2018**, *256*, 1173–1178. [CrossRef] [PubMed]
55. Lazaldin, M.A.M.; Iezhitsa, I.; Agarwal, R.; Agarwal, P.; Ismail, N.M. Neuroprotective effects of exogenous brain-derived neurotrophic factor on amyloid-beta 1-40-induced retinal degeneration. *Neural Regen. Res.* **2023**, *18*, 382–388. [CrossRef] [PubMed]
56. Binley, K.E.; Ng, W.S.; Barde, Y.A.; Song, B.; Morgan, J.E. Brain-derived neurotrophic factor prevents dendritic retraction of adult mouse retinal ganglion cells. *Eur. J. Neurosci.* **2016**, *44*, 2028–2039. [CrossRef]
57. Shpak, A.A.; Guekht, A.B.; Druzhkova, T.A.; Kozlova, K.I.; Gulyaeva, N.V. Ciliary neurotrophic factor in patients with primary open-angle glaucoma and age-related cataract. *Mol. Vis.* **2017**, *23*, 799–809.
58. Boulton, T.G.; Stahl, N.; Yancopoulos, G.D. Ciliary neurotrophic factor/leukemia inhibitory factor/interleukin 6/oncostatin M family of cytokines induces tyrosine phosphorylation of a common set of proteins overlapping those induced by other cytokines and growth factors. *J. Biol. Chem.* **1994**, *269*, 11648–11655. [CrossRef]
59. Escartin, C.; Pierre, K.; Colin, A.; Brouillet, E.; Delzescaux, T.; Guillermier, M.; Dhenain, M.; Deglon, N.; Hantraye, P.; Pellerin, L.; et al. Activation of astrocytes by CNTF induces metabolic plasticity and increases resistance to metabolic insults. *J. Neurosci.* **2007**, *27*, 7094–7104. [CrossRef]
60. Lee, K.; Choi, J.O.; Hwang, A.; Bae, H.W.; Kim, C.Y. Ciliary Neurotrophic Factor Derived From Astrocytes Protects Retinal Ganglion Cells Through PI3K/AKT, JAK/STAT, and MAPK/ERK Pathways. *Investig. Ophthalmol. Vis. Sci.* **2022**, *63*, 4. [CrossRef]
61. Leibinger, M.; Andreadaki, A.; Diekmann, H.; Fischer, D. Neuronal STAT3 activation is essential for CNTF- and inflammatory stimulation-induced CNS axon regeneration. *Cell Death Dis.* **2013**, *4*, e805. [CrossRef]
62. Muller, A.; Hauk, T.G.; Leibinger, M.; Marienfeld, R.; Fischer, D. Exogenous CNTF stimulates axon regeneration of retinal ganglion cells partially via endogenous CNTF. *Mol. Cell Neurosci.* **2009**, *41*, 233–246. [CrossRef] [PubMed]
63. Mey, J.; Thanos, S. Intravitreal injections of neurotrophic factors support the survival of axotomized retinal ganglion cells in adult rats in vivo. *Brain Res.* **1993**, *602*, 304–317. [CrossRef] [PubMed]
64. Cui, Q.; Yip, H.K.; Zhao, R.C.; So, K.F.; Harvey, A.R. Intraocular elevation of cyclic AMP potentiates ciliary neurotrophic factor-induced regeneration of adult rat retinal ganglion cell axons. *Mol. Cell Neurosci.* **2003**, *22*, 49–61. [CrossRef] [PubMed]

65. Bertrand, J.; Winton, M.J.; Rodriguez-Hernandez, N.; Campenot, R.B.; McKerracher, L. Application of Rho antagonist to neuronal cell bodies promotes neurite growth in compartmented cultures and regeneration of retinal ganglion cell axons in the optic nerve of adult rats. *J. Neurosci.* **2005**, *25*, 1113–1121. [CrossRef]
66. Lingor, P.; Tonges, L.; Pieper, N.; Bermel, C.; Barski, E.; Planchamp, V.; Bahr, M. ROCK inhibition and CNTF interact on intrinsic signalling pathways and differentially regulate survival and regeneration in retinal ganglion cells. *Brain* **2008**, *131*, 250–263. [CrossRef]
67. Kauper, K.; McGovern, C.; Sherman, S.; Heatherton, P.; Rapoza, R.; Stabila, P.; Dean, B.; Lee, A.; Borges, S.; Bouchard, B.; et al. Two-year intraocular delivery of ciliary neurotrophic factor by encapsulated cell technology implants in patients with chronic retinal degenerative diseases. *Investig. Ophthalmol. Vis. Sci.* **2012**, *53*, 7484–7491. [CrossRef]
68. Talcott, K.E.; Ratnam, K.; Sundquist, S.M.; Lucero, A.S.; Lujan, B.J.; Tao, W.; Porco, T.C.; Roorda, A.; Duncan, J.L. Longitudinal study of cone photoreceptors during retinal degeneration and in response to ciliary neurotrophic factor treatment. *Investig. Ophthalmol. Vis. Sci.* **2011**, *52*, 2219–2226. [CrossRef]
69. Zhang, K.; Hopkins, J.J.; Heier, J.S.; Birch, D.G.; Halperin, L.S.; Albini, T.A.; Brown, D.M.; Jaffe, G.J.; Tao, W.; Williams, G.A. Ciliary neurotrophic factor delivered by encapsulated cell intraocular implants for treatment of geographic atrophy in age-related macular degeneration. *Proc. Natl. Acad. Sci. USA* **2011**, *108*, 6241–6245. [CrossRef]
70. Tirassa, P.; Rosso, P.; Iannitelli, A. Ocular Nerve Growth Factor (NGF) and NGF Eye Drop Application as Paradigms to Investigate NGF Neuroprotective and Reparative Actions. *Methods Mol. Biol.* **2018**, *1727*, 19–38. [CrossRef]
71. Chen, Q.; Wang, H.; Liao, S.; Gao, Y.; Liao, R.; Little, P.J.; Xu, J.; Feng, Z.P.; Zheng, Y.; Zheng, W. Nerve growth factor protects retinal ganglion cells against injury induced by retinal ischemia-reperfusion in rats. *Growth. Factors* **2015**, *33*, 149–159. [CrossRef]
72. Patapoutian, A.; Reichardt, L.F. Trk receptors: Mediators of neurotrophin action. *Curr. Opin. Neurobiol.* **2001**, *11*, 272–280. [CrossRef] [PubMed]
73. Colafrancesco, V.; Parisi, V.; Sposato, V.; Rossi, S.; Russo, M.A.; Coassin, M.; Lambiase, A.; Aloe, L. Ocular application of nerve growth factor protects degenerating retinal ganglion cells in a rat model of glaucoma. *J. Glaucoma* **2011**, *20*, 100–108. [CrossRef] [PubMed]
74. Guo, L.; Davis, B.M.; Ravindran, N.; Galvao, J.; Kapoor, N.; Haamedi, N.; Shamsher, E.; Luong, V.; Fico, E.; Cordeiro, M.F. Topical recombinant human Nerve growth factor (rh-NGF) is neuroprotective to retinal ganglion cells by targeting secondary degeneration. *Sci. Rep.* **2020**, *10*, 3375. [CrossRef] [PubMed]
75. Lebrun-Julien, F.; Bertrand, M.J.; De Backer, O.; Stellwagen, D.; Morales, C.R.; Di Polo, A.; Barker, P.A. ProNGF induces TNFalpha-dependent death of retinal ganglion cells through a p75NTR non-cell-autonomous signaling pathway. *Proc. Natl. Acad. Sci. USA* **2010**, *107*, 3817–3822. [CrossRef]
76. Mesentier-Louro, L.A.; Rosso, P.; Carito, V.; Mendez-Otero, R.; Santiago, M.F.; Rama, P.; Lambiase, A.; Tirassa, P. Nerve Growth Factor Role on Retinal Ganglion Cell Survival and Axon Regrowth: Effects of Ocular Administration in Experimental Model of Optic Nerve Injury. *Mol. Neurobiol.* **2019**, *56*, 1056–1069. [CrossRef]
77. Ferrari, M.P.; Mantelli, F.; Sacchetti, M.; Antonangeli, M.I.; Cattani, F.; D'Anniballe, G.; Sinigaglia, F.; Ruffini, P.A.; Lambiase, A. Safety and pharmacokinetics of escalating doses of human recombinant nerve growth factor eye drops in a double-masked, randomized clinical trial. *BioDrugs* **2014**, *28*, 275–283. [CrossRef]
78. Parsons, C.G.; Stoffler, A.; Danysz, W. Memantine: A NMDA receptor antagonist that improves memory by restoration of homeostasis in the glutamatergic system–too little activation is bad, too much is even worse. *Neuropharmacology* **2007**, *53*, 699–723. [CrossRef]
79. Osborne, N.N. Recent clinical findings with memantine should not mean that the idea of neuroprotection in glaucoma is abandoned. *Acta Ophthalmol.* **2009**, *87*, 450–454. [CrossRef]
80. Hare, W.A.; WoldeMussie, E.; Lai, R.K.; Ton, H.; Ruiz, G.; Chun, T.; Wheeler, L. Efficacy and safety of memantine treatment for reduction of changes associated with experimental glaucoma in monkey, I: Functional measures. *Investig. Ophthalmol. Vis. Sci.* **2004**, *45*, 2625–2639. [CrossRef]
81. Seki, M.; Lipton, S.A. Targeting excitotoxic/free radical signaling pathways for therapeutic intervention in glaucoma. *Prog. Brain Res.* **2008**, *173*, 495–510. [CrossRef]
82. Yucel, Y.H.; Gupta, N.; Zhang, Q.; Mizisin, A.P.; Kalichman, M.W.; Weinreb, R.N. Memantine protects neurons from shrinkage in the lateral geniculate nucleus in experimental glaucoma. *Arch. Ophthalmol.* **2006**, *124*, 217–225. [CrossRef] [PubMed]
83. Kim, T.W.; Kim, D.M.; Park, K.H.; Kim, H. Neuroprotective effect of memantine in a rabbit model of optic nerve ischemia. *Korean J. Ophthalmol.* **2002**, *16*, 1–7. [CrossRef] [PubMed]
84. Burgoyne, C.F. The non-human primate experimental glaucoma model. *Exp. Eye Res.* **2015**, *141*, 57–73. [CrossRef] [PubMed]
85. Wamsley, S.; Gabelt, B.T.; Dahl, D.B.; Case, G.L.; Sherwood, R.W.; May, C.A.; Hernandez, M.R.; Kaufman, P.L. Vitreous glutamate concentration and axon loss in monkeys with experimental glaucoma. *Arch. Ophthalmol.* **2005**, *123*, 64–70. [CrossRef]
86. Levkovitch-Verbin, H.; Quigley, H.A.; Kerrigan-Baumrind, L.A.; D'Anna, S.A.; Kerrigan, D.; Pease, M.E. Optic nerve transection in monkeys may result in secondary degeneration of retinal ganglion cells. *Investig. Ophthalmol. Vis. Sci.* **2001**, *42*, 975–982.
87. Vianna, J.R.; Chauhan, B.C. How to detect progression in glaucoma. *Prog. Brain Res.* **2015**, *221*, 135–158. [CrossRef]
88. Eckert, A.; Keil, U.; Scherping, I.; Hauptmann, S.; Muller, W.E. Stabilization of mitochondrial membrane potential and improvement of neuronal energy metabolism by Ginkgo biloba extract EGb 761. *Ann. N. Y. Acad. Sci.* **2005**, *1056*, 474–485. [CrossRef]

89. Ritch, R. Potential role for Ginkgo biloba extract in the treatment of glaucoma. *Med. Hypotheses* **2000**, *54*, 221–235. [CrossRef]
90. Kang, J.M.; Lin, S. Ginkgo biloba and its potential role in glaucoma. *Curr. Opin. Ophthalmol.* **2018**, *29*, 116–120. [CrossRef]
91. Hirooka, K.; Tokuda, M.; Miyamoto, O.; Itano, T.; Baba, T.; Shiraga, F. The Ginkgo biloba extract (EGb 761) provides a neuroprotective effect on retinal ganglion cells in a rat model of chronic glaucoma. *Curr. Eye Res.* **2004**, *28*, 153–157. [CrossRef]
92. Ma, K.; Xu, L.; Zhan, H.; Zhang, S.; Pu, M.; Jonas, J.B. Dosage dependence of the effect of Ginkgo biloba on the rat retinal ganglion cell survival after optic nerve crush. *Eye* **2009**, *23*, 1598–1604. [CrossRef]
93. Ma, K.; Xu, L.; Zhang, H.; Zhang, S.; Pu, M.; Jonas, J.B. The effect of ginkgo biloba on the rat retinal ganglion cell survival in the optic nerve crush model. *Acta Ophthalmol.* **2010**, *88*, 553–557. [CrossRef] [PubMed]
94. Harris, A.; Gross, J.; Moore, N.; Do, T.; Huang, A.; Gama, W.; Siesky, B. The effects of antioxidants on ocular blood flow in patients with glaucoma. *Acta Ophthalmol.* **2018**, *96*, e237–e241. [CrossRef] [PubMed]
95. Park, J.W.; Kwon, H.J.; Chung, W.S.; Kim, C.Y.; Seong, G.J. Short-term effects of Ginkgo biloba extract on peripapillary retinal blood flow in normal tension glaucoma. *Korean J. Ophthalmol.* **2011**, *25*, 323–328. [CrossRef]
96. Faiq, M.A.; Wollstein, G.; Schuman, J.S.; Chan, K.C. Cholinergic nervous system and glaucoma: From basic science to clinical applications. *Prog. Retin. Eye Res.* **2019**, *72*, 100767. [CrossRef]
97. Roberti, G.; Tanga, L.; Michelessi, M.; Quaranta, L.; Parisi, V.; Manni, G.; Oddone, F. Cytidine 5′-Diphosphocholine (Citicoline) in Glaucoma: Rationale of Its Use, Current Evidence and Future Perspectives. *Int. J. Mol. Sci.* **2015**, *16*, 28401–28417. [CrossRef]
98. Adibhatla, R.M.; Hatcher, J.F.; Dempsey, R.J. Effects of citicoline on phospholipid and glutathione levels in transient cerebral ischemia. *Stroke* **2001**, *32*, 2376–2381. [CrossRef] [PubMed]
99. van der Merwe, Y.; Yang, X.; Ho, L.C.; Yu, Y.; Chau, Y.; Leung, C.K.-S.; Conner, I.P.; Steketee, M.B.; Wollstein, G.; Schuman, J.S.; et al. Citicoline preserves optic nerve integrity and visuomotor function following chronic intraocular pressure elevation. *Investig. Ophthalmol. Vis. Sci.* **2016**, *57*.
100. Virno, M.; Pecori-Giraldi, J.; Liguori, A.; De Gregorio, F. The protective effect of citicoline on the progression of the perimetric defects in glaucomatous patients (perimetric study with a 10-year follow-up). *Acta Ophthalmol. Scand. Suppl.* **2000**, *78*, 56–57. [CrossRef]
101. Parisi, V. Electrophysiological assessment of glaucomatous visual dysfunction during treatment with cytidine-5′-diphosphocholine (citicoline): A study of 8 years of follow-up. *Doc. Ophthalmol.* **2005**, *110*, 91–102. [CrossRef]
102. Rejdak, R.; Toczolowski, J.; Kurkowski, J.; Kaminski, M.L.; Rejdak, K.; Stelmasiak, Z.; Grieb, P. Oral citicoline treatment improves visual pathway function in glaucoma. *Med. Sci. Monit.* **2003**, *9*, PI24-8. [PubMed]
103. Parisi, V.; Coppola, G.; Centofanti, M.; Oddone, F.; Angrisani, A.M.; Ziccardi, L.; Ricci, B.; Quaranta, L.; Manni, G. Evidence of the neuroprotective role of citicoline in glaucoma patients. *Prog. Brain Res.* **2008**, *173*, 541–554. [CrossRef] [PubMed]
104. Ottobelli, L.; Manni, G.L.; Centofanti, M.; Iester, M.; Allevena, F.; Rossetti, L. Citicoline oral solution in glaucoma: Is there a role in slowing disease progression? *Ophthalmologica* **2013**, *229*, 219–226. [CrossRef]
105. Rossetti, L.; Iester, M.; Tranchina, L.; Ottobelli, L.; Coco, G.; Calcatelli, E.; Ancona, C.; Cirafici, P.; Manni, G. Can Treatment With Citicoline Eyedrops Reduce Progression in Glaucoma? The Results of a Randomized Placebo-controlled Clinical Trial. *J. Glaucoma* **2020**, *29*, 513–520. [CrossRef]
106. Parisi, V.; Centofanti, M.; Ziccardi, L.; Tanga, L.; Michelessi, M.; Roberti, G.; Manni, G. Treatment with citicoline eye drops enhances retinal function and neural conduction along the visual pathways in open angle glaucoma. *Graefes Arch. Clin. Exp. Ophthalmol.* **2015**, *253*, 1327–1340. [CrossRef]
107. Parisi, V.; Oddone, F.; Roberti, G.; Tanga, L.; Carnevale, C.; Ziccardi, L.; Manni, G. Enhancement of Retinal Function and of Neural Conduction Along the Visual Pathway Induced by Treatment with Citicoline Eye Drops in Liposomal Formulation in Open Angle Glaucoma: A Pilot Electrofunctional Study. *Adv. Ther.* **2019**, *36*, 987–996. [CrossRef]
108. Roberti, G.; Tanga, L.; Parisi, V.; Sampalmieri, M.; Centofanti, M.; Manni, G. A preliminary study of the neuroprotective role of citicoline eye drops in glaucomatous optic neuropathy. *Indian J. Ophthalmol.* **2014**, *62*, 549–553. [CrossRef] [PubMed]
109. Grieb, P.; Junemann, A.; Rekas, M.; Rejdak, R. Citicoline: A Food Beneficial for Patients Suffering from or Threated with Glaucoma. *Front. Aging Neurosci.* **2016**, *8*, 73. [CrossRef]
110. Anton, A.; Garcia, V.; Munoz, M.; Gonzales, K.; Ayala, E.; Del Mar Sanchez, E.; Morilla-Grasa, A. The Effect of Oral Citicoline and Docosahexaenoic Acid on the Visual Field of Patients with Glaucoma: A Randomized Trial. *Life* **2022**, *12*, 1481. [CrossRef]
111. Marino, P.F.; Rossi, G.C.M.; Campagna, G.; Capobianco, D.; Costagliola, C.; on behalf of QUALICOS Study Group. Effects of Citicoline, Homotaurine, and Vitamin E on Contrast Sensitivity and Visual-Related Quality of Life in Patients with Primary Open-Angle Glaucoma: A Preliminary Study. *Molecules* **2020**, *25*, 5614. [CrossRef]
112. Bonechi, C.; Mahdizadeh, F.F.; Talarico, L.; Pepi, S.; Tamasi, G.; Leone, G.; Consumi, M.; Donati, A.; Magnani, A. Liposomal Encapsulation of Citicoline for Ocular Drug Delivery. *Int. J. Mol. Sci.* **2023**, *24*, 16864. [CrossRef] [PubMed]
113. Verdin, E. NAD(+) in aging, metabolism, and neurodegeneration. *Science* **2015**, *350*, 1208–1213. [CrossRef] [PubMed]
114. Cimaglia, G.; Votruba, M.; Morgan, J.E.; Andre, H.; Williams, P.A. Potential Therapeutic Benefit of NAD(+) Supplementation for Glaucoma and Age-Related Macular Degeneration. *Nutrients* **2020**, *12*, 2871. [CrossRef] [PubMed]
115. Zhang, M.; Ying, W. NAD(+) Deficiency Is a Common Central Pathological Factor of a Number of Diseases and Aging: Mechanisms and Therapeutic Implications. *Antioxid Redox Signal* **2019**, *30*, 890–905. [CrossRef]
116. Araie, M.; Mayama, C. Use of calcium channel blockers for glaucoma. *Prog. Retin. Eye Res.* **2011**, *30*, 54–71. [CrossRef]

117. Pasquale, L.R. Vascular and autonomic dysregulation in primary open-angle glaucoma. *Curr. Opin. Ophthalmol.* **2016**, *27*, 94–101. [CrossRef]
118. Resch, H.; Garhofer, G.; Fuchsjager-Mayrl, G.; Hommer, A.; Schmetterer, L. Endothelial dysfunction in glaucoma. *Acta Ophthalmol.* **2009**, *87*, 4–12. [CrossRef]
119. Williams, P.A.; Harder, J.M.; Foxworth, N.E.; Cardozo, B.H.; Cochran, K.E.; John, S.W.M. Nicotinamide and WLD(S) Act Together to Prevent Neurodegeneration in Glaucoma. *Front. Neurosci.* **2017**, *11*, 232. [CrossRef]
120. Williams, P.A.; Harder, J.M.; Foxworth, N.E.; Cochran, K.E.; Philip, V.M.; Porciatti, V.; Smithies, O.; John, S.W. Vitamin B(3) modulates mitochondrial vulnerability and prevents glaucoma in aged mice. *Science* **2017**, *355*, 756–760. [CrossRef] [PubMed]
121. Tribble, J.R.; Otmani, A.; Sun, S.; Ellis, S.A.; Cimaglia, G.; Vohra, R.; Joe, M.; Lardner, E.; Venkataraman, A.P.; Dominguez-Vicent, A.; et al. Nicotinamide provides neuroprotection in glaucoma by protecting against mitochondrial and metabolic dysfunction. *Redox Biol* **2021**, *43*, 101988. [CrossRef]
122. Yu, N.; Wu, X.; Zhang, C.; Qin, Q.; Gu, Y.; Ke, W.; Liu, X.; Zhang, Q.; Liu, Z.; Chen, M.; et al. NADPH and NAC synergistically inhibits chronic ocular hypertension-induced neurodegeneration and neuroinflammation through regulating p38/MAPK pathway and peroxidation. *Biomed. Pharmacother.* **2024**, *175*, 116711. [CrossRef] [PubMed]
123. Kim, M.; Kim, J.Y.; Rhim, W.K.; Cimaglia, G.; Want, A.; Morgan, J.E.; Williams, P.A.; Park, C.G.; Han, D.K.; Rho, S. Extracellular vesicle encapsulated nicotinamide delivered via a trans-scleral route provides retinal ganglion cell neuroprotection. *Acta Neuropathol. Commun.* **2024**, *12*, 65. [CrossRef] [PubMed]
124. Felici, R.; Lapucci, A.; Ramazzotti, M.; Chiarugi, A. Insight into molecular and functional properties of NMNAT3 reveals new hints of NAD homeostasis within human mitochondria. *PLoS ONE* **2013**, *8*, e76938. [CrossRef]
125. Tribble, J.R.; Joe, M.; Varricchio, C.; Otmani, A.; Canovai, A.; Habchi, B.; Daskalakis, E.; Chaleckis, R.; Loreto, A.; Gilley, J.; et al. NMNAT2 is a druggable target to drive neuronal NAD production. *Nat. Commun.* **2024**, *15*, 6256. [CrossRef]
126. Kouassi Nzoughet, J.; Chao de la Barca, J.M.; Guehlouz, K.; Leruez, S.; Coulbault, L.; Allouche, S.; Bocca, C.; Muller, J.; Amati-Bonneau, P.; Gohier, P.; et al. Nicotinamide Deficiency in Primary Open-Angle Glaucoma. *Investig. Ophthalmol. Vis. Sci.* **2019**, *60*, 2509–2514. [CrossRef]
127. Hui, F.; Tang, J.; Williams, P.A.; McGuinness, M.B.; Hadoux, X.; Casson, R.J.; Coote, M.; Trounce, I.A.; Martin, K.R.; van Wijngaarden, P.; et al. Improvement in inner retinal function in glaucoma with nicotinamide (vitamin B3) supplementation: A crossover randomized clinical trial. *Clin. Exp. Ophthalmol.* **2020**, *48*, 903–914. [CrossRef]
128. De Moraes, C.G.; John, S.W.M.; Williams, P.A.; Blumberg, D.M.; Cioffi, G.A.; Liebmann, J.M. Nicotinamide and Pyruvate for Neuroenhancement in Open-Angle Glaucoma: A Phase 2 Randomized Clinical Trial. *JAMA Ophthalmol.* **2022**, *140*, 11–18. [CrossRef] [PubMed]
129. Leung, C.K.S.; Ren, S.T.; Chan, P.P.M.; Wan, K.H.N.; Kam, A.K.W.; Lai, G.W.K.; Chiu, V.S.M.; Ko, M.W.L.; Yiu, C.K.F.; Yu, M.C.Y. Nicotinamide riboside as a neuroprotective therapy for glaucoma: Study protocol for a randomized, double-blind, placebo-control trial. *Trials* **2022**, *23*, 45. [CrossRef]
130. Mehmel, M.; Jovanovic, N.; Spitz, U. Nicotinamide Riboside-The Current State of Research and Therapeutic Uses. *Nutrients* **2020**, *12*, 1616. [CrossRef]
131. Di Polo, A. Dendrite pathology and neurodegeneration: Focus on mTOR. *Neural Regen. Res.* **2015**, *10*, 559–561. [CrossRef]
132. Agostinone, J.; Alarcon-Martinez, L.; Gamlin, C.; Yu, W.Q.; Wong, R.O.L.; Di Polo, A. Insulin signalling promotes dendrite and synapse regeneration and restores circuit function after axonal injury. *Brain* **2018**, *141*, 1963–1980. [CrossRef] [PubMed]
133. Jhanwar-Uniyal, M.; Wainwright, J.V.; Mohan, A.L.; Tobias, M.E.; Murali, R.; Gandhi, C.D.; Schmidt, M.H. Diverse signaling mechanisms of mTOR complexes: mTORC1 and mTORC2 in forming a formidable relationship. *Adv. Biol. Regul.* **2019**, *72*, 51–62. [CrossRef] [PubMed]
134. Avgerinos, K.I.; Kalaitzidis, G.; Malli, A.; Kalaitzoglou, D.; Myserlis, P.G.; Lioutas, V.A. Intranasal insulin in Alzheimer's dementia or mild cognitive impairment: A systematic review. *J. Neurol.* **2018**, *265*, 1497–1510. [CrossRef] [PubMed]
135. Craft, S.; Baker, L.D.; Montine, T.J.; Minoshima, S.; Watson, G.S.; Claxton, A.; Arbuckle, M.; Callaghan, M.; Tsai, E.; Plymate, S.R.; et al. Intranasal insulin therapy for Alzheimer disease and amnestic mild cognitive impairment: A pilot clinical trial. *Arch. Neurol.* **2012**, *69*, 29–38. [CrossRef]
136. Wennberg Smith, Z.; Beykin, G.; Saludares, M.; Nunez, M.; Wang, Q.; Di Polo, A.; Goldberg, J.L. A Phase 1 Trial of Topical Insulin for Patients with Glaucoma. *Investig. Ophthalmol. Vis. Sci.* **2024**, *65*, 668.
137. Pezzuto, J.M. Resveratrol: Twenty Years of Growth, Development and Controversy. *Biomol. Ther.* **2019**, *27*, 1–14. [CrossRef]
138. Pang, Y.; Qin, M.; Hu, P.; Ji, K.; Xiao, R.; Sun, N.; Pan, X.; Zhang, X. Resveratrol protects retinal ganglion cells against ischemia induced damage by increasing Opa1 expression. *Int. J. Mol. Med.* **2020**, *46*, 1707–1720. [CrossRef]
139. Wu, Y.; Pang, Y.; Wei, W.; Shao, A.; Deng, C.; Li, X.; Chang, H.; Hu, P.; Liu, X.; Zhang, X. Resveratrol protects retinal ganglion cell axons through regulation of the SIRT1-JNK pathway. *Exp. Eye Res.* **2020**, *200*, 108249. [CrossRef]
140. Ji, K.; Li, Z.; Lei, Y.; Xu, W.; Ouyang, L.; He, T.; Xing, Y. Resveratrol attenuates retinal ganglion cell loss in a mouse model of retinal ischemia reperfusion injury via multiple pathways. *Exp. Eye Res.* **2021**, *209*, 108683. [CrossRef]
141. Luo, H.; Zhuang, J.; Hu, P.; Ye, W.; Chen, S.; Pang, Y.; Li, N.; Deng, C.; Zhang, X. Resveratrol Delays Retinal Ganglion Cell Loss and Attenuates Gliosis-Related Inflammation From Ischemia-Reperfusion Injury. *Investig. Ophthalmol. Vis. Sci.* **2018**, *59*, 3879–3888. [CrossRef]

142. Cao, K.; Ishida, T.; Fang, Y.; Shinohara, K.; Li, X.; Nagaoka, N.; Ohno-Matsui, K.; Yoshida, T. Protection of the Retinal Ganglion Cells: Intravitreal Injection of Resveratrol in Mouse Model of Ocular Hypertension. *Investig. Ophthalmol. Vis. Sci.* **2020**, *61*, 13. [CrossRef]
143. Chen, S.; Zhi, Z.; Ruan, Q.; Liu, Q.; Li, F.; Wan, F.; Reinach, P.S.; Chen, J.; Qu, J.; Zhou, X. Bright Light Suppresses Form-Deprivation Myopia Development With Activation of Dopamine D1 Receptor Signaling in the ON Pathway in Retina. *Investig. Ophthalmol. Vis. Sci.* **2017**, *58*, 2306–2316. [CrossRef]
144. Lin, J.; Xue, J.; Xu, Q.; Liu, Z.; Zhao, C.; Tang, J.; Han, J.; A, S.; Wang, W.; Zhuo, Y.; et al. In situ-crosslinked hydrogel-induced experimental glaucoma model with persistent ocular hypertension and neurodegeneration. *Biomater. Sci.* **2022**, *10*, 5006–5017. [CrossRef]
145. Zhang, Q.; Xue, J.; Tang, J.; Wu, S.; Liu, Z.; Wu, C.; Liu, C.; Liu, Y.; Lin, J.; Han, J.; et al. Modulating amacrine cell-derived dopamine signaling promotes optic nerve regeneration and preserves visual function. *Sci. Adv.* **2024**, *10*, eado0866. [CrossRef] [PubMed]
146. Ohta, Y.; Takaseki, S.; Yoshitomi, T. Effects of ripasudil hydrochloride hydrate (K-115), a Rho-kinase inhibitor, on ocular blood flow and ciliary artery smooth muscle contraction in rabbits. *Jpn. J. Ophthalmol.* **2017**, *61*, 423–432. [CrossRef] [PubMed]
147. Tanna, A.P.; Johnson, M. Rho Kinase Inhibitors as a Novel Treatment for Glaucoma and Ocular Hypertension. *Ophthalmology* **2018**, *125*, 1741–1756. [CrossRef]
148. Nakamura, N.; Honjo, M.; Yamagishi-Kimura, R.; Sakata, R.; Watanabe, S.; Aihara, M. Neuroprotective effect of omidenepag on excitotoxic retinal ganglion cell death regulating COX-2-EP2-cAMP-PKA/Epac pathway via Neuron-Glia interaction. *Neuroscience* **2024**, *553*, 145–159. [CrossRef]
149. Kapic, A.; Zaman, K.; Nguyen, V.; Neagu, G.C.; Sumien, N.; Prokai, L.; Prokai-Tatrai, K. The Prodrug DHED Delivers 17beta-Estradiol into the Retina for Protection of Retinal Ganglion Cells and Preservation of Visual Function in an Animal Model of Glaucoma. *Cells* **2024**, *13*, 1126. [CrossRef] [PubMed]
150. Zhu, Y.; Li, S.Y.; Zhang, L.J.; Lei, B.; Wang, Y.C.; Wang, Z. Neuroprotection of the P2X7 receptor antagonist A740003 on retinal ganglion cells in experimental glaucoma. *Neuroreport* **2024**, *35*, 822–831. [CrossRef]
151. Zhang, Q.; Xiang, S.; Chen, X.; Rong, Y.; Huang, L.; Chen, Z.; Yao, K.; Chen, W.; Deng, C.; Wang, J. Irisin attenuates acute glaucoma-induced neuroinflammation by activating microglia-integrin alphaVbeta5/AMPK and promoting autophagy. *Int. Immunopharmacol.* **2024**, *138*, 112545. [CrossRef]
152. Wu, L.H.; Cheng, Y.W.; Lin, F.L.; Hsu, K.C.; Wang, M.H.; Yen, J.L.; Wang, T.J.; Lin, T.E.; Liu, Y.C.; Huang, W.J.; et al. A novel HDAC8 inhibitor H7E exerts retinoprotective effects against glaucomatous injury via ameliorating aberrant Muller glia activation and oxidative stress. *Biomed. Pharmacother.* **2024**, *174*, 116538. [CrossRef]
153. Di Pierdomenico, J.; Gallego-Ortega, A.; Norte-Munoz, M.; Vidal-Villegas, B.; Bravo, I.; Boluda-Ruiz, M.; Bernal-Garro, J.M.; Fernandez-Bueno, I.; Pastor-Jimeno, J.C.; Villegas-Perez, M.P.; et al. Evaluation of the neuroprotective efficacy of the gramine derivative ITH12657 against NMDA-induced excitotoxicity in the rat retina. *Front. Neuroanat.* **2024**, *18*, 1335176. [CrossRef]
154. Dahlmann-Noor, A.; Vijay, S.; Jayaram, H.; Limb, A.; Khaw, P.T. Current approaches and future prospects for stem cell rescue and regeneration of the retina and optic nerve. *Can. J. Ophthalmol.* **2010**, *45*, 333–341. [CrossRef]
155. Fu, L.; Kwok, S.S.; Chan, Y.K.; Ming Lai, J.S.; Pan, W.; Nie, L.; Shih, K.C. Therapeutic Strategies for Attenuation of Retinal Ganglion Cell Injury in Optic Neuropathies: Concepts in Translational Research and Therapeutic Implications. *Biomed. Res. Int.* **2019**, *2019*, 8397521. [CrossRef]
156. Guo, X.; Zhou, J.; Starr, C.; Mohns, E.J.; Li, Y.; Chen, E.P.; Yoon, Y.; Kellner, C.P.; Tanaka, K.; Wang, H.; et al. Preservation of vision after CaMKII-mediated protection of retinal ganglion cells. *Cell* **2021**, *184*, 4299–4314.e4212. [CrossRef]
157. Bull, N.D.; Johnson, T.V.; Martin, K.R. Stem cells for neuroprotection in glaucoma. *Prog. Brain Res.* **2008**, *173*, 511–519. [CrossRef] [PubMed]
158. Hu, B.Y.; Xin, M.; Chen, M.; Yu, P.; Zeng, L.Z. Mesenchymal stem cells for repairing glaucomatous optic nerve. *Int. J. Ophthalmol.* **2024**, *17*, 748–760. [CrossRef] [PubMed]
159. Harrell, C.R.; Fellabaum, C.; Arsenijevic, A.; Markovic, B.S.; Djonov, V.; Volarevic, V. Therapeutic Potential of Mesenchymal Stem Cells and Their Secretome in the Treatment of Glaucoma. *Stem Cells Int.* **2019**, *2019*, 7869130. [CrossRef] [PubMed]
160. Johnson, T.V.; DeKorver, N.W.; Levasseur, V.A.; Osborne, A.; Tassoni, A.; Lorber, B.; Heller, J.P.; Villasmil, R.; Bull, N.D.; Martin, K.R.; et al. Identification of retinal ganglion cell neuroprotection conferred by platelet-derived growth factor through analysis of the mesenchymal stem cell secretome. *Brain* **2014**, *137*, 503–519. [CrossRef] [PubMed]
161. Park, S.S.; Moisseiev, E.; Bauer, G.; Anderson, J.D.; Grant, M.B.; Zam, A.; Zawadzki, R.J.; Werner, J.S.; Nolta, J.A. Advances in bone marrow stem cell therapy for retinal dysfunction. *Prog. Retin. Eye Res.* **2017**, *56*, 148–165. [CrossRef]
162. Crigler, L.; Robey, R.C.; Asawachaicharn, A.; Gaupp, D.; Phinney, D.G. Human mesenchymal stem cell subpopulations express a variety of neuro-regulatory molecules and promote neuronal cell survival and neuritogenesis. *Exp. Neurol.* **2006**, *198*, 54–64. [CrossRef] [PubMed]
163. Emre, E.; Yuksel, N.; Duruksu, G.; Pirhan, D.; Subasi, C.; Erman, G.; Karaoz, E. Neuroprotective effects of intravitreally transplanted adipose tissue and bone marrow-derived mesenchymal stem cells in an experimental ocular hypertension model. *Cytotherapy* **2015**, *17*, 543–559. [CrossRef]

164. Harper, M.M.; Grozdanic, S.D.; Blits, B.; Kuehn, M.H.; Zamzow, D.; Buss, J.E.; Kardon, R.H.; Sakaguchi, D.S. Transplantation of BDNF-secreting mesenchymal stem cells provides neuroprotection in chronically hypertensive rat eyes. *Investig. Ophthalmol. Vis. Sci.* **2011**, *52*, 4506–4515. [CrossRef] [PubMed]
165. Johnson, T.V.; Bull, N.D.; Hunt, D.P.; Marina, N.; Tomarev, S.I.; Martin, K.R. Neuroprotective effects of intravitreal mesenchymal stem cell transplantation in experimental glaucoma. *Investig. Ophthalmol. Vis. Sci.* **2010**, *51*, 2051–2059. [CrossRef]
166. Mead, B.; Amaral, J.; Tomarev, S. Mesenchymal Stem Cell-Derived Small Extracellular Vesicles Promote Neuroprotection in Rodent Models of Glaucoma. *Investig. Ophthalmol. Vis. Sci.* **2018**, *59*, 702–714. [CrossRef] [PubMed]
167. Roubeix, C.; Godefroy, D.; Mias, C.; Sapienza, A.; Riancho, L.; Degardin, J.; Fradot, V.; Ivkovic, I.; Picaud, S.; Sennlaub, F.; et al. Intraocular pressure reduction and neuroprotection conferred by bone marrow-derived mesenchymal stem cells in an animal model of glaucoma. *Stem Cell Res. Ther.* **2015**, *6*, 177. [CrossRef]
168. Wang, Y.; Lv, J.; Huang, C.; Li, X.; Chen, Y.; Wu, W.; Wu, R. Human Umbilical Cord-Mesenchymal Stem Cells Survive and Migrate within the Vitreous Cavity and Ameliorate Retinal Damage in a Novel Rat Model of Chronic Glaucoma. *Stem Cells Int.* **2021**, *2021*, 8852517. [CrossRef]
169. Osborne, A.; Sanderson, J.; Martin, K.R. Neuroprotective Effects of Human Mesenchymal Stem Cells and Platelet-Derived Growth Factor on Human Retinal Ganglion Cells. *Stem Cells* **2018**, *36*, 65–78. [CrossRef]
170. Vilela, C.A.P.; Messias, A.; Calado, R.T.; Siqueira, R.C.; Silva, M.J.L.; Covas, D.T.; Paula, J.S. Retinal function after intravitreal injection of autologous bone marrow-derived mesenchymal stromal cells in advanced glaucoma. *Doc. Ophthalmol.* **2021**, *143*, 33–38. [CrossRef]
171. Lopez Sanchez, M.I.; Crowston, J.G.; Mackey, D.A.; Trounce, I.A. Emerging Mitochondrial Therapeutic Targets in Optic Neuropathies. *Pharmacol. Ther.* **2016**, *165*, 132–152. [CrossRef]
172. Chao, J.R.; Lamba, D.A.; Klesert, T.R.; Torre, A.; Hoshino, A.; Taylor, R.J.; Jayabalu, A.; Engel, A.L.; Khuu, T.H.; Wang, R.K.; et al. Transplantation of Human Embryonic Stem Cell-Derived Retinal Cells into the Subretinal Space of a Non-Human Primate. *Transl. Vis. Sci. Technol.* **2017**, *6*, 4. [CrossRef] [PubMed]
173. Sluch, V.M.; Davis, C.H.; Ranganathan, V.; Kerr, J.M.; Krick, K.; Martin, R.; Berlinicke, C.A.; Marsh-Armstrong, N.; Diamond, J.S.; Mao, H.Q.; et al. Differentiation of human ESCs to retinal ganglion cells using a CRISPR engineered reporter cell line. *Sci. Rep.* **2015**, *5*, 16595. [CrossRef] [PubMed]
174. Venugopalan, P.; Wang, Y.; Nguyen, T.; Huang, A.; Muller, K.J.; Goldberg, J.L. Transplanted neurons integrate into adult retinas and respond to light. *Nat. Commun.* **2016**, *7*, 10472. [CrossRef] [PubMed]
175. Ma, J.; Guo, C.; Guo, C.; Sun, Y.; Liao, T.; Beattie, U.; Lopez, F.J.; Chen, D.F.; Lashkari, K. Transplantation of Human Neural Progenitor Cells Expressing IGF-1 Enhances Retinal Ganglion Cell Survival. *PLoS ONE* **2015**, *10*, e0125695. [CrossRef]
176. Zhou, X.; Xia, X.B.; Xiong, S.Q. Neuro-protection of retinal stem cells transplantation combined with copolymer-1 immunization in a rat model of glaucoma. *Mol. Cell Neurosci.* **2013**, *54*, 1–8. [CrossRef]
177. Oswald, J.; Kegeles, E.; Minelli, T.; Volchkov, P.; Baranov, P. Transplantation of miPSC/mESC-derived retinal ganglion cells into healthy and glaucomatous retinas. *Mol. Ther. Methods Clin. Dev.* **2021**, *21*, 180–198. [CrossRef] [PubMed]
178. Suen, H.C.; Qian, Y.; Liao, J.; Luk, C.S.; Lee, W.T.; Ng, J.K.W.; Chan, T.T.H.; Hou, H.W.; Li, I.; Li, K.; et al. Transplantation of Retinal Ganglion Cells Derived from Male Germline Stem Cell as a Potential Treatment to Glaucoma. *Stem Cells Dev.* **2019**, *28*, 1365–1375. [CrossRef]
179. Qin, Y.; Ge, G.; Yang, P.; Wang, L.; Qiao, Y.; Pan, G.; Yang, H.; Bai, J.; Cui, W.; Geng, D. An Update on Adipose-Derived Stem Cells for Regenerative Medicine: Where Challenge Meets Opportunity. *Adv. Sci.* **2023**, *10*, e2207334. [CrossRef]
180. Greco, S.J.; Rameshwar, P. Microenvironmental considerations in the application of human mesenchymal stem cells in regenerative therapies. *Biologics* **2008**, *2*, 699–705. [CrossRef]
181. Kasetty, M.A.; Hedges, T.R., 3rd; Witkin, A.J. Bilateral Epiretinal Membrane Formation after Intravitreal Injections of Autologous Mesenchymal Stem Cells. *Retin. Cases Brief Rep.* **2022**, *16*, 561–564. [CrossRef]
182. Jain, A.; Zode, G.; Kasetti, R.B.; Ran, F.A.; Yan, W.; Sharma, T.P.; Bugge, K.; Searby, C.C.; Fingert, J.H.; Zhang, F.; et al. CRISPR-Cas9-based treatment of myocilin-associated glaucoma. *Proc. Natl. Acad. Sci. USA* **2017**, *114*, 11199–11204. [CrossRef] [PubMed]
183. Souma, T.; Tompson, S.W.; Thomson, B.R.; Siggs, O.M.; Kizhatil, K.; Yamaguchi, S.; Feng, L.; Limviphuvadh, V.; Whisenhunt, K.N.; Maurer-Stroh, S.; et al. Angiopoietin receptor TEK mutations underlie primary congenital glaucoma with variable expressivity. *J. Clin. Investig.* **2016**, *126*, 2575–2587. [CrossRef]
184. Thomson, B.R.; Heinen, S.; Jeansson, M.; Ghosh, A.K.; Fatima, A.; Sung, H.K.; Onay, T.; Chen, H.; Yamaguchi, S.; Economides, A.N.; et al. A lymphatic defect causes ocular hypertension and glaucoma in mice. *J. Clin. Investig.* **2014**, *124*, 4320–4324. [CrossRef] [PubMed]
185. Osborne, A.; Khatib, T.Z.; Songra, L.; Barber, A.C.; Hall, K.; Kong, G.Y.X.; Widdowson, P.S.; Martin, K.R. Neuroprotection of retinal ganglion cells by a novel gene therapy construct that achieves sustained enhancement of brain-derived neurotrophic factor/tropomyosin-related kinase receptor-B signaling. *Cell Death Dis.* **2018**, *9*, 1007. [CrossRef]
186. Osborne, A.; Wang, A.X.Z.; Tassoni, A.; Widdowson, P.S.; Martin, K.R. Design of a Novel Gene Therapy Construct to Achieve Sustained Brain-Derived Neurotrophic Factor Signaling in Neurons. *Hum. Gene Ther.* **2018**, *29*, 828–841. [CrossRef] [PubMed]
187. Khatib, T.Z.; Osborne, A.; Yang, S.; Ali, Z.; Jia, W.; Manyakin, I.; Hall, K.; Watt, R.; Widdowson, P.S.; Martin, K.R. Receptor-ligand supplementation via a self-cleaving 2A peptide-based gene therapy promotes CNS axonal transport with functional recovery. *Sci. Adv.* **2021**, *7*, eabd2590. [CrossRef]

188. Wojcik-Gryciuk, A.; Gajewska-Wozniak, O.; Kordecka, K.; Boguszewski, P.M.; Waleszczyk, W.; Skup, M. Neuroprotection of Retinal Ganglion Cells with AAV2-BDNF Pretreatment Restoring Normal TrkB Receptor Protein Levels in Glaucoma. *Int. J. Mol. Sci.* **2020**, *21*, 6262. [CrossRef]
189. Alqawlaq, S.; Sivak, J.M.; Huzil, J.T.; Ivanova, M.V.; Flanagan, J.G.; Beazely, M.A.; Foldvari, M. Preclinical development and ocular biodistribution of gemini-DNA nanoparticles after intravitreal and topical administration: Towards non-invasive glaucoma gene therapy. *Nanomedicine* **2014**, *10*, 1637–1647. [CrossRef]
190. Cen, L.P.; Liang, J.J.; Chen, J.H.; Harvey, A.R.; Ng, T.K.; Zhang, M.; Pang, C.P.; Cui, Q.; Fan, Y.M. AAV-mediated transfer of RhoA shRNA and CNTF promotes retinal ganglion cell survival and axon regeneration. *Neuroscience* **2017**, *343*, 472–482. [CrossRef]
191. Hellstrom, M.; Pollett, M.A.; Harvey, A.R. Post-injury delivery of rAAV2-CNTF combined with short-term pharmacotherapy is neuroprotective and promotes extensive axonal regeneration after optic nerve trauma. *J. Neurotrauma* **2011**, *28*, 2475–2483. [CrossRef]
192. Do Rhee, K.; Wang, Y.; Ten Hoeve, J.; Stiles, L.; Nguyen, T.T.T.; Zhang, X.; Vergnes, L.; Reue, K.; Shirihai, O.; Bok, D.; et al. Ciliary neurotrophic factor-mediated neuroprotection involves enhanced glycolysis and anabolism in degenerating mouse retinas. *Nat. Commun.* **2022**, *13*, 7037. [CrossRef] [PubMed]
193. Hakim, A.; Guido, B.; Narsineni, L.; Chen, D.W.; Foldvari, M. Gene therapy strategies for glaucoma from IOP reduction to retinal neuroprotection: Progress towards non-viral systems. *Adv. Drug. Deliv. Rev.* **2023**, *196*, 114781. [CrossRef]
194. Bosco, A.; Anderson, S.R.; Breen, K.T.; Romero, C.O.; Steele, M.R.; Chiodo, V.A.; Boye, S.L.; Hauswirth, W.W.; Tomlinson, S.; Vetter, M.L. Complement C3-Targeted Gene Therapy Restricts Onset and Progression of Neurodegeneration in Chronic Mouse Glaucoma. *Mol. Ther.* **2018**, *26*, 2379–2396. [CrossRef] [PubMed]
195. Shen, J.; Xiao, R.; Bair, J.; Wang, F.; Vandenberghe, L.H.; Dartt, D.; Baranov, P.; Ng, Y.S.E. Novel engineered, membrane-localized variants of vascular endothelial growth factor (VEGF) protect retinal ganglion cells: A proof-of-concept study. *Cell Death Dis.* **2018**, *9*, 1018. [CrossRef]
196. Wang, Q.; Zhuang, P.; Huang, H.; Li, L.; Liu, L.; Webber, H.C.; Dalal, R.; Siew, L.; Fligor, C.M.; Chang, K.C.; et al. Mouse gamma-Synuclein Promoter-Mediated Gene Expression and Editing in Mammalian Retinal Ganglion Cells. *J. Neurosci.* **2020**, *40*, 3896–3914. [CrossRef] [PubMed]
197. Donahue, R.J.; Fehrman, R.L.; Gustafson, J.R.; Nickells, R.W. BCLX(L) gene therapy moderates neuropathology in the DBA/2J mouse model of inherited glaucoma. *Cell Death Dis.* **2021**, *12*, 781. [CrossRef] [PubMed]
198. Fang, F.; Zhuang, P.; Feng, X.; Liu, P.; Liu, D.; Huang, H.; Li, L.; Chen, W.; Liu, L.; Sun, Y.; et al. NMNAT2 is downregulated in glaucomatous RGCs, and RGC-specific gene therapy rescues neurodegeneration and visual function. *Mol. Ther.* **2022**, *30*, 1421–1431. [CrossRef]
199. Lani-Louzada, R.; Marra, C.; Dias, M.S.; de Araujo, V.G.; Abreu, C.A.; Ribas, V.T.; Adesse, D.; Allodi, S.; Chiodo, V.; Hauswirth, W.; et al. Neuroprotective Gene Therapy by Overexpression of the Transcription Factor MAX in Rat Models of Glaucomatous Neurodegeneration. *Investig. Ophthalmol. Vis. Sci.* **2022**, *63*, 5. [CrossRef]
200. Visuvanathan, S.; Baker, A.N.; Lagali, P.S.; Coupland, S.G.; Miller, G.; Hauswirth, W.W.; Tsilfidis, C. XIAP gene therapy effects on retinal ganglion cell structure and function in a mouse model of glaucoma. *Gene Ther.* **2022**, *29*, 147–156. [CrossRef]
201. Jiang, W.; Tang, L.; Zeng, J.; Chen, B. Adeno-associated virus mediated SOD gene therapy protects the retinal ganglion cells from chronic intraocular pressure elevation induced injury via attenuating oxidative stress and improving mitochondrial dysfunction in a rat model. *Am. J. Transl. Res.* **2016**, *8*, 799–810.
202. Petrova, V.; Pearson, C.S.; Ching, J.; Tribble, J.R.; Solano, A.G.; Yang, Y.; Love, F.M.; Watt, R.J.; Osborne, A.; Reid, E.; et al. Protrudin functions from the endoplasmic reticulum to support axon regeneration in the adult CNS. *Nat. Commun.* **2020**, *11*, 5614. [CrossRef] [PubMed]
203. Mackiewicz, J.; Tomczak, J.; Lisek, M.; Sakowicz, A.; Guo, F.; Boczek, T. NFATc4 Knockout Promotes Neuroprotection and Retinal Ganglion Cell Regeneration After Optic Nerve Injury. *Mol. Neurobiol.* **2024**. *ahead of print*. [CrossRef]
204. Thananthirige, K.P.M.; Chitranshi, N.; Basavarajappa, D.; Rajput, R.; Abbasi, M.; Palanivel, V.; Gupta, V.B.; Paulo, J.A.; Koronyo-Hamaoui, M.; Mirzaei, M.; et al. Tau modulation through AAV9 therapy augments Akt/Erk survival signalling in glaucoma mitigating the retinal degenerative phenotype. *Acta Neuropathol. Commun.* **2024**, *12*, 89. [CrossRef]
205. Brown, M.D.; Starikovskaya, E.; Derbeneva, O.; Hosseini, S.; Allen, J.C.; Mikhailovskaya, I.E.; Sukernik, R.I.; Wallace, D.C. The role of mtDNA background in disease expression: A new primary LHON mutation associated with Western Eurasian haplogroup. *J. Hum. Genet.* **2002**, *110*, 130–138. [CrossRef]
206. Alexander, C.; Votruba, M.; Pesch, U.E.; Thiselton, D.L.; Mayer, S.; Moore, A.; Rodriguez, M.; Kellner, U.; Leo-Kottler, B.; Auburger, G.; et al. OPA1, encoding a dynamin-related GTPase, is mutated in autosomal dominant optic atrophy linked to chromosome 3q28. *Nat. Genet.* **2000**, *26*, 211–215. [CrossRef] [PubMed]
207. Osborne, N.N.; Nunez-Alvarez, C.; Joglar, B.; Del Olmo-Aguado, S. Glaucoma: Focus on mitochondria in relation to pathogenesis and neuroprotection. *Eur. J. Pharmacol.* **2016**, *787*, 127–133. [CrossRef] [PubMed]
208. Harun-Or-Rashid, M.; Pappenhagen, N.; Palmer, P.G.; Smith, M.A.; Gevorgyan, V.; Wilson, G.N.; Crish, S.D.; Inman, D.M. Structural and Functional Rescue of Chronic Metabolically Stressed Optic Nerves through Respiration. *J. Neurosci.* **2018**, *38*, 5122–5139. [CrossRef] [PubMed]

209. Oddone, F.; Rossetti, L.; Parravano, M.; Sbardella, D.; Coletta, M.; Ziccardi, L.; Roberti, G.; Carnevale, C.; Romano, D.; Manni, G.; et al. Citicoline in Ophthalmological Neurodegenerative Disease: A Comprehensive Review. *Pharmaceuticals* **2021**, *14*, 281. [CrossRef]
210. Williams, P.A.; Harder, J.M.; Cardozo, B.H.; Foxworth, N.E.; John, S.W.M. Nicotinamide treatment robustly protects from inherited mouse glaucoma. *Commun. Integr. Biol.* **2018**, *11*, e1356956. [CrossRef]
211. Lanza, M.; Gironi Carnevale, U.A.; Mele, L.; Bifani Sconocchia, M.; Bartollino, S.; Costagliola, C. Morphological and Functional Evaluation of Oral Citicoline Therapy in Chronic Open-Angle Glaucoma Patients: A Pilot Study With a 2-Year Follow-Up. *Front. Pharmacol.* **2019**, *10*, 1117. [CrossRef]
212. Gabelein, C.G.; Feng, Q.; Sarajlic, E.; Zambelli, T.; Guillaume-Gentil, O.; Kornmann, B.; Vorholt, J.A. Mitochondria transplantation between living cells. *PLoS Biol.* **2022**, *20*, e3001576. [CrossRef]
213. Nascimento-Dos-Santos, G.; de-Souza-Ferreira, E.; Lani, R.; Faria, C.C.; Araujo, V.G.; Teixeira-Pinheiro, L.C.; Vasconcelos, T.; Goncalo, T.; Santiago, M.F.; Linden, R.; et al. Neuroprotection from optic nerve injury and modulation of oxidative metabolism by transplantation of active mitochondria to the retina. *Biochim. Biophys. Acta Mol. Basis Dis.* **2020**, *1866*, 165686. [CrossRef] [PubMed]
214. Li, S.F.; Han, Y.; Wang, F.; Su, Y. Progress in exosomes and their potential use in ocular diseases. *Int. J. Ophthalmol.* **2020**, *13*, 1493–1498. [CrossRef] [PubMed]
215. Lener, T.; Gimona, M.; Aigner, L.; Borger, V.; Buzas, E.; Camussi, G.; Chaput, N.; Chatterjee, D.; Court, F.A.; Del Portillo, H.A.; et al. Applying extracellular vesicles based therapeutics in clinical trials—An ISEV position paper. *J. Extracell Vesicles* **2015**, *4*, 30087. [CrossRef] [PubMed]
216. Mead, B.; Tomarev, S. Bone Marrow-Derived Mesenchymal Stem Cells-Derived Exosomes Promote Survival of Retinal Ganglion Cells Through miRNA-Dependent Mechanisms. *Stem Cells Transl. Med.* **2017**, *6*, 1273–1285. [CrossRef]
217. Pan, D.; Chang, X.; Xu, M.; Zhang, M.; Zhang, S.; Wang, Y.; Luo, X.; Xu, J.; Yang, X.; Sun, X. UMSC-derived exosomes promote retinal ganglion cells survival in a rat model of optic nerve crush. *J. Chem. Neuroanat.* **2019**, *96*, 134–139. [CrossRef] [PubMed]
218. Yu, B.; Wang, K.; Hao, H.; Liu, Y.; Yue, Y.; Li, X.; Xing, X.; Zhang, X. Small extracellular vesicles derived from microRNA-22-3p-overexpressing mesenchymal stem cells protect retinal ganglion cells by regulating MAPK pathway. *Commun. Biol.* **2024**, *7*, 807. [CrossRef]
219. Wang, J.J.; Zeng, Z.W.; Xiao, R.Z.; Xie, T.; Zhou, G.L.; Zhan, X.R.; Wang, S.L. Recent advances of chitosan nanoparticles as drug carriers. *Int. J. Nanomed.* **2011**, *6*, 765–774. [CrossRef]
220. Giannaccini, M.; Usai, A.; Chiellini, F.; Guadagni, V.; Andreazzoli, M.; Ori, M.; Pasqualetti, M.; Dente, L.; Raffa, V. Neurotrophin-conjugated nanoparticles prevent retina damage induced by oxidative stress. *Cell Mol Life Sci.* **2018**, *75*, 1255–1267. [CrossRef]
221. Jiang, C.; Moore, M.J.; Zhang, X.; Klassen, H.; Langer, R.; Young, M. Intravitreal injections of GDNF-loaded biodegradable microspheres are neuroprotective in a rat model of glaucoma. *Mol Vis* **2007**, *13*, 1783–1792.
222. Checa-Casalengua, P.; Jiang, C.; Bravo-Osuna, I.; Tucker, B.A.; Molina-Martinez, I.T.; Young, M.J.; Herrero-Vanrell, R. Retinal ganglion cells survival in a glaucoma model by GDNF/Vit E PLGA microspheres prepared according to a novel microencapsulation procedure. *J. Control Release* **2011**, *156*, 92–100. [CrossRef]
223. Jiang, W.; Xiao, D.; Wu, C.; Yang, J.; Peng, X.; Chen, L.; Zhang, J.; Zha, G.; Li, W.; Ju, R.; et al. Circular RNA-based therapy provides sustained and robust neuroprotection for retinal ganglion cells. *Mol. Ther. Nucleic Acids* **2024**, *35*, 102258. [CrossRef] [PubMed]
224. Davis, B.M.; Pahlitzsch, M.; Guo, L.; Balendra, S.; Shah, P.; Ravindran, N.; Malaguarnera, G.; Sisa, C.; Shamsher, E.; Hamze, H.; et al. Topical Curcumin Nanocarriers are Neuroprotective in Eye Disease. *Sci. Rep.* **2018**, *8*, 11066. [CrossRef] [PubMed]
225. Rai, A.; Mhatre, S.; Chandler, C.; Opere, C.; Singh, S. Application of Quality by Design in the Development of Hydrogen Sulfide Donor Loaded Polymeric Microparticles. *AAPS PharmSciTech* **2024**, *25*, 132. [CrossRef] [PubMed]

pharmaceuticals

Review

Review: Neuroprotective Nanocarriers in Glaucoma

Kun Pei [1], Maria Georgi [2,3], Daniel Hill [1], Chun Fung Jeffrey Lam [4], Wei Wei [3,5] and Maria Francesca Cordeiro [1,3,5,6,*]

1 UCL Institute of Ophthalmology, London EC1V 9EL, UK; kun.pei.21@ucl.ac.uk (K.P.); d.hill.13@ucl.ac.uk (D.H.)
2 St Mary's Hospital, Imperial College Healthcare NHS Trust, London W2 1NY, UK; m.georgi@imperial.ac.uk
3 Department of Surgery & Cancer, Imperial College London, London SW7 5NG, UK; w.wei20@imperial.ac.uk
4 GKT Medical School, Guy's Campus, Great Maze Pond, London SE1 1UL, UK; chun.lam@kcl.ac.uk
5 Imperial College Ophthalmic Research Group (ICORG) Unit, Imperial College, London NW1 5QH, UK
6 Western Eye Hospital, London NW1 5QH, UK
* Correspondence: m.cordeiro@imperial.ac.uk

Abstract: Glaucoma stands as a primary cause of irreversible blindness globally, characterized by the progressive dysfunction and loss of retinal ganglion cells (RGCs). While current treatments primarily focus on controlling intraocular pressure (IOP), many patients continue to experience vision loss. Therefore, the research focus has shifted to therapeutic targets aimed at preventing or delaying RGC death and optic nerve degeneration to slow or halt disease progression. Traditional ocular drug administration, such as eye drops or oral medications, face significant challenges due to the eye's unique structural and physiological barriers, which limit effective drug delivery. Invasive methods like intravitreal injections can cause side effects such as bleeding, inflammation, and infection, making non-invasive delivery methods with high bioavailability very desirable. Nanotechnology presents a promising approach to addressing these limitations in glaucoma treatment. This review summarizes current approaches involving neuroprotective drugs combined with nanocarriers, and their impact for future use.

Keywords: glaucoma; ocular neurodegenerative disease; nanomedicine; nanocarriers; neuroprotection; neuroprotective; retinal ganglion cell

1. Introduction

Glaucoma, a principal cause of irreversible blindness globally [1,2], remains a multifactorial neurodegenerative condition with an unclear pathogenesis. It is marked by the gradual degeneration and structural damage to RGCs, primarily manifesting as progressive visual field loss [3]. Due to neural compensatory mechanisms for visual deficits, patients typically present late with gradual and often asymmetrical vision loss [4]. The current estimated global prevalence of glaucoma in a population aged 40–80 years of age is 3.54%. In 2013, it was estimated that 64.3 million people worldwide between the ages of 40 and 80 had glaucoma. This number increased to 76.0 million by 2020 and is projected to reach 111.8 million by 2040 [5]. However, these figures are likely an underestimation with over 90% of glaucoma cases in low- and middle-income countries remaining undiagnosed [6,7]. Glaucoma also poses significant direct and indirect financial loses, with an estimated annual cost of USD 2.9 billion in the United States [6].

The Glaucoma Research Foundation recognizes multiple glaucoma subtypes, including primary and secondary open-angle, primary and secondary closed-angle, and normal tension glaucoma. Risk factors include elevated intraocular pressure, genetic predisposition, myopia, ethnicity (such as Afro-Caribbean), and hypertension [7,8]. Currently, elevated IOP remains the only modifiable risk factor that medical intervention can target [9,10]. Most current treatments therefore try to lower IOP through surgery, laser, or medication to prevent or delay the progression of RGC degeneration.

Although most current treatments have resulted in effectively controlled IOP, these approaches are not optimal as they fail to address the persistent loss of RGCs [11]. RGCs serve a pivotal function in visual processing by transmitting information from the retina to the brain and are the primary cellular component of the ganglion cell layer (GCL) [11,12]. The adult human eye contains approximately 1.2 million RGCs [13]. Cell bodies, intricate dendritic structures, and distinctive axons make up the structural components of RGCs [11]. Dendrites process photoreceptor impulses and receive synaptic input mostly from amacrine and bipolar cells, which act as intermediary neurons in the retina to transmit and modulate information from the photoreceptors (Figure 1). These interactions form a critical hierarchical network essential for the integration and processing of visual signals into nerve impulses. The action potentials generated then travel via the axons of RGCs, which merge to form the optic nerve. These axons pass through the optic disc and optic canal, forming synapses in the lateral geniculate nucleus (LGN) of the thalamus and partially in the superior colliculus (SC), thereby linking the visual pathway of the eye with that of the brain [14]. The LGN processes visual information, relaying it to the primary visual cortex for further complex analysis and integration, highlighting the crucial role of RGCs in visual perception [13,15].

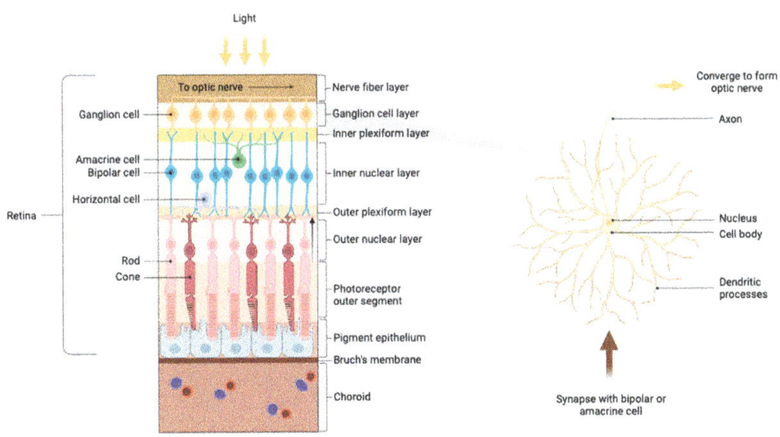

Figure 1. Anatomy of the retina and ganglion cells.

Mechanisms leading to RGC apoptosis in glaucoma include ischemia [16], oxidative stress [17,18], mitochondrial dysfunction [17], diminished levels of neurotrophic factor [18,19], accumulation of misfolded proteins [20], neuroinflammation [21], and mechanical stress due to elevated intraocular pressure [22,23]. Despite concerted efforts by ophthalmologists to use pharmacological, laser, and surgical methods to lower intraocular pressure, these strategies are not always successful in protecting retinal ganglion cells [24]. Furthermore, the side effects associated with these treatments present significant concerns, particularly adverse reactions to the medications or complications from surgical procedures [25]. This underscores the complexity of treating glaucoma and highlights the need for developing more effective and safer therapeutic approaches. The neuroprotective approach aims to preserve RGCs and their axons independently of IOP, lowering the number of interventions [26]. This approach has gained prominence as it seeks to directly address the cellular and molecular pathways involved in RGC damage, providing a potentially more comprehensive treatment paradigm for glaucoma [26]. Neuroprotection focuses on protecting neurons from predominantly secondary neurodegeneration to improve health outcomes [27,28]. The primary target of neuroprotection in glaucoma is the retinal ganglion cell. RGC loss is irreversible [15]; thus, the protection of these cells is essential. Neuroprotection in glaucoma provides a supplementary therapy for those whose progressive visual loss is not sufficiently mitigated by IOP reduction alone [25,27]. In addition, it has potential to serve as an alternative and complementary therapy for patients who are intolerant or unresponsive to IOP-lowering treatments, thus broadening the scope of potential treatment options in glaucoma management [25,27].

Numerous studies have investigated neuroprotective therapies for glaucoma, identifying several promising avenues. These include neurotrophic factors [29], dietary supplements [27,30,31], anti-inflammatory therapies [32,33], α2-adrenoceptor agonists [34], N-methyl-D-aspartate (NMDA) receptor antagonists [35], calcium channel blockers [36], and metabolic-targeting compounds [37]. Table 1 provides a summary of the contemporary neuroprotective treatments for glaucoma. These therapies, which are either undergoing preclinical evaluation or are currently being tested in clinic, have demonstrated promising results in terms of efficacy. Each therapeutic avenue targets a unique pathway, offering diverse mechanisms to mitigate the neurodegenerative processes in glaucoma. The diversity of neuroprotective therapeutic strategies reflects the evolving understanding of the complex pathophysiology of glaucoma, with each pathway offering a unique potential therapeutic target.

Among the various challenges faced by neuroprotective therapies, a significant barrier to their effectiveness is the blood–retinal barrier (BRB), which plays a crucial role in the eye's posterior segment, tightly regulating and isolating the retina's microenvironment from systemic circulation [38]. The BRB is composed of two main components: the internal BRB (iBRB) and the external BRB (oBRB) [38]. The iBRB is formed by tight junctions between endothelial cells of the retinal capillaries, which originate from the central retinal artery. It serves the inner retinal layers through three vascular plexuses, effectively controlling the internal environment. The oBRB, on the other hand, includes tight junctions between retinal pigment epithelial cells, the choroid, BM, and the RPE [38]. The choroid's capillaries are crucial for supplying oxygen and nutrients to the outer retinal layers, removing waste products, and maintaining retinal temperature and volume [39]. The BM, located between these capillaries and the RPE, allows for selective molecular diffusion and blocks larger molecules, which helps stabilize the RPE and absorb physical stress [40]. These barriers prevent water, plasma, and toxic compounds from penetrating the eye. They may also impede the administration of therapeutic drugs [38].

The effectiveness of common treatment methods is significantly compromised by the barrier effect of the BRB. Currently, the most prevalent methods of ocular drug delivery include topical administration or intracameral, suprachoroidal, subconjunctival, intravitreal, and subretinal injections, and finally, systemic administration (Figure 2) [41]. Table 2 presents the advantages and disadvantages of these modes of administration. These methods achieve varying degrees of therapeutic success, especially as their bioavailability is generally low, typically less than 5% (largely due to the BRB [42]). As a result, it is important to develop innovative techniques overcoming or bypassing the challenges imposed by the BRB.

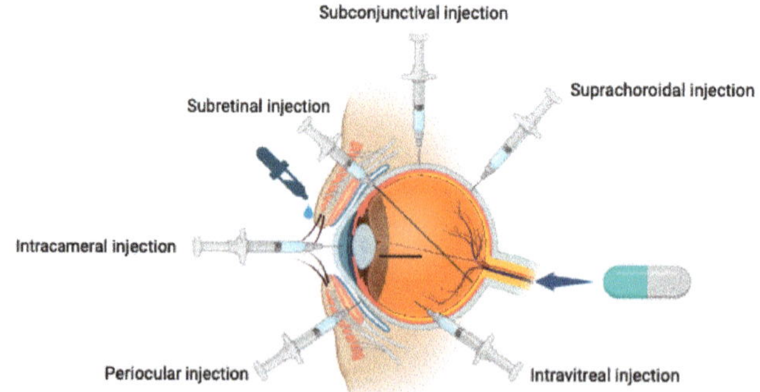

Figure 2. Ocular drug delivery systems.

The BRB's effects on drug bioavailability limit the beneficial effects of neuroprotective agents using conventional drug delivery methods. However, advances in nanotechnology have significantly pushed the field of ocular drug delivery forward [43]. Nanocarriers including micelles, nanoemulsions, solid lipid nanoparticles, dendrimers, liposomes, and exosomes have shown promise compared to the established methods of drug delivery [44]. The design of these novel carriers aims to overcome ocular barriers, improve drug stability, and boost therapeutic efficacy [44]. They also extend the half-life of the drugs in ocular tissues, and reduce the frequency of dosing; these factors improve patient compliance and decrease drug-related side effects [45]. Furthermore, nanocarriers enable sustained and controlled release, targeted drug delivery, and the combination of multiple drugs, resulting in improved treatment outcomes for a variety of ocular conditions [45] (Figure 3).

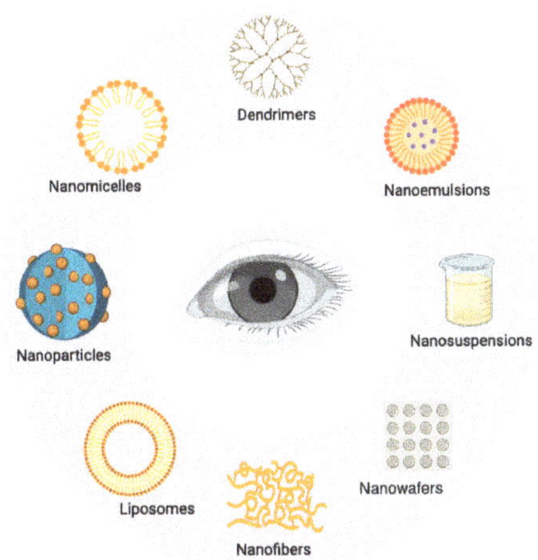

Figure 3. Nanocarrier formulations for ocular drug delivery.

Nanocarriers 1.2 exhibit a range of mechanisms of action which enhance the delivery and efficacy of neuroprotective agents.

Dendrimers are symmetrical synthetic polymers which consist of three architectural domains: a central core, a hyperbranched mantle, and a corona with peripheral reactive functional groups (dendrons). Their unique properties, including hyperbranching and high biocompatibility with biological systems, make them valuable in a wide range of medical and biomedical applications [46]. These three-dimensional nanopolymeric structures are particularly promising as vectors for drug and gene delivery. Enhanced drug delivery is achieved by increasing the solubility of hydrophobic drugs through the entrapment of neuroprotective agents within the intramolecular cavities of dendrimers. Additionally, the exterior functional groups allow for the attachment of targeting moieties, improving the specificity of drug delivery to certain diseases [47]. However, dendrimers can interact with biological membranes, potentially resulting in apoptosis and hemolysis. In vivo studies have demonstrated hepatotoxicity and renal toxicity of higher-generation dendrimers [48]; further in vivo investigations into the surface modification and safety characterization of dendrimers are required.

Liposomes are spherical phospholipid vesicles with a cell-like membrane which provides low immunogenicity, low toxicity, and high biocompatibility [49]. They can encapsulate either hydrophobic or hydrophilic agents, protecting drugs from hydrolysis and prolonging their biological half-life, while controlling drug release [50]. Liposomes are limited by the low entrapment efficiency of water-soluble molecules, rapid hemofiltration and excretions, and fast drug release (Gireesh, T., C. Kuldeep, and K. Pramod. "Liposomal current status, evaluation and recent advances" [51]).

Nanoparticles (NPs) are drug carriers ranging from 1 to 1000 nm in size. They can be divided into polymer and lipid nanoparticles [52]. Ocular NPs are primarily composed of lipids, proteins, and either natural or synthetic polymers [53]. Nanoparticles (NPs) can be classified into nanocapsules and nanospheres based on their morphological structure. Nanospheres possess a solid polymeric structure, while nanocapsules feature a thin polymeric shell, about 5 nm thick, encasing an oily core. These NPs have the capability to encapsulate both hydrophobic and hydrophilic drugs, protecting them from degradation and enhancing targeted delivery. This results in more efficient drug absorption, controlled release, and improved bioavailability [54].

Their absorption efficiency is significantly affected by their surface charge [55]; catatonic NPs, for instance, exhibit higher retention times on ocular surfaces when compared to anionic NPs due to the negatively charged surfaces on corneal and conjunctival tissues [55]. NPs have several advantages, specifically regarding ocular drug delivery; (1) their small size reduces ocular irritation; (2) their superior absorption abilities enable the sustained drug release and therefore reduce dosing intervals; (3) they feature improved cellular penetration due to their better absorption; (4) they enable targeted drug delivery to desired tissues [56–58].

Although NPs hold significant potential for the management of ocular diseases, severe challenges hinder their widespread use in clinical practice. These challenges include insufficient drug loading, premature drug release during storage, difficulties in achieving uniform particle dispersion, and toxic effects associated with surfactant concentrations [59]. Further research is needed to advance the clinical application of NPs.

Table 3 summarizes recent research focused on neuroprotective drugs combined with nanocarriers, specifically for the protection of RGCs in glaucoma treatment. This review categorizes and discusses these studies, as well as addresses the safety of nanocarriers and their impact on therapeutic efficacy. The existing literature explores the effects of neuroprotective drugs combined with nanocarriers in retinal diseases, as well as the use of neuroprotective drugs in treating glaucomatous conditions. This review comprehensively summarizes the current impact of neuroprotective drugs combined with nanocarriers, specifically on glaucoma.

Table 1. Neuroprotective drugs for anti-glaucoma or retinal ganglion cell protection.

Neuroprotectant Class	Drug (Mode of Delivery)	Mechanism
α2-adrenoceptor agonists	Brimonidine [60] (topical, SC)	It decreases the production of aqueous humor and increases uveoscleral outflow, meaning it is commonly used to manage glaucoma due to these hypotensive properties. Alpha2-adrenoceptors have been found in the retinal ganglion layer and inner nuclear layer, indicating potential neuroprotective advantages. They can upregulate retinal expression of anti-apoptotic proteins, including bcl-2 and bcl-x, thereby protecting the RGCs.
N-methyl-D-aspartate (NMDA) receptor antagonists	Memantine [61] (oral)	A non-competitive NMDA receptor antagonist that specifically targets activated glutamatergic receptors, reducing toxic calcium influx and excessive glutamatergic activity, while preserving normal neurotransmission. It has demonstrated protective effects on RGCs in rat models.
Calcium channel blockers	Brovincamine [62] (topical) Flunarizine [63] Nilvadipine [64]	Prevent calcium-mediated apoptosis and increase ocular blood flow by inhibiting calcium influx into vascular smooth muscle cells. This leads to peripheral vasodilation, reduced vascular resistance, and increased blood flow to the optic nerve.
Metabolic-targeting compounds	Metformin [65] (oral)	Reduces oxidative stress by targeting fibrotic signaling, oxidation of nicotinamide adenine dinucleotide (NAD), and mitochondrial energetics.
	Insulin (SC) [66]	A peptide hormone with receptors distributed throughout the central nervous system, including the retina, which operates through the PI3K/Akt signaling pathway to protect retinal neurons, glial cells, and vasculature from excessive glucose flux and apoptosis.
Neurotrophins	NGF [67] (topical, IVT)	Neurotrophins are transported to the retina in a retrograde manner and regulate neuronal growth, function, and survival. Elevated IOP in glaucomatous conditions obstructs this retrograde transport. NGF is expressed in both the retina and target brain cells. In the retina, NGF binds to Tyrosine kinase receptor A (TrkA) on RGCs to promote neural differentiation and prevent apoptosis.

Table 1. *Cont.*

Neuroprotectant Class	Drug (Mode of Delivery)	Mechanism
	BDNF [68]	BDNF, through its receptor TrkB, demonstrates strong neuroprotective effects by reducing dendritic degeneration in mouse models.
Dietary supplements	Nicotinamide (oral) [69]	A dietary supplement and redox reaction coenzyme which increases extracellular NAD+ levels, thereby increasing mitochondrial NAD+ concentration. Animal glaucoma models have demonstrated a declining capacity to retain NAD during early disease.
	Citicoline [70] (oral, IM, topical)	A structural and functional precursor for essential components cell membranes, which also reduces oxidative stress in the central nervous system through the promotion of glutathione production.
	Phosphoserine [71]	A structural matrix of cellular membranes which plays a fundamental role in the synthesis of neurotransmitters.
Anti-inflammatory therapies	Anti-complement: ANX005 [72] (IV) ANX007 [73] (IVT)	A recombinant antibody humanized against complement C1q. C1q, along with C1s2 and C1r2, constitutes the C1 complex, which helps initiate the complement pathway. In glaucoma animal models, C1q is upregulated in the retina. It contributes to the formation of the membrane attack complex, which phagocytoses weakened synapses and RGCs.
	Anti-fas–ONL1204 [74] (IVT)	Fas ligand (fragment apoptosis stimulator), a member of the TNF family, exhibits distinct properties in ocular inflammation. Soluble FasL does not trigger inflammation in the eye, whereas membrane-bound FasL induces potent inflammation. Additionally, soluble FasL inhibits corneal inflammation, which is exacerbated by the pro-inflammatory effects of membrane-bound FasL in mice models.
	MicroRNA-124 [75]	Regulates target genes expression through post-transcriptional regulation.

Table 2. Advantages and disadvantages of ocular drug deliveries.

Method of Delivery	Advantages	Disadvantages
Topical eye drops [76]	Non-invasive.	Ocular surface side effects. Low bioavailability especially for posterior segment tissues.
Subconjunctival injections [27]	Bypasses ocular surface.	Invasive. Limited for anterior segment pathologies.
Intracameral injections [27]	Commonly for delivering antibiotics post-operatively, especially cataract surgery, to reduce endophthalmitis rates.	Invasive. Risk of toxic anterior segment syndrome, toxic endothelial cell destruction syndrome.
Intra-vitreal injections [76]	Higher bioavailability. Relatively less technically demanding compared to other delivery methods of posterior segment pathologies.	Invasive. Risk of endophthalmitis, hemorrhage, retinal detachment.
Subretinal injections [27]	Higher bioavailability. Targeted treatment for RPE and outer retina, especially useful in gene therapy.	Invasive, requires vitrectomy. Technically challenging. Effects confined to site of injection with limited distribution.
Suprachoroidal injections [77]	Higher bioavailability. Targeted treatment for choroid, RPE, outer retina.	Technically challenging.
Systemic [27]	Non-invasive for oral administration.	Higher risk of systemic side effects. Low bioavailability.

Table 3. Neuroprotective nanocarrier in glaucoma studies.

Neuroprotective Strategy	Year	Nanocarrier	Neuroprotective Agent	Administration	Size	PDI	Zeta	EE	Subject	Assesment	Outcome
Antioxident	2024	GelCA, a gelatin-modified hydrogel (Cur@PDA NPs)	Curcumin	Intravitreal	328.6 nm	less than 0.3	−26.7 ± 1.1 mV	70.5 ± 10%	ONC ICR mice, RGC-5 (H_2O_2 induced)	Vitro: AmplexTM Red reagent (10-acetyl-3,7-dihydroxyphenoxazine), Vivo: H&E staining (Bio-Compatibility), tissue staining (Antioxidative Performance)	Cur@PDA@GelCA shows strong biocompatibility, protects retinal tissue from oxidative stress, and offers lasting adhesion.
Antioxident, anti-inflammation	2023	chitosan-hyaluronic acid (CS/HA)	EPOβ	Topical	330 ± 15 nm	0.174 ± 0.016	+28 ± 1 mV	38.4 ± 0.3%	Wistar Hannover Rat (cauterizing three episcleral veins)	ERG, IOP, Histologic Evaluation	The nanoparticle reaches the retina through topical use, significantly improving ERG and retinal thickness earlier in treated animals.
Antioxident	2023	hydroxyl PAMAM dendrimers	N-acetylcysteine (NAC) (D-NAC)	intravitreal, intravenous	5.8 nm	NA	+6.5 mV	NA	Wistar rat model of laser-induced, Translimbal laser (TLL)	IOP, Immunostaining, transcriptomic studies were performed on RNA isolated from retina	Hydroxyl PAMAM dendrimers target activated microglia/macrophages quickly post-injury and remain for 28 days, while NAC conjugation offers neuroprotection.
Antioxident, anti-inflammation	2022	chitosan-hyaluronic acid nanoparticles (CS/HA)	epoetin beta (EPOβ)	Subconjunctival	289 ± 3 nm	0.126 ± 0.085	39 ± 1 mV	38.4 ± 0.3%	Wistar Hannover rat	IOP, ERG and microhematocrit evaluations; histological evaluation (immunofluorescence and HE)	CS/HA nanoparticles enhance mucoadhesion and retention on the ocular surface, safely delivering EPOβ to the retina.
Antioxident	2018	Pluronic-F127 stabilised TPGS	Curcumin	Topical	17.9 nm	0.002	+18.69 mV	94.20%	R28 cell line, DA rat (OHT, pONT)	IOP, Immunostaining	Boosts the drug's solubility by nearly 400,000 times, significantly preserving RGC density.
a2-adrenergic agonist	2022	Polydopamine (PDA)	brimonidine	Intravitreal	223.9 ± 4.7 nm	NA	−28.5 ± 0.58 mV	20%	vitro: Human umbilical vein endothelial cells (HUVECs), Raw 264.7, N2a, 661W and ARPE-19; vivo: C57BL/6 mice (ONC)	Nissl staining, Propidium iodide (PI) uptake and cell death analysis, RT-PCR, transcriptome analysis, Immunostaining, vivo: Optomotor Test; Light/Dark Transition Test	PDA nanoparticles effectively eliminate reactive species and reduce cellular ROS. Brimonidine-loaded PDA (Br@PDA) offers superior protection against RGC loss and visual impairment.
a2-adrenergic agonist	2016	Alkoxylphenacyl-based polycarbonates	Brimonidine tartrate (BRT)	Intravitreal	189.9–199.8 nm	NA	around −0.2 mV	less than 20%	human trabecular meshwork (HTM) cell, Wistar rats	Immunostaining, cytotoxicity studies	AP-PCL microfilms sustained BRT release for over 90 days due to slow degradation and microporous formation, showing good biocompatibility in cell studies and rat models.

Table 3. *Cont.*

Neuroprotective Strategy	Year	Nanocarrier	Neuroprotective Agent	Administration	Size	PDI	Zeta	EE	Subject	Assesment	Outcome
a2-adrenergic agonist, IOP lowering	2015	nanosponge (NS)	Brimonidine, Travoprost, Bimatoprost	Intravitreal, Topical	Bimatoprost: 400 nm and 700 nm; Brimonidine, Travoprost: 50 nm	NA	NA	NA	C57BL/6 (C57) mice	IOP, qualification of Neuro-DiO	A single NS injection delivers ocular hypotensive drugs continuously for up to 32 days and may effectively target RGCs.
a2-adrenergic agonist	2015	human serum albumin nanoparticle (HSA-NP)	Brimonidine	Intravitreal	152.8 ± 51.1 nm	NA	−29.7 ± 7.5 mV	NA	RGC-5 cells; Sprague–Dawley (SD) rat (ONC)	Immunostaining	Br-loaded HSA-NPs provide a prolonged therapeutic effect and enhanced neuroprotection synergistically.
Neurotrophic factors	2023	Cell adhesion peptide (CAP)-gemini surfactants (18-7N(p1-5)-18)	BDNF	Intravitreal	180–320 nm	0.2–0.5	+53.7 ± 1.15 or 13.1 ± 0.5 mV	NA	CD1 mice, Rat A7 astrocyte, 3D retinal neurosphere model from CD 1–4 multipotent retinal stem cells (MRSC)	Vitro: Flow cytometry (Transfection efficiency and viability studies), Transfection study, Vivo: stained with Syto™ 13 nucleic acid stain observed CLSM, BDNF-ELISA (for BDNF expression)	IgSF CAPs were successfully used to improve the adhesion and delivery of gemini NPXs to retinal cells through conjugation with gemini surfactant gene vectors.
Neurotrophic factors	2016	K2® nanoparticle gene delivery system (K2-NPs)	BDNF	NA	83.9 ± 0.4 nm	0.17 ± 0.01	+57.3 ± 2.8 mV	NA	two-layer contact-independent 3D neuronal co-culture model	flow cytometry, and enzyme-linked immunosorbent assay (ELISA)	Quantifying neurite growth in astrocyte-SH-SY5Y cell co-cultures serves as an effective bioassay model for evaluating non-viral gene delivery systems.
NMDA receptor antagonists	2018	memantine-loaded PLGA-PEG nanoparticles (MEM-NP)	Memantine	Topical	141.8 nm	0.078 ± 0.018	−26.5 mV	80.60%	retinoblastoma (Y-79) and keratinocytes (HaCaT) cells. Dark Agouti rat (OHT). New Zealand rabbits. (DA) rats	IOP, Immunostaining, Corneal and scleral permeation	In vitro epithelial and neuronal cell cultures, this formulation was better tolerated than free memantine and effectively preserved RGC density.
blockade of glutamate transmission	2022	PLGA nanoparticles in situ gelling system (NPs-Gel)	Riluzole	Topical	below 200 nm	below 0.2	−30.0 mV	94%	C57BL6 mice	Immunostaining	Optimized PLGA nanoparticles penetrate the blood-retinal barrier, delivering RLZ for 24 h, while RLZ NP gels enhance retention, clear vision, and comfort for glaucoma and dry eye patients.

Table 3. *Cont.*

Neuroprotective Strategy	Year	Nanocarrier	Neuroprotective Agent	Administration	Size	PDI	Zeta	EE	Subject	Assesment	Outcome
IOP lowering/antioxidant	2020	Soluplus	Melatonin/Agomelatine	Topical	61.78 ± 1.61 nm	0.068 ± 0.022	NA	NA	Vivo: Sprague–Dawley RAT (Methylcellulose (MCE) model)	IOP, ERG, Western Blot Analysis, Immunostaining	In the MCE model, melatonin/agomelatine significantly reduced IOP elevation more effectively than timolol or brimonidine.
IOP lowering, anti-inflammatory	2020	The diblock poly ethylene glycol-co-polysebacic acid (PSA-PEG)	brinzolamide and mi-RNA-124	Intravitreal	250–400 nm	NA	0.015–0.020 mV	90%	dutch-belted rabbits and 69 C57/BL6 mice	IOP, Immunostaining, real time PCR	The miRNA/NP-BRZ system is safe, non-toxic, and effectively reduces IOP while providing neuroprotection. The novel nanoparticles offer a slow release and accurately deliver BRZ to the target site, ensuring a long-lasting effect.

2. Therapeutic Agents

2.1. Brimonidine

Brimonidine, a selective α2-adrenergic receptor agonist, is primarily used in clinical settings to lower IOP by reducing aqueous humor production and enhancing uveoscleral outflow [78]. Additionally, the presence of α2-adrenergic receptors in the GCL and the INL is associated with brimonidine having neuroprotective actions [34]. These effects manifest indirectly through interactions with glial cells and directly on RGCs [79]. Research has shown that brimonidine can prevent RGC apoptosis by reducing excitotoxicity, which inhibits the release of neurotransmitters like glutamate and aspartate, and by elevating the concentration of neurotrophic growth factors [34,80,81]. Commonly available as topical eye drops (Alphagan® P at concentrations of 0.1% and 0.15%, and Alphagan® at 0.2%) [82], brimonidine's effectiveness is limited by its low bioavailability (1–7%) and quick clearance from the eye, necessitating frequent application to achieve the desired outcomes [83]. For this reason, various studies have utilized nanocarriers with brimonidine with the objective of improving the efficacy and duration of their action (Table 4).

Table 4. Efficacy and safety of different nanocarrier formulations of brimonidine.

Neuroprotective Drug	Nanocarrier	Efficacy	Safety	Key Findings
Brimonidine	Standard eye drops–Topical concentrations: 0.1%, 0.15%, 0.2%	Limited bioavailability (1–7%) and rapid clearance necessitate frequent dosing.	Common side effects include asthenia; drowsiness; eye discomfort (dry eye, eye inflammation); hyperemia; hypersensitivity; sensation of foreign body; taste altered.	Limited by frequent administration needs due to low bioavailability.
	Polydopamine (PDA)	Enhances RGC protection and significantly boosts axon regeneration compared to brimonidine alone; superior efficacy in reducing microglial activation and increasing visual function.	Demonstrates good ocular biocompatibility; intravitreal injections show no adverse effects.	PDA nanoparticles provide a synergistic neuroprotective effect, improve RGC density, and boost visual outcomes.
	Brimonidine with human serum albumin (HSA) nanoparticles	Prolongs RGC survival, extending treatment efficacy compared to brimonidine alone.	High biocompatibility: no toxic effects reported.	HSA improves brimonidine's duration of action, suggesting potential for more sustained neuroprotection.
	alkoxylphenacyl-based polycarbonates copolymerized with polycaprolactone (AP-PCL)	Sustained release up to 90 days with consistent therapeutic levels; effective long-term IOP management.	Well tolerated; no retinal detachment observed in vivo; no toxic effects on human trabecular meshwork cells.	Demonstrates potential for extended-release formulations, reducing dosing frequency.
	Brimonidine with nanosponges	Effective IOP reduction sustained for up to 3 weeks, depending on nanoparticle size; promising for long-term management.	No major adverse effects reported in animal studies.	Enhanced IOP control with longer-lasting effects compared to standard formulations, though further testing is needed for functional RGC protection assessment.

Luo et al. explored the use of polydopamine (PDA), a biocompatible melanin-like polymer derived from dopamine, as a nanocarrier for brimonidine in a glaucoma model [83]. Their research demonstrated that an intravitreal injection of brimonidine in PDA effectively reduced damaged RGCs and diminished microglial activation, while protecting the RPE [83]. Moreover, the PDA nanoparticles significantly enhanced axon regeneration compared to brimonidine alone [83]. Notably, the brimonidine–PDA formulation increased axon density tenfold from the lesion site compared to other treatments, indicating a synergistic effect. This formulation also more effectively increased RGC density and reduced microglial numbers compared to either component alone, likely due to PDA's antioxidant properties [83]. Visual function assessments through optomotor response tests and light/dark conversion assays revealed that brimonidine–PDA markedly improved visual function. Thus, brimonidine encapsulated in PDA nanoparticles served as an effective neuroprotective nanotherapeutic, significantly enhancing ROS elimination and improving RGC protection.

A study by Kim et al. demonstrated an enhancement in the efficacy of brimonidine using human serum albumin (HSA) as carriers, which independently showed great improvements in RGC survival in the optic nerve crush model [3]. This neuroprotective effect of HSA on RGC survival is likely mediated by modulating the aggregation of amyloid-beta (Aβ) [84]. On day 5, the RGC survival rates were similar for HSA–NPs with and without brimonidine, at 52.6 ± 3.3% and 63.5 ± 7.1%, respectively. By day 14, these rates were 30.7 ± 11.7% for HSA–NPs alone and 38.1 ± 3.6% for HSA–NPs with brimonidine. The group receiving only brimonidine had a higher survival rate on day 5 (58.0 ± 4.2%), but this dropped to 18.6 ± 3.9% by day 14, indicating that the treatment effects lasted longer in both HSA groups [3]. Notably, the HSA with brimonidine (HSA–Br) group exhibited the best treatment efficacy at 14 days [3]. This enhanced outcome may be due to HSA prolonging the availability of brimonidine by supporting its continuous release, or possibly the initial neuroprotective effects of brimonidine synergistically enhancing the therapeutic impact of HSA, resulting in a more extended and potent neuroprotective effect than either HSA–NPs or brimonidine alone. Compared to Luo et al.'s research, Kim et al.'s study not only showed improved efficacy of nanocarriers but also highlighted their potential for extended therapeutic durations. However, unlike Luo et al.'s experiments, Kim et al.'s did not test for functionality.

In exploring advanced drug delivery systems for extended durations of action, alkoxylphenacyl-based polycarbonates copolymerized with polycaprolactone (AP–PCL) were used to develop both injectable nanoparticles and implantable microfilms [82]. The AP–PCL microfilms enabled an extended release of brimonidine tartrate (BRT), starting with an initial burst and then gradually releasing the drug consistently up to approximately 90 days. Importantly, the viability of human trabecular meshwork (HTM) cells exposed to these microfilms was not compromised, indicating no toxic effects. In a separate application, AP–PCL nanoparticles were injected intravitreally into Wistar rats. Observations up to 40 days post-injection showed no signs of retinal detachment, suggesting good ocular tolerance and biocompatibility of the nanoparticles. Additionally, a nanosponge technology encapsulating brimonidine was developed to extend the duration of drug release [85]. Smaller nanosponges (around 50 nm) proved as effective in reducing IOP as the topical application immediately after treatment or intravitreal injection, with effects lasting up to about a week. Larger nanosponges (about 500 nm) showed a 27% reduction in IOP lasting nearly 3 weeks post-injection, demonstrating their potential for long-term IOP management [85]. While nanosponges demonstrate enhanced capability for sustained IOP management, Lambert et al. did not perform a quantitative evaluation of RGC survival. However, considering brimonidine's known protective effects on RGCs, this approach still holds promise for long-lasting RGC protection. It should be noted that neither study conducted in vivo functional testing, which would provide a more comprehensive and convincing assessment of the therapeutic benefits of long-term drug release.

2.2. Brain-Derived Neurotrophic Factor (BDNF)

BDNF is a protein produced by ganglion cells and inner nuclear layer cells that provides neuroprotection directly through the pro-myosin receptor kinase B (TrkB) expressed in RGCs or indirectly via TrkB on glial cells [86]. Increased IOP impairs the vital retrograde transport of BDNF, necessary for RGC survival, by inhibiting dynamin, a protein that assists in the transport of molecules like TrkB and BDNF [86]. In a rat model of glaucoma, the upregulation of the BDNF gene has shown protective effects on RGCs, suggesting its potential as an adjunct therapy alongside IOP reduction [87]. Although injections of neurotrophic factors like BDNF can inhibit or even reverse cell damage, their effects are transient and typically need multiple injection. Naked nucleic acids are highly susceptible to degradation without a delivery system and struggle to escape endosomes for successful nuclear translocation. Gene delivery vectors, essential for addressing these challenges, are divided into viral and non-viral systems. While viral vectors are effective at delivering nucleic acids, their potential to cause adverse immune reactions due to their mobility and immunogenicity is a concern [88]. While many viral systems have progressed to the clinical stage, nonviral systems are still under development [89]. The challenge of effectively and reliably delivering these therapies to the retina remains significant. The development of a non-viral, nano-based BDNF gene delivery vector system holds promise for addressing these issues by enabling sustained therapeutic delivery (Table 5).

Table 5. Efficacy and safety of different nanocarrier formulations of brain-derived neurotrophic factor (BDNF).

Neuroprotective Drug	Formulation/Approach	Efficacy	Safety	Key Findings	References
Brain-Derived Neurotrophic Factor (BDNF)	**BDNF (Standard Injections)**	Provides neuroprotection to RGCs by acting directly through TrkB receptors or indirectly via glial cells. Effective in improving RGC survival.	Generally safe, but repeated injections are needed due to the transient nature of BDNF, which can raise safety concerns over time.	Challenges include rapid degradation, limited nuclear translocation, and difficulty sustaining therapeutic levels.	[68]
	K2® Nanoparticle-Based BDNF Gene Delivery	Superior retinal transfection and increased gene expression levels compared to conventional methods; 3.4× higher BDNF levels in animal models.	Good biocompatibility: no major adverse reactions noted in studies.	Effective BDNF gene delivery via nanoparticles demonstrated in co-culture models; promising for sustained gene therapy.	[70]
	Combined BDNF and TrkB Gene Therapy	Provides superior neuroprotection and axonal transport compared to individual therapies; prevents receptor downregulation, ensuring sustained activation.	Safe in experimental settings.	Synergistic effects improve therapeutic outcomes, offering enhanced and prolonged RGC protection.	[72]
	PACAP38-Linked Exosomes (Non-Glaucoma Model)	Markedly improves survival rates, nerve layer thickness, and visual recovery in traumatic optic neuropathy models.	High biocompatibility observed; no significant adverse effects reported.	Demonstrates the effectiveness of neurotrophic factor delivery via exosomes for enhanced neuroprotection.	[73]

Chen et al. developed an astrocyte and SH-SY5Y cell co-culture model to assess neurite outgrowth, thus providing a valuable bioassay for evaluating non-viral systems [88]. They employed the K2® nanoparticle-based delivery system (K2-NPs) loaded with BDNF-encoding plasmids to transfect A7 astrocytes. Enhanced neurite outgrowth was consistently observed in trA7/SH-SY5Y and trA7/oxSH-SY5Y conditions compared to controls [88]. Additionally, the team synthesized dual surfactants modified with integrin-targeting RGD and neuroimmune immunoglobulin superfamily (IgSF) peptides to transfect genes into retinal cells in vitro and vivo [89]. In animal studies, a CD1 rodent model demonstrated that intravitreal injection of 18-7N(pFASNKL)-18pNPX resulted in superior retinal cell transfection and elevated gene expression levels of tdTomato/BDNF compared to its RGD counterpart and the parental compound, with BDNF levels 3.4 times greater than the control [89]. These findings suggest that pNPX modified with integrins and IgSF CAM-binding peptides has the potential to deliver therapeutic DNA to the target site, thereby significantly increasing BDNF secretion.

Recent research suggests that gene therapy delivering both BDNF and its receptor TrkB together produces superior results compared to administering each component alone [90]. This synergistic approach enhances axonal transport and provides superior protection for RGCs compared to isolated gene therapies. The likely mechanism for these enhanced therapeutic outcomes is the prevention of receptor downregulation, ensuring sustained receptor activation and prolonged therapeutic effects. Khatib et al.'s research supports this, demonstrating that the combination of TrkB and BDNF stimulates axonal transport more effectively than administering the receptor or ligand alone, highlighting the superior efficacy of combined treatments.

In terms of other neurotrophic factors, combination with nanocarriers significantly enhances therapeutic outcomes for RGC survival in non-glaucoma models. For instance, Wang et al. linked Pituitary adenylate cyclase-activating polypeptide 38 (PACAP38) to rat RGC exosomes, markedly improving survival rates, nerve layer thickness, axonal regeneration, and visual function recovery in a rat model of traumatic optic neuropathy (TON) [91]. Another study found that proteins like BDNF and neurotrophies, when conjugated to magnetic nanoparticles (MNPs), effectively protected against RGC loss in an oxidative stress model, while equivalent doses of free proteins were ineffective [92]. These results show the potential and benefits of combining neurotrophic factors with nanocarriers to enhance therapeutic efficacy.

2.3. *Memantine*

Memantine selectively binds to activated glutamate receptors and has been shown to protect ganglion cells from glutamate toxicity by inhibiting overactive NMDA receptors, without affecting normal receptor activity [61]. This protection stems from preventing excitotoxic damage, which is often due to excessive calcium influx through NMDA receptors, leading to free radical formation—a process implicated in the pathophysiology of glaucoma [93]. However, preclinical studies reveal that memantine offers limited benefits in glaucoma models. For example, in one study, oral memantine improved early multifocal electroretinography (ERG) responses and visual evoked cortical potential (VECP) amplitudes in monkeys at 3 and 5 months following an increase in IOP but showed no benefits at 16 months [94]. While safe, the neuroprotective effects of memantine are modest, likely due to its low bioavailability [94]. To address this, Sánchez et al. developed an eye drop using memantine-loaded nanoparticles (MEM–NPs), demonstrating good efficacy and safety [95].

In comparison with the untreated control group, the number of RGCs was significantly lower in the group with health control. Treatment with MEM–NPs effectively prevented OHT-induced RGC damage, indicating that its protective mechanism does not depend on IOP, as IOP levels in the MEM–NPs and ocular hypertension (OHT) groups were similar [95]. MEM–NPs showed no toxicity at any tested concentration, caused less irritation, and provided better penetration than the free drug [95]. The detection of memantine in scleral tissue suggests that a drug aggregate may have formed within the sclera, allowing memantine to diffuse into intraocular tissues and improving the drug's efficacy. By reducing the dosage and focusing on localized administration, the potential for systemic side effects from memantine could be lessened, ensuring that effective drug concentrations reach the targeted tissues. However, a vehicle control group, which would have clarified the effect of the carrier alone on RGC health, was lacking in this study. Additionally, the absence of a group treated with free memantine alone precludes a direct comparison that could demonstrate the potential enhancement of therapeutic efficacy by the nanoparticle carrier. These omissions are important because they limit the final assessment of whether the observed neuroprotective effects can be specifically attributed to the nanoparticle formulation rather than to the drug itself or the vehicle.

Memantine has been shown to reduce amyloid-beta peptide levels in vitro and in Alzheimer's disease (AD) transgenic mouse models [96]. There is increasing evidence linking amyloid-beta accumulation to the pathophysiology of glaucoma and age-related macular degeneration (AMD) [97]. Since amyloid-beta is also implicated in AMD progression, therapies like brimonidine and memantine that inhibit amyloid-beta pathways may be valuable for treating AMD [97]. Memantine has also shown promising results in Alzheimer's disease patients by reducing excitotoxic damage to neurons [98], suggesting it could be very suitable for elderly patients with multiple neurodegenerative disorders. Current evidence does not validate memantine's effectiveness in slowing the progression of glaucoma; hence, its off-label use for this condition is not advised. Nevertheless, employing nanoparticle carriers could introduce new therapeutic opportunities [99]. This method might improve memantine's bioavailability and effectiveness, potentially enhancing treatment outcomes for glaucoma.

2.4. Riluzole (RLZ)

RLZ is an FDA-approved drug that is primarily used in the treatment of amyotrophic lateral sclerosis (ALS). Its neuroprotective effects could be attributed to several factors, although the exact mechanism of action is not fully understood. These include inhibition of voltage-dependent sodium channels, which reduces sustained sodium currents, partial protection of axons in white matter from hypoxia, and facilitation of excess glutamate removal via the glutamate transporter EAAT2. RLZ has shown neuroprotective properties in neuronal axon models and has been effective in slowing RGC death in experimental glaucoma models. However, its low water solubility could cause serious adverse effects such as pancreatitis, angioedema, and hepatitis. Using nanoparticle carriers to enhance RLZ's solubility and enable localized administration might be a valuable strategy to reduce these adverse reactions and improve its therapeutic impact. This approach could allow for the targeted delivery of RLZ directly to affected areas, potentially minimizing systemic exposure, and maximizing its neuroprotective effects.

Esteruelas and colleagues developed a nanocarrier system for RLZ using poly (lactic-co-glycolic acid) (PLGA) nanoparticles and a nanogel formulation. Initial tests showed that both the RLZ nanoparticles (RLZ–NPs) and the RLZ–NPs gel were non-irritating within the first five minutes of application, whereas free RLZ caused mild vasoconstriction and moderate irritancy. Drug distribution studies revealed that, 24 h after administration, both formulations were primarily located in the eye posterior area, with the gel formulation more prevalent in the anterior structures compared to the nanoparticles [100]. Although RLZ–NPs reached therapeutic targets faster, they were also cleared more quickly than the RLZ–NPs gel. The gel's greater mucoadhesive properties allow it to retain the RLZ–NPs within ocular tissues for a longer period, potentially offering more consistent dosing. Additionally, the gel possesses suitable rheological and viscosity properties that prevent visual blurriness [100]. However, this study did not include efficacy tests; it only demonstrated that the drug delivery systems could reach targeted areas within the eye. There were no evaluations of therapeutic effectiveness, such as RGC survival rates or functional tests like ERG, nor comparisons to determine whether encapsulation in carriers improved therapeutic outcomes. This focus on drug delivery without assessing actual treatment efficacy marks a significant limitation in evaluating the potential benefits of these systems for treating glaucoma.

2.5. Curcumin

Curcumin is a polyphenol pigment produced from herb, also known as natural Yellow 3 or E100 under the European Food Additive Directive [101]. Curcumin has shown potential in mitigating pathways related to the development of common ophthalmic diseases, including oxidative stress linked to mitochondrial activity, β-amyloid aggregation, inflammation via PPAR-γ agonist activity, inhibition of astrocyte proliferation through iNOS and JAK2-STAT3 pathways, and anti-angiogenic effects via the VEGF/VEGFR/K-ras pathway [102]. Despite its relatively low toxicity and high tolerability, curcumin faces several major obstacles that limit its clinical use. Its poor bioavailability and low water solubility (approximately 11 ng/mL) pose significant challenges [101]. Although side effects are typically mild and often limited to gastrointestinal disturbances, curcumin undergoes rapid metabolism in the liver and intestines, where it is quickly converted into water-soluble metabolites (glucuronides and sulfates) that are rapidly excreted from the body. This rapid breakdown and elimination result in low bioavailability and limit its therapeutic effectiveness [102].

A novel hydrogel system was developed by partially grafting cinnamic acid onto gelatin to create a GelCA macromonomer with a high substitution degree (91%) [103]. This GelCA hydrogel is crosslinked under UV light via photoinitiated dimerization of the olefinic groups in cinnamic acid, providing π-π stacking capabilities for drug adsorption. Polydopamine (PDA) nanoparticles were synthesized using these π-π stacking and hydrogen bonding properties to effectively load curcumin, enhancing biocompatibility. Cytotoxicity studies showed no significant differences in cell viability among GelCA, PDA-GelCA, and Cur-PDA-GelCA hydrogels, nor were there signs of tissue atrophy or cell loss in retinal tissues injected with these hydrogels after two weeks, indicating good biocompatibility. In vivo experiments using an ONC model revealed that GelCA alone did not significantly affect antioxidative properties compared to the control ONC group. However, both groups showed potential in mitigating oxidative damage after ONC, with significant improvements over the control [103]. The similar antioxidative results between two groups, as indicated by DHE staining, suggest the need for more specific in vivo tests to discern differences in their therapeutic effects [103]. Despite Cur-PDA nanoparticles demonstrating superior efficacy in inhibiting DPPH• radicals and other ROS compared to PDA alone, the distinct therapeutic benefits of curcumin remain to be fully established. Given the comparable therapeutic outcomes and the invasive nature of intravitreal injections, further investigations related to drug release duration and minimizing injection frequency are required to reduce the burden of frequent injections for patients.

Davis developed an eye drop formulation using curcumin-loaded nanoparticles by TPGS/Pluronic F127, which enhances curcumin solubility by nearly 400,000 times [102]. This formulation remains stable in both liquid and lyophilized forms at room temperature for at least two months. Alarmable cell viability studies showed that incubating immortalized R28 cells in both the treatment and vehicle groups dramatically decreased glutamate-induced toxicity. The inclusion of α-tocopherol (TPGS) in the formulation may play a protective role against this toxicity, although TPGS alone did not protect against damage from cobalt chloride, suggesting potential additive effects of curcumin and TPGS. Previous research indicated that a 10 µL intravitreal injection of 0.5% (*w*/*v*) TPGS provided neuroprotection against ischemia/reperfusion injury in a rat model, an effect not observed with topical application, likely due to lower concentrations reaching the retina [30]. However, the current study shows that topical TPGS loaded with curcumin enhances RGC survival, possibly by modulating P-glycoprotein (P-gp) activity to enhance curcumin's transocular transport [104]. There were no significant changes in IOP between the experimental and the OHT groups, indicating that the neuroprotective effects are independent of IOP changes [102].

Functional testing to distinguish between the effects of the carrier and the active drug made it possible to clarify their potential synergistic interactions and contribute to optimizing treatment strategies for eye diseases wherein IOP reduction alone is insufficient. Such tests could be crucial for understanding the mechanisms underlying the observed therapeutic benefits and refining approaches to glaucoma treatment.

2.6. Epoetin Beta (EPOβ)

Erythropoietin (Epo) is a glycoprotein hormone primarily known for stimulating erythropoiesis, which involves activating the bone marrow to produce red blood cells [105]. In addition to its hematopoietic role, Epo also has protective effects in the retina, such as resistance to inflammation, oxidative damage, ischemia, degeneration, and permeability problems [106]. It has been found to inhibit apoptosis in RGCs, maintaining visual function in various glaucoma models. Epo's anti-apoptotic actions are linked to a variety of pathways, including caspase-3 downregulation via the PI3K/AKT pathway, mitochondrial cytochrome release inhibition, and calcium in the cell regulation [107]. Different methods of Epo administration to the retina, such as intravitreal, systemic, and subretinal, have been investigated. Different administration modalities may have varying risk profiles and side-effects. Systemic treatment could result in systemic adverse effects including increased hematopoiesis, hypertension, and thrombosis [108], whereas intravitreal injections may lead to ocular issues such cataract, vitritis, choroidoretinitis, retinal detachment, and endophthalmitis [79]. Table 2 summarizes the general advantages and disadvantages of various modalities for ocular drug delivery. To harness the therapeutic potential of Epo while minimizing these risks, Silva et al. developed a formulation binding with chitosan. This formulation is designed for the intraocular and topical administration of recombinant EPOβ, with the aim of improving retinal health through localized, targeted therapy, thereby reducing the likelihood of both systemic and severe local complications.

Sliva et al. designed a nanoparticle-based system for sub-conjunctival administration carrying EPOβ and conducted safety and release profile evaluations for both the carrier and the drug-loaded groups [109]. The cytotoxicity test cell lines confirmed the formulation's safety, as no toxicity was observed. The study also monitored hematocrit levels before and after administering the EPOβ encapsulated in chitosan and hyaluronic acid, with results showing no significant changes, further affirming the safety of the formulation. ERG assessments performed pre- and post-injection demonstrated no detrimental effects on retinal health, indicating good biocompatibility of the vehicle group, regardless of whether they contained encapsulated EPOβ. In terms of drug release, lacrimal fluid simulation experiments showed that about 65% of EPOβ was released within the initial 16 min, with around 90% released over six hours. Immunofluorescence assays on ocular tissues revealed that the subconjunctivally administered drug-loaded group effectively transported EPOβ to the retina [109]. The detectability of EPOβ in the retina up to 21 days post-administration, with complete clearance by day 28, underscores the formulation's capability to deliver and maintain therapeutic levels of EPOβ in the retina for an extended period.

Silva et al. conducted further studies about this formulation in a rat model topically [107]. They used ERG to measure retinal function at intervals of 3, 7, 14, and 21 days after inducing glaucoma. On day 21 under scotopic conditions, the treatment group demonstrated significantly higher b-wave amplitudes (198 $\pm$ 51 μV) compared to the control group (145 $\pm$ 39 μV), although the overall differences between the groups did not reach statistical significance. The treatment group showed a quicker recovery trend with both a-wave and b-wave amplitudes nearing baseline over time. Measurements of retinal thickness near the optic nerve indicated a marked improvement in the treatment group by day 21, with an average thickness of 145.6 μm versus 120.2 μm in the control group [107]. This reflects an improvement in retinal function. EPOβ presence was detected in the GCL, inner nuclear layer (INL), and outer nuclear layer (ONL) on days 7, 14, and 21. By day 14, EPOβ was found in the cornea and, by day 21, in the ciliary body and sclera, indicating that it infiltrated the cornea. The nanocarrier efficiently transported the active drug into the eyeball, as shown by the persistence of fluorescence signals 21 days after delivery. In addition, there was a significant difference in retinal thickness between the treatment group and the control group on day 21: 120.2 $\pm$ 10.6 μm and 145.6 $\pm$ 22 μm.

In the control group of the study, a decrease in IOP was noted following cauterization, which possibly aided in retinal recovery, with IOP normalizing by the seventh day post-operation. Previous studies suggest that temporary retinal damage caused by increased IOP in rats can fully reverse in three weeks [110]. To better mimic chronic conditions and extend the period for assessing the neuroprotective and neurodegenerative effects of the CS/HA-EPOβ nanoparticles, it might be beneficial to cauterize additional episcleral veins beyond the three used in this experiment. Furthermore, the experimental setup included only treatment and empty carrier groups, lacking a pure drug group for comparative analysis. Chitosan, known for its adhesive properties due to positively charged amino groups, enhances the corneal residence time of eye drops. Chitosan nanoparticles can bind to negatively charged sialic acid residues on the conjunctival and corneal mucosa, extending the retention time of the drug [111]. The absence of a pure drug group makes it challenging to determine whether the nanoparticle carrier indeed improves the drug's bioavailability, targeting, stability, or other therapeutic effects.

2.7. N-Acetylcysteine (NAC)

NAC, an acetylated form of the amino acid cysteine, was first patented in 1960 and introduced to medicine in 1967 [112]. It is known for its mucolytic properties and acts as a powerful anti-inflammatory and antioxidant agent. NAC is used as a dietary supplement and has several medical applications [113]. It serves as a mucolytic agent in chronic pulmonary conditions and is frequently employed to manage paracetamol overdose [113]. It is also used to protect the kidneys in contrast-induced nephropathy and to prevent atrial fibrillation. NAC is also being investigated for its potential to modulate various pathophysiological processes in neurological and psychiatric disorders, acting on factors such as neuroinflammation, mitochondrial dysfunction, oxidative stress, apoptosis, and neurotransmitter systems involving glutamate and dopamine [114].

Pitha et al. present a PAMAM dendrimer (D-NAC) covalently coupled to the drug NAC, which lowers microglia activation, and RGC death following an individual salvage administration one week after laser therapy [115]. The carrier hydroxyl PAMAM dendrimer targeted activated retinal microglia/macrophages and was retained for up to about one month after administration. Compared to intravitreal injections, wherein D-Cy5 (PMAM dendritic polymer fluorescent marker) uptake was seen only in the retina, after intravenous injection, D-Cy5 was seen only in the optic nerve, and D-Cy5 was not present in the retina as a whole outside of the ONH [115]. This suggests that systemic administration of PAMAM dendrimer is unable to cross the outer retinal barrier. A single intravitreal dose of D-NAC given 1 week after IOP elevation significantly reduced the transcription of pro-inflammatory (IL-1β, IL-6, MCP-1) and A1 astrocyte (Serping1, Amigo2, Fkbp5) markers and increased retinal ganglion cell survival (39 $\pm$ 12%) compared to free NAC- (26 $\pm$ 14%) and BSS-treated eyes (20 $\pm$ 15%, p = 0.02) which had lower survival rates [115]. This highlights the neuroprotective effect of the combination of dendritic polymers with the antioxidant NAC and demonstrates the potential of PMAM dendritic polymers for early glaucoma detection and as a vehicle for neuroprotective therapy.

It is worth noting that this experiment did not include a vehicle-only group; had a vehicle group been established, it might have been more straightforward to observe whether there was a superimposed or antagonistic effect of NAC on dendritic combination. Both the treatment group and the drug-only group showed a decrease in IOP after administration, and it is challenging to determine whether the increase in RGC survival was attributable to the decrease in IOP. After 6 weeks, it might be a little late for enucleation, and it may be more appropriate to remove the lenses and perform immunostaining if the IOP has not returned to normal. In addition, incorporating functional in vivo testing could also strengthen efficacy evaluations, offering direct evidence of the treatment's impact on visual function.

2.8. Melatonin

Melatonin is a neurohormone produced by the pineal gland and ocular structure [116]. It possesses significant antioxidant properties, lending itself well to potential therapeutic applications [116], with the demonstration that several of its metabolites help to scavenge free radicals. Melatonin modulates the activity of antioxidant enzymes such as superoxide dismutase (SOD), catalase, and glutathione peroxidase in the retina, which are enzymes protecting the retina from free-radical-induced damage [117]. In addition, melatonin effectively inhibits the nitric oxide pathway, affects glutamate uptake, and modulates the activity of key enzymes in the retina—increasing glutamine synthetase (GS) activity and decreasing glutaminase activity in golden hamsters [116]. These properties suggest that melatonin or its analogues could be effective in glaucoma treatment through the protection of retinal ganglion cells [118].

Dal et al. designed a topical formulation (MelAgo) combining melatonin with its analogue agomelatine, and compared it with two glaucoma medications, brimonidine and timolol [118]. They found that MelAgo had a better IOP-lowering effect than either timolol or brimonidine, and it was about twice as effective as either drug. Electrophysiological examination revealed that neither the vehicle nor timolol affected the Photopic Negative Response (PhNR) amplitude, with brimonidine, melatonin, and agomelatine almost restoring it to control level. The PERG test also showed that neither the vehicle nor timolol improved the amplitude or implicit time. However, brimonidine, melatonin, and agomelatine were able to restore the amplitude and implicit time. Conversely, RGC survival aligned with electrophysiological tests, indicating that RGC density nearly returned to normal levels after MelAgo treatment, but not completely following treatment with brimonidine, melatonin, or agomelatine. Although timolol and brinzolamide had comparable IOP-lowering effects, brimonidine demonstrated superior neuroprotective efficacy. Importantly, due to its potent antihypertensive effects, the formulation of melatonin and agomelatine effectively counteracted RGC loss and restored RGC function by inhibiting inflammation associated with gliosis [118]. Table 6 below summarizes the effects on IOP, retinal function, anti-inflammatory, and RGC density for MelAgo, timolol, and brimonidine for methylcellulose-treated eyes, as described by Dal et al.

Table 6. Summarizing findings of the effects of MelAgo, timolol, and brimonidine on intraocular pressure (IOP) and retinal ganglion cell (RGC) survival.

Agents	IOP Reduction	Photopic Negative Response	Pattern ERG	Gliosis-Related Inflammation	RGC Density (% of Healthy Control)
MelAgo	60%	PhNR amplitude close to normal.	Reduced MCE-induced prolonged implicit time.	Reduced Iba1 and GFAP by 2.3- and 3.2-fold. Reduced levels of TNF-α, IL-1β and IL-6 (by 2.0-, 2.1-, and 2.3-fold). Increased levels of IL-4 and IL-10 (by about 2.0- and 2.5-fold).	Central 92 ± 5 Middle 94 ± 5 Peripheral 94 ± 5
Timolol	32%	No effects.	No effects.	No effects.	Central 78 ± 4 Middle 83 ± 3 Peripheral 81 ± 4
Brimonidine	34%	PhNR amplitude close to normal.	Reduced MCE-induced prolonged implicit time.	Reduced GFAP by 1.6-fold.	Central 88 ± 3 Middle 91 ± 6 Peripheral 91 ± 3

However, it is plausible that the efficacy of MelAgo on ganglion cell survival corresponds to decreased eye pressure, albeit direct neuroprotection cannot be ruled out. This model involves elevated IOP, and it could be more effective to establish a glaucoma model with normal IOP to distinguish between direct neuroprotection and indirect neuroprotection resulting from a decrease in IOP. In addition, setting up a drug-only group (melatonin + agomelatine) might better demonstrate the effectiveness of the vehicle on drug bioavailability.

2.9. MicroRNA (miRNA)

MiRNAs are short non-coding RNA sequences that modulate gene expression post-transcriptionally, affecting processes such as apoptosis, cell growth, and differentiation [119]. Research indicates that miRNAs could be valuable therapeutic agents for diseases like glaucoma, with specific miRNAs targeted by synthetic analogs or inhibitors. miR-124, abundantly present in the central nervous system, encoded by three distinct genomic loci, named miR-124-1, miR-124-2, and miR-124-3, and has shown potential in early therapeutic tests [119,120]. It influences neuronal differentiation and has anti-inflammatory properties by promoting an M2 phenotype in microglial cells and macrophages, providing neuroprotection [121]. These properties are particularly beneficial [121] in reducing apoptosis in RGCs, crucial for combating the progressive degeneration seen in glaucoma.

A study delivered miRNA-124 and brinzolamide (BRZ) encapsulated in PEG–PSA nanoparticles to the retina via intravitreal injection, demonstrating good efficacy and neuroprotective effects [122]. Two animal models were used: a high IOP model and an optic nerve crush (ONC) model. In the high IOP model, the microbeads + miRNA/NP–BRZ group showed the most significant reduction in IOP, surpassing both the microbeads + PBS group and the microbeads-only group [122]. The ONC model was utilized to evaluate the neuroprotective efficacy of the formulation, and the results indicate that co-administration of miRNA/NP–BRZ significantly inhibited RGC apoptosis [122]. Both the fluorogold labeling method and TUNEL staining confirmed the significant reduction in RGC apoptosis with the combined miRNA/NP–BRZ treatment [122].

While it is well known that BRZ effectively lowers IOP [123], this study did not include a comparison group with BRZ + microbeads to determine whether the addition of miRNA affects the IOP-lowering efficacy. Additionally, the ONC model's control group only included PBS, lacking a comparison with pure drug and pure carrier groups. Including these groups would have provided a more comprehensive assessment of the formulation's potential and the synergistic effects of miRNA and BRZ.

3. Limitations

Despite the progress in the development of ocular drug delivery systems, a number of barriers still exist. Rodent models contribute much valuable data in glaucoma research, yet they fall short in replicating the specific anatomical features, progressive progression, and pathophysiology of the disease in humans. One major limitation is that rodents do not have a macula, a critical aspect when studying diseases like glaucoma that primarily affect human vision [124]. Additionally, the rapid progression of disease in these models, due to rodents' shorter lifespans, does not accurately represent the typically slow progression of glaucoma in humans, wherein retinal and RGC damage unfolds over years rather than the weeks to months observed in rodents. There is a clear need for new animal models that more closely mirror the chronic neurodegenerative nature of human glaucoma [124].

Larger animals like rabbits and pigs might be more suitable owing to their longer lifespans and larger eye sizes, which are more comparable to humans. However, the use of larger animals is often limited by higher costs, greater maintenance requirements, and facility constraints. Rabbits, in particular, share key anatomical similarities with the human eye, such as the lamina cribrosa and astrocytes [125]. The composition of the aqueous humor in rabbits, which includes amino acids and free peptides, closely parallels that of humans. This makes the rabbit eye an excellent model for studying ocular biochemistry. Additionally, this animal model offers a straightforward, cost-effective, and reproducible option for preclinical eye disease research. Rabbits are gentle, have appropriately sized eyes for various techniques, and possess a relatively longer lifespan compared to other animal species [126].

However, rabbit models have limitations, including thinner scleral thickness, higher choroidal blood flow, and a lower vitreous volume, approximately 1.5 mL less than that of humans [127]. Additionally, rabbits have a larger and thicker lens that occupies more of the vitreous cavity [128], which may complicate the collection of samples from the posterior segment for biochemical analysis.

The retina of pigs and mini-pigs shares more similarities with the human retina than that of other mammals, including dogs, cats, goats, and cows [129]. In terms of ocular anatomy, pigs have a retina that closely resembles the human eye, featuring holangiotic retinal vasculature, a lack of tapetum, cone photoreceptors in the outer retina, similar scleral thickness, and three types of RGCs. These characteristics aid in studying the mechanisms behind selective RGC death [130]. Pigs are also considered easy to handle, grow slowly, and have appropriately sized eyes for diagnostic tools such as OCT, corneal topography imaging, and ERG compared to non-human primates.

Additionally, given that glaucoma is a multifactorial disease, most existing models are insufficient as they typically only simulate one aspect of the disease's progression. The variability among different animal models means that the effects of the same drug can differ markedly, making it essential to choose an animal model that aligns well with the drug's mechanism of action. Furthermore, if a drug has an IOP-lowering effect, it is crucial to determine whether its neuroprotective actions are due to lowering IOP or a direct effect on neural tissues. This differentiation could be better understood by employing two models: one with elevated IOP and another with normal IOP, to isolate the effects of the drug on neuroprotection independent of IOP changes.

The current safety assessment of nanocarriers heavily relies on cell culture models due to their cost-effectiveness and simplicity [131]. However, these models are limited in terms of evaluating prolonged toxicological exposure as they do not sustain cell viability long enough. Additionally, nanocarriers can encounter stability issues during long-term storage, potentially skewing toxicity evaluation results. While there is no universally accepted method for assessing nanocarrier toxicity, factors such as surface charge, size, and solubility are considered potential indicators, yet extensive research is needed to validate these parameters. Most toxicity studies investigating nanocarriers have been conducted in animal models, with a significant gap in long-term clinical trials involving humans. Furthermore, for topically administered therapies, the frequent dosing required raises concerns about the accumulation of drugs and carriers, which could pose safety risks over time [131].

Some studies have used the RGC-5 line in cell culture. However, this is controversial as it is now known to be derived from mice rather than rats, and additional cytogenetic tests, karyotype analysis, as well as genetic and protein analyses, have revealed that these cells are not actually retinal RGCs but rather the 661 W cell line, which is a photoreceptor cell line transformed with the SV-40 T antigen from mice [132]. This indicates that RGC-5 cells are not a good representative model for RGCs. In contrast, the ARPE-19 cell line, derived from the human retinal pigment epithelium, should serve as a better alternative for studies involving retinal cells.

A key aim of neuroprotective treatments for glaucoma is to preserve RGCs during the initial phases of the disease, preventing their complete degeneration. Consequently, advanced techniques for the early diagnosis of glaucoma are essential. However, most in vivo experiments primarily focus on measuring IOP without conducting functional tests like ERG or VEP. Even when ERG is used, protocols that specifically isolate RGC responses from general retinal activity may be more appropriate than standard full-field ERG measurements of the a-wave, b-wave, and their implicit times [133]. For instance, pattern ERG (PERG) and PhNR, which is the negative wave after b-wave, could provide a better assessment of the RGC's function [133]. Additionally, a novel technique called Detection of Apoptosing Retinal Cells (DARC) utilizes the fluorophore DY-776 conjugated to an Annexin A5 protein, termed ANX776. This molecule binds to apoptotic cells and can be imaged using confocal scanning laser ophthalmoscopy (cSLO). By monitoring changes in the number of labeled spots, the survival of RGCs can be assessed more accurately [134]. Implementing these advanced in vivo diagnostic and monitoring techniques can significantly enhance the evaluation of neuroprotective therapies' effectiveness in preserving RGC function and viability in glaucoma research.

Finally, in studies exploring nanocarriers for delivering neuroprotective drugs, some experiments lack control groups with pure drugs or pure vehicles to compare with treatment groups. A comprehensive intergroup design, including these control groups, could better demonstrate the improvements that nanocarriers provide in therapeutic efficacy. This approach would offer clearer insights into the specific contributions of the nanocarriers and the drugs they deliver, thereby strengthening the validity and applicability of the findings.

4. Future Work

It is well-known that, while intravitreal injections can bypass the BRB and likely offer better efficacy, their invasive nature may lead to side effects such as infection, inflammation, and bleeding, thereby making non-invasive eye drop administration more attractive. Combining nanocarriers can overcome the issues of poor bioavailability and limited delivery to the retina associated with topical administration. However, current studies using eye drops often require very high administration frequencies, nearly 1–2 times daily, raising concerns about drug accumulation, as previously mentioned. Eye drops incorporating chitosan offer potential advantages, as chitosan can relax tight junctions on the corneal epithelium, provide mucoadhesiveness due to its positive charge, and extend drug release time [135]. These benefits suggest a reduced administration frequency and improved safety for topical delivery. However, the safety of chitosan nanoparticles themselves has not been extensively studied in clinical research, necessitating further clinical research to confirm their safety and efficacy. Future studies should also try to elucidate the optimal physicochemical properties of chitosan for delivering glaucoma drugs, such as the best zeta potential, amination level, molecular weight, and degree of deacetylation.

Another non-invasive delivery method with potential is intranasal administration. Khan et al. demonstrated that, after intranasal administration, drugs rapidly accumulate in the eyes and optic nerves of rodents within 30 min, likely by directly entering the central nervous system and traveling along the optic nerve to the retina, thus bypassing the blood–brain barrier (BBB) and the BRB [136]. This method offers the advantages of low dosing and reduced systemic risk compared to oral administration [137]. Comparing the efficacy of the same formulation via eye drops and intranasal delivery could be a novel direction for glaucoma treatment, addressing the issue of poor patient compliance with simple administration methods.

The complicated pathophysiology of glaucoma comprises several routes and processes. To prevent RGC apoptosis and delay disease progression, targeting a specific pathway or cause may not be sufficient. Multi-drug therapies, which simultaneously target multiple pathways and mechanisms or regulate key modulators affecting multiple downstream effectors, offer a promising approach to neuroprotection and could be more effective. Delivering multiple drug formulations with different physicochemical properties using conventional solvents (e.g., PBS and glucose solutions) is challenging, but feasible with nanocarriers. For example, studies on ONC mouse models have shown that combined AAV2 gene therapy for BDNF and TrkB provides more sustained and effective results than either therapy alone [138]. Topical administration of melatonin/agomelatine nanomicelles significantly reduces elevated IOP, and the combined use of both melatonin compounds prolongs the hypotensive effect compared to using either drug alone [139]. Understanding innate compensatory mechanisms is critical to designing combination medicines. As a result, personalized multi-drug therapies based on nanocarriers adapted to individual patient physiology could become the new mainstream treatment. This may involve developing novel nanocarriers that simultaneously deliver multiple active drugs targeting different mechanisms or distinct pathways within the same mechanism to delay RGC apoptosis. By combining multifunctionality and targeted delivery, these nanocarriers could address multiple pathways involved in RGC degeneration, providing a more comprehensive neuroprotective strategy for retinal diseases.

The majority of novel drug delivery techniques were predominantly tested in vitro studies or on animals, thereby lacking detailed in vivo experimentation in human eyes. Their systemic and metabolic side effects remain unclear. Current experiments remain in the preclinical stage; therefore, further clinical translational studies are required due to the anatomical differences between animal models and human eyes. More research is required to fully understand the safety and repeatability of combining carriers with neuroprotective drugs. Large-scale, long-term clinical studies are anticipated. However, transitioning these nanotechnology-based drug delivery systems from the experimental stage to practical clinical use faces significant challenges, particularly in scaling up production and maintaining consistent quality control scale. Additionally, the impact of different intraocular environments, including history of eye surgery, repeated implants, or vitrectomy patient, on various characteristics of nanocarriers also warrants investigation [10]. For instance, nanoparticle movement post-operation in cataract or vitrectomy with silicone oil [140], which animal models cannot simulate, are important considerations. Thus, surgical history is a significant factor that merits attention in further clinic studies.

In the future, greater emphasis should be placed on developing novel non-invasive ocular drug delivery systems that can effectively overcome ocular barriers, extend drug release duration, and maintain therapeutic concentrations at the targeted lesion sites. To achieve this, optimizing key properties of nanocarriers—such as size, zeta potential, refractive index, safety, stability, pH, surface tension, and osmotic pressure—is crucial. Additionally, more extensive in vitro and in vivo studies are needed, alongside the development of animal models that more closely mimic human eye diseases.

5. Conclusions

Glaucoma is a vision-threatening disease affecting patients of all ages globally. While current treatments primarily focus on lowering IOP to slow disease progression, they are not curative. Neuroprotective therapies are gaining increasing attention, as they aim to directly prevent or delay the apoptosis of irreplaceable neuronal cells, particularly RGCs. Such treatments can be used alone or in conjunction with IOP-lowering therapies to achieve complementary or synergistic effects and better therapeutic outcomes. As research into neuroprotective therapies advances, it enhances our understanding of the mechanisms underlying glaucoma onset and progression. This knowledge could eventually lead to preventative strategies and potentially the restoration of neural cell function, greatly benefiting both healthcare systems and patients' quality of life. Nanocarriers loaded with neuroprotective agents can address several limitations, such as poor penetration of the BRB, suboptimal pharmacokinetics, and short half-life in vivo. By improving drug bioavailability, targeted delivery, and efficacy, nanocarriers can reduce the required drug dosage, thereby minimizing adverse reactions, provided the carriers themselves are safe. Additionally, nanocarriers can simultaneously deliver both hydrophilic and hydrophobic drugs by encapsulating them within cavities or by using covalent bonds to bind the drugs. This capability allows for a single formulation to carry multiple active drugs, facilitating combination therapy. Although preclinical investigations have shown positive results, few neuroprotective nanocarrier treatments have made it into clinical trials. More clinical research is required to confirm their therapeutic effects. The path to effective neuroprotection in glaucoma remains long, but neuroprotective nanocarriers have demonstrated great potential toward becoming the next generation of mainstream treatments for glaucoma.

Author Contributions: Conceptualization, K.P. and M.F.C.; writing—original draft preparation, K.P.; writing—review and editing, K.P., M.F.C., D.H., M.G., C.F.J.L. and W.W.; supervision, M.F.C. All authors have read and agreed to the published version of the manuscript.

Funding: This research received no external funding.

Conflicts of Interest: The authors declare no conflicts of interest.

References

1. Quigley, H.A.; Broman, A.T. The number of people with glaucoma worldwide in 2010 and 2020. *Br. J. Ophthalmol.* **2006**, *90*, 262–267. [CrossRef] [PubMed]
2. Flaxman, S.R.; Bourne, R.R.A.; Resnikoff, S.; Ackland, P.; Braithwaite, T.; Cicinelli, M.V.; Das, A.; Jonas, J.B.; Keeffe, J.; Kempen, J.H.; et al. Global causes of blindness and distance vision impairment 1990–2020: A systematic review and meta-analysis. *Lancet Glob. Health* **2017**, *5*, e1221–e1234. [CrossRef] [PubMed]
3. Kim, K.E.; Jang, I.; Moon, H.; Kim, Y.J.; Jeoung, J.W.; Park, K.H.; Kim, H. Neuroprotective Effects of Human Serum Albumin Nanoparticles Loaded with Brimonidine on Retinal Ganglion Cells in Optic Nerve Crush Model. *Investig. Ophthalmol. Vis. Sci.* **2015**, *56*, 5641–5649. [CrossRef] [PubMed]
4. Ramulu, P.Y.; Hochberg, C.; Maul, E.A.; Chan, E.S.; Ferrucci, L.; Friedman, D.S. Glaucomatous visual field loss associated with less travel from home. *Optom. Vis. Sci.* **2014**, *91*, 187–193. [CrossRef] [PubMed]

5. Tham, Y.C.; Li, X.; Wong, T.Y.; Quigley, H.A.; Aung, T.; Cheng, C.Y. Global prevalence of glaucoma and projections of glaucoma burden through 2040: A systematic review and meta-analysis. *Ophthalmology* **2014**, *121*, 2081–2090. [CrossRef]
6. Rein, D.B.; Zhang, P.; Wirth, K.E.; Lee, P.P.; Hoerger, T.J.; McCall, N.; Klein, R.; Tielsch, J.M.; Vijan, S.; Saaddine, J. The economic burden of major adult visual disorders in the United States. *Arch. Ophthalmol.* **2006**, *124*, 1754–1760. [CrossRef]
7. Jayaram, H.; Kolko, M.; Friedman, D.S.; Gazzard, G. Glaucoma: Now and beyond. *Lancet* **2023**, *402*, 1788–1801. [CrossRef]
8. Coleman, A.L.; Miglior, S. Risk factors for glaucoma onset and progression. *Surv. Ophthalmol.* **2008**, *53* (Suppl. S1), S3–S10. [CrossRef]
9. Belamkar, A.; Harris, A.; Zukerman, R.; Siesky, B.; Oddone, F.; Verticchio Vercellin, A.; Ciulla, T.A. Sustained release glaucoma therapies: Novel modalities for overcoming key treatment barriers associated with topical medications. *Ann. Med.* **2022**, *54*, 343–358. [CrossRef]
10. Shen, Y.; Sun, J.; Sun, X. Intraocular nano-microscale drug delivery systems for glaucoma treatment: Design strategies and recent progress. *J. Nanobiotechnology* **2023**, *21*, 84. [CrossRef]
11. Masland, R.H. The neuronal organization of the retina. *Neuron* **2012**, *76*, 266–280. [CrossRef] [PubMed]
12. Xiang, M.; Zhou, H.; Nathans, J. Molecular biology of retinal ganglion cells. *Proc. Natl. Acad. Sci. USA* **1996**, *93*, 596–601. [CrossRef] [PubMed]
13. Kim, U.S.; Mahroo, O.A.; Mollon, J.D.; Yu-Wai-Man, P. Retinal Ganglion Cells-Diversity of Cell Types and Clinical Relevance. *Front. Neurol.* **2021**, *12*, 661938. [CrossRef] [PubMed]
14. Ju, W.K.; Perkins, G.A.; Kim, K.Y.; Bastola, T.; Choi, W.Y.; Choi, S.H. Glaucomatous optic neuropathy: Mitochondrial dynamics, dysfunction and protection in retinal ganglion cells. *Prog. Retin. Eye Res.* **2023**, *95*, 101136. [CrossRef] [PubMed]
15. Feng, K.M.; Tsung, T.H.; Chen, Y.H.; Lu, D.W. The Role of Retinal Ganglion Cell Structure and Function in Glaucoma. *Cells* **2023**, *12*, 2797. [CrossRef]
16. Shiga, Y.; Kunikata, H.; Aizawa, N.; Kiyota, N.; Maiya, Y.; Yokoyama, Y.; Omodaka, K.; Takahashi, H.; Yasui, T.; Kato, K.; et al. Optic Nerve Head Blood Flow, as Measured by Laser Speckle Flowgraphy, Is Significantly Reduced in Preperimetric Glaucoma. *Curr. Eye Res.* **2016**, *41*, 1447–1453. [CrossRef]
17. Tezel, G. Multifactorial Pathogenic Processes of Retinal Ganglion Cell Degeneration in Glaucoma towards Multi-Target Strategies for Broader Treatment Effects. *Cells* **2021**, *10*, 1372. [CrossRef]
18. Halliwell, B. Oxidative stress and neurodegeneration: Where are we now? *J. Neurochem.* **2006**, *97*, 1634–1658. [CrossRef]
19. Kimura, A.; Namekata, K.; Guo, X.; Harada, C.; Harada, T. Neuroprotection, Growth Factors and BDNF-TrkB Signalling in Retinal Degeneration. *Int. J. Mol. Sci.* **2016**, *17*, 1584. [CrossRef]
20. Baltmr, A.; Duggan, J.; Nizari, S.; Salt, T.E.; Cordeiro, M.F. Neuroprotection in glaucoma—Is there a future role? *Exp. Eye Res.* **2010**, *91*, 554–566. [CrossRef]
21. Russo, R.; Varano, G.P.; Adornetto, A.; Nucci, C.; Corasaniti, M.T.; Bagetta, G.; Morrone, L.A. Retinal ganglion cell death in glaucoma: Exploring the role of neuroinflammation. *Eur. J. Pharmacol.* **2016**, *787*, 134–142. [CrossRef] [PubMed]
22. Downs, J.C.; Roberts, M.D.; Burgoyne, C.F. Mechanical environment of the optic nerve head in glaucoma. *Optom. Vis. Sci.* **2008**, *85*, 425–435. [CrossRef] [PubMed]
23. Burgoyne, C.F.; Downs, J.C.; Bellezza, A.J.; Suh, J.K.; Hart, R.T. The optic nerve head as a biomechanical structure: A new paradigm for understanding the role of IOP-related stress and strain in the pathophysiology of glaucomatous optic nerve head damage. *Prog. Retin. Eye Res.* **2005**, *24*, 39–73. [CrossRef] [PubMed]
24. Boccaccini, A.; Cavaterra, D.; Carnevale, C.; Tanga, L.; Marini, S.; Bocedi, A.; Lacal, P.M.; Manni, G.; Graziani, G.; Sbardella, D.; et al. Novel frontiers in neuroprotective therapies in glaucoma: Molecular and clinical aspects. *Mol. Aspects Med.* **2023**, *94*, 101225. [CrossRef]
25. Cantor, L. Achieving low target pressures with today's glaucoma medications. *Surv. Ophthalmol.* **2003**, *48* (Suppl. S1), S8–S16. [CrossRef]
26. Akopian, A.; Kumar, S.; Ramakrishnan, H.; Roy, K.; Viswanathan, S.; Bloomfield, S.A. Targeting neuronal gap junctions in mouse retina offers neuroprotection in glaucoma. *J. Clin. Investig.* **2017**, *127*, 2647–2661. [CrossRef]
27. Naik, S.; Pandey, A.; Lewis, S.A.; Rao, B.S.S.; Mutalik, S. Neuroprotection: A versatile approach to combat glaucoma. *Eur. J. Pharmacol.* **2020**, *881*, 173208. [CrossRef]
28. Schober, M.S.; Chidlow, G.; Wood, J.P.; Casson, R.J. Bioenergetic-based neuroprotection and glaucoma. *Clin. Exp. Ophthalmol.* **2008**, *36*, 377–385. [CrossRef]
29. Teng, K.K.; Hempstead, B.L. Neurotrophins and their receptors: Signaling trios in complex biological systems. *Cell. Mol. Life Sci.* **2004**, *61*, 35–48. [CrossRef]
30. Nucci, C.; Tartaglione, R.; Cerulli, A.; Mancino, R.; Spanò, A.; Cavaliere, F.; Rombolà, L.; Bagetta, G.; Corasaniti, M.T.; Morrone, L.A. Retinal damage caused by high intraocular pressure-induced transient ischemia is prevented by coenzyme Q10 in rat. *Int. Rev. Neurobiol.* **2007**, *82*, 397–406. [CrossRef]
31. Cheung, Z.H.; So, K.F.; Lu, Q.; Yip, H.K.; Wu, W.; Shan, J.J.; Pang, P.K.; Chen, C.F. Enhanced survival and regeneration of axotomized retinal ganglion cells by a mixture of herbal extracts. *J. Neurotrauma* **2002**, *19*, 369–378. [CrossRef] [PubMed]
32. Dunkelberger, J.R.; Song, W.C. Complement and its role in innate and adaptive immune responses. *Cell Res.* **2010**, *20*, 34–50. [CrossRef] [PubMed]

33. Wallach, D. The Tumor Necrosis Factor Family: Family Conventions and Private Idiosyncrasies. *Cold Spring Harb. Perspect. Biol.* **2018**, *10*, a028431. [CrossRef] [PubMed]
34. Wheeler, L.; WoldeMussie, E.; Lai, R. Role of alpha-2 agonists in neuroprotection. *Surv. Ophthalmol.* **2003**, *48* (Suppl. S1), S47–S51. [CrossRef]
35. Murtas, G.; Marcone, G.L.; Sacchi, S.; Pollegioni, L. L-serine synthesis via the phosphorylated pathway in humans. *Cell. Mol. Life Sci.* **2020**, *77*, 5131–5148. [CrossRef]
36. Hill, D.; Compagnoni, C.; Cordeiro, M.F. Investigational neuroprotective compounds in clinical trials for retinal disease. *Expert. Opin. Investig. Drugs* **2021**, *30*, 571–577. [CrossRef]
37. Lusthaus, J.; Goldberg, I. Current management of glaucoma. *Med. J. Aust.* **2019**, *210*, 180–187. [CrossRef]
38. O'Leary, F.; Campbell, M. The blood-retina barrier in health and disease. *FEBS J.* **2023**, *290*, 878–891. [CrossRef]
39. Guymer, R.H.; Bird, A.C.; Hageman, G.S. Cytoarchitecture of choroidal capillary endothelial cells. *Investig. Ophthalmol. Vis. Sci.* **2004**, *45*, 1660–1666. [CrossRef]
40. Booij, J.C.; Baas, D.C.; Beisekeeva, J.; Gorgels, T.G.; Bergen, A.A. The dynamic nature of Bruch's membrane. *Prog. Retin. Eye Res.* **2010**, *29*, 1–18. [CrossRef]
41. Gaudana, R.; Ananthula, H.K.; Parenky, A.; Mitra, A.K. Ocular drug delivery. *AAPS J.* **2010**, *12*, 348–360. [CrossRef] [PubMed]
42. Akhter, M.H.; Ahmad, I.; Alshahrani, M.Y.; Al-Harbi, A.I.; Khalilullah, H.; Afzal, O.; Altamimi, A.S.A.; Najib Ullah, S.N.M.; Ojha, A.; Karim, S. Drug Delivery Challenges and Current Progress in Nanocarrier-Based Ocular Therapeutic System. *Gels* **2022**, *8*, 82. [CrossRef] [PubMed]
43. Kang-Mieler, J.J.; Rudeen, K.M.; Liu, W.; Mieler, W.F. Advances in ocular drug delivery systems. *Eye* **2020**, *34*, 1371–1379. [CrossRef] [PubMed]
44. Li, S.; Chen, L.; Fu, Y. Nanotechnology-based ocular drug delivery systems: Recent advances and future prospects. *J. Nanobiotechnology* **2023**, *21*, 232. [CrossRef]
45. Srinivasarao, D.A.; Lohiya, G.; Katti, D.S. Fundamentals, challenges, and nanomedicine-based solutions for ocular diseases. *Wiley Interdiscip. Rev. Nanomed. Nanobiotechnology* **2019**, *11*, e1548. [CrossRef]
46. Palmerston Mendes, L.; Pan, J.; Torchilin, V.P. Dendrimers as Nanocarriers for Nucleic Acid and Drug Delivery in Cancer Therapy. *Molecules* **2017**, *22*, 1401. [CrossRef]
47. Mittal, P.; Saharan, A.; Verma, R.; Altalbawy, F.M.A.; Alfaidi, M.A.; Batiha, G.E.; Akter, W.; Gautam, R.K.; Uddin, M.S.; Rahman, M.S. Dendrimers: A New Race of Pharmaceutical Nanocarriers. *Biomed. Res. Int.* **2021**, *2021*, 8844030. [CrossRef]
48. Chis, A.A.; Dobrea, C.; Morgovan, C.; Arseniu, A.M.; Rus, L.L.; Butuca, A.; Juncan, A.M.; Totan, M.; Vonica-Tincu, A.L.; Cormos, G.; et al. Applications and Limitations of Dendrimers in Biomedicine. *Molecules* **2020**, *25*, 3982. [CrossRef]
49. Wu, P.T.; Lin, C.L.; Lin, C.W.; Chang, N.C.; Tsai, W.B.; Yu, J. Methylene-Blue-Encapsulated Liposomes as Photodynamic Therapy Nano Agents for Breast Cancer Cells. *Nanomaterials* **2018**, *9*, 14. [CrossRef]
50. Sousa, I.; Rodrigues, F.; Prazeres, H.; Lima, R.T.; Soares, P. Liposomal therapies in oncology: Does one size fit all? *Cancer Chemother. Pharmacol.* **2018**, *82*, 741–755. [CrossRef]
51. Kube, S.; Hersch, N.; Naumovska, E.; Gensch, T.; Hendriks, J.; Franzen, A.; Landvogt, L.; Siebrasse, J.P.; Kubitscheck, U.; Hoffmann, B.; et al. Fusogenic Liposomes as Nanocarriers for the Delivery of Intracellular Proteins. *Langmuir* **2017**, *33*, 1051–1059. [CrossRef] [PubMed]
52. Peng, C.; Kuang, L.; Zhao, J.; Ross, A.E.; Wang, Z.; Ciolino, J.B. Bibliometric and visualized analysis of ocular drug delivery from 2001 to 2020. *J. Control Release* **2022**, *345*, 625–645. [CrossRef] [PubMed]
53. Xu, J.; Zheng, S.; Hu, X.; Li, L.; Li, W.; Parungao, R.; Wang, Y.; Nie, Y.; Liu, T.; Song, K. Advances in the Research of Bioinks Based on Natural Collagen, Polysaccharide and Their Derivatives for Skin 3D Bioprinting. *Polymers* **2020**, *12*, 1237. [CrossRef] [PubMed]
54. Zielińska, A.; Carreiró, F.; Oliveira, A.M.; Neves, A.; Pires, B.; Venkatesh, D.N.; Durazzo, A.; Lucarini, M.; Eder, P.; Silva, A.M.; et al. Polymeric Nanoparticles: Production, Characterization, Toxicology and Ecotoxicology. *Molecules* **2020**, *25*, 3731. [CrossRef]
55. Akhter, S.; Anwar, M.; Siddiqui, M.A.; Ahmad, I.; Ahmad, J.; Ahmad, M.Z.; Bhatnagar, A.; Ahmad, F.J. Improving the topical ocular pharmacokinetics of an immunosuppressant agent with mucoadhesive nanoemulsions: Formulation development, in-vitro and in-vivo studies. *Colloids Surf. B Biointerfaces* **2016**, *148*, 19–29. [CrossRef]
56. Yetisgin, A.A.; Cetinel, S.; Zuvin, M.; Kosar, A.; Kutlu, O. Therapeutic Nanoparticles and Their Targeted Delivery Applications. *Molecules* **2020**, *25*, 2193. [CrossRef]
57. Sánchez-López, E.; Espina, M.; Doktorovova, S.; Souto, E.B.; García, M.L. Lipid nanoparticles (SLN, NLC): Overcoming the anatomical and physiological barriers of the eye—Part I—Barriers and determining factors in ocular delivery. *Eur. J. Pharm. Biopharm.* **2017**, *110*, 70–75. [CrossRef]
58. Meng, T.; Kulkarni, V.; Simmers, R.; Brar, V.; Xu, Q. Therapeutic implications of nanomedicine for ocular drug delivery. *Drug Discov. Today* **2019**, *24*, 1524–1538. [CrossRef]
59. Gorantla, S.; Rapalli, V.K.; Waghule, T.; Singh, P.P.; Dubey, S.K.; Saha, R.N.; Singhvi, G. Nanocarriers for ocular drug delivery: Current status and translational opportunity. *RSC Adv.* **2020**, *10*, 27835–27855. [CrossRef]
60. Scuteri, D.; Bagetta, G.; Nucci, C.; Aiello, F.; Cesareo, M.; Tonin, P.; Corasaniti, M.T. Evidence on the neuroprotective properties of brimonidine in glaucoma. *Prog. Brain Res.* **2020**, *257*, 155–166. [CrossRef]
61. Vorwerk, C.K.; Lipton, S.A.; Zurakowski, D.; Hyman, B.T.; Sabel, B.A.; Dreyer, E.B. Chronic low-dose glutamate is toxic to retinal ganglion cells. Toxicity blocked by memantine. *Investig. Ophthalmol. Vis. Sci.* **1996**, *37*, 1618–1624.

62. Sawada, A.; Kitazawa, Y.; Yamamoto, T.; Okabe, I.; Ichien, K. Prevention of visual field defect progression with brovincamine in eyes with normal-tension glaucoma. *Ophthalmology* **1996**, *103*, 283–288. [CrossRef] [PubMed]
63. Mansouri, K.; Silva, S.E.; Shaarawy, T. Efficacy and tolerability of topical 0.05% flunarizine in patients with open-angle glaucoma or ocular hypertension-a pilot study. *J. Glaucoma* **2011**, *20*, 519–522. [CrossRef] [PubMed]
64. Tsuruga, H.; Murata, H.; Araie, M.; Aihara, M. Neuroprotective effect of the calcium channel blocker nilvadipine on retinal ganglion cell death in a mouse ocular hypertension model. *Heliyon* **2023**, *9*, e13812. [CrossRef] [PubMed]
65. Kumar, N. Commentary—Association of metformin use among diabetics and the incidence of primary open-angle glaucoma—The Chennai Eye Disease Incidence Study. *Indian. J. Ophthalmol.* **2021**, *69*, 3339–3340. [CrossRef]
66. Faiq, M.A.; Sengupta, T.; Nath, M.; Velpandian, T.; Saluja, D.; Dada, R.; Dada, T.; Chan, K.C. Ocular manifestations of central insulin resistance. *Neural Regen. Res.* **2023**, *18*, 1139–1146. [CrossRef]
67. Johnson, T.V.; Bull, N.D.; Martin, K.R. Neurotrophic factor delivery as a protective treatment for glaucoma. *Exp. Eye Res.* **2011**, *93*, 196–203. [CrossRef]
68. Binley, K.E.; Ng, W.S.; Barde, Y.A.; Song, B.; Morgan, J.E. Brain-derived neurotrophic factor prevents dendritic retraction of adult mouse retinal ganglion cells. *Eur. J. Neurosci.* **2016**, *44*, 2028–2039. [CrossRef]
69. Hui, F.; Tang, J.; Williams, P.A.; McGuinness, M.B.; Hadoux, X.; Casson, R.J.; Coote, M.; Trounce, I.A.; Martin, K.R.; van Wijngaarden, P.; et al. Improvement in inner retinal function in glaucoma with nicotinamide (vitamin B3) supplementation: A crossover randomized clinical trial. *Clin. Exp. Ophthalmol.* **2020**, *48*, 903–914. [CrossRef]
70. Anton, A.; Garcia, V.; Muñoz, M.; Gonzales, K.; Ayala, E.; Del Mar Sanchez, E.; Morilla-Grasa, A. The Effect of Oral Citicoline and Docosahexaenoic Acid on the Visual Field of Patients with Glaucoma: A Randomized Trial. *Life* **2022**, *12*, 1481. [CrossRef]
71. Scalinci, S.Z.; Lugaresi, M.; Scorolli, L.; Ralli, M.; Greco, A.; Pantaleone, V.; Taurone, S.; Franzone, F. Neuroprotective role of phosphoserine in primary open-angle glaucoma patients. *Eur. Rev. Med. Pharmacol. Sci.* **2020**, *24*, 9780–9786. [CrossRef] [PubMed]
72. Lansita, J.A.; Mease, K.M.; Qiu, H.; Yednock, T.; Sankaranarayanan, S.; Kramer, S. Nonclinical Development of ANX005: A Humanized Anti-C1q Antibody for Treatment of Autoimmune and Neurodegenerative Diseases. *Int. J. Toxicol.* **2017**, *36*, 449–462. [CrossRef] [PubMed]
73. Sun, Y.; Wirta, D.; Murahashi, W.; Mathur, V.; Sankaranarayanan, S.; Taylor, L.K.; Yednock, T.; Fong, D.S.; Goldberg, J.L. Safety and Target Engagement of Complement C1q Inhibitor ANX007 in Neurodegenerative Eye Disease: Results from Phase I Studies in Glaucoma. *Ophthalmol. Sci.* **2023**, *3*, 100290. [CrossRef] [PubMed]
74. Krishnan, A.; Kocab, A.J.; Zacks, D.N.; Marshak-Rothstein, A.; Gregory-Ksander, M. A small peptide antagonist of the Fas receptor inhibits neuroinflammation and prevents axon degeneration and retinal ganglion cell death in an inducible mouse model of glaucoma. *J. Neuroinflammation* **2019**, *16*, 184. [CrossRef] [PubMed]
75. Guo, J.; Liu, H.; Fu, L. MicroRNA-124 ameliorates autophagic dysregulation in glaucoma via regulation of P2X7-mediated Akt/mTOR signaling. *Cutan. Ocul. Toxicol.* **2022**, *41*, 43–48. [CrossRef]
76. Ghate, D.; Edelhauser, H.F. Ocular drug delivery. *Expert. Opin. Drug Deliv.* **2006**, *3*, 275–287. [CrossRef]
77. Wu, K.Y.; Fujioka, J.K.; Gholamian, T.; Zaharia, M.; Tran, S.D. Suprachoroidal Injection: A Novel Approach for Targeted Drug Delivery. *Pharmaceuticals* **2023**, *16*, 1241. [CrossRef]
78. Barnes, S.D.; Campagna, J.A.; Dirks, M.S.; Doe, E.A. Control of intraocular pressure elevations after argon laser trabeculoplasty: Comparison of brimonidine 0.2% to apraclonidine 1.0%. *Ophthalmology* **1999**, *106*, 2033–2037. [CrossRef]
79. Shih, G.C.; Calkins, D.J. Secondary neuroprotective effects of hypotensive drugs and potential mechanisms of action. *Expert. Rev. Ophthalmol.* **2012**, *7*, 161–175. [CrossRef]
80. Gao, H.; Qiao, X.; Cantor, L.B.; WuDunn, D. Up-regulation of brain-derived neurotrophic factor expression by brimonidine in rat retinal ganglion cells. *Arch. Ophthalmol.* **2002**, *120*, 797–803. [CrossRef]
81. Lai, R.K.; Chun, T.; Hasson, D.; Lee, S.; Mehrbod, F.; Wheeler, L. Alpha-2 adrenoceptor agonist protects retinal function after acute retinal ischemic injury in the rat. *Vis. Neurosci.* **2002**, *19*, 175–185. [CrossRef] [PubMed]
82. Manickavasagam, D.; Wehrung, D.; Chamsaz, E.A.; Sanders, M.; Bouhenni, R.; Crish, S.D.; Joy, A.; Oyewumi, M.O. Assessment of alkoxylphenacyl-based polycarbonates as a potential platform for controlled delivery of a model anti-glaucoma drug. *Eur. J. Pharm. Biopharm.* **2016**, *107*, 56–66. [CrossRef] [PubMed]
83. Lou, X.; Hu, Y.; Zhang, H.; Liu, J.; Zhao, Y. Polydopamine nanoparticles attenuate retina ganglion cell degeneration and restore visual function after optic nerve injury. *J. Nanobiotechnology* **2021**, *19*, 436. [CrossRef]
84. Stanyon, H.F.; Viles, J.H. Human serum albumin can regulate amyloid-β peptide fiber growth in the brain interstitium: Implications for Alzheimer disease. *J. Biol. Chem.* **2012**, *287*, 28163–28168. [CrossRef] [PubMed]
85. Lambert, W.S.; Carlson, B.J.; van der Ende, A.E.; Shih, G.; Dobish, J.N.; Calkins, D.J.; Harth, E. Nanosponge-Mediated Drug Delivery Lowers Intraocular Pressure. *Transl. Vis. Sci. Technol.* **2015**, *4*, 1. [CrossRef] [PubMed]
86. Lambuk, L.; Mohd Lazaldin, M.A.; Ahmad, S.; Iezhitsa, I.; Agarwal, R.; Usković, V.; Mohamud, R. Brain-Derived Neurotrophic Factor-Mediated Neuroprotection in Glaucoma: A Review of Current State of the Art. *Front. Pharmacol.* **2022**, *13*, 875662. [CrossRef]
87. Martin, K.R.; Quigley, H.A.; Zack, D.J.; Levkovitch-Verbin, H.; Kielczewski, J.; Valenta, D.; Baumrind, L.; Pease, M.E.; Klein, R.L.; Hauswirth, W.W. Gene therapy with brain-derived neurotrophic factor as a protection: Retinal ganglion cells in a rat glaucoma model. *Investig. Ophthalmol. Vis. Sci.* **2003**, *44*, 4357–4365. [CrossRef]

88. Chen, D.W.; Foldvari, M. In vitro bioassay model for screening non-viral neurotrophic factor gene delivery systems for glaucoma treatment. *Drug Deliv. Transl. Res.* **2016**, *6*, 676–685. [CrossRef]
89. Narsineni, L.; Chen, D.W.; Foldvari, M. BDNF gene delivery to the retina by cell adhesion peptide-conjugated gemini nanoplexes in vivo. *J. Control Release* **2023**, *359*, 244–256. [CrossRef]
90. Khatib, T.Z.; Osborne, A.; Yang, S.; Ali, Z.; Jia, W.; Manyakin, I.; Hall, K.; Watt, R.; Widdowson, P.S.; Martin, K.R. Receptor-ligand supplementation via a self-cleaving 2A peptide-based gene therapy promotes CNS axonal transport with functional recovery. *Sci. Adv.* **2021**, *7*, eabd2590. [CrossRef]
91. Wang, T.; Li, Y.; Guo, M.; Dong, X.; Liao, M.; Du, M.; Wang, X.; Yin, H.; Yan, H. Exosome-Mediated Delivery of the Neuroprotective Peptide PACAP38 Promotes Retinal Ganglion Cell Survival and Axon Regeneration in Rats with Traumatic Optic Neuropathy. *Front. Cell Dev. Biol.* **2021**, *9*, 659783. [CrossRef] [PubMed]
92. Giannaccini, M.; Usai, A.; Chiellini, F.; Guadagni, V.; Andreazzoli, M.; Ori, M.; Pasqualetti, M.; Dente, L.; Raffa, V. Neurotrophin-conjugated nanoparticles prevent retina damage induced by oxidative stress. *Cell. Mol. Life Sci.* **2018**, *75*, 1255–1267. [CrossRef] [PubMed]
93. Shen, J.; Wang, Y.; Yao, K. Protection of retinal ganglion cells in glaucoma: Current status and future. *Exp. Eye Res.* **2021**, *205*, 108506. [CrossRef] [PubMed]
94. Hare, W.; WoldeMussie, E.; Lai, R.; Ton, H.; Ruiz, G.; Feldmann, B.; Wijono, M.; Chun, T.; Wheeler, L. Efficacy and safety of memantine, an NMDA-type open-channel blocker, for reduction of retinal injury associated with experimental glaucoma in rat and monkey. *Surv. Ophthalmol.* **2001**, *45* (Suppl. S3), S284–S289, discussion S295–S286. [CrossRef]
95. Sánchez-López, E.; Egea, M.A.; Davis, B.M.; Guo, L.; Espina, M.; Silva, A.M.; Calpena, A.C.; Souto, E.M.B.; Ravindran, N.; Ettcheto, M.; et al. Memantine-Loaded PEGylated Biodegradable Nanoparticles for the Treatment of Glaucoma. *Small* **2018**, *14*, 1701808. [CrossRef] [PubMed]
96. Ito, K.; Tatebe, T.; Suzuki, K.; Hirayama, T.; Hayakawa, M.; Kubo, H.; Tomita, T.; Makino, M. Memantine reduces the production of amyloid-β peptides through modulation of amyloid precursor protein trafficking. *Eur. J. Pharmacol.* **2017**, *798*, 16–25. [CrossRef] [PubMed]
97. Nizari, S.; Guo, L.; Davis, B.M.; Normando, E.M.; Galvao, J.; Turner, L.A.; Bizrah, M.; Dehabadi, M.; Tian, K.; Cordeiro, M.F. Non-amyloidogenic effects of α2 adrenergic agonists: Implications for brimonidine-mediated neuroprotection. *Cell Death Dis.* **2016**, *7*, e2514. [CrossRef]
98. Reuben, D.B.; Kremen, S.; Maust, D.T. Dementia Prevention and Treatment: A Narrative Review. *JAMA Intern. Med.* **2024**, *184*, 563–572. [CrossRef]
99. Bou Ghanem, G.O.; Wareham, L.K.; Calkins, D.J. Addressing neurodegeneration in glaucoma: Mechanisms, challenges, and treatments. *Prog. Retin. Eye Res.* **2024**, *100*, 101261. [CrossRef]
100. Esteruelas, G.; Halbaut, L.; García-Torra, V.; Espina, M.; Cano, A.; Ettcheto, M.; Camins, A.; Souto, E.B.; Luisa García, M.; Sánchez-López, E. Development and optimization of Riluzole-loaded biodegradable nanoparticles incorporated in a mucoadhesive in situ gel for the posterior eye segment. *Int. J. Pharm.* **2022**, *612*, 121379. [CrossRef]
101. López-Malo, D.; Villarón-Casares, C.A.; Alarcón-Jiménez, J.; Miranda, M.; Díaz-Llopis, M.; Romero, F.J.; Villar, V.M. Curcumin as a Therapeutic Option in Retinal Diseases. *Antioxidants* **2020**, *9*, 48. [CrossRef]
102. Davis, B.M.; Pahlitzsch, M.; Guo, L.; Balendra, S.; Shah, P.; Ravindran, N.; Malaguarnera, G.; Sisa, C.; Shamsher, E.; Hamze, H.; et al. Topical Curcumin Nanocarriers are Neuroprotective in Eye Disease. *Sci. Rep.* **2018**, *8*, 11066. [CrossRef] [PubMed]
103. Liu, Y.C.; Lin, Y.K.; Lin, Y.T.; Lin, C.W.; Lan, G.Y.; Su, Y.C.; Hu, F.R.; Chang, K.H.; Chen, V.; Yeh, Y.C.; et al. Injectable, Antioxidative, and Tissue-Adhesive Nanocomposite Hydrogel as a Potential Treatment for Inner Retina Injuries. *Adv. Sci.* **2024**, *11*, e2308635. [CrossRef] [PubMed]
104. Davis, B.M.; Tian, K.; Pahlitzsch, M.; Brenton, J.; Ravindran, N.; Butt, G.; Malaguarnera, G.; Normando, E.M.; Guo, L.; Cordeiro, M.F. Topical Coenzyme Q10 demonstrates mitochondrial-mediated neuroprotection in a rodent model of ocular hypertension. *Mitochondrion* **2017**, *36*, 114–123. [CrossRef] [PubMed]
105. Schuster, S.J.; Koury, S.T.; Bohrer, M.; Salceda, S.; Caro, J. Cellular sites of extrarenal and renal erythropoietin production in anaemic rats. *Br. J. Haematol.* **1992**, *81*, 153–159. [CrossRef]
106. Luo, W.; Hu, L.; Wang, F. The protective effect of erythropoietin on the retina. *Ophthalmic Res.* **2015**, *53*, 74–81. [CrossRef]
107. Silva, B.; Gonçalves, L.M.; Braz, B.S.; Delgado, E. Topical Administration of a Nanoformulation of Chitosan-Hyaluronic Acid-Epoetin Beta in a Rat Model of Glaucoma. *Pharmaceuticals* **2023**, *16*, 164. [CrossRef]
108. Brar, S.K.; Perveen, S.; Chaudhry, M.R.; AlBabtain, S.; Amreen, S.; Khan, S. Erythropoietin-Induced Hypertension: A Review of Pathogenesis, Treatment, and Role of Blood Viscosity. *Cureus* **2021**, *13*, e12804. [CrossRef]
109. Silva, B.; Gonçalves, L.M.; Braz, B.S.; Delgado, E. Chitosan and Hyaluronic Acid Nanoparticles as Vehicles of Epoetin Beta for Subconjunctival Ocular Delivery. *Mar. Drugs* **2022**, *20*, 151. [CrossRef]
110. Abbott, C.J.; Choe, T.E.; Lusardi, T.A.; Burgoyne, C.F.; Wang, L.; Fortune, B. Evaluation of retinal nerve fiber layer thickness and axonal transport 1 and 2 weeks after 8 hours of acute intraocular pressure elevation in rats. *Investig. Ophthalmol. Vis. Sci.* **2014**, *55*, 674–687. [CrossRef]
111. Wadhwa, S.; Paliwal, R.; Paliwal, S.R.; Vyas, S.P. Hyaluronic acid modified chitosan nanoparticles for effective management of glaucoma: Development, characterization, and evaluation. *J. Drug Target.* **2010**, *18*, 292–302. [CrossRef]

112. Lillibridge, C.B.; Docter, J.M.; Eidelman, S. Oral administration of n-acetyl cysteine in the prophylaxis of "meconium ileus equivalent". *J. Pediatr.* **1967**, *71*, 887–889. [CrossRef] [PubMed]
113. Schwalfenberg, G.K. N-Acetylcysteine: A Review of Clinical Usefulness (an Old Drug with New Tricks). *J. Nutr. Metab.* **2021**, *2021*, 9949453. [CrossRef] [PubMed]
114. Deepmala; Slattery, J.; Kumar, N.; Delhey, L.; Berk, M.; Dean, O.; Spielholz, C.; Frye, R. Clinical trials of N-acetylcysteine in psychiatry and neurology: A systematic review. *Neurosci. Biobehav. Rev.* **2015**, *55*, 294–321. [CrossRef] [PubMed]
115. Pitha, I.; Kambhampati, S.; Sharma, A.; Sharma, R.; McCrea, L.; Mozzer, A.; Kannan, R.M. Targeted Microglial Attenuation through Dendrimer-Drug Conjugates Improves Glaucoma Neuroprotection. *Biomacromolecules* **2023**, *24*, 1355–1365. [CrossRef] [PubMed]
116. Belforte, N.A.; Moreno, M.C.; de Zavalía, N.; Sande, P.H.; Chianelli, M.S.; Keller Sarmiento, M.I.; Rosenstein, R.E. Melatonin: A novel neuroprotectant for the treatment of glaucoma. *J. Pineal Res.* **2010**, *48*, 353–364. [CrossRef]
117. Monteiro, K.; Shiroma, M.E.; Damous, L.L.; Simões, M.J.; Simões, R.D.S.; Cipolla-Neto, J.; Baracat, E.C.; Soares-Jr., J.M. Antioxidant Actions of Melatonin: A Systematic Review of Animal Studies. *Antioxidants* **2024**, *13*, 439. [CrossRef]
118. Dal Monte, M.; Cammalleri, M.; Amato, R.; Pezzino, S.; Corsaro, R.; Bagnoli, P.; Rusciano, D. A Topical Formulation of Melatoninergic Compounds Exerts Strong Hypotensive and Neuroprotective Effects in a Rat Model of Hypertensive Glaucoma. *Int. J. Mol. Sci.* **2020**, *21*, 9267. [CrossRef]
119. Sun, Y.; Luo, Z.M.; Guo, X.M.; Su, D.F.; Liu, X. An updated role of microRNA-124 in central nervous system disorders: A review. *Front. Cell Neurosci.* **2015**, *9*, 193. [CrossRef]
120. Wang, P.; Chen, L.; Zhang, J.; Chen, H.; Fan, J.; Wang, K.; Luo, J.; Chen, Z.; Meng, Z.; Liu, L. Methylation-mediated silencing of the miR-124 genes facilitates pancreatic cancer progression and metastasis by targeting Rac1. *Oncogene* **2014**, *33*, 514–524. [CrossRef]
121. Dobrzycka, M.; Sulewska, A.; Biecek, P.; Charkiewicz, R.; Karabowicz, P.; Charkiewicz, A.; Golaszewska, K.; Milewska, P.; Michalska-Falkowska, A.; Nowak, K.; et al. miRNA Studies in Glaucoma: A Comprehensive Review of Current Knowledge and Future Perspectives. *Int. J. Mol. Sci.* **2023**, *24*, 4699. [CrossRef]
122. Li, T.; Wang, Y.; Chen, J.; Gao, X.; Pan, S.; Su, Y.; Zhou, X. Co-delivery of brinzolamide and miRNA-124 by biodegradable nanoparticles as a strategy for glaucoma therapy. *Drug Deliv.* **2020**, *27*, 410–421. [CrossRef] [PubMed]
123. Iester, M. Brinzolamide ophthalmic suspension: A review of its pharmacology and use in the treatment of open angle glaucoma and ocular hypertension. *Clin. Ophthalmol.* **2008**, *2*, 517–523. [CrossRef] [PubMed]
124. Pang, I.H.; Clark, A.F. Inducible rodent models of glaucoma. *Prog. Retin. Eye Res.* **2020**, *75*, 100799. [CrossRef] [PubMed]
125. Perlman, I. Testing retinal toxicity of drugs in animal models using electrophysiological and morphological techniques. *Doc. Ophthalmol.* **2009**, *118*, 3–28. [CrossRef] [PubMed]
126. Fiore, T.; Iaccheri, B.; Pietrolucci, F.; Giansanti, F.; Cavaliere, A.; Coltella, R.; Mameli, M.G.; Androudi, S.; Brazitikos, P.; Cagini, C. Retinal toxicity of intravitreal genistein in a rabbit model. *Retina* **2010**, *30*, 1536–1541. [CrossRef] [PubMed]
127. Bill, A.; Stjernschantz, J. Cholinergic vasoconstrictor effects in the rabbit eye: Vasomotor effects of pentobarbital anesthesia. *Acta Physiol. Scand.* **1980**, *108*, 419–424. [CrossRef]
128. Werner, L.; Chew, J.; Mamalis, N. Experimental evaluation of ophthalmic devices and solutions using rabbit models. *Vet. Ophthalmol.* **2006**, *9*, 281–291. [CrossRef]
129. Ruiz-Ederra, J.; García, M.; Hicks, D.; Vecino, E. Comparative study of the three neurofilament subunits within pig and human retinal ganglion cells. *Mol. Vis.* **2004**, *10*, 83–92.
130. Ruiz-Ederra, J.; García, M.; Hernández, M.; Urcola, H.; Hernández-Barbáchano, E.; Araiz, J.; Vecino, E. The pig eye as a novel model of glaucoma. *Exp. Eye Res.* **2005**, *81*, 561–569. [CrossRef]
131. Patel, C.; Pande, S.; Sagathia, V.; Ranch, K.; Beladiya, J.; Boddu, S.H.S.; Jacob, S.; Al-Tabakha, M.M.; Hassan, N.; Shahwan, M. Nanocarriers for the Delivery of Neuroprotective Agents in the Treatment of Ocular Neurodegenerative Diseases. *Pharmaceutics* **2023**, *15*, 837. [CrossRef]
132. Krishnamoorthy, R.R.; Clark, A.F.; Daudt, D.; Vishwanatha, J.K.; Yorio, T. A forensic path to RGC-5 cell line identification: Lessons learned. *Investig. Ophthalmol. Vis. Sci.* **2013**, *54*, 5712–5719. [CrossRef] [PubMed]
133. Wilsey, L.J.; Fortune, B. Electroretinography in glaucoma diagnosis. *Curr. Opin. Ophthalmol.* **2016**, *27*, 118–124. [CrossRef] [PubMed]
134. Cordeiro, M.F.; Hill, D.; Patel, R.; Corazza, P.; Maddison, J.; Younis, S. Detecting retinal cell stress and apoptosis with DARC: Progression from lab to clinic. *Prog. Retin. Eye Res.* **2022**, *86*, 100976. [CrossRef] [PubMed]
135. Albarqi, H.A.; Garg, A.; Ahmad, M.Z.; Alqahtani, A.A.; Walbi, I.A.; Ahmad, J. Recent Progress in Chitosan-Based Nanomedicine for Its Ocular Application in Glaucoma. *Pharmaceutics* **2023**, *15*, 681. [CrossRef] [PubMed]
136. Khan, R.S.; Dine, K.; Bauman, B.; Lorentsen, M.; Lin, L.; Brown, H.; Hanson, L.R.; Svitak, A.L.; Wessel, H.; Brown, L.; et al. Intranasal Delivery of a Novel Amnion Cell Secretome Prevents Neuronal Damage and Preserves Function in a Mouse Multiple Sclerosis Model. *Sci. Rep.* **2017**, *7*, 41768. [CrossRef]
137. Keller, L.A.; Merkel, O.; Popp, A. Intranasal drug delivery: Opportunities and toxicologic challenges during drug development. *Drug Deliv. Transl. Res.* **2022**, *12*, 735–757. [CrossRef]
138. Osborne, A.; Khatib, T.Z.; Songra, L.; Barber, A.C.; Hall, K.; Kong, G.Y.X.; Widdowson, P.S.; Martin, K.R. Neuroprotection of retinal ganglion cells by a novel gene therapy construct that achieves sustained enhancement of brain-derived neurotrophic factor/tropomyosin-related kinase receptor-B signaling. *Cell Death Dis.* **2018**, *9*, 1007. [CrossRef]

139. Dal Monte, M.; Cammalleri, M.; Pezzino, S.; Corsaro, R.; Pescosolido, N.; Bagnoli, P.; Rusciano, D. Hypotensive Effect of Nanomicellar Formulation of Melatonin and Agomelatine in a Rat Model: Significance for Glaucoma Therapy. *Diagnostics* **2020**, *10*, 138. [CrossRef]
140. Moritera, T.; Ogura, Y.; Honda, Y.; Wada, R.; Hyon, S.H.; Ikada, Y. Microspheres of biodegradable polymers as a drug-delivery system in the vitreous. *Investig. Ophthalmol. Vis. Sci.* **1991**, *32*, 1785–1790.

pharmaceuticals

MDPI

Review

Drug Delivery Systems for Glaucoma: A Narrative Review

Antonio M. Fea, Veronica Vallino, Michela Cossu, Valentina Marica, Cristina Novarese, Michele Reibaldi and Francesco Petrillo *

Department of Ophthalmology, "City of Health and Science" Hospital, 10126 Turin, Italy; antoniofea@icloud.com (A.M.F.); veronica.vallino@gmail.com (V.V.); michela.cossu@hotmail.com (M.C.); vale.marica.vm@gmail.com (V.M.); cristina.novarese@edu.unito.it (C.N.); mreibaldi@libero.it (M.R.)
* Correspondence: francescopetrillo09@gmail.com

Abstract: Glaucoma is one of the world's leading causes of blindness, and its management is challenging. The main objective is to lower intraocular pressure through medical, para-surgical, and surgical therapy. Medical therapy often represents the first line of treatment. Although effective in many cases, the eye drops are accompanied by significant problems. They require high patient compliance and can be associated with various side effects, limiting their efficacy. Consequently, the research for new drug delivery systems trying to overcome these limitations is ongoing: numerous devices are developing and gradually entering clinical practice. These new therapeutic options may offer better control of the intraocular pressure, with fewer side effects, and are less dependent on patients' compliance. Hence, the research in this field continues to flourish. This review summarizes the most recent findings in the scientific literature, underlines the role and possible limitations of the new glaucoma drug delivery systems in clinical practice, and recognizes their new horizons and perspectives.

Keywords: glaucoma; punctal plug delivery systems; conjunctival fornix delivery systems; drug-laden contact lenses; periocular drug delivery systems; intracameral delivery systems

1. Introduction

Glaucoma is a leading cause of irreversible blindness worldwide and poses a significant public health problem. It has been estimated that about 80 million people are affected by this condition, and 11 million are bilaterally blind [1]. The disease consists of an optic neuropathy characterized by progressive degeneration of retinal ganglion cells, resulting in cupping (an excavated appearance of the optic disc) and visual loss [2]. The most common form of glaucoma is open-angle (OAG), and intraocular pressure (IOP) is the leading risk factor for its development and progression. Although there is no cure, glaucoma therapies intend to preserve visual function by lowering IOP to the target pressure [3]. The term target pressure refers to an IOP below which it is reasonable to estimate the rate of disease progression to be sufficiently slow to minimize the risk of further symptomatic visual loss. Available treatment options include medical therapy (both topical drops and oral medications), laser therapy, and incisional surgery. Topical medications are often the first-line treatment. Even when eye drops are adequate, patients' adherence to daily medication is a significant challenge. Patients often require multiple medications and doses, and older people have difficulty instilling eye drops. It has been estimated that up to 80% of patients do not use their IOP-lowering eye drops as prescribed [4,5]. In glaucoma, poor medication adherence has been associated with worse visual outcomes [6]. Besides adherence problems, significant barriers to drug delivery can result in suboptimal medication levels in the target tissues. Additionally, the eye drop delivery of drugs is characterized by a pulsatile fashion with a peak drug concentration followed by a through before the next dose, which could result in worse control of the IOP and a higher risk of adverse effects [7]. Sustained release drug delivery systems can potentially avoid many of the limitations of eye drops. A continuous drug release is associated with sustained lowering of the IOP, a reduced total dose, and a lower or slower systemic exposure, which reduces the risk of systemic side effects. Moreover, eliminating the need to remember to use medication for that period and bypassing the challenges of administering them at the right time helps to overcome adherence problems [8]. Several sustained-release drug delivery systems are in preclinical development or clinical trials to treat glaucoma and ocular hypertension (OHT). This narrative review aims to provide an overview of recent advancements in drug delivery systems for glaucoma. In particular, we will focus on five innovative approaches: drug-eluting punctal plugs, periocular inserts, contact lenses, subconjunctival implants, and intracameral implants. Through a critical analysis of the existing literature, we will examine the characteristics, advantages, limitations, and future perspectives of each of these systems, with the aim of offering a comprehensive understanding of their potential clinical applications and associated challenges.

2. Punctal Plugs Delivery System

Punctal plugs (Figure 1) are rod-shaped medical devices, usually made of a polymeric material, designed to be implanted into the tear ducts. In order to reduce tear drainage and increase the tears on the ocular surface, they block the punctum and the canaliculus [9]. These plugs can be loaded with drugs that are slowly released in a unidirectional manner toward the ocular surface, prolonging contact with the eye surface. Drug mixing with the tear fluid and subsequent delivery to the eye surface and intraocular tissues is thus enhanced. Punctal plugs can be perpetual (for dry eye) or temporary (for drug delivery). Plugs come in a variety of shapes and sizes. Each part of the structure can change, such as the core (polymer matrix or drug matrix permeable to tear fluid), the cap (semi- or impermeable membrane with one or more pores), the body (impermeable to drug and tear fluid), and the nose (assists the insertion process) [7]. Inside the core of the plug is a potentially possible place to load the drug solution, suspension, emulsion, nanoparticle or microparticle, or liposome suspensions [10].

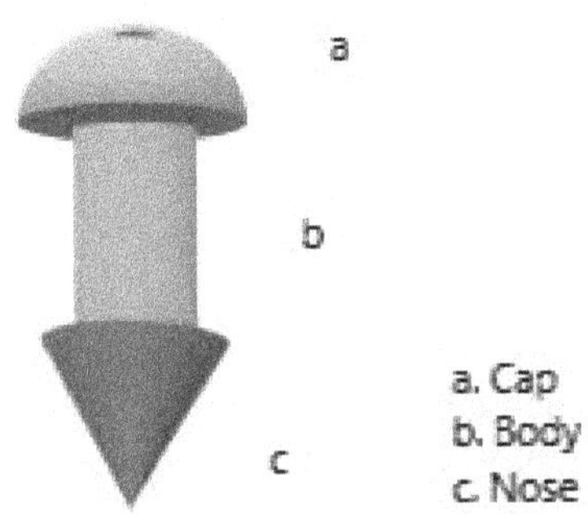

Figure 1. Punctal plug.

Glaucoma patients may benefit from punctal plugs for three main reasons. Ocular surface disease (OSD) frequently coexists with glaucoma and may be initiated or exacerbated by topical medications. Patients with OSD may have an aqueous deficiency and dysfunctional tear film circulation. Recent studies have underlined the utility of punctal plugs in curing these patients because they can prevent tears from draining through the nasolacrimal ducts, thereby maintaining a greater volume of tears on the surface of the eye and relieving dry eye symptoms [11–13]. Furthermore, punctal plugs could reduce the systemic absorption of drugs and improve their efficacy on the target tissue. Lastly, when loaded with antiglaucoma drugs, they represent an alternative to eye drops, thus reducing their burden and compliance issues.

Many studies supported the advantages of punctal plugs in patients with glaucoma (Table 1).

Table 1. Punctal plugs delivery systems.

Authors	Type of Study	Year	Device	Drugs	Results
Chen et al. [14]	Case-control study	2020	Temporary Collagen Punctal Plugs		A statistically significant IOP diminution in the case group compared with the control group. DED ameliorated significantly in the case group rather than in the control group.
Sherwin et al. [15]	Randomized controlled trial	2018	Punctal plugs in patients using prostaglandin analog monotherapy		Punctal plugs resulted in a significantly lowered IOP (MD 1.5 mm Hg, 95% CI 0.1–2.9, $p = 0.032$).
Perera et al. [16]	Initial feasibility, prospective, single-arm	2016	OTX-TP	Travoprost	OTX-TP can reduce IOP by 24% (day 10) and 15.6% (day 30).
Kompella et al. [17]	Phase 2 trial	2010	L-PPDS	Latanoprost	A mean reduction in IOP by 3.5 mm Hg at the end of 4 weeks, with 36% of patients showing a decrease in IOP of more than 5 mm Hg.

Chen et al. used a temporary collagen punctal plug as a pre-test before the decision on permanent punctal closure or long-term plug use for patients with dry eye disease (DED) and primary open-angle glaucoma (POAG). In order to predict the efficacy of the long-term ones and find possible complications, it was used a non-expensive temporary collagen plug. It has been demonstrated that in eyes responding to temporary plugs with fewer DED symptoms and reduced IOP without complications, a long-term punctal plug or permanent punctal closure could be considered. Using punctal occlusion can be safer because it decreases the absorption of glaucoma medications into nasal mucosae, such as beta-blockers, preventing heart rate disturbances [14].

Sherwin et al., in a randomized controlled study of punctal plug placement in glaucoma patients treated with a prostaglandin analog, found reduced OSDI scores, increased tear break-up time, reduced corneal staining, and decreased tear osmolarity. Curiously, IOP also decreased, probably because of increased holding and assimilation of prostaglandin [15]. Zimmermann et al. demonstrated that nasolacrimal occlusion is a simple technique that increases the ocular bioavailability of topically applied ocular drugs and reduces the probability of adverse systemic effects. Punctal occlusion can reduce systemic drug absorption of topical timolol by more than 60%. This technique seems safe, simple, and effective [18]. Also, Huang et al. found, after punctal plug occlusion, a statistically significant reduction in IOP with a variance of 1.82 mm Hg or 10.7% (baseline of 16.96 mm Hg) compared with before punctal occlusion [19].

Punctal plugs can be filled with medication and act as a conduit for its delivery. However, their capacity is restricted to accommodating low doses of medication, making corticosteroids and prostaglandins the primary drugs suitable for this purpose due to their effectiveness at low concentrations. The sole commercially available plug system is Dextenza™, specifically designed to administer dexamethasone for the management of post-operative inflammation and pain. Nevertheless, several research efforts have been directed towards developing systems for prostaglandin delivery as well [20,21].

Perera et al. developed travoprost tear plugs named OTX-TP (Ocular Therapeutix, Inc., Bedford, MA, USA) utilizing polylactic acid particles enclosed in polyethylene glycol absorbable hydrogel rods. These microparticles hydrolyze to deliver travoprost over a 90-day period. In their initial single-arm study, all plugs were directly visible post-insertion without slit lamp examination. Retention rates, assessed through visualization, remained at 100% for the first 10 days, declining to 42% by day 30 [16]. In a subsequent phase 2 trial, patients were divided into two groups and randomly prescribed to receive either two times a day artificial tears or timolol 0.5% twice daily with a drug-free punctal plug. Both groups showed clinically significant reductions in IOP from baseline, with the OTX-TP group demonstrating a reduction of 4.5–5.7 mm Hg compared with 6.4–7.6 mm Hg in the timolol group. Retention rates improved compared with earlier reports but remained relatively low in the mid-term (91% at day 60 and 48% at day 90). However, this may underestimate retention due to the lack of ability of transillumination to locate small remnants of OTX-TP by visualization [22]. Finally, in a multicenter randomized phase 3 trial comparing OTX-TP to a placebo vehicle insert, OTX-TP showed generally good tolerability, with no serious ocular adverse events observed. The most common ocular adverse events (AEs) included dacryocanaliculitis (approximately 7% in OTX-TP vs. 3% in placebo) and lacrimal structure disorder (around 6% in OTX-TP vs. 4% in placebo). The tolerability of the implant improved over time. However, the trial did not prove meaningful preponderance in mean IOP reduction, likely due to its sample size and short duration. According to personal communication with the manufacturer, the OTX-TP system is no longer being built up as it did not show superiority to eye drops [23]. In 2010, Kompella et al. and subsequently Goldberg et al. in 2011 investigated the Latanoprost punctal plug delivery system (PPDS) developed by QLT (Vancouver, BC, Canada). Later, the L-PPDS (Latanoprost Punctal Plug Delivery System) was acquired by Mati Therapeutics Inc. (Austin, TX, USA) They employed an L-shaped plug containing a nonbiodegradable Latanoprost core, insertable at the slit lamp, with two plugs placed into one eye. Each plug contained 70.5 mg of Latanoprost, exhibiting slow drug release over 30 days. After four weeks of L-PPDS treatment, a statistically significant mean reduction in IOP from baseline (−5.7 mm Hg) was observed. The L-PPDS demonstrated good tolerability throughout the study period, with AEs comparable to those reported for commercial punctal plugs. This extracellular and minimally invasive drug delivery system demonstrated clinically significant and sustained (4 weeks) reduction in IOP [17,24]. Although punctal plugs are believed to be relatively safe, their use is commonly associated with epiphora, conjunctival irritation, punctal plug extrusion, and, less commonly, with inflammatory conditions such as dacryocystitis. There can also be a slight irritation or itchiness in the tear duct area after the first insertion, which usually disappears after acclimatization. Furthermore, an external plug can be forced down into the canaliculus through insertion or lost afterwards. The same problem can occur with intracanalicular plugs [25,26]. The inconstant delivery from the punctum to the tear film and for the low drug doses that punctal plugs, as drug carriers, can handle may also limit the IOP-lowering effect.

3. Conjunctival Fornix Delivery System

Conjunctival fornix delivery systems include implants, gels, or particles designed for prolonged drug release (Table 2).

Table 2. Conjunctival fornix delivery system.

Authors	Type of Study	Year	Device	Drugs	Results
Alza Corporation [27]	Phase 3 trial, randomized, multicenter, masked (previously on the market)	1975	Pilocarpine Ocusert	Conjunctival fornix delivery system, pilocarpine	Positive outcomes in IOP lowering but also many side effects. Withdrawn from the market in 1993.
James D. Brandt et al. [28]	Double-masked, randomized, multicenter, phase 2 study	2016	Topical Bimatoprost Ocular Insert	Conjunctival fornix delivery system, Bimatoprost	Bimatoprost Ocular Ring did not reach the non-inferiority standard in lowering IOP compared with timolol eye drops.
James D. Brandt et al. [29]	A 13-month open-label extension (OLE) study	2017	Topical Bimatoprost Ocular Insert	Conjunctival fornix delivery system, Bimatoprost	A median reduction of 4 mm Hg was reached over a follow-up period of 19 months.
Nair et al. [30]	Preclinical study	2018	Timolol maleate-loaded inserts	Conjunctival fornix delivery system, timolol	Faster response and longer persistence of the drug in timolol maleate-loaded inserts compared with timolol maleate drop solution.
Kelly et al. [31]	Single-dose crossover study	1989	New Ophthalmic Delivery System (NODS)	Conjunctival fornix delivery system, pilocarpine	The bioavailability of NODS is eight times greater compared with conventional eye drops.

Table 2. *Cont.*

Authors	Type of Study	Year	Device	Drugs	Results
Greaves et al. [32]	Gamma scintigraphic study on 12 volunteers	1992	Radiolabeled New Ophthalmic Delivery System (NODS) loaded with pilocarpine nitrate	Conjunctival fornix delivery system, pilocarpine	IOP-lowering; more significant pupil diameter reduction and higher bioavailability compared with a 25 μL drop of 2% pilocarpine nitrate solution.
M. Diestelhorst [33]	Open-label, crossover study	1994	Radiolabeled New Ophthalmic Delivery System (NODS)	Conjunctival fornix delivery system	NODS were less tolerated than conventional eye drops.
Bigdeli et al. [34]	In vivo (rabbits)	2023		Timolol Maleate-Brimonidine tartrate-loaded liposomes	Both liposomal and aqueous formulations reduce IOP, although liposomal formulation is more efficient (rabbits).

The implant system comprises a non-toxic sterile device shaped to fit the conjunctival cul-de-sac. Like other delivery systems described in this review, it is developed to administer glaucoma therapy more efficiently, avoiding the patient's compliance problem. Many recent studies focus on the search for a new method to extend the drug permanence on the ocular surface, considering that the half-life of eyedrops is about 4 min after administration [35]. These devices can be distinguished as soluble, insoluble, and bioerodible. The drug release occurs by diffusion for the non-degradable inserts and by dissolution or erosion for the degradable ones. Their main advantages include precise dosage, increased bioavailability, lack of preservatives, continuous effectiveness without peaks for a lasting period ranging from hours to days, and reduced number of administrations. Unfortunately, some disadvantages were reported, such as foreign body sensation, eye discharge, occasional device loss, and difficulty placing and removing the insert.

The Pilocarpine Ocusert insert (developed by Alza corporation) [27], approved for clinical use in 1975, was a polymer-membrane system composed of two elliptical membranes 13.4 mm long and 13.7 mm wide, among which is positioned the pilocarpine core. Designed to be worn in the upper or lower conjunctival fornix, it supplied the drug for seven days through diffusion with two different rates: 20 μg per hour or 40 μg per hour. This system avoided the pulsatile fluctuations of the eye drops by maintaining a constant rate of the drug on the eye's surface. Many adverse events peculiar to pilocarpine drops, such as brow ache and blurred vision caused by myopic shift, were reduced with Ocusert. However, many side effects were reported, such as dislodgement, foreign body perception, burning sensation, conjunctival hyperemia, and high cost [36]. This device was used until 1993, when it was withdrawn from the market.

The Topical Bimatoprost Ocular Insert [28] is a preservative-free ocular ring with an inner polypropylene structure surrounded by a silicone matrix containing 13 mg of Bimatoprost. The diameter of this ring varies between 24 and 29 mm depending on the intercanthal distance. Drug delivery is based on passive diffusion of the drug in the tear film with a declining dose from day 0 (35 μg/day) to day 180 (6 μg/day). A multicenter, phase 2, randomized, prospective study assessed the efficacy and safety of a Bimatoprost insert compared to a 0.5% timolol ophthalmic solution administered twice a day in 130 patients with OAG and OHT over a six-month period. Ocular and non-ocular adverse events were more prevalent in the Bimatoprost insert group. The most frequent were eye discharge, conjunctival hyperemia, punctate keratitis, eye itching, and ocular discomfort. However, all the ocular AEs were mild to moderate, and patients fully recovered without consequences. A clinically relevant IOP reduction of −3.2 to −6.4 mm Hg was reached (more than 20% compared with the baseline) with the Bimatoprost insert compared with a decrease of −4.2 to −6.4 mm Hg for the timolol group. Although a considerable IOP reduction was reached, non-inferiority of the Bimatoprost insert was not demonstrated: non-inferiority was set as a mean difference lower than 1.5 mm Hg, and this goal was not reached in the Bimatoprost group. The authors attribute this result to the small sample size and the probable paradoxical response of prostaglandins, which are less effective when frequently administered. At the end of this study, 75 patients joined an open-label single-arm study for 13 months [29]. A median reduction of 4 mm Hg was observed, proving that the Bimatoprost insert can lead to a clinically relevant decrease in IOP over 19 months.

A combination ring containing Bimatoprost and timolol has been studied in a clinical trial. However, results are still unavailable (nct02742649).

Timolol maleate-loaded inserts designed to fit in the conjunctival cul-de-sac were investigated by Nair et al. [30] These devices are composed of a polymeric matrix with a solvent evaporation complex incorporating both a hydrophilic polymer (sodium alginate) and a hydrophobic polymer (ethyl cellulose), with polyethylene glycol (PEG) acting as a plasticizer. Different formulations (F1–F7) were developed by changing the concentrations of the polymer, which was found to influence drug release negatively. The formulation with the best pharmacologic properties (F4) was selected for use in vitro and then in vivo. In vivo studies on rabbits compared F4 formulation and marked timolol maleate drop solution. A faster response and longer persistence of the drug were observed in the former group.

Recent advancements in pharmaceutical research have explored the utilization of liposomes as a novel formulation to prolong the effects of eye drops and enhance their efficacy. In 2022, Bigdeli et al. found the optimized liposomal formulation containing timolol maleate (TM) and brimonidine tartrate (BT) and studied its effectiveness in 36 rabbits [34]. The IOP was successfully increased in the right eye of these rabbits by injecting Hydroxypropyl methylcellulose (HPMC) polymer into the anterior chamber. Then, the rabbits were divided into six groups: the control group, the ones that received drug-loaded liposomes (TM/BT, only TM, only BT) or a drug-free liposomal formulation, and finally, the one in which aqueous solutions of TM/BT were instilled. The IOP was measured before HPMC polymer injection in the AC, 24 h after, and 2 h after each eye drop instillation (27, 41, 51, 65, 75, 89, 99, 113, 123, and 137 h after starting the injection in the right eye AC). In each group, the study ended when eye pressure returned to its baseline value. The results showed that 27 h after the injection, both formulations were capable of substantially lowering the IOP compared with the control group ($p < 0.001$), and specifically, the liposomal formulation was more effective in lowering the IOP compared with the simple formulation of the aqueous solution. Specifically, the selected liposomal formulation made by thin layer hydration increased the efficacy of the timolol maleate and brimonidine tartrate combination could provide sustained release for 12 h.

Nanotechnology has led to the development of innovative systems for the administration of hypotensive drugs such as timolol. A study by Burcu Uner et al. [37] investigated timolol-loaded ethosomes, which are lipid-based vesicles characterized by their small size and elasticity. These ethosomes enhance drug permeation, and both in vitro and in vivo tests showed no significant difference ($p > 0.05$) between a single application and three applications of the conventional drug, suggesting potential benefits for glaucoma treatment.

Another study [38] examined the effectiveness of timolol encapsulated in nanogels made from N-isopropylacrylamide (NIPA) and acrylic acid (AAc). This in vivo study in animal models showed that a single application could reduce and stabilize intraocular pressure (IOP) for 48 h. Additionally, nanogels containing timolol maleate and biopolymers such as chitosan and alginate demonstrated sustained drug release and improved permeation, indicating a promising strategy for glaucoma therapy [39].

Furthermore, a nanofibrous drug delivery system (DDS) was developed as an alternative to eye drops for brinzolamide (BRZ) delivery [40]. This system utilized β-cyclodextrin (β-CD), hydroxypropyl cellulose (HPC), and polycaprolactone (PCL) and exhibited sustained drug permeation through sheep corneas, achieving therapeutic levels. This nanofibrous DDS offers a significant advantage over traditional eye drops by providing extended drug delivery primarily through the cornea.

The New Ophthalmic Delivery System (NODS) in a novel device where the drug is loaded into a water-soluble polyvinylalcohol (PVA) film, which dissolves when in contact with the tears in the lower conjunctival sac, detaching from the other parts of the device.

Kelly et al. [31] demonstrated that pilocarpine delivered in NODS is well-tolerated, and its dosage is eight times smaller compared with conventional eye drops with a consequent reduced systemic absorption.

Greaves et al. [32] utilized radiolabeled NODS containing pilocarpine nitrate for a scintigraphic study to evaluate precorneal residence and pharmacodynamics in human subjects. This study showed eight-fold precorneal bioavailability of the pilocarpine NODS compared with a 25 μL drop of 2% pilocarpine nitrate solution.

Later, Diestelhorst et al. [33] analyzed the tolerability of NODS, comparing it with the tolerability of Isopto-Naturale R eye drops. The study underlined that NODS were less tolerated than conventional drops through questionnaires. However, this problem was related not to the device itself but to the mechanism of delivery. In many cases, the medicated plaque did not detach from the rest of the delivery system.

Finally, numerous natural polymers have been integrated into drug delivery systems to produce drops that can persist longer on the eye's surface. An example of these polymers is collagen, which has been utilized extensively in ocular applications, including collagen shields. These devices have been loaded with different drugs like steroids or antibiotics. However, their use decreased due to limitations such as non-transparency and toxicity of the cross-linking agents.

Agban et al. [41] devised cross-linked collagen shields employing metal oxide nanoparticles (NPs) such as titanium dioxide (TiO_2), zinc oxide (ZnO), and zinc oxide encapsulated with polyvinylpyrrolidone (ZnO/PVP) as cross-linking agents for the controlled release of pilocarpine hydrochloride. The results of this study underlined a release of pilocarpine hydrochloride in vitro over 14 days.

4. Contact Lens for Ophthalmic Drug Delivery

Contact lenses serve a multifaceted role in ophthalmic care, being utilized not only for the correction of visual impairments but also for protecting compromised corneal surfaces. Ongoing investigations have unveiled their potential as an alternative method for the delivery of ocular medications (Table 3).

Table 3. Drug-laden contact lenses.

Authors	Type of Study	Year	Device	Drugs	Results
Hiratami and Alvarez-Lorenzo [42]	Preclinical trial	2002	Soft contact lenses of a cross-linked hydrogel loaded with molecular imprinting method	Timolol	This method improves the drug loading capacity and prolongs timolol release for more than 24 h
Kim et al. [43]	Preclinical trial	2007	Soaked silicone hydrogel soft contact lenses	Timolol	This method increases drug release for about 15–20 days
Peng et al. [44]	Preclinical trial (animal model)	2012	Silicone hydrogel contact lenses loaded with vitamin E	Timolol	This method increases the bioavailability of the drug and reduces systemic drug uptake
Braga et al. [45]	Preclinical trial	2011	Drug loaded into commercial soft contact lenses with a supercritical solvent impregnation process	Acetazolamide	With this method, it is possible to control acetazolamide-loaded amounts and adjust the drug release levels
Jung and Chauhan [46]	Preclinical trial	2012	Drug encapsulating highly cross-linked nanoparticles in contact lenses	Timolol	This method increases the duration of drug release to 2–4 weeks
Peng et al. [47]	Preclinical trial (animal model)	2012	Commercial soft contact lenses incorporating vitamin E	Timolol	This method increases the drug release duration and lowers the IOP by continuously wearing
Jung et al. [48]	Preclinical trial (animal model)	2013	Nanoparticle-loaded contact lenses	Timolol	This method allows a constant reduction in IOP
Hsu et al. [49]	Preclinical trial (animal model)	2015	Vitamin E-loaded contact lenses	Timolol and Dorzolamide	This method ensures prolonged release of the drug and further reduces IOP
Ciolino et al. [50]	Preclinical trial (animal model)	2016	Contact lenses with a drug-polymer film	Latanoprost	This method is effective as delivery with daily Latanoprost ophthalmic solution

In 1960, Wichterle [51] drew attention to drug-laden contact lenses, and the interest in this area has grown [8]. Drug-laden contact lenses are thin, curved lenses applied to the corneal surface. They can be made of different materials, but silicone hydrogel lenses are currently the best for drug carriage, having good oxygen permeability and being safely worn for 30 consecutive days [52]. When the drug is released from the lens, it maintains high concentrations for approximately 30 min into the tear film, compared with 4 min on average for eye drops [44,47]. For this reason, drug-laden contact lenses seem to be an excellent alternative to eye drops for managing anterior segment disease.

In 1965, Sedlacek pioneered the development of a contact lens soaked in a 1% homatropine aqueous solution to achieve prolonged mydriasis in patients [53]. Shortly thereafter, North devised a lens immersed in a 4% pilocarpine solution with the aim of reducing IOP [54]. Soaking is the first, simplest, and cheapest method of drug-loading hydrogel contact lenses. However, the loading capacity of the drug is limited, and the duration of its effect is short (about 24 h). In 2008, Kim and colleagues designed hydrogel soft contact lenses for extended wear, in which, although the drug-loading method is soaking, the particular material allows the drug (timolol) to be carried for a more extended period, ranging from weeks to months [43]. Moreover, Peng in 2010 demonstrated that silicone hydrogel lenses medicated with vitamin E (10–40%) significantly prolonged the release duration of hydrophilic drugs (e.g., timolol) while preserving lens transparency (vitamin E has a shorter wavelength than visible light) [32].

Subsequently, other authors have developed hydrogel contact lenses with embedded ionic components that interact with the drugs in which the lenses are soaked [55].

In 1998, a device incorporating two contact lenses fused with a central cavity for drug insertion was developed by Nakada and Sugiyama [56]. Following this, the concept was further refined by other researchers, who introduced the term "molecular imprinting" to describe the process of creating cavities within contact lenses with a high affinity for the drug [57]. This method garnered interest due to its simplicity of refilling: by immersing the lens in the drug solution, the drug is drawn into the contact lens, resulting in the formation of a polymer—a compound composed of numerous smaller molecules [44]. However, it is essential that the composition of the lens is tailored specifically to the drug when utilizing this technique. In 2002, Hiratani et al. demonstrated that contact lenses loaded with timolol, prepared using molecular imprinting, could sustain drug release for over 24 h in a saline solution [42].

Another promising approach is the "hypercritical fluid method", classically applied for bulk transport and extraction, which consists of separating one component from another using supercritical fluids as the extracting solvent. The drug is dispersed in the supercritical solvent and interacts with the contact lens. This method is effective for both hydrophilic and hydrophobic drugs and can be combined with molecular imprinting, allowing a more significant drug loading and extended release [45,58].

Another strategy is the encapsulation of drugs in colloidal nanoparticles, allowing more extended drug durability at the level of the cornea [59]. However, these nanoparticles form aggregates and reduce lens transparency and oxygen permeability [60]. Jung and Chauhan in 2012 published a study on the dispersion of timolol encapsulated in cross-linked polymeric nanoparticles within contact lenses, which extended the length of drug release to 2–4 weeks [46]. Liposomes are nanoparticles that can be embedded in the lens matrix or at the level of the lens surface, and these have been shown to increase the duration of drug release, enhance the stability of the tear film, and increase the thickness of the lipid layer [61]. In addition, microemulsions are promising nanoparticles. They have the advantage of easy preparation and sterilization. However, their use is still challenging because their stability is poor (for example, timolol dissolved in ethyl butyrate releases very rapidly) [60].

Finally, Ciolino et al. have developed contact lenses with a drug-polymer film. In this case, by altering either the type or molecular weight of the polymer, drug release characteristics can be modified, showing promising potential [50].

Unfortunately, the existing literature on drug-eluting contact lenses remains predominantly preclinical, with studies primarily conducted in animal models. Nonetheless, we draw attention to two noteworthy investigations that have underscored the safety and efficacy of such devices. In 2012, Jung pioneered the development of extended-wear contact lenses infused with dispersed nanoparticles of PGT (propoxylated glycerol triacrylate) loaded with timolol. This loading was achieved by immersing the lens in an ethanol solution containing the particles. Both in vitro and in vivo experiments conducted on Beagle dogs exhibited a consistent reduction in IOP [48]. Similarly, in 2015, Hsu demonstrated in laboratory and animal models that contact lenses loaded with Dorzolamide and timolol, when co-loaded with vitamin E, could sustain prolonged drug release for approximately two days. This extended release led to a further reduction in IOP compared with traditional Cosopt eye drops [49].

The advantages associated with contact lenses suggest its potential for broader acceptance in the future. They offer ease of use, maintain drug concentrations effectively over extended periods, reduce administration frequency compared with eye drops, and can simultaneously correct visual defects. However, contact lenses pose certain risks, including the potential for sight-threatening microbial keratitis. Challenges related to biocompatibility, patient comfort, optical clarity, drug loading, and delivery persist [62]. Moreover, medications used for glaucoma treatment necessitate continuous wear of contact lenses, even overnight, which may compromise oxygen permeability, leading to corneal edema [62]. Additionally, in the elderly population, where glaucoma prevalence is high, both insertion and wearing of contact lenses can be challenging due to reduced and poor-quality tear film, often causing discomfort [63]. Furthermore, prolonged use of drug-eluting contact lenses can result in systemic absorption of medications through the conjunctiva, posing risks, especially for patients with conditions such as asthma or heart disease [64]. Striking a balance between effective drug release, intraocular pressure reduction, and maintaining transparency and adequate oxygen permeability in contact lenses remains a challenge. Despite ongoing research, there is currently no FDA-approved device for these reasons. However, research on drug-eluting contact lenses persists due to their user-friendly nature and direct contact with the corneal surface, offering promising avenues for future development [8].

5. Periocular Drug Delivery Systems

The main periocular routes of drug delivery include subconjunctival, sub-Tenon, and posterior juxtascleral pathways.

However, the most recent studies in the field of glaucoma predominantly focused on the subconjunctival route due to its minimally invasive nature and proximity to the target tissues (Table 4) [65].

Table 4. Periocular drug delivery systems.

Authors	Type of Study	Year	Device	Drugs	Results
Wong et al. [66]	In vivo (rabbits)	2014		Subconjunctival Latanoprost injection	More significant IOP reduction than Latanoprost eye drops for about 80 days (2 injections-day one and day 50) (rabbits).
Wong et al. [67]	In vivo (rabbits)	2012		Latanoprost-loaded EggPC liposomes	A single injection of these liposomes lowers the IOP for up to 90 days, a more significant IOP reduction than Latanoprost eye drops (rabbits).
Wong et al. [68]	Open-label, non-comparative study	2014		Latanoprost-loaded nanoliposome	A single injection decreased IOP immediately in all six subjects; after three months, IOP was lower than baseline in all subjects except one (humans).
Kesav et al. [9]	Phase 1/2a clinical trial, multicenter	2017	Eye-D VD-101 insert	Latanoprost-releasing subconjunctival insert	The implant reduces IOP by 24% from baseline at 12 weeks. Non-inferior to Latanoprost eye drops (humans).
Kesav et al. [9]	In vivo (rabbits)	2018	GB-401 injection	Beta adrenergic prodrug subconjunctival injection	The implant reduces IOP within the first week and up to 20% from baseline at 2 months (rabbits).

The subconjunctival route enables, in fact, continuous drug delivery to the anterior segment tissues by positioning the drug in close proximity to the limbus, cornea, and ciliary body. Advantages of this route include the wide range of delivery systems available, such as implants, microspheres, nanospheres, liposomes, and gels, as well as the ability of particles to remain in place for extended periods, potentially facilitating sustained drug delivery for several months [69,70]. Larger particles have been found to be more suitable than micro- and nano-particles, with a more hydrophobic surface enhancing the retention of negatively charged nanoparticles [70]. However, these drug delivery systems may be visible and could induce visible hemorrhage on the eye surface. Most trials focusing on this route of administration have explored the use of liposomes.

Wong et al. proved that liposomal Latanoprost injected subconjunctivally in rabbits was more effective than daily eye drops in reducing IOP for approximately 50 days with good tolerability in all the subjects treated [66]. In 2012, the same group developed Latanoprost-loaded egg-phosphatidylcholine (EggPC) liposomes and evaluated their stability, tolerability, and efficacy in vivo using 16 rabbits divided into two groups: one group received injections of Latanoprost-loaded EggPC liposomes, while the other received Latanoprost eye drops. Results indicated that a single injection of these liposomes was not only effective in lowering IOP for up to 90 days but also achieved a greater reduction compared with daily topical administration of Latanoprost (4.8 ± 1.5 vs. 2.5 ± 0.9 mm Hg; $p < 0.001$). No signs of inflammation were observed in the eyes during slit-lamp examination [67]. However, it is noteworthy that these findings were obtained in rabbits, in which the prostaglandin analogs induce a significantly lower IOP reduction compared with humans due to the different sensitivity and localization of the receptors involved [71]. In 2014, the research team initiated an open-label safety and efficacy study involving six humans affected by ocular hypertension or primary open-angle glaucoma, focusing on both short-term and long-term (3 months) efficacy of a single subconjunctival injection of nanoliposomes (100 nm) containing Latanoprost [68]. One hour after the injection, intraocular pressure decreased from 27.55 ± 3.25 mm Hg to 14.52 ± 3.31 mm Hg (range 10–18 mm Hg), with an intraocular pressure reduction ranging from 37% to 63% and a mean decrease of 13.03 ± 2.88 mm Hg (range 9–17 mm Hg), which was statistically significant at each post-treatment visit ($p = 0.001$ to 0.049). All six subjects experienced an immediate IOP reduction of 9 mm Hg or more (≥20%). After three months, five out of six subjects showed lower IOP compared with baseline, especially those who had been using Latanoprost drops before enrollment; the one patient whose IOP after three months returned to baseline had previously been treated with timolol eye drops. Regarding safety, all patients tolerated the injection well, reporting only mild ocular discomfort immediately after the injection. The subconjunctival liposome delivery system is still not approved to date, and the long-term potential adverse effects of nanomedicine are still unknown.

In 2017, a multicenter 12-week phase 1/2a clinical trial was conducted to assess the effectiveness and tolerance of the Eye-D VD-101 insert, a subconjunctival implant built to release Latanoprost over a 12-month period. The study enrolled 77 patients and compared three different doses of the VS-101 insert with Latanoprost eye drops applied once daily. Results at the 12-week mark showed a 24% decrease in IOP from the baseline, accompanied by mostly mild adverse events and favorable tolerability [9].

Graybug Vision demonstrated that the subconjunctival injection of GB-401, an encapsulated microparticle formulation of a beta-adrenergic antagonist prodrug that was able to lower IOP by hydrolyzing into an active agent. The IOP reduction was significant within the first week after the injection and was sustained with maximum IOP lowering of ~20% for over two months. These results were obtained in pigmented rabbits both in vitro and in vivo, but human glaucomatous patients are being recruited for the upcoming first-in-human phase 1/2a study [9].

Although the results of periocular drug delivery systems studies are promising, their main limitation is that they were conducted mostly on rabbits; more studies on human patients are necessary to better define the efficacy and tolerability of these drugs.

6. Intracameral Delivery System

The corneal and precorneal barriers represent significant impediments to drug penetration into the eye, thereby diminishing intraocular bioavailability. The intracameral approach, involving direct injection of therapeutic agents into the anterior chamber, offers a solution to these challenges posed by extraocular routes (Table 5). This approach yields higher concentrations of drugs in the aqueous humor and adjacent ocular tissues, enhancing therapeutic efficacy while requiring smaller doses of the drug [72].

Table 5. Intracameral delivery systems.

Authors	Type of Study	Year	Device	Drugs	Results
J. Bacharach et al. [73]	Phase 3 trial, randomized, multicenter, masked	2021	Durysta	Intracameral sustained released Bimatoprost	BimSR 10 µg and 15 µg were non-inferior in lowering IOP to timolol eye drops.
F.A. Medeiros et al. [74]	Phase 3 trial, randomized, multicenter, masked	2020	Durysta	Intracameral sustained released Bimatoprost	BimSR 10 µg and 15 µg were non-inferior in lowering IOP to timolol eye drops.
John P. Berdahl [75]	Phase 2 trial, randomized, multicenter, masked	2023	Idose	Intracameral sustained released travoprost	The fast-release and slow-release formulations were non-inferior in lowering IOP to timolol eye drops.
Courtesy of the producer	Phase 2 trial, randomized, multicenter, masked		ENV515	Intracameral sustained released travoprost	ENV515 was non-inferior to topical travoprost on day 25 and to topical timolol 0.5% on the 11th month.

Currently, two sustained-release prostaglandin implants have received the U.S. Food and Drug Administration (FDA) approval, with several others undergoing development.

In March 2020, the U.S. FDA granted approval for "Durysta", marking the introduction of the first intracameral biodegradable implant containing Bimatoprost (BimSR), designed for single-use applications. This implant takes the form of a solid, rod-shaped structure comprising 10 µg of Bimatoprost encapsulated within a poly(D, L-lactide-co-glycolide) polymer matrix. Post-implantation, the matrix gradually degrades, facilitating the sustained, non-pulsatile release of the drug through zero-order kinetics [72]. Pharmacokinetic analysis revealed that 80.5% of the Bimatoprost payload was released by day 51, reaching 99.8% by day 80, with no detectable drug levels beyond 4.5 months post-injection. The intracameral administration route enables direct delivery of the drug to the iris and ciliary body, thereby reducing the overall medication requirement for achieving therapeutic effects. The 10 µg of drug released from the implant is equivalent to the dosage contained in a single drop of Bimatoprost 0.03% ophthalmic solution. Additionally, minimal to undetectable drug levels were observed in non-target extraocular tissues, mitigating the risk of adverse events such as orbital fat atrophy [34]. Prostaglandin analogs (PGAs) are commonly employed as first-line treatments due to their established efficacy and safety profile. While PGAs primarily lower IOP by enhancing aqueous humor outflow via the uveoscleral pathway, intracameral BimSR may offer additional mechanisms for IOP reduction. Pharmacokinetic investigations revealed that following BimSR injection, drug concentrations were 4400-fold higher compared with once-daily topical Bimatoprost 0.03% eye drops, with intact Bimatoprost being the predominant drug moiety detected in the iris-ciliary body, as opposed to Bimatoprost acid observed with topical eye drops. Furthermore, supplementary IOP reduction was demonstrated when BimSR was administered to normotensive cynomolgus monkeys already receiving topical Bimatoprost eye drops. Hence, it is plausible that the Bimatoprost implant provides IOP reduction through alternative mechanisms beyond those offered by Bimatoprost eye drops [76]. The implantation procedure necessitates the use of an aseptic technique conducted under an operating microscope. A 28-gauge needle is utilized to insert the implant into the anterior chamber, typically through the temporal or superotemporal clear cornea. Following the injection, it is recommended that the patient maintain an upright position for approximately one hour to facilitate the settling of the implant in the inferior angle [77]. A 24-month, paired-eye, phase 1/2 clinical trial was undertaken to evaluate the safety and effectiveness of the Bimatoprost implant in individuals with OAG. Participants were divided into four groups, each receiving different dosages of the drug (6, 10, 15, and 20 μg). An interim analysis published at the six-month mark revealed that each dosage of BimSR resulted in a comparable reduction in IOP compared to once-daily Bimatoprost 0.03% eye drops. Additionally, it was observed that the reduction in IOP was dose-dependent for at least 12 weeks, and subsequent administrations were equally effective as the initial one. Following this, the approval of Durysta was based on the findings from two identical 20-month phase 3 clinical trials, known as ARTEMIS 1 and ARTEMIS 2. These trials were characterized by their prospective, multicenter, randomized design and compared the effects of a 10 or 15 μg Bimatoprost implant with twice-daily topical timolol maleate 0.5% in patients with open-angle glaucoma or ocular hypertension. Primary endpoints included IOP and IOP change from baseline through week 12. Safety assessments encompassed various measures such as best-corrected visual acuity (BCVA), biomicroscopy examination, gonioscopy, fundus examinations, corneal endothelial cell density (CECD) evaluation, corneal thickness evaluation, and treatment of emergent adverse events (TEAEs). Particular attention was paid to corneal AEs (mainly corneal endothelial cell loss and/or corneal edema) and inflammatory adverse events (especially keratitis, iritis, and vitritis). Results from ARTEMIS 1 indicated that both 10 μg and 15 μg Bimatoprost implants were non-inferior to timolol in terms of IOP and IOP changes from baseline. Furthermore, subsequent administrations of the implants demonstrated sustained efficacy, with many patients requiring no additional IOP-lowering treatment, even during extended safety follow-ups. However, the incidence of TEAEs, including corneal changes and inflammatory reactions, was higher in the Bimatoprost implant groups compared with the timolol group, particularly with repeated dosing. The most frequent side effects were conjunctival hyperemia, eye irritation, foreign body sensation, and eye pain (some probably caused by the injection procedure). There were no reports of eyelash growth or periorbital fat atrophy [78]. Similarly, ARTEMIS 2 showed favorable outcomes for the 10 μg and 15 μg Bimatoprost implant groups compared with the timolol group in terms of mean IOP and mean change from baseline IOP. The efficacy of repeated administrations was also confirmed. Despite the promising efficacy, concerns were raised regarding adverse events, particularly corneal changes such as corneal endothelial cell loss, which appeared to be higher than expected, especially with repeated administrations and in the 15 μg implant group [78]. Another noteworthy finding from ARTEMIS 1 and 2 is the reduced necessity for additional IOP-lowering treatments among many patients treated with BimSR, persisting even after four months from the last injection. Stramer et al. have reported that BimSR injection may sustain outflow by inducing a significant upregulation in MMP1 and downregulation of fibronectin. Notably, the dramatic upregulation in MMP1 and downregulation of fibronectin are observed exclusively with Bimatoprost at high concentrations and not with BFA [76]. In conclusion, the risk/benefit evaluation favored the 10 μg Bimatoprost implant over the 15 μg Bimatoprost implant. However, many other BimSR phase 3 trials are underway or pending the release of results.

The iDose, developed by Glaukos Corporation based in San Clemente, California, United States, is a sustained-release intraocular implant containing a unique formulation of travoprost. This implant consists of three primary components: (1) a scleral anchor for insertion and secure attachment to the inner scleral wall via the trabecular meshwork; (2) a titanium body serving as a drug reservoir; and (3) a membrane designed to release travoprost gradually over a targeted duration ranging from six to twelve months. Insertion and removal of the implant are performed under an operating microscope. A phase 2 trial, conducted as a multicenter, randomized, double-masked study, compared the efficacy of iDose with two different rates of drug elution to twice daily topical timolol 0.5%. The trial enrolled 154 patients, with 51 receiving the fast-release formulation of iDose, 54 receiving the slow-release formulation, and 49 receiving timolol 0.5% eye drops. Results showed comparable mean IOP reductions at the end of the first 24 months among the groups. In addition to its efficacy, the iDose device demonstrated a favorable safety profile during phase 2 clinical trials, with no significant corneal or conjunctival adverse events reported. The FDA granted approval to iDose® TR in December 2023 for managing IOP in patients with open-angle glaucoma or ocular hypertension. However, it is contraindicated in individuals with active or suspected ocular infections, corneal endothelial cell dystrophy, prior corneal transplants, or hypersensitivity to its components. Caution is advised for patients with certain angle abnormalities. FDA approval was based on two phase 3 pivotal trials (GC-010 and GC-012) involving 1150 subjects across 89 clinical sites. Both trials successfully met primary efficacy endpoints over a 3-month period and demonstrated favorable safety profiles over 12 months. While iDose TR showed non-inferiority to timolol ophthalmic solution in reducing IOP during the initial 3 months, it did not maintain this non-inferiority over the subsequent 9 months. At the 12-month follow-up, a significant portion of iDose TR subjects were medication-free, with high tolerability and subject retention rates. Ocular adverse reactions were mostly mild and transient, including increases in IOP, iritis, dry eye, and visual field defects [79].

ENV515, developed by Envisia Therapeutics located in Morrisville, NC, USA, is a biodegradable intracameral implant containing travoprost. This implant, shaped like a rod, is produced using innovative particle replication in non-wetting templates technology, allowing for the creation of sterile nanoparticles with an extended-release formulation of travoprost. Injection of the implant into the anterior chamber is performed using a specialized injector under an operating microscope, typically resting stably in the inferior iridocorneal angle. Prior to human clinical trials, preclinical studies were conducted in normotensive and hypertensive beagles, varying in implant size, drug concentration, and number. These studies showed a reduction in intraocular pressure following implantation, with the device being well-tolerated and causing minimal adverse effects, primarily transient ocular irritation related to the surgical procedure. In a phase 2a clinical trial involving 21 patients with open-angle glaucoma, ENV515 was compared to once-daily topical travoprost and met the primary endpoint of non-inferiority by day 25. Another clinical trial evaluated ENV515 against topical timolol 0.5% eye drops in open-angle glaucoma patients, demonstrating an IOP reduction of 6.7 ± 3.7 mm Hg and confirming non-inferiority to timolol over 11 months [69,74,78].

7. Conclusions

The primary goal of managing glaucoma patients is to enhance their well-being and quality of life within a sustainable healthcare framework. Visual functionality plays a pivotal role in determining quality of life, with individuals experiencing minor visual impairment generally maintaining an acceptable standard of living. However, as visual impairment worsens, overall well-being declines. Additionally, factors such as treatment costs and drug side effects significantly impact patients' quality of life, including the inconvenience faced during chronic therapy. The sustained drug delivery systems (SDDSs) explored in this review can have a significant impact on glaucoma management and patients' quality of life. The advantages are substantial: by releasing medication gradually over an extended period, SDDSs can significantly reduce the frequency of dosing, thereby enhancing patient adherence and slowing disease progression. Improved adherence through SDDSs is critical for maintaining optimal IOP control, essential for preventing optic nerve damage and vision loss. SDDSs can be precisely engineered for targeted drug delivery to specific ocular tissues, such as the anterior chamber or ciliary body. This targeted delivery approach increases drug efficacy by concentrating medication at the site of action while reducing systemic exposure, thus minimizing side effects. This is particularly advantageous for patients at elevated risk of systemic side effects or those taking multiple medications. Moreover, SDDSs alleviate the burden of frequent daily dosing, simplifying the treatment regimen, and this could improve patient convenience and quality of life. Research indicates that traditional eye drop regimens often cause frustration and reduce the quality of life for glaucoma patients. By making the treatment regimen less demanding, SDDS can empower patients to manage their condition more effectively and enhance their overall well-being. However the long-term impacts of SDDS in glaucoma treatment need further exploration. Most clinical trials on SDDSs focus on short-term efficacy and safety. To fully understand potential side effects, drug interactions, and the overall effect on glaucoma progression over extended periods, long-term studies are essential. Monitoring potential issues like inflammation, irritation, or scarring is crucial in these studies. Ensuring a consistent and controlled release of medication throughout the intended duration is critical. Long-term research can uncover potential problems, such as drug degradation within the SDDS or changes in release rates over time. Moreover, a comprehensive analysis is required to balance the initial costs of SDDSs against potential long-term savings. This includes factors like improved adherence, which reduces medication waste, fewer complications leading to less need for additional treatments, and reduced office visits due to less frequent dosing. Finally, it is very important to consider the central role of the patient, who should receive clear and thorough information about the benefits, proper usage, and possible limitations of SDDSs. Providing a variety of SDDS formulations and delivery methods can cater to individual patient needs and preferences [80].

Author Contributions: A.M.F., V.V., M.C., V.M., C.N. and F.P. had full access to all the data in the study and took responsibility for the integrity of the data and the accuracy of the data analysis. Study concept and design: A.M.F., V.V., M.C., V.M., C.N. and F.P. Acquisition, analysis, or interpretation of data: all authors. Drafting of the manuscript: all authors. Critical revision of the manuscript for important intellectual content: all authors. Study supervision: A.M.F. and F.P. All authors have read and agreed to the published version of the manuscript.

Funding: This research received no external funding.

Conflicts of Interest: The authors declare no conflicts of interest.

References

1. Allison, K.; Patel, D.; Alabi, O. Epidemiology of Glaucoma: The Past, Present, and Predictions for the Future. *Cureus* **2020**, *12*, e11686. [CrossRef]
2. Schuster, A.K.; Erb, C.; Hoffmann, E.M.; Dietlein, T.; Pfeiffer, N. The Diagnosis and Treatment of Glaucoma. *Dtsch. Ärzteblatt Int.* **2020**, *117*, 225. [CrossRef] [PubMed]
3. Weinreb, R.N.; Aung, T.; Medeiros, F.A. The Pathophysiology and Treatment of Glaucoma: A Review. *JAMA* **2014**, *311*, 1901. [CrossRef] [PubMed]
4. Newman-Casey, P.A.; Robin, A.L.; Blachley, T.; Farris, K.; Heisler, M.; Resnicow, K.; Lee, P.P. The Most Common Barriers to Glaucoma Medication Adherence. *Ophthalmology* **2015**, *122*, 1308–1316. [CrossRef] [PubMed]
5. Olthoff, C.; Schouten, J.; Vandeborne, B.; Webers, C. Noncompliance with Ocular Hypotensive Treatment in Patients with Glaucoma or Ocular HypertensionAn Evidence-Based Review. *Ophthalmology* **2005**, *112*, 953–961.E7. [CrossRef]
6. Rossi, G.C.; Pasinetti, G.M.; Scudeller, L.; Radaelli, R.; Bianchi, P.E. Do adherence rates and glaucomatous visual field progression correlate? *Eur. J. Ophthalmol.* **2011**, *21*, 410–414. [CrossRef] [PubMed]
7. Kompella, U.B.; Hartman, R.R.; Patil, M.A. Extraocular, periocular and intraocular routes for sustained drug delivery for glaucoma. *Prog. Retin. Eye Res.* **2021**, *82*, 100901. [CrossRef] [PubMed] [PubMed Central]
8. Singh, R.B.; Ichhpujani, P.; Thakur, S.; Jindal, S. Promising Therapeutic Drug Delivery Systems for Glaucoma: A Comprehensive Review. *Ophthalmol. Eye Dis.* **2020**, *12*, 251584142090574. [CrossRef]
9. Kesav, N.P.; Young, C.E.C.; Ertel, M.K.; Seibold, L.K.; Kahook, M.Y. Sustained-Release Drug Delivery Systems for the Treatment of Glaucoma. *Int. J. Ophthalmol.* **2021**, *14*, 148–159. [CrossRef]
10. Jehangir, N.; Bever, G.; Mahmood, S.M.J.; Moshirfar, M. Comprehensive Review of the Literature on Existing Punctal Plugs for the Management of Dry Eye Disease. *J. Ophthalmol.* **2016**, *2016*, 9312340. [CrossRef]
11. Chen, X.; Shen, T. Advances in Innovative Delivery Systems for Antiglaucoma Drugs. *Curr. Opin. Ophthalmol.* **2023**, *34*, 123–128. [CrossRef] [PubMed]
12. Calonge, M. The Treatment of Dry Eye. *Surv. Ophthalmol.* **2001**, *45* (Suppl. S2), S227–S239. [CrossRef] [PubMed]
13. Balaram, M.; Schaumberg, D.A.; Dana, M.R. Efficacy and Tolerability Outcomes after Punctal Occlusion with Silicone Plugs in Dry Eye Syndrome. *Am. J. Ophthalmol.* **2001**, *131*, 30–36. [CrossRef] [PubMed]
14. Chen, M.; Yung Choi, S. Preliminary Outcomes of Temporary Collagen Punctal Plugs for Patients with Dry Eye and Glaucoma. *Med. Hypothesis Discov. Innov. Ophthalmol.* **2020**, *9*, 56–60.

15. Sherwin, J.C.; Ratnarajan, G.; Elahi, B.; Bilkiewicz-Pawelec, A.; Salmon, J.F. Effect of a Punctal Plug on Ocular Surface Disease in Patients Using Topical Prostaglandin Analogues: A Randomized Controlled Trial. *Clin. Exp. Ophthalmol.* **2018**, *46*, 888–894. [CrossRef]
16. Perera, S.A.; Ting, D.S.; Nongpiur, M.E.; Chew, P.T.; Aquino, M.C.D.; Sng, C.C.; Ho, S.-W.; Aung, T. Feasibility Study of Sustained-Release Travoprost Punctum Plug for Intraocular Pressure Reduction in an Asian Population. *Clin. Ophthalmol.* **2016**, *10*, 757–764. [CrossRef]
17. Kompella, U.B.; Kadam, R.S.; Lee, V.H.L. Recent Advances in Ophthalmic Drug Delivery. *Ther. Deliv.* **2010**, *1*, 435–456. [CrossRef]
18. Zimmerman, T.J.; Kooner, K.S.; Kandarakis, A.S.; Ziegler, L.P. Improving the Therapeutic Index of Topically Applied Ocular Drugs. *Arch. Ophthalmol.* **1984**, *102*, 551–553. [CrossRef]
19. Huang, T.C.; Lee, D.A. Punctal Occlusion and Topical Medications for Glaucoma. *Am. J. Ophthalmol.* **1989**, *107*, 151–155. [CrossRef]
20. Tyson, S.L.; Bafna, S.; Gira, J.P.; Goldberg, D.F.; Jones, J.J.; Jones, M.P.; Kim, J.K.; Martel, J.M.; Nordlund, M.L.; Piovanetti-Perez, I.K.; et al. Multicenter Randomized Phase 3 Study of a Sustained-Release Intracanalicular Dexamethasone Insert for Treatment of Ocular Inflammation and Pain after Cataract Surgery. *J. Cataract. Refract. Surg.* **2019**, *45*, 204–212. [CrossRef]
21. Gira, J.P.; Sampson, R.; Silverstein, S.M.; Walters, T.R.; Metzinger, J.L.; Talamo, J.H. Evaluating the Patient Experience after Implantation of a 0.4 mg Sustained Release Dexamethasone Intracanalicular Insert (Dextenza^{TM}): Results of a Qualitative Survey. *Patient Prefer. Adherence* **2017**, *11*, 487–494. [CrossRef] [PubMed]
22. Wilson, C.; Sall, K.N.; Bafna, S.; Gira, J.P.; McLaurin, E.B.; Protzko, E.; Sampson, R.; Tekwani, N.; Tepedino, M.; Vold, S.; et al. Results of A Randomized, Double-Masked, Parallel-Arm Phase 2b Study Evaluating the Safety and Efficacy of OTX-TP (Travoprost Insert) Compared to Timolol Drops for the Treatment of Patients with Open-Angle Glaucoma or Ocular Hypertension. *Investig. Ophthalmol. Vis. Sci.* **2017**, *58*, 2111.
23. Vantipalli, S.; Sall, K.N.; Stein, E.; Schenker, H.; Mulaney, J.; Smyth-Medina, R.; Day, D.; Benza, R.; Dixon, E.-R.; Rissman, N.; et al. Evaluation of the Safety and Efficacy of OTX-TP, an Intracanalicular Travoprost Insert, for the Treatment of Patients with Open-Angle Glaucoma or Ocular Hypertension: A Phase 3 Study. *Investig. Ophthalmol. Vis. Sci.* **2020**, *61*, 3488.
24. Goldberg, D.F.; Williams, R. A Phase 2 Study Evaluating Safety and Efficacy of the Latanoprost Punctal Plug Delivery System (L-PPDS) in Subjects with Ocular Hypertension (OH) or Open-Angle Glaucoma (OAG). *Investig. Ophthalmol. Vis. Sci.* **2012**, *53*, 5095.
25. Tost, F.H.; Geerling, G. Plugs for Occlusion of the Lacrimal Drainage System. *Surg. Dry Eye* **2008**, *41*, 193–212.
26. Stout, N.; Harsch, A.G.; Lighthizer, N. Plug the Drain with Lacrimal Occlusion: Keep Dry Eye Patients Flowing into Your Office with This Time-Honored Technique. *Rev. Optom.* **2016**, *153*, 70–75.
27. Macoul, K.L.; Pavan-Langston, D. Pilocarpine Ocusert System for Sustained Control of Ocular Hypertension. *Arch. Ophthalmol.* **1975**, *93*, 587–590. [CrossRef] [PubMed]
28. Brandt, J.D.; Sall, K.; DuBiner, H.; Benza, R.; Alster, Y.; Walker, G.; Semba, C.P.; Budenz, D.; Day, D.; Flowers, B.; et al. Six-Month Intraocular Pressure Reduction with a Topical Bimatoprost Ocular Insert. *Ophthalmology* **2016**, *123*, 1685–1694. [CrossRef]
29. Brandt, J.D.; DuBiner, H.B.; Benza, R.; Sall, K.N.; Walker, G.A.; Semba, C.P.; Budenz, D.; Day, D.; Flowers, B.; Lee, S.; et al. Long-Term Safety and Efficacy of a Sustained-Release Bimatoprost Ocular Ring. *Ophthalmology* **2017**, *124*, 1565–1566. [CrossRef]
30. Nair, R.V.; Shefrin, S.; Suresh, A.; Anoop, K.R.; Nair, S.C. Sustained Release Timolol Maleate Loaded Ocusert Based on Biopolymer Composite. *Int. J. Biol. Macromol.* **2018**, *110*, 308–317. [CrossRef]
31. Kelly, J.A.; Molyneux, P.D.; Smith, S.A.; Smith, S.E. Relative Bioavailability of Pilocarpine from a Novel Ophthalmic Delivery System and Conventional Eye Drop Formulations. *Br. J. Ophthalmol.* **1989**, *73*, 360–362. [CrossRef] [PubMed]
32. Greaves, J.; Wilson, C.; Birmingham, A.; Richardson, M.; Bentley, P. Scintigraphic Studies on the Corneal Residence of a New Ophthalmic Delivery System (NODS): Rate of Clearance of a Soluble Marker in Relation to Duration of Pharmacological Action of Pilocarpine. *Br. J. Clin. Pharmacol.* **1992**, *33*, 603–609. [CrossRef] [PubMed]
33. Diestelhorst, M.; Krieglstein, G.K. The Ocular Tolerability of a New Ophthalmic Drug Delivery System (NODS). *Int. Ophthalmol.* **1994**, *18*, 1–4. [CrossRef] [PubMed]
34. Bigdeli, A.; Makhmalzadeh, B.S.; Feghhi, M.; SoleimaniBiatiani, E. Cationic Liposomes as Promising Vehicles for Timolol/Brimonidine Combination Ocular Delivery in Glaucoma: Formulation Development and in Vitro/in Vivo Evaluation. *Drug Deliv. Transl. Res.* **2023**, *13*, 1035–1047. [CrossRef]
35. Awwad, S.; Mohamed Ahmed, A.H.A.; Sharma, G.; Heng, J.S.; Khaw, P.T.; Brocchini, S.; Lockwood, A. Principles of Pharmacology in the Eye: Principles of Pharmacology in the Eye. *Br. J. Pharmacol.* **2017**, *174*, 4205–4223. [CrossRef]
36. Pollack, I.P.; Quigley, H.A.; Harbin, T.S. The Ocusert Pilocarpine System: Advantages and Disadvantages. *South. Med. J.* **1976**, *69*, 1296–1298. [CrossRef]
37. Uner, B.; Ozdemir, S.; Nur Pilevne, S.; Cenk Celebi, A.R. Timolol-loaded ethosomes for ophthalmic delivery: Reduction of high intraocular pressure in vivo. *Int. J. Pharm.* **2023**, *640*, 123021. [CrossRef] [PubMed]
38. Cuggino, J.C.; Tártara, L.I.; Gugliotta, L.M.; Palma, S.D.; Alvarez Igarzabal, C.I. Mucoadhesive and responsive nanogels as carriers for sustainable delivery of timolol for glaucoma therapy. *Mater. Sci. Eng. C Mater. Biol. Appl.* **2021**, *118*, 111383. [CrossRef] [PubMed]
39. Ilka, R.; Mohseni, M.; Kianirad, M.; Naseripour, M.; Ashtari, K.; Mehravi, B. Nanogel-based natural polymers as smart carriers for the controlled delivery of Timolol Maleate through the cornea for glaucoma. *Int. J. Biol. Macromol.* **2018**, *109*, 955–962. [CrossRef] [PubMed]

40. Cegielska, O.; Sierakowski, M.; Sajkiewicz, P.; Lorenz, K.; Kogermann, K. Mucoadhesive brinzolamide-loaded nanofibers for alternative glaucoma treatment. *Eur. J. Pharm. Biopharm.* **2022**, *180*, 48–62. [CrossRef] [PubMed]
41. Agban, Y.; Lian, J.; Prabakar, S.; Seyfoddin, A.; Rupenthal, I.D. Nanoparticle Cross-Linked Collagen Shields for Sustained Delivery of Pilocarpine Hydrochloride. *Int. J. Pharm.* **2016**, *501*, 96–101. [CrossRef] [PubMed]
42. Hiratani, H.; Alvarez-Lorenzo, C. Timolol Uptake and Release by Imprinted Soft Contact Lenses Made of N,N-Diethylacrylamide and Methacrylic Acid. *J. Control. Release* **2002**, *83*, 223–230. [CrossRef]
43. Kim, J.; Conway, A.; Chauhan, A. Extended Delivery of Ophthalmic Drugs by Silicone Hydrogel Contact Lenses. *Biomaterials* **2008**, *29*, 2259–2269. [CrossRef] [PubMed]
44. Peng, C.-C.; Burke, M.T.; Carbia, B.E.; Plummer, C.; Chauhan, A. Extended Drug Delivery by Contact Lenses for Glaucoma Therapy. *J. Control. Release* **2012**, *162*, 152–158. [CrossRef] [PubMed]
45. Braga, M.E.M.; Costa, V.P.; Pereira, M.J.T.; Fiadeiro, P.T.; Gomes, A.P.; Duarte, C.M.M.; De Sousa, H.C. Effects of Operational Conditions on the Supercritical Solvent Impregnation of Acetazolamide in Balafilcon A Commercial Contact Lenses. *Int. J. Pharm.* **2011**, *420*, 231–243. [CrossRef]
46. Jung, H.J.; Chauhan, A. Temperature Sensitive Contact Lenses for Triggered Ophthalmic Drug Delivery. *Biomaterials* **2012**, *33*, 2289–2300. [CrossRef]
47. Peng, C.-C.; Ben-Shlomo, A.; Mackay, E.O.; Plummer, C.E.; Chauhan, A. Drug Delivery by Contact Lens in Spontaneously Glaucomatous Dogs. *Curr. Eye Res.* **2012**, *37*, 204–211. [CrossRef]
48. Jung, H.J.; Abou-Jaoude, M.; Carbia, B.E.; Plummer, C.; Chauhan, A. Glaucoma Therapy by Extended Release of Timolol from Nanoparticle Loaded Silicone-Hydrogel Contact Lenses. *J. Control. Release* **2013**, *165*, 82–89. [CrossRef]
49. Hsu, K.-H.; Carbia, B.E.; Plummer, C.; Chauhan, A. Dual Drug Delivery from Vitamin E Loaded Contact Lenses for Glaucoma Therapy. *Eur. J. Pharm. Biopharm.* **2015**, *94*, 312–321. [CrossRef]
50. Ciolino, J.B.; Ross, A.E.; Tulsan, R.; Watts, A.C.; Wang, R.-F.; Zurakowski, D.; Serle, J.B.; Kohane, D.S. Latanoprost-Eluting Contact Lenses in Glaucomatous Monkeys. *Ophthalmology* **2016**, *123*, 2085–2092. [CrossRef]
51. Wichterle, O.; Lím, D. Hydrophilic Gels for Biological Use. *Nature* **1960**, *185*, 117–118. [CrossRef]
52. Muntz, A.; Subbaraman, L.N.; Sorbara, L.; Jones, L. Tear Exchange and Contact Lenses: A Review. *J. Optom.* **2015**, *8*, 2–11. [CrossRef] [PubMed]
53. Sedlácek, J. Possibility of the application of ophthalmic drugs with the use of gel contact lenses. *Ceskoslovenska Oftalmol.* **1965**, *21*, 509–512.
54. North, D.P. Treatment of Acute Glaucoma. *Can. Med. Assoc. J.* **1971**, *105*, 561.
55. Uchida, R.; Sato, T.; Tanigawa, H.; Uno, K. Azulene incorporation and release by hydrogel containing methacrylamide propyltrimenthylammonium chloride, and its application to soft contact lens. *J. Control. Release* **2003**, *92*, 259–264. [CrossRef]
56. Lovrec-Krstič, T.; Orthaber, K.; Maver, U.; Sarenac, T. Review of Potential Drug-Eluting Contact Lens Technologies. *Materials* **2023**, *16*, 3653. [CrossRef]
57. Venkatesh, S.; Saha, J.; Pass, S.; Byrne, M. Transport and Structural Analysis of Molecular Imprinted Hydrogels for Controlled Drug Delivery. *Eur. J. Pharm. Biopharm.* **2008**, *69*, 852–860. [CrossRef]
58. Yañez, F.; Martikainen, L.; Braga, M.E.M.; Alvarez-Lorenzo, C.; Concheiro, A.; Duarte, C.M.M.; Gil, M.H.; De Sousa, H.C. Supercritical Fluid-Assisted Preparation of Imprinted Contact Lenses for Drug Delivery. *Acta Biomater.* **2011**, *7*, 1019–1030. [CrossRef]
59. Janagam, D.R.; Wu, L.; Lowe, T.L. Nanoparticles for Drug Delivery to the Anterior Segment of the Eye. *Adv. Drug Deliv. Rev.* **2017**, *122*, 31–64. [CrossRef]
60. Xu, J.; Xue, Y.; Hu, G.; Lin, T.; Gou, J.; Yin, T.; He, H.; Zhang, Y.; Tang, X. A Comprehensive Review on Contact Lens for Ophthalmic Drug Delivery. *J. Control. Release* **2018**, *281*, 97–118. [CrossRef]
61. Wang, M.T.M.; Ganesalingam, K.; Loh, C.S.; Alberquerque, T.; Al-Kanani, S.; Misra, S.L.; Craig, J.P. Compatibility of Phospholipid Liposomal Spray with Silicone Hydrogel Contact Lens Wear. *Contact Lens Anterior Eye* **2017**, *40*, 53–58. [CrossRef] [PubMed]
62. Lim, C.H.L.; Stapleton, F.; Mehta, J.S. Review of Contact Lens–Related Complications. *Eye Contact Lens Sci. Clin. Pract.* **2018**, *44*, S1–S10. [CrossRef]
63. Rouen, P.A.; White, M.L. Dry Eye Disease: Prevalence, Assessment, and Management. *Home Healthc. Now* **2018**, *36*, 74–83. [CrossRef]
64. Stewart, W.C.; Garrison, P.M. β-Blocker–Induced Complications and the Patient With Glaucoma: Newer Treatments to Help Reduce Systemic Adverse Events. *Arch. Intern. Med.* **1998**, *158*, 221. [CrossRef]
65. Raghava, S.; Hammond, M.; Kompella, U.B. Periocular routes for retinal drug delivery. *Expert. Opin. Drug Deliv.* **2004**, *1*, 99–114. [CrossRef] [PubMed]
66. Natarajan, J.V.; Chattopadhyay, S.; Ang, M.; Darwitan, A.; Foo, S.; Zhen, M.; Koo, M.; Wong, T.T.; Venkatraman, S.S. Sustained Release of an Anti-Glaucoma Drug: Demonstration of Efficacy of a Liposomal Formulation in the Rabbit Eye. *PLoS ONE* **2011**, *6*, e24513. [CrossRef] [PubMed]
67. Natarajan, J.V.; Ang, M.; Darwitan, A.; Chattopadhyay, S.; Wong, T.T.; Venkatraman, S.S. Nanomedicine for Glaucoma: Liposomes Provide Sustained Release of Latanoprost in the Eye. *Int. J. Nanomed.* **2012**, *7*, 123–131. [CrossRef]
68. Wong, T.T.; Novack, G.D.; Natarajan, J.V.; Ho, C.L.; Htoon, H.M.; Venkatraman, S.S. Nanomedicine for Glaucoma: Sustained Release Latanoprost Offers a New Therapeutic Option with Substantial Benefits over Eyedrops. *Drug Deliv. Transl. Res.* **2014**, *4*, 303–309. [CrossRef]

69. Amrite, A.C.; Ayalasomayajula, S.P.; Cheruvu, N.P.S.; Kompella, U.B. Single Periocular Injection of Celecoxib-PLGA Microparticles Inhibits Diabetes-Induced Elevations in Retinal PGE 2, VEGF, and Vascular Leakage. *Investig. Ophthalmol. Vis. Sci.* **2006**, *47*, 1149. [CrossRef]
70. Amrite, A.C.; Kompella, U.B. Size-Dependent Disposition of Nanoparticles and Microparticles Following Subconjunctival Administration. *J. Pharm. Pharmacol.* **2010**, *57*, 1555–1563. [CrossRef]
71. Nagai, N.; Tsukamoto, A.; Kotake, T.; Ito, Y.; Okamoto, N.; Shimomura, Y. Evaluation of Rabbit Model for Glaucoma Study: Drug Interaction in a Rabbit Model Instilled with Ophthalmic Preparation containing Latanoprost and Timolol. *Iryo Yakugaku (Jpn. J. Pharm. Health Care Sci.)* **2016**, *42*, 645–650. [CrossRef]
72. Gautam, M.; Gupta, R.; Singh, P.; Verma, V.; Verma, S.; Mittal, P.; Karkhur, S.; Sampath, A.; Mohan, R.R.; Sharma, B. Intracameral Drug Delivery: A Review of Agents, Indications, and Outcomes. *J. Ocul. Pharmacol. Ther.* **2023**, *39*, 102–116. [CrossRef] [PubMed]
73. Bacharach, J.; Tatham, A.; Ferguson, G.; Belalcázar, S.; Thieme, H.; Goodkin, M.L.; Chen, M.Y.; Guo, Q.; Liu, J.; Robinson, M.R.; et al. ARTEMIS 2 Study Group. Phase 3, Randomized, 20-Month Study of the Efficacy and Safety of Bimatoprost Implant in Patients with Open-Angle Glaucoma and Ocular Hypertension (ARTEMIS 2). *Drugs* **2021**, *81*, 2017–2033. [CrossRef] [PubMed]
74. Medeiros, F.A.; Walters, T.R.; Kolko, M.; Coote, M.; Bejanian, M.; Goodkin, M.L.; Guo, Q.; Zhang, J.; Robinson, M.R.; Weinreb, R.N.; et al. Phase 3, Randomized, 20-Month Study of Bimatoprost Implant in Open-Angle Glaucoma and Ocular Hypertension (ARTEMIS 1). *Ophthalmology* **2020**, *127*, 1627–1641. [CrossRef]
75. Berdahl, J.P.; Sarkisian, S.R., Jr.; Ang, R.E.; Doan, L.V.; Kothe, A.C.; Usner, D.W.; Katz, L.J.; Navratil, T.; Travoprost Intraocular Implant Study Group. Efficacy and Safety of the Travoprost Intraocular Implant in Reducing Topical IOP-Lowering Medication Burden in Patients with Open-Angle Glaucoma or Ocular Hypertension. *Drugs* **2024**, *84*, 83–97. [CrossRef]
76. Sirinek, P.E.; Lin, M.M. Intracameral Sustained Release Bimatoprost Implants (Durysta). *Semin. Ophthalmol.* **2022**, *37*, 385–390. [CrossRef]
77. Stamer, W.D.; Perkumas, K.M.; Kang, M.H.; Dibas, M.; Robinson, M.R.; Rhee, D.J. Proposed Mechanism of Long-Term Intraocular Pressure Lowering With the Bimatoprost Implant. *Investig. Ophthalmol. Vis. Sci.* **2023**, *64*, 15. [CrossRef]
78. Medeiros, F.A.; Sheybani, A.; Shah, M.M.; Rivas, M.; Bai, Z.; Werts, E.; Ahmed, I.I.K.; Craven, E.R. Single Administration of Intracameral Bimatoprost Implant 10 mg in Patients with Open-Angle Glaucoma or Ocular Hypertension. *Ophthalmol. Ther.* **2022**, *11*, 1517–1537. [CrossRef]
79. Belamkar, A.; Harris, A.; Zukerman, R.; Siesky, B.; Oddone, F.; Verticchio Vercellin, A.; Ciulla, T.A. Sustained release glaucoma therapies: Novel modalities for overcoming key treatment barriers associated with topical medications. *Ann. Med.* **2022**, *54*, 343–358. [CrossRef]
80. Miller, P.E.; Eaton, J.S. Medical Antiglaucoma Therapy: Beyond the Drop. *Vet. Ophthalmol.* **2021**, *24*, 2–15. [CrossRef]

MDPI AG
Grosspeteranlage 5
4052 Basel
Switzerland
Tel.: +41 61 683 77 34

Pharmaceuticals Editorial Office
E-mail: pharmaceuticals@mdpi.com
www.mdpi.com/journal/pharmaceuticals

www.ingramcontent.com/pod-product-compliance
Lightning Source LLC
Chambersburg PA
CBHW061823210926
51850CB00006B/788